Arithmetic Geometry

Springer
*New York
Berlin
Heidelberg
Barcelona
Budapest
Hong Kong
London
Milan
Paris
Santa Clara
Singapore
Tokyo*

Arithmetic Geometry

Edited by
Gary Cornell Joseph H. Silverman

With Contributions by
M. Artin C.-L. Chai T. Chinburg G. Faltings
B. H. Gross F. O. McGuinness J. S. Milne
M. Rosen S. S. Shatz J. H. Silverman P. Vojta

Springer

Gary Cornell
Department of Mathematics
University of Connecticut
Storrs, CT 06268
U.S.A.

Joseph H. Silverman
Department of Mathematics
Brown University
Providence, RI 02912
U.S.A.

Mathematics Subject Classification (1991): 12-02, 14K15, 14L15

With 5 Illustrations

Library of Congress Cataloging in Publication Data
Arithmetic geometry.
 Papers presented at an instructional conference on arithmetic geometry held July 30–Aug. 10, 1984, at the University of Connecticut in Storrs.
 Includes bibliographies.
 1. Geometry, Algebraic—Congresses. 2. Algebraic number theory—Congresses. I. Cornell, Gary.
II Silverman, Joseph H., 1955– . III. Artin, Michael.
QA564.A73 1986   516.3'5   86-17758

© 1986 by Springer-Verlag New York Inc.
All rights reserved. No part of this book may be translated or reproduced in any form without written permission from Springer-Verlag, 175 Fifth Avenue, New York, New York 10010, U.S.A.

Authorization to photocopy items for internal or personal use, or the internal or personal use of specific clients, is granted by Springer-Verlag New York, Inc., provided that the appropriate fee is paid directly to Copyright Clearance Center, 222 Rosewood Drive, Danvers, MA 01923, USA (Telephone: (508) 750-8400), stating the ISBN of the volume, the volume title and the first and last page numbers of each chapter copied. The copyright owner's consent does not include copying for general distribution, promotion, new works, or resale. In these cases, specific written permission must be first obtained from the publisher.

Typeset by Asco Trade Typesetting Ltd., Hong Kong.
Printed and bound by Braun-Brumfield, Inc., Ann Arbor, MI.
Printed in the United States of America.

9 8 7 6 5 4 3 2 (Revised second printing, 1998)

ISBN 0-387-96311-1 Springer-Verlag New York Berlin Heidelberg
ISBN 3-540-96311-1 Springer-Verlag Berlin Heidelberg New York SPIN 10669856

Preface

This volume is the result of a (mainly) instructional conference on arithmetic geometry, held from July 30 through August 10, 1984 at the University of Connecticut in Storrs. This volume contains expanded versions of almost all the instructional lectures given during the conference. In addition to these expository lectures, this volume contains a translation into English of Faltings' seminal paper which provided the inspiration for the conference. We thank Professor Faltings for his permission to publish the translation and Edward Shipz who did the translation.

We thank all the people who spoke at the Storrs conference, both for helping to make it a successful meeting and enabling us to publish this volume. We would especially like to thank David Rohrlich, who delivered the lectures on height functions (Chapter VI) when the second editor was unavoidably detained. In addition to the editors, Michael Artin and John Tate served on the organizing committee for the conference and much of the success of the conference was due to them—our thanks go to them for their assistance.

Finally, the conference was only made possible through generous grants from the Vaughn Foundation and the National Science Foundation.

December, 1985 G. Cornell
J. H. Silverman

Contents

Contributors	xiii
Introduction	xv

CHAPTER I
Some Historical Notes — 1
GERD FALTINGS

§1.	The Theorems of Mordell and Mordell–Weil	2
§2.	Siegel's Theorem About Integral Points	2
§3.	The Proof of the Mordell Conjecture for Function Fields, by Manin and Grauert	4
§4.	The New Ideas of Parshin and Arakelov, Relating the Conjectures of Mordell and Shafarevich	4
§5.	The Work of Szpiro, Extending This to Positive Characteristic	6
§6.	The Theorem of Tate About Endomorphisms of Abelian Varieties over Finite Fields	6
§7.	The Work of Zarhin	6
	Bibliographic Remarks	7

CHAPTER II
Finiteness Theorems for Abelian Varieties over Number Fields — 9
GERD FALTINGS

§1.	Introduction	9
§2.	Semiabelian Varieties	10
§3.	Heights	14
§4.	Isogenies	17
§5.	Endomorphisms	20
§6.	Finiteness Theorems	22
	References	26
	Erratum	27

CHAPTER III
Group Schemes, Formal Groups, and p-Divisible Groups   29
Stephen S. Shatz

§1. Introduction	29
§2. Group Schemes, Generalities	30
§3. Finite Group Schemes	37
§4. Commutative Finite Group Schemes	45
§5. Formal Groups	56
§6. p-Divisible Groups	60
§7. Applications of Groups of Type $(p, p, \ldots, p)$ to p-Divisible Groups	76
References	78

CHAPTER IV
Abelian Varieties over $\mathbb{C}$ 79
Michael Rosen
(Notes by F. O. McGuinness)

§0. Introduction	79
§1. Complex Tori	79
§2. Isogenies of Complex Tori	81
§3. Abelian Varieties	83
§4. The Néron–Severi Group and the Picard Group	92
§5. Polarizations and Polarized Abelian Manifolds	95
§6. The Space of Principally Polarized Abelian Manifolds	97
References	100

CHAPTER V
Abelian Varieties   103
J. S. Milne

§1. Definitions	104
§2. Rigidity	104
§3. Rational Maps into Abelian Varieties	105
§4. Review of the Cohomology of Schemes	108
§5. The Seesaw Principle	109
§6. The Theorems of the Cube and the Square	110
§7. Abelian Varieties Are Projective	112
§8. Isogenies	114
§9. The Dual Abelian Variety: Definition	117
§10. The Dual Abelian Variety: Construction	119
§11. The Dual Exact Sequence	120
§12. Endomorphisms	121
§13. Polarizations and the Cohomology of Invertible Sheaves	126
§14. A Finiteness Theorem	127
§15. The Étale Cohomology of an Abelian Variety	128
§16. Pairings	131
§17. The Rosati Involution	137
§18. Two More Finiteness Theorems	140

§19. The Zeta Function of an Abelian Variety	143
§20. Abelian Schemes	145
References	150

CHAPTER VI
The Theory of Height Functions
JOSEPH H. SILVERMAN

151

The Classical Theory of Heights	151
§1. Absolute Values	151
§2. Height on Projective Space	151
§3. Heights on Projective Varieties	153
§4. Heights on Abelian Varieties	156
§5. The Mordell–Weil Theorem	158
Heights and Metrized Line Bundles	161
§6. Metrized Line Bundles on Spec (R)	161
§7. Metrized Line Bundles on Varieties	161
§8. Distance Functions and Logarithmic Singularities	163
References	166

CHAPTER VII
Jacobian Varieties
J. S. MILNE

167

§1. Definitions	167
§2. The Canonical Maps from C to its Jacobian Variety	171
§3. The Symmetric Powers of a Curve	174
§4. The Construction of the Jacobian Variety	179
§5. The Canonical Maps from the Symmetric Powers of C to its Jacobian Variety	182
§6. The Jacobian Variety as Albanese Variety; Autoduality	185
§7. Weil's Construction of the Jacobian Variety	189
§8. Generalizations	192
§9. Obtaining Coverings of a Curve from its Jacobian; Application to Mordell's Conjecture	195
§10. Abelian Varieties Are Quotients of Jacobian Varieties	198
§11. The Zeta Function of a Curve	200
§12. Torelli's Theorem: Statement and Applications	202
§13. Torelli's Theorem: The Proof	204
Bibliographic Notes for Abelian Varieties and Jacobian Varieties	208
References	211

CHAPTER VIII
Néron Models
M. ARTIN

213

§1. Properties of the Néron Model, and Examples	214
§2. Weil's Construction: Proof	221
§3. Existence of the Néron Model: R Strictly Local	223

§4.	Projective Embedding	227
§5.	Appendix: Prime Divisors	229
	References	230

CHAPTER IX
Siegel Moduli Schemes and Their Compactifications over $\mathbb{C}$ 231
Ching-Li Chai

§0.	Notations and Conventions	231
§1.	The Moduli Functors and Their Coarse Moduli Schemes	232
§2.	Transcendental Uniformization of the Moduli Spaces	235
§3.	The Satake Compactification	238
§4.	Toroidal Compactification	243
§5.	Modular Heights	247
	References	250

CHAPTER X
Heights and Elliptic Curves   253
Joseph H. Silverman

§1.	The Height of an Elliptic Curve	253
§2.	An Estimate for the Height	256
§3.	Weil Curves	260
§4.	A Relation with the Canonical Height	263
	References	264

CHAPTER XI
Lipman's Proof of Resolution of Singularities for Surfaces   267
M. Artin

§1.	Introduction	267
§2.	Proper Intersection Numbers and the Vanishing Theorem	270
§3.	Step 1: Reduction to Rational Singularities	274
§4.	Basic Properties of Rational Singularities	278
§5.	Step 2: Blowing Up the Dualizing Sheaf	281
§6.	Step 3: Resolution of Rational Double Points	283
	References	287

CHAPTER XII
An Introduction to Arakelov Intersection Theory   289
T. Chinburg

§1.	Definition of the Arakelov Intersection Pairing	289
§2.	Metrized Line Bundles	292
§3.	Volume Forms	294
§4.	The Riemann–Roch Theorem and the Adjunction Formula	299
§5.	The Hodge Index Theorem	304
	References	307

CHAPTER XIII
Minimal Models for Curves over Dedekind Rings   309
T. Chinburg

§1.	Statement of the Minimal Models Theorem	309
§2.	Factorization Theorem	311

§3.	Statement of the Castelnuovo Criterion	314
§4.	Intersection Theory and Proper and Total Transforms	315
§5.	Exceptional Curves	317
	5A. Intersection Properties	317
	5B. Prime Divisors Satisfying the Castelnuovo Criterion	319
§6.	Proof of the Castelnuovo Criterion	321
§7.	Proof of the Minimal Models Theorem	323
	References	325

CHAPTER XIV
Local Heights on Curves 327
BENEDICT H. GROSS

§1.	Definitions and Notations	327
§2.	Néron's Local Height Pairing	328
§3.	Construction of the Local Height Pairing	329
§4.	The Canonical Height	331
§5.	Local Heights for Divisors with Common Support	332
§6.	Local Heights for Divisors of Arbitrary Degree	333
§7.	Local Heights on Curves of Genus Zero	334
§8.	Local Heights on Elliptic Curves	335
§9.	Green's Functions on the Upper Half-Plane	336
§10.	Local Heights on Mumford Curves	337
	References	339

CHAPTER XV
A Higher Dimensional Mordell Conjecture 341
PAUL VOJTA

§1.	A Brief Introduction to Nevanlinna Theory	341
§2.	Correspondence with Number Theory	344
§3.	Higher Dimensional Nevanlinna Theory	347
§4.	Consequences of the Conjecture	349
§5.	Comparison with Faltings' Proof	352
	References	353

Contributors

M. ARTIN
    Department of Mathematics, Massachusetts Institute of Technology, Cambridge, MA 02139, U.S.A., artin@math.mit.edu

CHING-LI CHAI
    Department of Mathematics, University of Pennsylvania, Philadelphia, PA 19104, U.S.A., chai@math.upenn.edu

T. CHINBURG
    Department of Mathematics, University of Pennsylvania, Philadelphia, PA 19104, U.S.A., ted@math.upenn.edu

GERD FALTINGS
    Max-Planck Institut für Mathematik, Gottfried-Claren-Strasse 26, 53225 Bonn, Germany, gerd@mpim-bonn.mpg.de

BENEDICT H. GROSS
    Department of Mathematics, Harvard University, Cambridge, MA 02138, U.S.A., gross@math.harvard.edu

F. O. MCGUINNESS
    Department of Mathematics, Columbia University, New York, NY 10027, U.S.A., om@math.colombia.edu

J. S. MILNE
    Department of Mathematics, University of Michigan, Ann Arbor, MI 48109, U.S.A., jmilne@math.lsa.umich.edu

MICHAEL ROSEN
    Department of Mathematics, Brown University, Providence, RI 02912, U.S.A., michael-rosen@brown.edu

STEPHEN S. SHATZ
   Department of Mathematics, University of Pennsylvania, Philadelphia, PA 19104, U.S.A., sss@math.upenn.edu

JOSEPH H. SILVERMAN
   Department of Mathematics, Brown University, Providence, RI 02912, U.S.A., jhs@math.brown.edu

PAUL VOJTA
   Department of Mathematics, University of California, Berkeley, CA 94720, U.S.A., vojta@math.berkeley.edu

Introduction

The chapters of this book, with the exception of Chapters II, XI and XII, are expanded versions of the lectures given at the Storrs Conference. They are intended, as was the conference, to introduce many of the ideas and techniques currently being used in arithmetic geometry; and in particular to explicate the tools used by Faltings in his proof of the Isogeny, Shafarevich and Mordell conjectures.

The first chapter is a brief overview, by Faltings himself, of the history leading up to the proof of the Mordell conjecture, and the second is a translation from the German of Faltings' paper in which he proved all three conjectures. The heart of this book, Chapters III through IX, contain (with varying amounts of detail) all of the results used in Faltings' paper. In particular, there is a thorough treatment of finite group schemes and p-divisible groups (Chapter III), Abelian and Jacobian varieties and schemes (Chapters IV, V, VII and VIII), their moduli spaces (Chapter IX) and height functions (Chapter VI). The prerequisites vary for each chapter, but in general, little is needed beyond what would normally be covered in one-year graduate courses in algebraic number theory and algebraic geometry.

After a brief chapter to illustrate the general theory for the particular case of elliptic curves (Chapter X), there are four chapters devoted to the theory of local height functions and arithmetic (Arakelov) intersection theory. Finally, Chapter XV contains an exposition of Vojta's far-reaching conjecture, whose consequences would include many of the standard finiteness theorems and outstanding conjectures in arithmetic geometry.

The editors hope that this volume will provide a path into the forest that is modern arithmetic geometry, wherein you will discover the beautiful flowers that blossom when arithmetic and geometry are intertwined, and there perchance, discover some new, exotic species heretofore unknown to the world of mathematics.

CHAPTER I

Some Historical Notes

GERD FALTINGS

The purpose of these notes is to give some information about the origin of the ideas used in the proofs of the conjectures of Tate, Shafarevich, and Mordell. They are not meant to be a complete historical treatment, and they present only the author's very personal opinion of how things evolved, and who contributed important ideas. He therefore apologizes in advance for the inaccuracies in them, and that he has omitted many who have contributed their share. He does not intend to offend them, and welcomes advice and remarks. Hopefully his remarks will encourage the reader to look into the original papers. The general strategy is to explain when and why the main ideas were invented. In explaining them we use the modern terminology, which usually makes it much easier to state them than it was at the time when they were first used. Of course, this does not mean that we intend to criticize those who invented them, which had to state them at a time when the technical means available were much weaker than those we have today.

The main topics are:

(1) The theorems of Mordell and Mordell–Weil.
(2) Siegel's theorem about integral points.
(3) The proof of the Mordell conjecture for function fields, by Manin and Grauert.
(4) The new ideas of Parshin and Arakelov, relating the conjectures of Mordell and Shafarevich.
(5) The work of Szpiro, extending this to positive characteristic.
(6) The theorem of Tate about endomorphisms of abelian varieties over finite fields.
(7) The work of Zarhin.
(8) Bibliographical remarks.

§1. The Theorems of Mordell and Mordell–Weil

Let us start (arbitrarily) with the theorem of L. J. Mordell, that the rational points on an elliptic curve are a finitely generated abelian group. He also conjectured that there are only finitely many rational points on curves of higher genus, but stated that this was only a guess, and that he had no argument for it. A. Weil, trying to prove this conjecture, extended his result to Jacobians of curves of arbitrary genus (and thus, in the modern terminology which is far more sophisticated than that which he had in his hands, to abelian varieties). Both Mordell and Weil used infinite descent. In short the argument is as follows:

Let A denote an abelian variety over a number field K. Show first that $A(K)/2 \cdot A(K)$ is finite, by injecting it into a cohomology-group $H^1(O_S, A[2])$. Then use the fact that the height function $h(P)$ is almost a quadratic form on $A(K)$.

A natural idea of how to use this result to prove the Mordell conjecture is the following:

Inject a curve X of genus $g > 1$ into its Jacobian J, and prove that any finitely generated subgroup of J has finite intersection with X. The latter property may be verified over the complex numbers, so that we may use analytic tools.

Unfortunately, this assertion is rather difficult and its only known proof consists in deriving it from the Mordell conjecture. Thus the theorem of Mordell–Weil does not help us in proving the Mordell conjecture, and in fact it is not used in the proof.

§2. Siegel's Theorem About Integral Points

This theorem of C. L. Siegel states that on an affine curve of positive genus there can be only finitely many integral points over any number field. (For affine curves there is a difference between integral and rational points, in contrast to complete curves.) The main tool is diophantine approximation, first used by Thue, refined by Siegel, and put into its final form by Roth.

The idea of Siegel's proof goes as follows:

Let K denote a number field, X a complete curve over K, and $\infty \in X(K)$ a rational point. For any other rational point $P \in X(K)$ the height $h(P)$ (for some projective embedding) can be computed as an Arakelov type intersection number between the divisors P and ∞. If P is an integral point on the affine curve $X - \{\infty\}$ only the infinite places of K contribute to this intersection, and we obtain $h(P)$ by summing over local contributions $h_v(P)$, for v an infinite place

$$h(P) = \sum_{v|\infty} h_v(P).$$

Diophantine approximation gives us a constant c such that for almost all P we have

$$h_v(P) \le c \cdot h(P) \quad \text{for all } v|\infty.$$

In fact, any $c > 2$ will do.

This estimate can be refined by using coverings:

If the genus g of X is positive, we can find unramified coverings $X' \to X$ of arbitrarily high degree n^g, n some natural number. After extending K we may assume that P lifts to a rational point $P' \in X'(K)$.

The construction of X' uses multiplication by n on the Jacobian of X, and gives a projective embedding of X' such that

$$h(P') = \frac{1}{n^2} \cdot h(P).$$

Furthermore, we obtain for the infinite contributions that $h_v(P')$ is approximately equal to $h_v(P)$, since at each infinte place P' can be very close only to one of the points lying above ∞, the covering being unramified.

If we apply the inequality above to X' instead of X, we obtain

$$h_v(P) \le \frac{c}{n^2} \cdot h(P)$$

for almost all P; that is, the inequality remains true for almost all integral points P if we take any positive $c > 0$ (instead of $c > 2$). As $h(P)$ is the sum over the finitely many $h_v(P)$, we obtain a contradiction for almost all P; that is, the number of P's must be finite.

After Siegel's theorem there was almost no progress for about 30 years, except that K. Mahler extended the results to S-integral points, S a finite set of places containing all the infinite ones. (He extended diophantine approximation to p-adic fields.) The results are ineffective because of the use of diophantine approximation, but we obtain effective bounds for the number of integral points. Nobody so far has been able to apply the technique to the study of rational points.

§3. The Proof of the Mordell Conjecture for Function Fields, by Manin and Grauert

In 1963 Yu Manin proved the Mordell conjecture for curves over a complex function field. It states that nonisotrivial curves have only finitely many rational points. (Some assumption is necessary to exclude the constant curves.) This proof was very ingenious and used the Gauss–Manin connection. Another proof was given by H. Grauert in 1966. Both times the essential use was made of the fact that the ground field has a nontrivial derivation. Thus the proofs do not carry over to number fields.

§4. The New Ideas of Parshin and Arakelov, Relating the Conjectures of Mordell and Shafarevich

A. N. Parshin introduced a new idea to relate Mordell's conjecture to the Shafarevich conjecture for curves. For this he introduced the "Parshin trick," which is amply explained elsewhere in this volume. He also proved the Shafarevich conjecture for families with good reduction everywhere, over a complex curve. The general idea is as follows:

Let $f: X \to B$ denote a smooth family of curves of genus $g \geq 2$, over the base B (a compact Riemann surface). Consider the relative differentials $\omega_{X/B} = \Omega^1_{X/B}$ and their direct images $f_*(\omega_{X/B})$, $f_*(\omega^2_{X/B})$. We can bound the degree of $f_*(\omega_{X/B})$, because

$$\Gamma(B, f_*(\omega_{X/B}) \otimes \Omega^1_B) = \Gamma(X, \Omega^2_X)$$

is a subspace of $H^2(X, \mathbb{C})$, and the dimension of the singular cohomology of X is bounded because of the Leray spectral sequence

$$E_2^{p,q} = H^q(B, R^p f_* \mathbb{C}) \Rightarrow H^{p+q}(X, \mathbb{C}).$$

This leads to bounds for the numerical invariants of the surface X, like c_1^2, $\omega^2_{X/B}$, etc.

We furthermore show that X is a surface of general type if the genus of B is bigger than one (which we can assume), and we derive that the possible X make up only finitely many algebraic families.

Finally, we show that X can be deformed only if it is isotrivial. This goes as follows:

The infinitesimal deformations are given by $H^1(X, \omega^{-1}_{X/B})$. By Kodaira it suffices to show that $\omega_{X/B}$ is ample if X is not isotrivial.

It is easy to see that $\omega_{X/B}$ has positive intersection with any curve on X. It remains to be seen that $\omega_{X/B}^2$ is positive (by Nakai's criterion). If $\omega_{X/B}^2 = 0$ we show that $\Lambda^g f_*(\omega_{X/B})$ has degree zero. But this is $\omega_{J/B}$, where $J = \text{Pic}^0(X/B)$ denotes the Jacobian of X, and this is the pullback of an ample line bundle on the moduli space of abelian varieties. Thus $\omega_{X/B}^2 = 0$ implies J/B isotrivial implies X/B isotrivial, by Torelli.

For the general case (that is, X may have bad reduction over some points of B) Parshin still proves quite a lot, namely:

Let X/B denote a stable family, with good reduction outside $S \subset B$.

(a) $\omega_{X/B}^2$ is bounded ($\omega_{X/B}$ = relative dualizing sheaf).
(b) X moves in only finitely many algebraic families.

(a) suffices to prove the Mordell conjecture for function fields, with an explicit bound for the height of a rational point. To conclude the proof of the Shafarevich conjecture we only need an extension of the deformation argument above.

This extension was given by Arakelov. His main new idea was the use of Weierstrass points to construct an injection

$$f^*(\Lambda^g f_*(\omega_{X/B})) \hookrightarrow \omega_{X/B}^{g(g+1)/2}.$$

As before we get that $\deg(\Lambda^g f_*(\omega_{X/B})) > 0$ for families which are not isotrivial, and the inclusion above can be used to show that $\omega_{X/B}^2 > 0$. The rest essentially goes as before.

Some further comments:

(i) S. Arakelov has developed an intersection theory for arithmetic surfaces, and some of the results above carry over (for example, $\omega_{X/B}^2 \geq 0$). But as the proof of the Shafarevich conjecture makes essential use of the differentiation in the ground field, it does not carry over. Similarly for the weaker boundedness statements which suffice for the Mordell conjecture:

They use Hodge theory, and even do not carry over to positive characteristics (Frobenius gives counterexamples).

(ii) The Shafarevich conjecture had been shown for curves. One part of its proof applies also to families of principally polarized abelian variety: The results about boundedness still hold.

But the deformation theory fails:

There are highly nonisotrivial families of abelian varieties which can be deformed. Therefore the stronger form of the Shafarevich conjecture does not hold for function fields. It thus came as a surprise that it can be shown for number fields.

(iii) The methods used for function fields leads to effective bounds for the heights of rational points. For number fields the situation is worse.

§5. The Work of Szpiro, Extending This to Positive Characteristic

L. Szpiro extended the results of Parshin and Arakelov to positive characteristics. The main new ideas are:

(a) A vanishing theorem in positive characteristic, which replaces Kodaira.
(b) Instead of nonisotrivial we demand that the Kodaira–Spencer class of a family is nonzero. We derive ampleness for $\omega_{X/B}$, as before.
(c) To get upper bounds for $\deg(\Lambda^g f_* \omega_{X/B})$ we use that the Kodaira–Spencer class $\kappa \in H^1(X, f^*(\Omega_B^1) \otimes \omega_{X/B}^{-1})$ does not vanish. Thus $\omega_{X/B}$ cannot be "too ample," because otherwise this would contradict the vanishing theorem mentioned above.

Furthermore, his work very much helped to clarify the problems remaining for number fields. Also, for some time he seemed to stand alone with his optimism and his belief that the ideas of Parshin and Arakelov were the right method to tackle the problems. He also convinced the author of these lines that his view was correct.

§6. The Theorem of Tate About Endomorphisms of Abelian Varieties over Finite Fields

A different line of investigation was started by J. Tate when he proved the Tate conjecture for endomorphisms of abelian varieties over finite fields (which is a special case of his much more general conjecture about algebraic cycles). For this he invented the idea of dividing the abelian variety by finite subgroups, and of using the fact that up to isomorphism there are only finitely many different quotients. (This was easy to prove, since the quotients are parametrized by some variety of finite type, which has only finitely many rational points over the finite base field.) He had some difficulty in treating polarizations, which forced him to use the fact that the Galois group in question is cyclic, generated by the Frobenius element.

His essential contribution was to show how to derive the Tate conjecture from the Shafarevich conjecture. It is thus quite amusing that the proof of the Shafarevich conjecture for number fields goes the opposite way, since the Tate conjecture is the first important step in it.

§7. The Work of Zarhin

J. G. Zarhin generalized Tate's work to function fields over finite fields. He contributed two main new ideas.

(i) He showed how to get rid of the polarizations (the "Zarhin trick").

(ii) He used the height on the moduli space of abelian varieties. (Here he could build on some ideas of Parshin.)

Point (ii) was made difficult by the fact that he had to use Mumford's theta functions to compactify the moduli space of abelian varieties. Today we might phrase his argument as follows:

The height of a semiabelian variety G over a base B ($=$ a curve over a finite field) is the degree of $\omega_{G/B}$. Any isogeny of degree prime to the characteristic is étale and induces an isomorphism for the ω's. Hence it does not change the height.

BIBLIOGRAPHICAL REMARKS

General Information

Talks of P. Deligne and L. Szpiro, in *Séminaire Bourbaki* 1983, No. 616/619.
Faltings, G. and Wüstholz, G. Rational points. *Seminar Bonn/Wuppertal* 1983/84. Vieweg: Wiesbaden 1984.
Faltings, G. Neure Fortschritte in der diophantischen Geometrie, in *Perspectives in Mathematics/Anniversary of Oberwolfach* 1984. Birkhäuser-Verlag: Basel, 1984.
Faltings, G. Die Vermutungen von Tate und Mordell. *Jber. d. Dt. Math.-Verein* **86** (1984), 1-13.
Szpiro, L. Séminaire sur les pinceaux arithmétiques: La conjecture de Mordell. *Astérisque* **127** (1985).

Section 1

Mordell, L. J. On the rational solutions of the indeterminate equation of third and fourth degrees. *Proc. Cambridge Phil. Soc.* **21** (1922), 179-192.
Mumford, D. *Abelian Varieties*. Oxford University Press: Oxford, 1974.
Weil, A. L'arithmétique sur les courbes algébriques. *Acta Math.* **52** (1928), 281-315.

Section 2

Lang, S. *Elements of Diophantine Geometry*. Springer-Verlag: New York, 1983.
Mahler, K. Über die rationalen Punkte auf Kurven vom Geschlecht 1. *J. Reine Angew. Math.* **170** (1934), 168-178.
Roth, K. F. Rational approximations to algebraic numbers. *Mathematika* **2** (1955), 1-20.
Siegel, C. L. Über einige Anwendungen diophantischer Approximationen. *Abh. Preuss. Akad. Wiss., Phys.-Math. Kl.* **1929**, No. 1 (also in collected works).

Section 3

Grauert, H. Mordell's Vermutung über rationale Punkte auf algebraischen Kurven und Funktionenkörper. *Publ. Math. IHES* **25** (1965), 131-149.
Manin, Yu. I. Rational points on an algebraic curve over function fields. *Trans. Amer. Math. Soc.* **50** (1966), 189-234.

Section 4

Arakelov, S. An intersection theory for divisors on an arithmetic surface. *Math. USSR Izv.* **8** (1974), 1167-1180.

Arakelov, S. Families of curves with fixed degeneracies. *Math. USSR Izv.* **5** (1979), 1277–1302.

Faltings, G. Arakelov's theorem for abelian varieties. *Invent. Math.* **73** (1983), 337–347.

Faltings, G. Calculus on arithmetic surfaces. *Ann. Math.* **119** (1984), 387–424.

Parshin, A. N. Algebraic curves over function fields I. *Math. USSR Izv.* **2** (1968), 1145–1170.

Parshin, A. N. Quelques conjectures de finitude en géometrie diophantienne. *Actes Congres Intern. Math. Nizza* (1970), I, 467–471.

Section 5

Szpiro, L. Sur le théorème de rigidité de Parsin et Arakelov. *Astérisque* **64** (1974), 169–202.

Szpiro, L. Séminaire sur les pinceaux de courbes de genre au moins deux. *Astérisque* **86** (1981).

Section 6

Tate, J. Algebraic cycles and poles of zeta functions, in *Arithmetical Algebraic Geometry*. Harper and Row: New York, 1965.

Tate, J. Endomorphisms of abelian varieties over finite fields. *Invent. Math.* **2** (1966), 134–144.

Section 7

Zarhin, J. G. Isogenies of abelian varieties over fields of finite characteristics. *Math. USSR Sbornik* **24** (1974), 451–461.

Zarhin, J. G. A remark on endomorphisms of abelian varieties over function fields of finite characteristics. *Math. USSR Izv.* **8** (1974), 477–480.

CHAPTER II

Finiteness Theorems for Abelian Varieties over Number Fields

GERD FALTINGS

§1. Introduction

Let K be a finite extension of $\mathbb{Q}$, A an abelian variety defined over K, $\pi = \operatorname{Gal}(\bar{K}/K)$ the absolute Galois group of K, and l a prime number. Then π acts on the (so-called) Tate module

$$T_l(A) = \varprojlim_n A[l^n](\bar{K}).$$

The goal of this chapter is to give a proof of the following results:

(a) The representation of π on $T_l(A) \otimes_{\mathbb{Z}_l} \mathbb{Q}_l$ is semisimple.
(b) The map

$$\operatorname{End}_K(A) \otimes_{\mathbb{Z}} \mathbb{Z}_l \to \operatorname{End}_\pi(T_l(A))$$

is an isomorphism.
(c) Let S be a finite set of places of K, and let $d > 0$. Then there are only finitely many isomorphism classes of abelian varieties over K with polarizations of degree d which have good reduction outside of S.

(a) and (b) are known as the Tate conjectures, (c) as the Shafarevich conjecture. Furthermore, one knows [9] that the Mordell conjecture follows from (c). The Tate conjectures for abelian varieties over finite fields have already been proven by Tate himself. Zarhin generalized this to function fields over such fields, [15], [16], and our proof is an adaptation of his method to the case of a number field. Arakelov supplied the dictionary necessary for this translation [2], and the author has built upon his methods [5]. In brief, what is needed is to provide "everything" with a hermitian metric.

The proof of (c) is achieved by first showing finiteness only for isogeny classes. The basic idea for this was communicated to me by a referee for *Inventiones* in connection with the publication of my paper [6], and I then had only to translate it from Hodge theory into étale cohomology. I would therefore like to heartily thank this referee, who is personally unknown to me, for his suggestion.

The rest of the proof of (c) uses a variant of the methods employed in proving (a) and (b).

The paper begins, first of all, with some technical details concerning heights. The complications arise because, at least to my knowledge, no good moduli space for semiabelian varieties over $\mathbb{Z}$ exists yet. (L. Moret-Bailly, who investigated the situation over functions fields, had to struggle with similar problems [7].) After that, we use the very beautiful results of Tate [13] on p-divisible groups. The conclusion is then again somewhat technical.

I have learned much about the subject from L. Szpiro, and I want to thank him here for introducing me to this circle of problems. P. Deligne called my attention to a discrepancy in an earlier version of this work.

§2. Semiabelian Varieties

Definition. Let S be a scheme (or an algebraic stack). A semiabelian variety of relative dimension g over S is a smooth algebraic group $p: G \to S$ whose fibres are connected of dimension g, and are extensions of an abelian variety by a torus.

EXAMPLE. Let $q: C \to S$ be a stable curve of genus g [4]. Then

$$J = \text{Pic}^\tau(C/S) \to S$$

is a semiabelian variety of relative dimension g.

We need the following:

Lemma 1. *Let S be normal, $U \subset S$ open and dense, $p_1: A_1 \to S$ and $p_2: A_2 \to S$ two semiabelian varieties, $\phi: A_1/U \to A_2/U$ a homomorphism of algebraic groups defined over U. Then ϕ can be extended uniquely over all of S.*

PROOF. This is well known in case S is the spectrum of a complete discrete valuation ring. In general, one reduces immediately to the case in which S is noetherian and excellent, and writes

$$X \subseteq A_1 \times_S A_2$$

for the closure of the graph of ϕ.

After base change with suitable valuation rings, one sees that the projection $\text{pr}_1: X \to A_1$ is proper, and that its fibres have only one point. Since A_1 is normal, pr_1 must be an isomorphism, and X the graph of the uniquely determined extension of ϕ. (Uniqueness follows, for example, by consideration of torsion points, or in a thousand other ways.) □

Definition. Let $p: A \to S$ be a semiabelian variety of relative dimension g, $s: S \to A$ the zero section.
Let
$$\omega_{A/S} = s^*(\Omega^g_{A/S}),$$
$\omega_{A/S}$ is a line bundle on S.

Remarks. (a) If p is proper, then $\omega_{A/S} \cong p_*(\Omega^g_{A/S})$.
(b) $\omega_{A/S}$ commutes with a change of base.
(c) If $A = \text{Pic}^\tau(C/S)$ for a stable curve $q: C \to S$, then $\omega_{A/S} \cong \Lambda^g q_*(\omega_{C/S})$, where $\omega_{C/S}$ denotes the relative dualizing module.
(d) If $S = \text{Spec}(\mathbb{C})$ and p is proper (i.e., $A/\mathbb{C}$ is a complex abelian variety), then $\omega_{A/S} \cong \Gamma(A, \Omega^g_{A/\mathbb{C}})$ admits a canonical hermitian scalar product, namely:
If α, β are holomorphic differential forms on A, then

$$\langle \alpha, \beta \rangle = \left(\frac{i}{2}\right)^g \int_A \alpha \wedge \bar{\beta}.$$

We need some facts about the moduli spaces for stable curves and abelian varieties. For this the language of algebraic stacks seems to be the most appropriate. Should this notation appear too abstract to the reader, he might think through the following considerations:

We are really concerned with finiteness statements. If $\mathfrak{G}$ is one of the stacks to be introduced below, and S denotes the corresponding coarse moduli space, there is always an open covering

$$S = \bigcup_{i=1}^{r} U_i$$

and finite surjective maps $V_i \to U_i$, such that over V_i the "universal object for $\mathfrak{G}$" exists. One can then carry out all calculations in the V_i.

Now for the algebraic stacks to be used here.

(1) $\overline{\mathfrak{M}}_g$ classifies stable curves of genus g [4]. $\overline{\mathfrak{M}}_g$ is proper over $\text{Spec}(\mathbb{Z})$, and the coarse moduli variety belonging to it is called $\overline{M}_g$.
(2) $\mathfrak{A}_g$ classifies the principally polarized abelian varieties of relative dimension g, and A_g the corresponding moduli variety.

$\mathfrak{A}_g$ is not proper over $\text{Spec}(\mathbb{Z})$, but the following facts are known:

(a) If
$$p: A \to \mathfrak{A}_g$$
denotes the universal abelian variety over $\mathfrak{A}_g$, then there exists an $r > 0$ for which $(\omega_{A/\mathfrak{A}_g})^{\otimes r}$ defines a very ample line bundle on $A_g/\mathbb{Q}$ [3].

Let $\bar{A}_g/\mathbb{Q}$ be the Zariski closure in the corresponding projective space $\mathbb{P}^N_\mathbb{Q}$, $\bar{A}_g/\mathbb{Z}$ the Zariski closure in $\mathbb{P}^N_\mathbb{Z}$, and $\mathcal{M}$ the line bundle $\mathcal{O}(1)$ on $\bar{A}_g/\mathbb{Z}$. ($\mathcal{M}$ extends $(\omega_{A/A_g})^{\otimes r}$ on $\bar{A}_g/\mathbb{Q}$.)

(b) Over $\mathbb{C}$ there is a proper dominating morphism

$$\phi\colon \mathfrak{N} \to \bar{A}_g/\mathbb{C},$$

such that there exists a semiabelian variety over $\mathfrak{N}$ which extends the universal variety over $\mathfrak{A}_g$ (see [8, §9]). Moreover, it is known that the $\omega^{\otimes r}$ of this semiabelian variety is isomorphic to $\phi^*(\mathcal{M})$. (This is proven by a direct calculation; see my exposition in [6, §2]).

Lemma 2. *Over $\mathrm{Spec}(\mathbb{Z})$, there exists a proper algebraic stack $\mathfrak{Z}$, an open subset $\mathfrak{U} \subset \mathfrak{Z}$, and a proper morphism $\psi\colon \mathfrak{U} \to \mathfrak{A}_g$, which extends to a $\bar{\psi}\colon \mathfrak{Z}/\mathbb{Q} \to \bar{A}_g/\mathbb{Q}$, such that the following objects exist:*

(a) *A stable curve $q\colon C \to \mathfrak{Z}$.*
(b) *A sub-line-bundle ($=$ local direct summand) $\mathcal{L} \subseteq \Lambda^g q_*(\omega_{C/\mathbb{Z}})$.*
(c) *A pair of group homomorphisms over $\mathfrak{U}$*

$$\alpha\colon \mathrm{Pic}^\tau(C/\mathfrak{Z}) \to \psi^*(A),$$

$$\beta\colon \psi^*(A) \to \mathrm{Pic}^\tau(C/\mathfrak{Z}),$$

with

$$\alpha \circ \beta = \text{multiplication by } d \quad d \in \mathbb{N}, d \neq 0.$$

(Here A is again the universal abelian variety over $\mathfrak{A}_g$.)

(d) *There exists an isomorphism $\mathcal{L}^{\otimes r} = \bar{\psi}^*(\mathcal{M})$ over $\mathfrak{Z} \otimes_\mathbb{Z} \mathbb{Q}$ and $\mathcal{L}$ is the image of*

$$\alpha^*\colon \psi^*(\omega_{A/A_g}) \to \Lambda^g q_*(\omega_{C/\mathbb{Z}})$$

over $\mathfrak{U}/\mathbb{Q}$. The resulting isomorphism (over $\mathfrak{U}/\mathbb{Q}$)

$$\psi^*(\omega_{A/A_g})^{\otimes r} \cong \psi^*(\mathcal{M})$$

is the ψ^*-pullback of the isomorphism over A_g resulting from the construction of $\mathcal{M}$.

PROOF. The abelian variety associated to the generic point of $\mathfrak{A}_g$ is the quotient of a Jacobian. The curve thus obtained corresponds to a rational map from $\mathfrak{A}_g$ to $\overline{\mathfrak{M}}_{\tilde{g}}$, for some $\tilde{g}$.

If one considers the graph of this map, one obtains (with the help of some trivial additional considerations) a first candidate $\mathfrak{Z}$, such that conditions (a) and (by Lemma 1) (c) are already fulfilled. $\mathcal{L}$ is then already determined over $\mathfrak{U} \otimes_\mathbb{Z} \mathbb{Q}$ by (d), and furnishes a rational map from $\mathfrak{U} \otimes_\mathbb{Z} \mathbb{Q}$ to a certain projective bundle over $\mathfrak{Z}$. One replaces $\mathfrak{Z}$ by the normalization of the closure of the graph of this map and then (b) and the second part of (d) are also fulfilled. For the rest of (d), one notes that we have already constructed the required isomorphism over $\mathfrak{U} \otimes_\mathbb{Z} \mathbb{Q}$, and one need now only prove the existence of an extension to $\mathfrak{Z} \otimes_\mathbb{Z} \mathbb{Q}$. To do this, one can extend the ground field from $\mathbb{Q}$ to $\mathbb{C}$, and it suffices to prove the existence of an extension for a $\mathfrak{Z}/\mathbb{C}$ which is dominant and proper over $\mathfrak{Z}$.

With the help of the map $\phi: \mathfrak{N} \to \bar{A}_g/\mathbb{C}$ introduced above, one constructs a normal $\mathfrak{Z}$, such that $\psi^*(A)$ extends to a semiabelian variety on $\mathfrak{Z}$. By Lemma 1 one can also extend α and β, and this furnishes the desired isomorphism over $\mathfrak{Z}$. □

Corollary. *There exists a natural number e with the following property:*
Let K be a number field, R its ring of integers,

$$p: A \to \mathrm{Spec}(R)$$

a semiabelian variety such that the generic fibre A/K is proper over K and possesses a principal polarization. Then the corresponding map $\rho: \mathrm{Spec}(K) \to A_g/\mathbb{Q}$ extends to a $\rho: \mathrm{Spec}(R) \to \bar{A}_g/\mathbb{Z}$.

By construction, there exists an isomorphism

$$\rho^*(\mathcal{M}) \otimes_R K \cong (\omega_{A/R})^{\otimes r} \otimes_R K.$$

Using this isomorphism one gets:

$$e \cdot \rho^*(\mathcal{M}) \subseteq (\omega_{A/R})^{\otimes r} \subseteq e^{-1} \cdot \rho^*(\mathcal{M}) \quad (\subseteq \rho^*(\mathcal{M}) \otimes_R K).$$

PROOF. We may assume that $\bar{\psi}: \mathfrak{Z}/\mathbb{Q} \to \bar{A}_g/\mathbb{Q}$ can be extended to a proper $\bar{\psi}: \mathfrak{Z}/\mathbb{Z} \to \bar{A}_g/\mathbb{Z}$. Then there is a finite field extension $K' \supseteq K$ (with integers $R' \subseteq K'$), such that ρ can be lifted to

$$\tilde{\rho}: \mathrm{Spec}(R') \to \mathfrak{Z}.$$

Since $\bar{\psi}^*(\mathcal{M})$ and $\mathcal{L}^{\otimes r}$ are isomorphic over $\mathfrak{Z} \otimes_\mathbb{Z} \mathbb{Q}$, there is an $e_1 > 0$ such that over $\mathfrak{Z}$

$$e_1 \cdot \mathcal{L}^{\otimes r} \subseteq \bar{\psi}^*(\mathcal{M}) \subseteq e_1^{-1} \cdot \mathcal{L}^{\otimes r}.$$

It suffices to prove the claim after a change of base to R', and then we need only compare $\omega_{A/R'}$ and $\tilde{\rho}^*(\mathcal{L})$.

By pullback one obtains a stable curve

$$q: C \to \mathrm{Spec}(R')$$

and

$$\alpha: \mathrm{Pic}^\tau(C/R') \to A/R', \qquad \beta: A/R' \to \mathrm{Pic}^\tau(C/R')$$

with $\alpha \circ \beta = d \cdot id$ (use Lemma 2 over R'), such that $\tilde{\rho}^*(R)$ is a subbundle of $\Lambda^g q_*(\omega_{C/R'})$, which is generated by the image of

$$\alpha^*: \omega_{A/R'} \to \Lambda^g q_*(\omega_{C/R'}).$$

From this, it follows immediately that $d^g \cdot \bar{\rho}^*(\mathcal{L}) \subseteq \omega_{A/R'} \subseteq \bar{\rho}^*(\mathcal{L})$, and we are done. □

§3. Heights

Again let K be a number field, R the ring of integers in K. By analogy to [5], we define a metrized line bundle on $\operatorname{Spec}(R)$ to be a projective R-module P of rank 1, together with norms $\| \ \|_v$ on $P \otimes_R K_v$ for all infinite places of K. K_v denotes the completion of K at v, and we define $\varepsilon_v = 1$ or 2 according to whether $K_v \cong \mathbb{R}$ or $K_v \cong \mathbb{C}$. The degree of the metrized line bundle is defined as ("#" = order)

$$\operatorname{Deg}(P, \| \ \|) = \log(\#(P/R \cdot p)) - \sum_v \varepsilon_v \log \|p\|_v,$$

where p is a nonzero element of P, the sum runs over all infinite places of K, and the right-hand side is of course independent of p.

Remark. The idea of metrized line bundle was introduced by Arakelov ([2]). The degree of P is naturally also connected with the volume of P.

We are especially interested in the metrized line bundle $\omega_{A/R}$, where

$$p\colon A \to \operatorname{Spec}(R)$$

is a semiabelian variety, with proper generic fibre A/K. The metrics at the infinite places come from the scalar product mentioned above

$$\|\alpha\|_v^2 = \left(\frac{i}{2}\right)^g \cdot \int_{A(\bar{K}_v)} \alpha \wedge \bar{\alpha}.$$

Definition. The moduli-theoretic height $h(A)$ is

$$h(A) = \frac{1}{[K:\mathbb{Q}]} \deg(\omega_{A/R}).$$

One sees immediately that $h(A)$ is invariant under extension of the ground field. The name "height" is justified by the following:

In general one defines the height of a point $x \in \mathbb{P}^n(K)$ by associating to x a morphism $\rho\colon \operatorname{Spec}(R) \to \mathbb{P}^n_{\mathbb{Z}}$, providing the bundle $\mathcal{O}(1)$ on $\mathbb{P}^n_{\mathbb{C}}$ with a metric, and then defining the height of x to be

$$\frac{1}{[K:\mathbb{Q}]} \cdot \deg(\rho^*\mathcal{O}(1)).$$

Changing the hermitian metric only changes the height function by a bounded amount, and it is known that for every c, there are only finitely many K-rational points of $\mathbb{P}^n$ with height $\leq c$. Corresponding considerations apply to closed subvarieties of $\mathbb{P}^n$. In our situation, one embeds A_g in $\mathbb{P}^n_{\mathbb{Z}}$ as above, by means of $\mathcal{M}$. We have already defined a metric $\| \ \|$ on the line bundle induced by $\mathcal{M}$ on $A_g(\mathbb{C})$. Should this metric admit an extension to $\bar{A}_g(\mathbb{C})$, one could use it to define the height, and then the corollary to Lemma 2 would show for a semiabelian variety A over R (as above), which has principal polarization over K and so defines an $x \in A_g(K)$, that $h(x)$ and $r \cdot h(A)$ differ only by a bounded amount.

Unfortunately, the metric $\| \ \|$ has singularities along $\bar{A}_g(\mathbb{C}) - A_g(\mathbb{C})$; however, these are so mild that the fundamental finiteness property of heights remains true.

Definition. Let $X/\mathbb{C}$ be a compact complex variety, $Y \subseteq X$ a closed subvariety, $\mathscr{M}$ a line bundle of X, $\| \ \|$ a hermitian metric on $\mathscr{M} | X - Y$. The metric has logarithmic singularities along Y if the following holds:

There is a proper dominant map

$$\phi: \tilde{X} \to X,$$

such that $\tilde{X}$ is smooth and $\phi^{-1}(Y)$ is a divisor with normal crossings, and such that for a local generator h of $\phi^*(\mathscr{M})$ and a local equation f of $\phi^{-1}(Y)$,

$$\text{Sup}\{\|h\|, \|h\|^{-1}\} \leq c_1 \cdot |(\log |f|)|^{c_2} \qquad \text{(with constants } c_1, c_2 > 0\text{) holds.}$$

EXAMPLE.

$$X = \bar{A}_g(\mathbb{C}), \qquad Y = \bar{A}_g(\mathbb{C}) - A_g(\mathbb{C}), \qquad \mathscr{M} \text{ and } \| \ \| \text{ as before.}$$

Indeed this was already proven in [6, end of §2], but at the request of the referees we present a short sketch of the proof here:

More generally, it is true that for a smooth X, a divisor Y on X with normal crossings and a semiabelian variety

$$p: A \to X,$$

such that p is proper and A is principally polarized over $X - Y$, the canonical metric on $\omega_{A/X}$ has logarithmic singularities along Y.

To see this one considers $p_*(\Omega^1_{A/X})$ instead of $\omega_{A/X}$ ("logarithmic singularities" can also be defined for vector bundles), and by the methods of Section 2 one reduces the problem to the case in which A is the Jacobian of a semistable curve $q: C \to X$. We will treat briefly the case of a semistable curve over the unit disc $\mathbb{D}$. The general case goes exactly the same way. If

$$q: C \to \mathbb{D} = \{t | |t| < 1\}$$

is a semistable curve, with good reduction except at 0, then C admits a covering

$$C = \bigcup_{i=1}^{l} U_i$$

such that either

(a) $\qquad U_i = \{(z, t) | |z| < 1, |t| < 1\}, \qquad q|U_i: U_i \to \mathbb{D}$

is smooth, and z furnishes coordinates on all fibres; or

(b) $\qquad U_i = \{(z_1, z_2, t) | |z_1| < \varepsilon, |z_2| < \varepsilon, z_1 z_2 = t^m\}, \qquad q(z_1, z_2, t) = t.$

If α is a local section of $q_*(\omega_{C/\mathbb{D}})$, then on the U_i's is of the form

(a) $\qquad\qquad\qquad \alpha = \text{(holomorph)} \cdot dz \quad \text{resp.}$

(b) $\qquad\qquad\qquad \alpha = \text{(holomorph)} \dfrac{dz_1}{z_1}.$

An explicit calculation shows that for $t \to 0$

$$\frac{i}{2} \int_{U_i \cap q^{-1}(t)} \alpha \wedge \bar{\alpha}$$

either remains bounded, or grows at most like $|\log|t||$. Further, one sees immediately that

$$\| \ \| \geq (\text{pos. const.}) \| \ \|_1,$$

where $\| \ \|_1$ denotes a hermitian metric on $q_*(\omega_{C/X})$ defined on all of X.

Lemma 3. *Let $X \subseteq \mathbb{P}^n_\mathbb{Z}$ be Zariski-closed, $Y \subseteq X$ closed, $\| \ \|$ a hermitian metric on $\mathcal{O}(1)|(X(\mathbb{C}) - Y(\mathbb{C}))$, with logarithmic singularities along Y. For a number field K, and $x \in X(K) - Y(K)$ one defines $h(x)$ as before. Then for every c, there are only finitely many $x \in X(K) - Y(K)$ with $h(x) \leq c$.*

PROOF. Let $\| \ \|_1$ be a hermitian metric for $\mathcal{O}(1)|X(\mathbb{C})$, h_1 the corresponding height function, and choose an $s > 0$ and

$$f_1, \ldots, f_i \in \Gamma(X/\mathbb{Z}, \mathcal{O}(s)),$$

whose set of common zeros is exactly Y. Then $\| \ \|_1$ defines a metric on $\mathcal{O}(s)$ (which is also called $\| \ \|_1$), and from the hypotheses it follows immediately that there exist constants $c_1, c_2 > 0$ with

$$\log \left| \frac{\| \ \|}{\| \ \|_1}(z) \right| \leq c_1 + c_2 \cdot \inf \{\log(|\log \| f_i(z) \|_1|)\}$$

for $z \in X(\mathbb{C})$.

If $x \in X(K) - Y(K)$, there corresponds a

$$\rho: \text{Spec}(R) \to X,$$

and then the f_i define sections $\rho^*(f_i)$ of $\rho^*(\mathcal{O}(s))$, with whose help the height $h_1(x)$ can be calculated. Since $\| f_i(z) \|_1$ is bounded above on $X(\mathbb{C})$, one immediately obtains constants $c_3, c_4 > 0$ with

$$|h(x) - h_1(x)| \leq c_3 + c_4 \log(h_1(x)).$$

The claim follows directly. $\qquad\square$

We can now reap the fruits of our labors. The following result is almost already proven.

Theorem 1. *Let c be given. Then there are only finitely many isomorphism classes of pairs of:*

(i) *a semiabelian variety of relative dimension g*

$$p: A \to \mathrm{Spec}(R)$$

with proper generic fibre A/K.
(ii) *a principal polarization of A/K for which*

$$h(A) \leq c$$

holds.

PROOF. According to the corollary to Lemma 2 the difference between $h(x)$ and $r \cdot h(A)$ is bounded ($x \in A_g(X)$ corresponding to A). According to Lemma 3, the A for which $h(A) \leq c$ provide only finitely many different $x \in A_g(K)$. We must now note that only finitely many K-isomorphism classes can induce the same isomorphism class over the algebraic closure $\bar{K}$. Thus, we fix such a class over $\bar{K}$ and consider the A/K belonging to it. It is known that all of these have bad reduction at the same places of K. It follows immediately from Lemma 4 below that there exists a finite extension $K' \supseteq K$, which for some $n \geq 3$ contains the nth division points of all A/K. It is known that our A's are already isomorphic over K', and the rest follows from basic general theorems of Galois cohomology. □

There remains to be added the

Lemma 4. *Let K be a number field, S a finite set of places of K. Then there are only finitely many field extensions of a given degree which are unramified outside of S.*

PROOF. Well known (Hermite–Minkowski). □

§4. Isogenies

We examine the behavior of $h(A)$ under isogeny. As always K is a number field, $R \subset K$ its ring of integers.

Let

$$p_1: A_1 \to \mathrm{Spec}(R)$$

and

$$p_2: A_2 \to \mathrm{Spec}(R)$$

be semiabelian varieties with proper generic fibre,

$$s: \mathrm{Spec}(R) \to A_1$$

the zero section, and $\phi: A_1 \to A_2$ an isogeny. (Of course, by Lemma 1, it is enough that ϕ be defined over K.)

We set $G = \text{Ker}(\phi) \subseteq A_1$. Since ϕ is automatically flat, G is a quasi-finite flat group scheme over $\text{Spec}(R)$.

ϕ induces an injection

$$\phi^*: \omega_{A_2/R} \to \omega_{A_1/R},$$

and one sees at once that

$$\#(\omega_{A_1/R}/\phi^*(\omega_{A_2/R})) = \#s^*(\Omega^1_{A_1/A_2}) = \#s^*(\Omega^1_{G/R}).$$

Since moreover ϕ^* changes the norms at the infinite places by $(\deg(\phi))^{1/2}$, there follows directly

Lemma 5.

$$h(A_2) = h(A_1) + \frac{1}{2}\log(\deg(\phi)) - \frac{1}{[K:\mathbb{Q}]} \cdot \log(\#s^*(\Omega^1_{G/R})).$$

Remark. If G is annihilated by a number $n \in \mathbb{N}$, then n also annihilates $\Omega^1_{G/R}$. It follows that

$$\exp(2[K:\mathbb{Q}] \cdot (h(A_2) - h(A_1)))$$

is a rational number in whose numerator and denominator only the prime factors of $\deg(\phi)$ appear. The exponents of these primes can be bounded by their exponents in $\deg(\phi)$.

We now investigate the behavior of the $h(A_n)$, in the case $A_n = A/G_n$, where G_n runs through the levels of an l-divisible group $G \subseteq A[l^\infty]$.

Theorem 2. *Let $p: A \to \text{Spec}(R)$ be a semiabelian variety with proper generic fibre, l a prime number, and $G/K \subseteq A[l^\infty]/K$ an l-divisible subgroup.*

Furthermore, let G_n be the kernel of l^n in G, and A_n the semiabelian variety $A_n = A/G_n$. Then

$$h(A_n) = h(A).$$

PROOF. Let $v_1, \ldots, v_n$ be the places of k lying over l, $K_i = K_{v_i}$ the corresponding local fields, $R_i \subseteq K_i$ the valuation rings, $m_i = [K_i : \mathbb{Q}_l]$, so that

$$m = [K:\mathbb{Q}] = \sum_{i=1}^{r} m_i.$$

We fix an i, and consider the formal group scheme $\hat{A}$ over $\text{Spf}(R_i)$, the completion of A/R_i along the fibre A_s over the closed point s of $\text{Spec}(R_i)$.

A_s is an extension $0 \to T_s \to A_s \to B_s \to 0$, with T_s a torus, B_s an abelian variety.

According to general fundamental theorems, one can lift T_s to a torus T over $\text{Spec}(R_i)$, and $\hat{T}$ is a closed formal subscheme of $\hat{A}$. (Morphisms from T_s into smooth group schemes can be lifted.)

Let
$$\hat{H}_i = \hat{A}[l^\infty]$$
be the associated l-divisible group. $\hat{H}_i$ is the formal completion of an l-divisible group H_i over R_i, and H_i/K_i is an l-divisible subgroup of $A[l^\infty]/K_i$. The same holds for $T[l^\infty]$, and to these subgroups, there correspond $\mathbb{Z}_l$-sublattices
$$T_l(T) \subseteq T_l(H_i) \subseteq T_l(A).$$

Lemma 6. *Let $D_i = \text{Gal}(\bar{K}_i/K_i)$ be the absolute Galois group of K_i, $I_i \subseteq D_i$ the ramification group.*

Then I_i acts trivially on $T_l(A)/T_l(H_i)$, and the induced action of $D_i/I_i \cong \hat{\mathbb{Z}}$ factors through a finite quotient of $\hat{\mathbb{Z}}$.

PROOF. Let
$$\langle \, , \, \rangle : T_l(A) \times T_l(A) \to \mathbb{Z}_l(1) = T_l(\mathbb{G}_m)$$
be the symplectic form induced by a polarization of A/K. $\langle \, , \, \rangle$ is not degenerate, and it is known [SGA VII, Exp IX, §7] that
$$\langle T_l(T), T_l(H_i) \rangle = 0.$$
By a dimension argument, $T_l(H_i) = T_l(T)^\perp$, and we have an injection
$$T_l(A)/T_l(H_i) \hookrightarrow \text{Hom}_{\mathbb{Z}_l}(T_l(T), \mathbb{Z}_l(1)).$$
This injection is D_i linear, and D_i acts on $\text{Hom}_{\mathbb{Z}_l}(T_l(T), \mathbb{Z}_l(1))$ in the required way. $\square$

Now, once again, back to our $G/K \subseteq A[l^\infty]/K$. After base extension $K \subseteq K_i$, we can form the intersection $G_i = G \cap H_i$. This is the maximal l-divisible subgroup of G_i/K_i which can be extended over R_i, and we have
$$\#(s^*\Omega^1_{A/A_n} \otimes_R R_i) = \#s^*(\Omega^1_{(G_i)_n/R_i}).$$
By [13, Prop. 2], one can calculate this immediately: Let d_i be the dimension of the maximal formal subgroup of G_i. Then
$$\#s^*(\Omega^1_{(G_i)_n/R_i}) = l^{n \cdot m_i \cdot d_i}.$$
If C_i denotes the completion of the algebraic closure of K_i, then it is known furthermore [13, Theorem 3, Cor. 2], that as D_i-modules
$$T_l(G_i) \otimes_{\mathbb{Z}_l} C_i \cong C_i^{n_i - d_i} \oplus C_i^{d_i}(+1) \qquad (h_i = \text{height}(G_i), \text{``}(+1)\text{''} = \text{Tate twist}).$$
Together with Lemma 6, this implies that D_i acts on
$$\Lambda^h T_l(G) \otimes_{\mathbb{Z}_l} C_i \subseteq \Lambda^h T_l(A) \otimes_{\mathbb{Z}_l} C_i \qquad (h = \text{height}(G))$$
as on
$$C_i(\chi_0^{d_i}) = C_i(d_i) \qquad (\chi_0 = \text{cyclotomic character}).$$
We now carry this over to the global case.

We have
$$\# s^*(\Omega^1_{A/A_n}) = l^{n \sum_{i=1}^r m_i d_i} \qquad (l^n \cdot \Omega^1_{A/A_n} = 0)$$
and
$$h(A_n) - h(A) = n \cdot \log(l) \cdot \left(\frac{h}{2} - \sum_{i=1}^r \frac{m_i}{m} \cdot d_i\right).$$

We must therefore show that $\sum_{i=1}^r m_i d_i = \frac{1}{2} mh$. For this purpose, we consider the absolute Galois group $\tilde{\pi} = \text{Gal}(\bar{\mathbb{Q}}/\mathbb{Q})$, and the $\tilde{\pi}$-module
$$\tilde{V} = \text{Ind}_\pi^{\tilde{\pi}}(T_l(A)) \qquad (\pi = \text{Gal}(\bar{K}/K)).$$

This contains the submodule
$$\tilde{W} = \text{Ind}_\pi^{\tilde{\pi}}(T_l(G))$$
of rank mh, and $\tilde{\pi}$ acts on the line
$$L = \Lambda^{mh}(\tilde{W}) \subseteq \Lambda^{mh}(\tilde{V})$$
via a character $\chi: \tilde{\pi} \to \mathbb{Z}_l^*$.

From class field theory it follows that χ is of the form
$$\chi = (l\text{-adic power of } \chi_0) \cdot (\text{character of finite order}).$$

The above l-adic power of χ_0 is determined as follows:
Let C be the completion of the algebraic closure of $\mathbb{Q}_l$,
$$D \cong \text{Gal}(\bar{\mathbb{Q}}_l/\mathbb{Q}_l) \subseteq \tilde{\pi}$$
the decomposition group of l. Then as D-modules we have
$$L \otimes_{\mathbb{Z}_l} C \cong C\left(+ \sum_{i=1}^r m_i d_i\right)$$
(this follows from our previous calculations), and hence by [13, Theorem 2],
$$\chi \cdot \chi_0^{-\sum_{i=1}^r m_i d_i}$$
is a character of finite order of D and also of $\tilde{\pi}$. Finally, from the part of the Weil conjectures already proven by Weil, together with some local considerations, it follows that, for almost all p, $\chi(F_p)$ (p = a prime number, F_p = Frobenius) is an algebraic number, all of whose conjugates have absolute value $p^{mh/2}$. Since $\chi_0(F_p) = p$, we have, as desired
$$\sum_{i=1}^r m_i d_i = \frac{mh}{2}.$$

§5. Endomorphisms

Let K be a number field, A/K an abelian variety of dimension g, l a prime number, $T_l = T_l(A)$ the Tate module, on which $\pi = \text{Gal}(\bar{K}/K)$ acts.

Theorem 3. *The action of π on $T_\ell \otimes_{\mathbb{Z}_\ell} \mathbb{Q}_\ell$ is semisimple.*

Theorem 4. *The map*
$$\mathrm{End}_K(A) \otimes_{\mathbb{Z}} \mathbb{Z}_l \to \mathrm{End}_\pi(T_l)$$
is an isomorphism.

PROOF. The two theorems are proven together. It is well known that it suffices, instead of Theorem 4, to prove the somewhat weaker statement that the map
$$\mathrm{End}_K(A) \otimes_{\mathbb{Z}} \mathbb{Q}_l \to \mathrm{End}_\pi(T_l \otimes_{\mathbb{Z}_l} \mathbb{Q}_l)$$
is bijective.

For the proof one may extend the ground field, or replace A by an isogenous abelian variety. We can also assume that A/K is principally polarized, and that A extends to a semiabelian variety over $\mathrm{Spec}(R)$. Then T_l admits a nondegenerate skew-symmetric bilinear form. Let
$$W \subseteq T_l \otimes_{\mathbb{Z}_l} \mathbb{Q}_l$$
be a π-invariant maximal isotropic subspace. This corresponds to an l-divisible subgroup $G \subseteq A[l^\infty]$, and the semiabelian varieties $A_n = A/G_n$ again admit principal polarizations.

By Theorem 2, $h(A_n) = h(A)$, and by Theorem 1, infinitely many A_n's are isomorphic.

As in [16], it follows that W is the image of an idempotent in $\mathrm{End}_K(A) \otimes_{\mathbb{Z}} \mathbb{Q}_l$. The rest of the proof goes exactly as in [16], and it will only be sketched here:

Choose $a, b, c, d \in \mathbb{Q}_l$ with $a^2 + b^2 + c^2 + d^2 = 1$.
Set
$$v = \begin{pmatrix} a & -b & -c & -d \\ b & a & d & -c \\ c & -d & a & b \\ d & c & -b & a \end{pmatrix}$$
(corresponding to the quaternion $a + bi + cj + dk$), so $v \cdot {}^t v = -1$. If W is an arbitrary π-invariant subspace of $T_l \otimes_{\mathbb{Z}_l} \mathbb{Q}_l$, then one applies the above considerations to the maximal isotropic subspace.
$$W_1 = \{(x, vx) | x \in W^4\} \oplus \{(y, -vy) | y \in (W^\perp)^4\} \subseteq T_l(A)^8 \otimes_{\mathbb{Z}_l} \mathbb{Q}_l.$$

Corollary 1. *Let A_1 and A_2 be abelian varieties over K. Then*
$$\mathrm{Hom}_K(A_1, A_2) \otimes_{\mathbb{Z}} \mathbb{Z}_l \to \mathrm{Hom}_\pi(T_l(A_1), T_l(A_2))$$
is an isomorphism. □

PROOF. Theorem 4 applied to $A_1 \times A_2$.

The *L*-series of A is defined, as is well known, as

$$L(A, s) = \prod_v \frac{1}{\det(1 - (N_v)^{-s} \cdot F_v | T_l(A))} = \prod_v L_v(A, s),$$

where the product runs over almost all places of K. The local *L*-factors are independent of l.

Corollary 2. *Let A_1, A_2 be as in Corollary 1. The following are equivalent:*

(i) A_1 *and* A_2 *are isogenous.*
(ii) $T_l(A_1) \otimes_{\mathbb{Z}_l} \mathbb{Q}_l \cong T_l(A_2) \otimes_{\mathbb{Z}_l} \mathbb{Q}_l$ *as π-modules.*
(iii) $L_v(s, A_1) = L_v(s, A_2)$ *for almost all places v of K.*
(iv) $L_v(s, A_1) = L_v(s, A_2)$ *for all v.*

PROOF. The equivalence of (i) and (ii) follow from Theorem 4, that of (ii) and (iii) from Theorem 3 (+ Čebotarev), and that (ii) implies (iv) implies (iii) is trivial. □

Corollary 3. *Let A/K be an abelian variety, $d > 0$. Then there are only finitely many isomorphism classes of abelian varieties B/K, with polarization of degree d, such that, for all l, $T_l(A) \cong T_l(B)$.*

PROOF. The assumptions imply that for every l there exists an isogeny between A and B which is of degree prime to l. Furthermore, for the purpose of the proof, we may extend the ground field, and then assume that A and all B's extend to semiabelian varieties over Spec(R), and that for all B's, there exists an isogeny of degree $\sqrt{d}$ with a principally polarized abelian variety. One then comes easily to the following assumptions:

(a) all B's have semistable reduction;
(b) all B/K are principally polarized;
(c) there exists an N such that for every prime number l and all B, there exist isogenies $\phi: A \to B$, for which the greatest power of l in deg(ϕ) divides N.

The remark after Lemma 5 then shows that $\exp(2[K : \mathbb{Q}](h(B) - h(A)))$ is a rational number, whose numerator and denominator divide a certain power of N.

Thus $h(B)$ is bounded, and one can apply Theorem 1. □

§6. Finiteness Theorems

Theorem 5. *Let S be a finite set of places of K. Then there are only finitely many isogeny classes of abelian varieties over K of a given dimension which have good reduction outside S.*

PROOF. Let A be such an abelian variety. According to the Weil conjectures, for $v \notin S$ there are only finitely many possibilities for the local L-factors $L_v(A, s)$. We will construct finitely many places $v_1, \ldots, v_r$, such that two A's are isogenous if they have the same local L-factor at these places. For this purpose, one chooses a prime number l. By Lemma 4, there exists a finite Galois extension $K' \supseteq K$ that contains all field extensions of K of degree $\leq l^{8g^2}$ which are unramified outside l and S ($g = \dim(A)$).

Let $G = \text{Gal}(K'/K)$; and choose $v_1, \ldots, v_r$ such that every conjugacy class in G contains the image of a Frobenius F_v for $v \in \{v_1, \ldots, v_r\}$ (Čebotarev). Then $v_1, \ldots, v_r$ fulfills our condition: Let A_1 and A_2 be two abelian varieties over K which have the same local L-factors at $v_1, \ldots, v_r$.

Let

$$M \subseteq \text{End}_{\mathbb{Z}_l}(T_l(A_1)) \times \text{End}_{\mathbb{Z}_l}(T_l(A_2))$$

be the $\mathbb{Z}_l$-subalgebra which is generated by the image of π.

Then M is a free $\mathbb{Z}_l$-module of rank $\leq 8g^2$, and M has representations on $T_l(A_1)$ and $T_l(A_2)$.

We must show that for every $m \in M$

$$\text{Tr}(m | T_l(A_1)) = \text{Tr}(m | T_l(A_2)).$$

It naturally suffices to prove this for m in a $\mathbb{Z}_l$-module basis of M, and by assumption the equality already holds if m is the image of an element of the conjugacy class of F_v, for $v \in \{v_1, \ldots, v_r\}$. We show that these images generate M over $\mathbb{Z}_l$. By Nakayama it is enough that they generate M/lM. This holds for the following reason:

We have a representation

$$\rho: \pi \to (M/lM)^* = \text{units of } M/lM,$$

whose image generates M/lM.

Since $\#(M/lM)^* \leq l^{8g^2}$, ρ factors through G, and $\rho(\pi)$ is the union of the images of the conjugacy classes of F_v, $v \in \{v_1, \ldots, v_r\}$. □

Theorem 6 (Shafarevich Conjecture). *Let S be a finite set of places of K, $d > 0$. Then there are only finitely many isomorphism classes of abelian varieties over K of a given dimension, with polarization of degree d, which have good reduction outside S.*

PROOF. By Theorem 5, we may assume that all the abelian varieties under consideration are isogenous to a fixed A/K. As in the proof of Corollary 3 to Theorem 4, we may further assume that all the B's extend to semiabelian varieties over $\text{Spec}(R)$, and that $d = 1$. We already know that

$$\exp(2[K : \mathbb{Q}](h(B) - h(A)))$$

is a rational number. We will construct a number N such that the numerator and denominator of this rational number have no prime factor $l > N$, and so that the powers of l dividing them are bounded for the prime numbers $l \le N$.

The latter is very easy: If, for two abelian varieties, B_1/K and B_2/K, $T_l(B_1)$ and $T_l(B_1)$ are isomorphic as π-modules, then by Theorem 4 there exists an isogeny between B_1 and B_2, of degree prime to l, and l does not occur in

$$\exp(2[K:\mathbb{Q}](h(B_1) - h(B_2))).$$

It thus suffices to show that there are only finitely many isomorphism classes of π invariant lattices in $T_l(A) \otimes_{\mathbb{Z}_l} \mathbb{Q}_l$. For this let M_l be the $\mathbb{Z}_l$-subalgebra of $\operatorname{End}_{\mathbb{Z}_l}(T_l(A))$ generated by π. Everything follows then from the fact that $M_l \otimes_{\mathbb{Z}_l} \mathbb{Q}_l$ is semisimple (Theorem 3).

We now come to the choice of N. For this, let n be the product of prime numbers l, for which either the extension $K \supseteq \mathbb{Q}$ is ramified at l, or A does not have good reduction at all of the places of characteristic l.

Choose any prime number p which does not divide n. Again let

$$\tilde{\pi} = \operatorname{Gal}(\bar{\mathbb{Q}}/\mathbb{Q}) \supseteq \pi = \operatorname{Gal}(\bar{K}/K),$$

and, for $0 \le h \le 2gm$ ($g = \dim(A)$, $m = [K:\mathbb{Q}]$), let

$$P_h(T) = \det[T - F_p | \Lambda^h(\operatorname{Ind}_\pi^{\tilde{\pi}}(T_l(A)))].$$

Here l is a prime number, prime to pn, and F_p denotes the Frobenius at the place p.

The $P_h(T)$ are independent of l, have coefficients in $\mathbb{Z}$, and their zeros have absolute value $p^{+h/2}$ (Weil conjecture, or better, theorem).

We now choose $N \ge 2$ so large that no prime number $l > N$ divides $P_h(\pm p^j)$ for

$$0 \le h \le 2gm,$$
$$0 \le j \le gm,$$
$$j \ne \tfrac{1}{2}h.$$

In addition, choose $N \ge np$.

We will show for every isogeny

$$\phi: B_1 \to B_2$$

of abelian varieties isogenous to A, whose degree is a power of l for a prime number $l > N$, that $h(B_1)$ and $h(B_2)$ are equal. This argument is similar to that in the proof of Theorem 2: We may assume that l annihilates the kernel G of ϕ. Let

$$V_l = T_l(B_1)/l \cdot T_l(B_1) \cong B_1[l](\bar{K}),$$
$$\tilde{V}_l = \operatorname{Ind}_\pi^{\tilde{\pi}}(V_l),$$
$$W_l = G(\bar{K}) \subseteq V_l$$
$$\tilde{W}_l = \operatorname{Ind}_\pi^{\tilde{\pi}}(W_l) \subseteq \tilde{V}_l.$$

If ϕ has degree l^h, then $\tilde{\pi}$ acts on
$$L = \Lambda^{mh}(\tilde{W}_l) \subseteq \Lambda^{mh}(\tilde{V}_l)$$
via a character $\chi: \tilde{\pi} \to (\mathbb{Z}/l\mathbb{Z})^*$.

If $\varepsilon: \tilde{\pi} \to \{\pm 1\}$ denotes the character through which $\tilde{\pi}$ acts on $\Lambda^m \operatorname{Ind}_\pi^{\tilde{\pi}}(\mathbb{Z})$, then $\chi \cdot \varepsilon^h$ is unramified outside l, because the inertia groups of places of K which do not divide l act unipotently on V_l (semistable reduction). By class field theory $\chi \cdot \varepsilon^h$ is a power of the cyclotomic character χ_0. The exponent of this power can be determined with the help of [10, Theorem 4.11] (instead of Tate's theory [13]) as follows:

Let
$$l^d = \# s^*(\Omega^1_{G/R}),$$
$$0 \le d \le gm.$$

Then $\chi \cdot \varepsilon^h = \chi_0^{+d}$ (according to Raynaud). Thus $\chi_0^d(F_p) = \pm p^d$ is a zero of $P_{mh}(T)$ modulo l, and by our choice of N, $d = hm/2$ must hold.

Since again
$$h(B_2) - h(B_1) = \log(l)\left(\frac{h}{2} - \frac{d}{m}\right),$$
our claim is proved, and it follows that the $h(B)$'s of the B's under consideration are bounded. Thus Theorem 6 follows from Theorem 1. $\square$

Corollary 1. *There are only finitely many isomorphism classes of smooth curves of genus $g \ge 2$ which have good reduction outside of S.*

PROOF. Torelli. $\square$

Theorem 7 (Mordell Conjecture). *Let X/K be a smooth curve of genus $g \ge 2$. Then $X(K)$ is finite.*

PROOF. This argument is in [9]: After extending the ground field if necessary, there is an unramified covering of degree $m > 2$:
$$\phi: X_1 \to X.$$

Lemma 4 furnishes a finite extension field $K_1 \supseteq K$ such that for every $x \in X(K)$, $\phi^{-1}(x)$ consists of m different K_1-rational points. Choose one of these points, say $y \in p^{-1}(x)$.

Let $D = \phi^{-1}(x) - \{y\}$ and A/K_1 the generalized Jacobian of the pair (X_1, D). With the help of y one constructs a map from $X_1 - D$ to A.

Multiplication by 2 on A then induces a covering $Y(X) \to X_1$, ramified exactly over D, for which the curve $Y(x)$ can have bad reduction only at those places v of K_1, and for which one of the following three conditions hold:

(a) v divides 2.
(b) X_1 has bad reduction at v.
(c) ϕ ramifies in the fibre mod v.

There are only finitely many such places, and thus only finitely many possibilities for $Y(x)$.

The same holds for the map $Y(x) \to X_1 \to X$, which ramifies exactly over x. The claim follows. □

Remarks. (1) In this way one also obtains a proof of the Siegel theorem about integral points, which makes no use of diophantine approximation.

(2) With the help of the methods of [16], one can conclude from Theorem 6, that for almost all prime numbers l, the subalgebra M_l of $\text{End}_{\mathbb{Z}_l}(T_l(A))$ generated by π is the full commutator of $\text{End}_K(A) \otimes_{\mathbb{Z}} \mathbb{Z}_l$.

REFERENCES

[1] Arakelov, S. Families of curves with fixed degeneracies. *Math. USSR Izv.*, **5** (1971), 1277–1302.
[2] Arakelov, S. An intersection theory for divisors on an arithmetic surface. *Math. USSR Izv.*, **8** (1974), 1167–1180.
[3] Baily, W. L. and Borel, A. Compactification of arithmetic quotients of bounded symmetric domains. *Ann. Math.*, **84** (1966), 442–528.
[4] Deligne, P. and Mumford, D. The irreducibility of the space of curves of a given genus. *Publ. Math. I.H.E.S.*, **36** (1969), 75–110.
[5] Faltings, G. Calculus on arithmetic surfaces. *Ann. Math.*
[6] Faltings, G. Arakelov's theorem for abelian varieties. *Invent. Math.*, **73** (1983), 337–347.
[7] Moret-Bailly, L. Variétés abéliennes polarisées sur les corps de fonctions. *C. R. Acad. Sci, Paris*, **296** (1983), 267–270.
[8] Namikawa, Y. *Toroidal Compactification of Siegel Spaces*. Lecture Notes in Mathematics, 812. Springer-Verlag: Berlin, Heidelberg, New York, 1980.
[9] Parshin, A. N. Algebraic curves over function fields, I. *Math. USSR Izv.*, **2** (1968), 1145–1170.
[10] Raynaud, M. Schémas en groupes de type $(p, \ldots, p)$. *Bull. Soc. Math. Fr.*, **102** (1974), 241–280.
[11] Szpiro, L. Sur le théorème de rigidité de Parsin et Arakelov. *Astérisque*, **64** (1979), 169–202.
[12] Szpiro, L. Séminaire sur les pinceaux de courbes de genre au moins deux. *Astérisque*, **86** (1981).
[13] Tate, J. p-divisible groups. *Proceedings of a Conference on Local Fields*, Driebergen, 1966. Springer-Verlag: Berlin, Heidelberg, New York, 1967, pp. 158–183.
[14] Tate, J. Endomorphisms of abelian varieties over finite fields. *Invent. Math.*, **2** (1966), 134–144.
[15] Zarhin, J. G. Isogenies of abelian varieties over fields of finite characteristics. *Math. USSR Sb.*, **24** (1974), 451–461.
[16] Zarhin, J. G. A remark on endomorphisms of abelian varieties over function fields of finite characteristics. *Math. USSR Izv.*, **8** (1974), 477–480.

ERRATUM

N. Katz has remarked that Theorem 2 in the above work is not completely correct. (O. Gabber constructed a counterexample.) The statement of the theorem should be replaced by the following which suffices for what comes after it:

The sequence $h(A_n)$ becomes stationary.

The mistake was in overlooking two subtle points. However, the original proof works if one replaces from the beginning, $A = A_0$ by A_m, for large enough m.

The problems are as follows:

(a) If $W \subseteq T_l(\hat{A})$ is a D_i-invariant sublattice, then of course there is a corresponding l-divisible subgroup of A/K_i, and by forming the Zariski closures, one obtains a system of finite flat group schemes over $\mathrm{Spf}(R_i)$ or also over $\mathrm{Spec}(R_i)$. However, these form an l-divisible group only when the mappings

$$G_{i,n+1}/G_{i,n} \to G_{i,n}/G_{i,n-1}$$

are isomorphisms for $n \geq 1$. In general, one cannot expect this. Nevertheless, a consideration of the discriminant shows that this is the case for large n. Passing from $A = A_0$ to A_m means that one need only consider these mappings for $n > m$. This argument is already found in Tate [1].

(b) In general, the intersection $G_i = G \cap H_i$ of l-divisible groups over $\mathrm{Spec}(R_i)$ does not define an l-divisible group even over K_i. This problem also disappears if we go to a suitable A_m. One may then continue as in (a).

REFERENCE

[1] Tate, J. *p*-divisible groups. *Proceedings of a Conference on local Fields*, Driebergen 1966. Springer-Verlag: Berlin, Heidelberg, New York, 1967, pp. 158–163.

CHAPTER III

Group Schemes, Formal Groups, and p-Divisible Groups

STEPHEN S. SHATZ

D. S. Rim, in memoriam

§1. Introduction

When the editors of this volume and organizers of the conference asked me to lecture on group schemes with an eye to applications in arithmetic, they gave me—with characteristic forethought—a nearly impossible task. I was to cover group schemes in general, finite group schemes in particular, sketch an acquaintance with formal groups, and study p-divisible groups—all in the compass of some six hours of lectures!

The audience was to consist of young graduate students, senior graduate students, professional research mathematicians of varying ages, and the leaders of the subject. I paid no mind to the latter and this article is not meant for them. Still the diversity of my listeners was staggering, and in this write-up of my lectures, the reader will notice many places where I have foundered against the implacable conditions of the task. The tempo is uneven: there are leisurely and terse arguments in the same paragraph, I have assumed no knowledge and large knowledge in the same proof, and I have omitted or sketched proofs after an idiosyncratic fashion. Also, personal choice has been exercised as to what was included and what omitted.

One could repair these defects by going to one of two extremes: The article could be shortened to one-half or one-third its size; this would render it more homogeneously concise and probably unreadable. Or, the article might be expanded to near book length; this would render it more pedagogical but unacceptable to the editors. So it stands as it is.

There is little that is original in what I have done. Perhaps the material on Sylow group schemes has not appeared before, and perhaps the arrangement of topics is somewhat novel. I have attempted to emphasize material which has so far only been in journal articles, and I have laid stress on examples and classes of examples.

It remains only for me to thank the organizers of the conference for a job well done and for the confidence they showed in me by tendering me their invitation to deliver the lectures. I would also like to remember my late colleague D. S. Rim, who had a lively interest in all of these matters and who would have been astonished and overjoyed at the way the Šafarevič, Tate, and Mordell conjectures were finally proved.

§2. Group Schemes, Generalities

In the sequel, all schemes will be locally noetherian, even noetherian, unless explicit mention is made to the contrary. This is the most important case for the applications, and our exposition is slanted towards these—especially the arithmetic applications. In addition, when dealing with categories, functors and the like, I shall neglect all mention of "universes" and such logical niceties. For one thing, such matters are not to my taste; for another, the knowledgeable reader will easily fix the exposition to fit correct foundations.

A *group functor over S* is a cofunctor, F, from schemes over S to the category of groups. Of course, this means each $F(X)$, for X a scheme over S, has the structure of group, and for each S morphism: $Y \to X$, we get a homomorphism of groups $F(X) \to F(Y)$. Many examples can be given, here are a few rather well-known ones:

(1) The *additive group scheme*, $\mathbb{G}_a$: For each X over S, put

$$\mathbb{G}_a(X) = \text{additive group of } \Gamma(X, \mathcal{O}_X).$$

(2) The *general linear group scheme*, $\mathbb{GL}(n)$:

$$\mathbb{GL}(n)(X) = \text{invertible } n \times n \text{ matrices with entries in } \Gamma(X, \mathcal{O}_X), \text{ under matrix multiplication.}$$

When $n = 1$, $\mathbb{GL}(n)$ has a special designation: $\mathbb{G}_m$, and is called the *multiplicative group scheme*. Of course,

$$\mathbb{G}_m(X) = \Gamma(X, \mathcal{O}_X)^* = \text{group of units of } \Gamma(X, \mathcal{O}_X).$$

(3) The *special linear group scheme*, $\mathbb{SL}(n)$:

$$\mathbb{SL}(n)(X) = \{A \in \mathbb{GL}(n)(X) | \det A = 1\}.$$

(4) The rth *roots of unity*, μ_r:

$$\mu_r(X) = \{A \in \mathbb{G}_m(X) | A^r = 1\}.$$

(5) If S is a scheme of characteristic $p > 0$, that is each stalk, $\mathcal{O}_{S,s}$, is a vector space over the prime field of characteristic p, then we define *the p^rth*

roots of zero, α_{p^r}:

$$\alpha_{p^r}(X) = \{A \in \mathbb{G}_a(X) | A^{p^r} = 0\}.$$

Since $(A + B)^p = A^p + B^p$ in $\Gamma(X, \mathcal{O}_X)$, it follows that $\alpha_{p^r}(X)$ is a subgroup of $\mathbb{G}_a(X)$.

(6) The *constant group functor*, $\mathscr{H}$: Let H be any group. Define a functor, $\mathscr{H}$, by

$\mathscr{H}(X) = H$, and for $Y \to X$, let $\mathscr{H}(X) \to \mathscr{H}(Y)$ be the identity map.

Other examples will appear below. If our group functor, F, is representable (by a scheme) and if G is the representing object, we call G a *group scheme*. By abuse of language, already indulged in, we also call the functor F a group scheme. This is the case with examples (1)–(5) above; example (6), however, is a non-representable functor. (A representable functor must take a disjoint union of schemes into the product of the functor evaluated on each component of the union.)

We assume F is representable, say by G, so that G is a group scheme over S. In the usual way, by Yoneda's lemma, the group axioms (functorial on each $F(X)$) translate into commutative diagrams for G over S. Here they are:

(1) There exists an S-morphism $m: G \times_S G \to G$, m is the group multiplication. The diagram

(A)
$$\begin{array}{ccc} G \times_S G \times_S G & \xrightarrow{1 \times m} & G \times_S G \\ \downarrow m \times 1 & & \downarrow m \\ G \times_S G & \xrightarrow{m} & G \end{array}$$

commutes (associative law for m).

(2) There exists a section $\varepsilon: S \to G$ for the structure morphism $\pi: G \to S$, so that the diagrams

(E)
$$\begin{array}{ccccccc} G \times_S G & \xrightarrow{m} & G & & G \times_S G & \xrightarrow{m} & G \\ \uparrow \varepsilon \times 1 & & \uparrow 1 & & \uparrow 1 \times \varepsilon & & \uparrow 1 \\ S \times_S G & \xrightarrow{1} & G & & G \times_S S & \xrightarrow{1} & G \end{array}$$

commute. (This means ε plays the role of identity.)

(3) There exists an S-morphism inv: $G \to G$, so that the diagrams

(I)
$$\begin{array}{ccccccccc} G & \xrightarrow{\Delta} & G \times_S G & \xrightarrow{1 \times \text{inv}} & G \times_S G & & G & \xrightarrow{\Delta} & G \times_S G & \xrightarrow{\text{inv} \times 1} & G \times_S G \\ \pi \downarrow & & & & \downarrow m & & \pi \downarrow & & & & \downarrow m \\ S & & \xrightarrow{\varepsilon} & & G & & S & & \xrightarrow{\varepsilon} & & G \end{array}$$

commute. Here, Δ is the diagonal. (This means inv is the inverse map.)

The concepts of *homomorphism* of group functors or group schemes and *subgroup functors* or *subgroup schemes* are trivial to formulate. We shall assume all our schemes are separated unless we explicitly mention to the contrary. Thus, $\Delta: G \to G \times_S G$ is a closed immersion; so, (E) shows that ε is a closed immersion and a homomorphism. Of course, S is the "trivial group scheme" over S.

Note that base extension of a group scheme G over S to *any* new base, T, over S, gives a new group scheme $G_T = G \times_S T$, *over T*. Hence, the correct way to think of a group scheme over S is as a family of group schemes, G_s, one for each $s \in S$. Here, G_s is $G \times_S \operatorname{Spec} \kappa(s)$ and $\kappa(s)$ is the residue field at s. Using the locution "group scheme over A" for a group scheme over $\operatorname{Spec} A$, where A is a commutative ring, we see that each G_s is a group scheme over a field (the field $\kappa(s)$). Of course, this implies one should have a theory for group schemes over a field as a first step. Here is an instructive example:

$$S = \operatorname{Spec} \mathbb{Z} \quad \text{and} \quad G = \operatorname{Spec} A, \quad \text{with} \quad A = \mathbb{Z}[X]/(X^2 + 2X).$$

Since, $G \times_S G = \operatorname{Spec}(A \otimes_{\mathbb{Z}} A)$, the map m corresponds to a $\mathbb{Z}$-algebra homomorphism

$$m^*: A \to A \otimes_{\mathbb{Z}} A.$$

In a similar way, the maps ε and inv, correspond to $\mathbb{Z}$-algebra maps

$$\varepsilon^*: A \to \mathbb{Z} \quad \text{and} \quad \operatorname{inv}^*: A \to A.$$

In this example, we choose

$$m^*(X) = 1 \otimes X + X \otimes 1 + X \otimes X,$$
$$\varepsilon^*(X) = 0,$$
$$\operatorname{inv}^*(X) = X.$$

The scheme G is then a group scheme over $\mathbb{Z}$. Let us examine it as a family of group schemes, G_p, each G_p being a group scheme over the field $\mathbb{Z}/p\mathbb{Z}$.

If p is odd, then $1/2$ exists in $\mathbb{Z}/p\mathbb{Z}$. If we write $Y = -X/2$, then as G_p is $\operatorname{Spec}(A/pA)$, we find

$$G_p = \operatorname{Spec}(\mathbb{Z}/p\mathbb{Z}[Y]/(Y^2 - Y)) = \operatorname{Spec}((\mathbb{Z}/p\mathbb{Z}) \cdot e_1) \coprod \operatorname{Spec}((\mathbb{Z}/p\mathbb{Z}) \cdot e_2),$$

where $e_1 = Y$, and $e_2 = 1 - Y$. The elements e_1 and e_2 are orthogonal idempotents; so G_p is the disjoint union of two connected components, each being a point. The group scheme structure in terms of e_1, e_2 is given by

$$m^*(e_1) = e_1 \otimes e_2 + e_2 \otimes e_1, \qquad m^*(e_2) = e_1 \otimes e_1 + e_2 \otimes e_2,$$
$$\varepsilon^*(e_1) = 0, \qquad \varepsilon^*(e_2) = 1,$$
$$\operatorname{inv}^*(e_1) = e_1, \qquad \operatorname{inv}^*(e_2) = e_2.$$

As a group functor, $G_p(X) = \{0, 1\}$ if X is connected (0 is the image of e_1, and 1 is the image of e_2), and $\{0, 1\}$ forms an additive group in the usual way. Should X be a disjoint union, $\bigcup_\alpha X_\alpha$, of connected components, we have $G_p(X) = \prod_\alpha G_p(X_\alpha)$. The group scheme G_p is disconnected over $\mathbb{Z}/p\mathbb{Z}$; indeed it is étale over $\mathbb{Z}/p\mathbb{Z}$.

If, instead of an odd prime p, we choose the generic point of Spec $\mathbb{Z}$, then in the above $\mathbb{Z}/p\mathbb{Z}$ is replaced by $\mathbb{Q}$—the rational numbers—and everything else goes through unchanged.

Now let $p = 2$. Then G_2 is Spec $A/2A$, and we find

$$A/2A = \mathbb{Z}/2\mathbb{Z}[X]/(X^2) = \mathbb{Z}/2\mathbb{Z}[Y]/(Y^2 - 1),$$

where $Y = X + 1$. The maps m^*, ε^*, inv* are given by

$$m^*(Y) = Y \otimes Y; \qquad \varepsilon^*(Y) = 1; \qquad \text{inv}^*(Y) = Y.$$

An easy check shows that G_2 is isomorphic to μ_2 over $\mathbb{Z}/2\mathbb{Z}$ and that G_2 is connected as a scheme. The transformation $Y = X + 1$ works over $\mathbb{Z}$ as well; it shows that our G is just μ_2 over $\mathbb{Z}$. The reason G was presented in its original form rather than as μ_2 is that this is the form in which it arises in the classification theory of group schemes.

Over $\mathbb{Z}$, the group scheme G is connected as $\mathbb{Z}$-scheme. Here is a sketch of it over Spec $\mathbb{Z}$

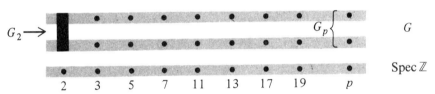

The shaded bands are the generic points ("continuous spectrum"), the dots are the special points ("discrete spectrum"). Observe that base extension of μ_2 to the open set $\mathbb{Z}[\frac{1}{2}]$, obtained by removing 2 from Spec $\mathbb{Z}$, yields an étale group scheme.

If a given group scheme, G, is affine, say $G = \text{Spec } A$, then (as in the example above) the diagrams (A), (E), (I) above are replaced by similar "dual" diagrams involving A, $k(= \Gamma(S, \mathcal{O}_S))$, $A \otimes_k A$, etc. All this is trivial. The notations m^*, ε^*, inv* will be used for the ring maps corresponding to m, ε, inv. Observe that if G is affine, diagram (E) shows that S is also affine. Also, the ring, A, of an affine group scheme is a *Hopf algebra* [20].

A more important case arises when G is *affine over* S. In this case, $\mathcal{O}_G$ is an $\mathcal{O}_S$-Hopf algebra, that is, $\mathcal{O}_G$ is a sheaf of Hopf algebras over $\mathcal{O}_S$. Of the examples given at the beginning of this section, all save (6) (which is *not* a group scheme) are affine over their base scheme S. As an exercise, one should compute explicitly the maps m^*, ε^*, inv* in these examples.

Group schemes non-affine over their bases are harder to come by. Essentially, they all involve abelian varieties. To give an example, we proceed as follows: Take $S = \text{Spec } \mathbb{C}$, and consider the projective plane cubic curve, E,

given by
$$y^2z = 4x^3 - axz^2 - bz^3; \quad a^3 - 27b^2 \neq 0.$$

Here, a and b are complex numbers, and the non-vanishing of $a^3 - 27b^2$ is necessary to guarantee that E is smooth. The scheme E is an algebraic variety which we shall make into a group scheme. We do this by the following geometric construction.

The line $z = 0$ is a flex tangent to E; we call its triple intersection with E the point O. This point will play the role of neutral group element of E. If P, Q are points on E, the line PQ cuts E in a third point R and the line OR cuts E in a third point. This last point is $P + Q$. If we use the standard model of $\mathbb{P}^2$ over $\mathbb{R}$ as a disc with anti-podal boundary points identified, then a picture of the real points of E and the construction above is:

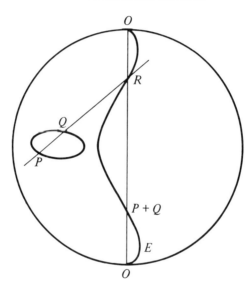

Real points of E and $\mathbb{P}^2$, with group law.

While it is not immediately obvious that we have defined a group law, this is easy to check. Moreover, by elementary analytic geometry, the coordinates of $P + Q$ may be computed as rational functions of the coordinates of P and Q, the coefficients of these rational functions lying in the field $\mathbb{Q}(a, b)$. Hence, for any field L over $\mathbb{Q}(a, b)$, and for any pair of points P, Q on E rational over L, we can write down expressions for the coordinates of $P + Q$ and these expressions are rational functions of the data given by P and Q. This means that E forms a group scheme over Spec $\mathbb{Q}(a, b)$ which is also a projective variety (of dimension 1). Such an object is an *elliptic curve* or *abelian variety* of dimension 1. (There can be no confusion here, since a theorem of Chevalley guarantees that the group law is commutative on an abelian variety.)

The scheme $E \times \cdots \times E$ (n times) is an abelian variety of dimension n. One can make abelian varieties which are not products of elliptic curves yet which still arise from curves (of genera greater than 1)—these are the *Jacobians* of curves. One can further make abelian varieties which are not Jacobians. All of these matters are thoroughly discussed in the articles by Rosen (classical case) [16] and Milne (abstract case) [11] in this volume.

If the base scheme, S, is not a field, one defines the notion of abelian scheme over S. A scheme, $\mathscr{A}$, over S is an *abelian scheme over S* if and only if: (0) $\mathscr{A}$ is a group scheme over S, (1) $\mathscr{A}$ is proper over S, (2) $\mathscr{A}$ is smooth over S, (3) all the fibres $\mathscr{A}_s$ are abelian varieties over the respective residue fields $\kappa(s)$. That is, an abelian scheme is a group scheme which is a relatively compact, continuously varying family of abelian varieties. It may be surprising that, given S, there may be no abelian scheme over S. For example, if $S = \operatorname{Spec} \mathbb{Z}$, there is no abelian scheme over S! For relative dimension 1, this is an old theorem of Tate [18], while for relative dimension > 1, it is a very recent theorem of Fontaine reported on in Ribet's article [15], [21].

We return to the general case. Suppose $f: H \to G$ is a homomorphism of group schemes (over the base S). We can easily make the group scheme $\ker f$ as the pull-back of the identity section, ε_G, of G. More specifically, $\ker f$ is the fibre product $H \times_G S$, indicated in the diagram:

$$\begin{array}{ccc} H & \xleftarrow{\mathrm{pr}_1} & H \times_G S = \ker f \\ f\downarrow & & \downarrow \mathrm{pr}_2 \\ G & \xleftarrow{\varepsilon_G} & S \end{array}$$

The morphism pr_2 makes $\ker f$ an S-group scheme, and pr_1 is a closed immersion because ε_G is one. From the functorial point of view, $\ker f$ is just what one expects. It assigns to the S-scheme, T, the group $\ker(H(T) \to G(T))$. Moreover, $\ker f$ is a normal subgroup scheme of H in the sense that $(\ker f)(T)$ is normal in $H(T)$ for every T over S.

What about $\operatorname{coker} f$, even as a homogeneous space (= coset space)? There are at least two reasons why this must be a much more subtle question than that for $\ker f$.

In the first place, the naive idea: "Why not represent the functor $T \rightsquigarrow G(T)/f(H(T))$?" fails because almost always this is *not* representable. However, this can be fixed as we shall now sketch.

If T is a scheme over S, we can consider a family $\{T_\alpha \to T\}_\alpha$ of T-schemes. One example of such a family is if each T_α is an open subscheme of T. We might call the family a "Zariski family" in this case. If we are thinking of covering T by the T_α, we should demand that the morphism $\coprod_\alpha T_\alpha \to T$ be surjective. In our example of a Zariski family, we would then get a *Zariski covering* of T (that is, an ordinary covering in the Zariski topology). When using such coverings for Čech cohomology and the like, it is not necessary that the T_α really be Zariski-opens of T, rather that certain formal properties of the maps $T_\alpha \to T$ (which make up the "covering") should hold. Why not throw in more "coverings" $\{T_\alpha \to T\}$ than just the Zariski-opens as long as the necessary properties hold? This seemingly innocuous idea is tremendously fruitful. It is the basic idea behind *sites* or Grothendieck topologies. In our case, the appropriate maps to add are indicated by descent theory.

Remember that we are interested in the representability of a certain functor. Descent theory gives a necessary condition for representability. Namely, if F is representable and if the morphism $\coprod_\alpha T_\alpha \to T$ is *faithfully flat* and *quasi-compact* (fpqc), then the diagram

(SHF) $$F(T) \xrightarrow{a} \prod_\alpha F(T_\alpha) \underset{c}{\overset{b}{\rightrightarrows}} \prod_{\beta,\gamma} F(T_\beta \times_T T_\gamma)$$

is an exact sequence (i.e., $F(T)$ is isomorphic *via* the map a to the equalizer of b and c in $\prod_\alpha F(T_\alpha)$).

So, we want to add those "coverings" $\{T_\alpha \to T\}$ in which each map $T_\alpha \to T$ is flat and quasi-compact and for which the map $\coprod_\alpha T_\alpha \to T$ is faithfully flat (so is surjective) and quasi-compact. The necessary condition for representability—exactness of (SHF) for fpqc coverings—is not met by the naive "coker f" functor. This functor does satisfy the weaker condition that the map a of (SHF) is injective. Under these conditions, one can show [1], [2] that the new functor

$$(\text{coker} f)(T) = \varinjlim \left(\text{Equalizer} \prod_\alpha G(T_\alpha)/fH(T_\alpha) \right.$$
$$\left. \rightrightarrows \prod_{\beta\gamma} G(T_\beta \times_T T_\gamma)/fH(T_\beta \times_T T_\gamma) \right)$$

in which the limit is taken over all fpqc coverings of T, does satisfy (SHF). Consequently, this new functor, coker f, which is called the *sheaf cokernel of* f, is a candidate for what we want.

The second reason is perhaps more surprising—it is best illustrated by an example. We let $G = \mathbb{GL}(2)$ over a field, and let H be the subgroup of upper triangular matrices,

$$H = \left\{ \begin{pmatrix} a & b \\ 0 & c \end{pmatrix} \middle| ac \neq 0 \right\}.$$

Clearly, H is the stabilizer of the line through the first basis vector; so, G/H—if it is to have any real sense—ought to be the orbit of this line under G. But then, G/H is the set of lines through the origin in the plane; that is, G/H is the projective line even though both G and H are affine. Any hope of constructing G/H in general by finding global H-invariant functions on G is thereby dashed, for such functions would be global functions on $\mathbb{P}^1$; hence, constant.

Nevertheless, the object representing the coker sheaf is the correct one. The problem is: When does it exist and how does one construct it? We will need only a special case of the following result proved by Grothendieck [8].

Theorem. *Let G be a finite type S-group scheme and let H be a closed subgroup scheme of G. If H is proper and flat over S and if G is quasi-projective over S, then the quotient sheaf G/H is representable.*

The theory of group schemes over S gives rise to a subtle interplay between the algebro-geometric properties of S and the group theory in the fibres of G. To simplify possible complications, one should make preliminary reductions of an easy nature. For example, one may always consider group schemes where the base scheme, S, is connected. To see this, just write S as the disjoint union of its components S_α. If $\pi: G \to S$, write G_α for $\pi^{-1}(S_\alpha)$, then the G_α are group schemes and G is the disjoint union of the G_α.

§3. Finite Group Schemes

Abelian schemes are of primary importance in the arithmetic applications. To study them, one examines group schemes finite over the base S and limits of these. Such group schemes have simpler structure than abelian schemes, yet knowledge of them is crucial for the study of abelian schemes. It scarcely makes any sense to study finite group schemes unless they are flat over the base S.

The group schemes μ_r and α_{p^r} are finite and flat over their base schemes. Another class of examples is given by the following construction.

(7) Let $S = \operatorname{Spec} R$, where R is a complete (more generally—Hensel) noetherian local ring with residue field k. Write π for the Galois group of k_s/k (k_s is a separable closure of k)—so, π is the étale fundamental group of S. An abstract group G will be called a π-group if π acts on it in such a way that the action is continuous when G is given the discrete topology. If G is a π-group, we write $\mathscr{A}(G)$ for the set of all functions on G to R_h, where R_h is the strict Henselization (= Hensel closure) of R.* Then $R_h \otimes_R k \cong k_s$, and if $f \in \mathscr{A}(G)$, we write $\tilde{f}$ for the function from G to k_s given by

$$\tilde{f}(\sigma) = f(\sigma) \otimes_R 1 \in R_h \otimes_R k \cong k_s.$$

Let $\tilde{\mathscr{A}}(G)$ be the set of such $\tilde{f}$, then $\tilde{\mathscr{A}}(G)$ has the usual π-action

$$(\forall\ t \in \pi)(\forall\ \sigma \in G)((t\tilde{f})(\sigma) = t \cdot \tilde{f}(t^{-1}\sigma)).$$

In this action, $\tilde{f}$ is fixed by π if and only if $\tilde{f}$ is π-equivariant, i.e., $t \cdot \tilde{f}(\sigma) = \tilde{f}(t\sigma)$.

Now assume G is a finite group, then

$$\mathscr{A}(G \times G) = \mathscr{A}(G) \otimes_{R_h} \mathscr{A}(G),$$

---

* Recall that R_h is the limit of "all" unramified local rings over R.

and so we can make $\mathscr{A}(G)$ into a Hopf algebra over R_h by setting

$$((m^*)f)(\sigma, \tau) = f(\sigma\tau),$$
$$\varepsilon^* f = f(1),$$
$$(\text{inv}^* f)(\sigma) = f(\sigma^{-1}).$$

As an R_h-module, $\mathscr{A}(G)$ is free of rank $\#(G)$ with a basis consisting of orthogonal idempotents, e_σ, defined by

$$e_\sigma(\tau) = \delta_{\sigma\tau},$$

this means Spec $\mathscr{A}(G)$ is a finite, flat (indeed étale) group scheme over R_h.

Finally, let $A(G)$ be the set of all $f \in \mathscr{A}(G)$ for which $\tilde{f}$ is π-equivariant. Then $A(G)$ is a Hopf algebra over R, and by continuity of the π-action, one sees that

$$A(G) \otimes_R R_h = \mathscr{A}(G).$$

It follows (by faithfully flat descent) that Spec $A(G)$ *is a finite, étale group scheme over R*, and $A(G)$ has rank $\#(G)$. We write $\mathscr{G}$ for Spec $A(G)$. A simple argument shows that *the functor*

$$G \rightsquigarrow \mathscr{G} = \text{Spec } A(G)$$

establishes a full embedding of the category of abstract, finite π-groups into the category of finite, flat group schemes over R. (The image consists exactly of the étale group schemes over R, and one recovers G from $\mathscr{G}$ (as π-group) from the equation $G = \mathscr{G}(R_h)$.)

Of course, every abstract group is a π-group (trivial action); so, the category of finite group schemes over S contains the category of ordinary groups. Consequently, a full theory of finite, flat group schemes over S contains every theorem about finite groups.

If G is a finite, flat group scheme over a connected base S, then the $\mathcal{O}_S$-module, $\mathcal{O}_G$, is locally free of constant rank called the *order of G* and denoted $\#(G)$. This terminology is consistent with the older notion of the order of a group as its cardinality, for in the case that G is étale of the form $\mathscr{H}$, we have $\#(G) = \#(H)$. *From now on all base schemes will be connected and finite group scheme means finite, flat group scheme.*

Here is a sketch of how to make the quotient group scheme in the case of interest to us, and a theorem giving some of its properties.

Theorem. *Let G be a finite group scheme over S, and let N be a flat subgroup scheme of G. Then the quotient scheme G/N exists and is flat and finite over S. We have*

$$\#(G) = \#(N)\#(G/N).$$

If N is normal, then G/N is a finite group scheme over S.

PROOF [13]. N is finite as $\mathcal{O}_N$ is a quotient of the $\mathcal{O}_S$-module $\mathcal{O}_G$; so, N is proper over S. Now G is quasi-projective because it is finite. Therefore, Grothendieck's theorem tells us that G/N is representable. In this case, however, we can say how to construct G/N. We know N is given by a quasi-coherent ideal, $\mathcal{J}$, of $\mathcal{O}_G$, and we consider the $\mathcal{O}_S$-subalgebra of $\mathcal{O}_G$ given by

(**) $$\{x \in \mathcal{O}_G \mid m^*(x) \equiv x \otimes 1 \bmod \mathcal{O}_G \otimes \mathcal{J}\}.$$

It turns out that the Spec of this $\mathcal{O}_S$-subalgebra is just G/N.

We still must show G/N is flat over S and that Lagrange's theorem is true. From the definition of $\mathcal{O}_Q$ by (**), where Q now denotes G/N, one finds that the equivalence relation on G induced by N is exactly the fibred product

$$\mathcal{R} = G \times_Q G.$$

This is a closed subscheme of $G \times_S G$, and one has the isomorphism

$$N \times_S G \xrightarrow{\sim} \mathcal{R} = G \times_Q G;$$

so, by base extension (N being faithfully flat over S), we find that $\operatorname{pr}_2: \mathcal{R} \to G$ is a faithfully flat morphism. But, the cartesian diagram

(*) $$\begin{array}{ccc} G & \xleftarrow{\operatorname{pr}_1} & \mathcal{R} = G \times_Q G \\ {\scriptstyle p}\downarrow & & \downarrow{\scriptstyle \operatorname{pr}_2} \\ Q & \xleftarrow{p} & G \end{array}$$

and its symmetry show that pr_1 is also faithfully flat. Now $G \to Q$ is a surjection, consequently it is dominated by a covering in the fpqc topology. We may suppose this is of the form

$$X \xrightarrow{r} G \xrightarrow{p} Q$$

and that $p \circ r$ is fpqc. Base extend (*) by X and get

(*)' $$\begin{array}{ccccc} G & \xleftarrow{\operatorname{pr}_1} & \mathcal{R} & \xleftarrow{\rho_1} & \mathcal{R} \times_G X \\ {\scriptstyle p}\downarrow & & \downarrow{\scriptstyle \operatorname{pr}_2} & & \downarrow{\scriptstyle \rho_2} \\ Q & \xleftarrow{p} & G & \xleftarrow{r} & X \end{array}$$

in which ρ_2 (being pr_2 base extended) is faithfully flat. But then, p, base extended by the fpqc morphism $p \circ r$ is faithfully flat. Descent theory [5] shows p to be faithfully flat. Since G is flat over S, the faithful flatness of p implies that Q is flat over S.

When one counts ranks in the isomorphism

$$N \times_S G \xrightarrow{\sim} G \times_Q G,$$

one finds that

$$ng = g^2/q, \quad g = \#(G), \quad q = \#(Q), \quad n = \#(N),$$

as required. That Q is a group when N is normal follows from the functorial description of Q as the sheaf cokernel. □

What about the deeper theorems of group theory in this context? For example, Sylow's theorems. When the base scheme, S, is sufficiently nice, there are positive results and they reveal the fact that some base extension of S is necessary. We shall assume that our base

(†) S is integral, excellent, and of dimension one.

Suppose we can prove the following lemma:

Lemma. *If G is a finite group scheme over S and if S satisfies* (†), *then*:

(1) *each subgroup scheme, $\mathscr{H}$, of the generic fibre, $\mathscr{G}$, of G over S, extends uniquely to a (flat) subgroup scheme, H, of G having the same order as $\mathscr{H}$;*
(2) *$\mathscr{H}$ is normal in $\mathscr{G}$ if and only if H is normal in G;*
(3) *$\mathscr{H}$ is conjugate to another subgroup scheme, $\mathscr{H}'$, of $\mathscr{G}$ if and only if their extensions H and H' are conjugate in G.*

Then, on the basis of this lemma, we can prove a version of the Sylow theorems. For notation, let us agree that $p^{\mathrm{ord}_p(G)}$ means the highest power of p which divides $\#(G)$.

Theorem (Sylow Theorems). *Let G be a finite group scheme over a base, S, which satisfies* (†). *If p is a prime number, then there exists a scheme, T, which satisfies* (†) *and is finite and faithfully flat over S, so that $G_T = G \times_S T$ possesses a subgroup scheme of order $p^{\mathrm{ord}_p(G)}$ over T. If p is not the characteristic of S, then the same statement is true for all exponents a with $0 \leq a \leq \mathrm{ord}_p(G)$. Again, if p is not the characteristic of S, then the number of such p-Sylow subgroup schemes divides $\#(G)$ and is congruent to one modulo p. Also, any two such p-Sylow subgroup schemes are conjugate in G_T, and any p-subgroup scheme of G_T is contained in one of them.*

PROOF. In this proof, we will use certain facts about group schemes over fields to be established shortly. The generic fibre, $\mathscr{G}$, of G is a group scheme over the field, K, which is the local ring of the generic point of S. We will show later, that each such $\mathscr{G}$ possesses a canonical exact sequence

$$0 \to \mathscr{G}^0 \to \mathscr{G} \to \mathscr{G}^{\mathrm{et}} \to 0$$

in which $\mathscr{G}^0$ is the connected component of identity and has order q^t, where q is the characteristic of K. The group scheme $\mathscr{G}^{\mathrm{et}}$ is étale over K; so, by example (7) above, it is $\mathscr{J}$ for some π-group J. Here, π is the Galois group of the separable closure of K over K. Since J is finite as π-group, there is some finite separable extension, L, of K, so that J is a π' ($= \mathrm{Gal}(K_s/L)$)-group with trivial action; that is $\mathscr{J} \otimes_K L$ is an étale group scheme coming from an ordinary group (with no Galois action) over L.

Now let p be a prime number, *not* equal to the characteristic of S (hence, arbitrary in characteristic zero). It is known that the exact sequence above splits over a perfect field; so, as $\mathscr{G}$ is a finite group scheme, it splits at some finite level below the perfect closure. By adjusting L upwards, if necessary, we may assume this splitting already takes place over L. Let T be the normalization of S in L; the excellence of S implies that T is finite, faithfully flat over S, and satisfies (†). By Lagrange's theorem, p^a divides the order of $\mathscr{I} \otimes_K L$ if $0 \leq a \leq \text{ord}_p(G)$; so, the ordinary Sylow theorem yields a subgroup, $\tilde{H}$, of $\mathscr{I} \otimes_K L$ of order p^a. By the splitting, $\tilde{H}$ gives a subgroup scheme, $\mathscr{H}$, of $\mathscr{G}$; the lemma extends $\mathscr{H}$ to the desired subgroup scheme, H, of G_T. The remaining statements are now clear. They correspond to the splitting $\mathscr{G}^{\text{et}} \otimes_K L \hookrightarrow \mathscr{G} \otimes_K L$, to the fact that all p-subgroup schemes of $\mathscr{G}$ must arise from $\mathscr{G}^{\text{et}}$ (as $(p,q) = 1$), to the ordinary Sylow statements applied in $\mathscr{I} \otimes_K L$, and to the uniqueness and conjugation statements of the lemma.

We now consider $p = q = \text{char}(S)$. Here, as above, we pick L so that $\mathscr{I} \otimes_K L$ becomes trivial as a π'-group; and we define T as the normalization of S in L. (Also, we require the exact sequence to split over L.) Now some power of p may divide $\#(\mathscr{G}^{\text{et}})$, and then $\mathscr{I} \otimes_K L$ possesses a p-Sylow subgroup. We pull this back in the exact sequence

$$0 \to \mathscr{G}^0 \otimes_K L \to \mathscr{G} \otimes_K L \to \mathscr{G}^{\text{et}} \otimes_K L \to 0$$

to a subgroup scheme of clear order $p^{\text{ord}_p(G)}$ of $\mathscr{G} \otimes_K L$. The lemma extends it over T as a subgroup scheme of G_T. The last two statements are now very easy. First of all, $\mathscr{G}^0 \otimes_K L$ is normal in $\mathscr{G} \otimes_K L$; so, conjugation will map it to itself. Thus, our p-Sylow subgroup schemes of $\mathscr{G} \otimes_K L$ are conjugate one to the other, and the lemma carries this over to G_T. Lastly, if Z is a p-subgroup scheme of G_T, then its generic fibre, $\mathscr{Z}$, is certainly contained in one of the p-Sylow subgroup schemes we have constructed for $\mathscr{G} \otimes_K L$; so, one more application of the lemma finishes our proof.

There remains only the

PROOF OF THE LEMMA. We can give two proofs of the existence of an extension across all of S. Fix G over S, and consider the scheme

$\text{Hilb}^r(G/S)(T)$

$= \left\{ \text{subschemes } H \text{ of } G_T \,\middle|\, \begin{array}{l} (1) \ H \text{ is flat over } T; \\ (2) \ \mathscr{O}_H \text{ is locally free of rank } r \text{ over } \mathscr{O}_T. \end{array} \right\}$

It is known [9] that $\text{Hilb}^r(G/S)$ is a projective scheme over S. Now, under our hypotheses, the subgroup scheme, $\mathscr{H}$, is a rational section of $\text{Hilb}^r(G/S)$ for some r (which divides $\#(G)$), and by (†) it extends to an honest section of $\text{Hilb}^r(G/S)$; that is, to a flat subscheme, H, of G. That this H is a subgroup scheme is seen as well from the second construction which we now give. We let H be the smallest closed subscheme ($=$ scheme-theoretic closure) of G which contains $\mathscr{H}$. Now $H \otimes_S K = \mathscr{H}$ because $\mathscr{H}$ is closed in $\mathscr{G}$. Also, if U is an affine open of G, with ring of global sections A, then $U \cap \mathscr{G} = \mathscr{U}$ is affine open in $\mathscr{G}$, with ring $\mathscr{A} = A \otimes_S K$. But then, $\mathscr{H} \cap \mathscr{U}$ is defined by an ideal $\mathscr{Q}$ of $\mathscr{A}$ and the pull-back of $\mathscr{Q}$ under $A \to \mathscr{A} = A \otimes_S K$, say I, is by definition the ideal defining $H \cap U$. Since A/I is a subring of $\mathscr{A}/\mathscr{Q}$, this persists under localization. As $\mathscr{A}/\mathscr{Q}$ is torsion-free, so then is A/I; it follows that all its localizations are flat, that is $\mathrm{Spec}(A/I)$ is flat over S. (We have used the one dimensionality of S in this argument.) Therefore, we now know H is flat over S. Furthermore, H is faithfully flat over S, and this gives the uniqueness at once.

The flatness of the scheme-theoretic closure shows that this operation preserves fibred products over S. That is, the extension of $\mathscr{H} \otimes_K \mathscr{H}$ to all of S is merely $H \times_S H$. Therefore, an easy argument involving the continuity of the multiplication, m, on G shows that H is a subgroup scheme of G whenever $\mathscr{H}$ is one in $\mathscr{G}$. It also shows that normality extends, as does conjugation (here uniqueness is invoked). The lemma is proved. □

The proof of Sylow's theorems just given is clearly the "wrong" one (even though correct). It uses too crudely the geometry of S and the group theory in the fibres of G—it does not mix them effectively. My best guess as to a plausible statement of the Sylow theorem is

If G is a finite group scheme over an integral scheme S and if p^a divides $\#(G)$, for some prime p, then there is a scheme, T, faithfully flat and quasi-compact over S, such that G_T possesses a finite subgroup scheme over T of order p^a.

I would suggest an attempt at a proof along the lines of Wielandt's proof: Look at $\mathrm{Hilb}^{p^a}(G/S)$. It is easy to see that G acts on $\mathrm{Hilb}^{p^a}(G/S)$, but then the difficulties begin—most especially with the combinatorial aspects of the classical Wielandt proof.

A remarkable suggestion, due to B. Gross (oral communication), is that one consider a finite group over the generic point of a (discrete) valuation ring. Assume the residue field of the ring is of characteristic $p > 0$ and that p divides $\#(G)$. Then the structure of the set of extensions of G to a group scheme over the ring should be closely related to the Brauer theory of modular characters of G over the residue field.

By the way, it is easy to use the methods of our proof of Sylow's theorems over base schemes S satisfying condition (†) to give a proof of the

Feit–Thompson Theorem. *Let G be a finite group scheme over a base scheme S which satisfies* (†), *we have*:

(1) *If G has odd order, then after a finite, faithfully flat base extension T, the group scheme G_T is solvable; and*
(2) *If G is non-abelian and geometrically simple (G_T is simple for all T as above), then $\#(G)$ is even.*

In all of the above, we based results on the theory over a field. We need to examine this now. First, the connected component sequence—done for the case of a complete local ring.

GROUP SCHEMES, FORMAL GROUPS, AND p-DIVISIBLE GROUPS

Proposition. *Let R be a complete local ring and let G be an affine group scheme, flat and of finite type over R. Then, there is an exact sequence*

$$0 \to G^0 \to G \to G^{et} \to 0$$

in which G^0 is a connected, flat affine group scheme over R normal in G and G^{et} is an étale finite group scheme over R.

PROOF. Let G^0 be the connected component of identity in G. Of course, G^0 is a connected normal subgroup scheme of G. In terms of rings, if A is the ring of global sections of G, let A^{et} be the maximal separable subalgebra of A. This exists because the completeness of R allows us to lift idempotents. It is not hard to see that Spec A^{et} is a group scheme over R; it is flat, finite and étale. Let G^{et} be Spec A^{et}. Now write $e_0, \ldots, e_t$ for the idempotent generators of A^{et}, then ε^* vanishes on all but one of them, say $\varepsilon^*(e_0) = 1$. (We may and do assume, by an easy reduction, that the residue field of R is separably closed.) But then,

$$G^0 = \text{Spec } A/(\ker \varepsilon^* \cap A^{et})A = \text{Spec } e_0 A,$$

and we see G^0 is flat over R.

In the case that G is a finite group scheme over R, the group G^0 is Spec $A_{\mathfrak{m}_0}$, where $\mathfrak{m}_0 = \ker \varepsilon^*$. As a functor, G^0 is given by

$$G^0(T) = \ker(G(T) \to G(T_{\text{red}}))$$

whenever T is the Spec of a local, finite R-algebra. □

To go further, we examine the special case that R is a field k. Then our proposition yields the following.

Corollary. *If G is a finite type affine group scheme over the field k, then the one-forms on G over k satisfy*

$$\Omega^1_{G/K} \cong A \otimes_k (\ker \varepsilon^*/(\ker \varepsilon^*)^2).$$

To see this, we merely remark that $\Omega^1_{G^0/k}$ is $A^0 \otimes_k (\ker \varepsilon^*/(\ker \varepsilon^*)^2)$. One should also check that the universal derivation is given as follows: Say

$$m^*(a) = \sum_i a_i \otimes b_i \in A \otimes_k A.$$

Consider the projection

$$A = k \cdot 1 \oplus \ker \varepsilon^* \to \ker \varepsilon^* \to \ker \varepsilon^*/(\ker \varepsilon^*)^2$$

and write $\bar{z}$ for the image of z under this projection. Then,

$$d(a) = \sum_i a_i \otimes \bar{b}_i \quad (d: A \to \Omega^1_{G/k} = A \otimes_k (\ker \varepsilon^*/(\ker \varepsilon^*)^2)). \quad \square$$

This corollary implies a fundamental result due to P. Cartier [3], [20].

Theorem (Cartier). *If k is a field of characteristic zero, and if $G = \text{Spec } A$ is an affine group scheme over k, then A is reduced. In particular, every finite group scheme is étale, and so has the form $\mathcal{H}$ for a π-group H.*

PROOF. We may assume (by taking direct limits) that A is finitely generated as k-algebra. Write $\mathfrak{m}_0$ for ker ε^*, then $\mathfrak{m}_0/\mathfrak{m}_0^2$ is a finite-dimensional k-space. Assume we have proved

(∗) If $x_1, \ldots, x_t$ are a k-basis for $\mathfrak{m}_0/\mathfrak{m}_0^2$, the monomials $x_1^{a_1} \ldots x_t^{a_t}$ in which $a_1 + \cdots + a_t = z$ form a k-basis for $\mathfrak{m}_0^z/\mathfrak{m}_0^{z+1}$, that is, $\text{gr}_{\mathfrak{m}_0}(A)$ is a polynomial ring in $x_1, \ldots, x_t$,

then we are done. For, we need only show that if $y^2 = 0$, the element y must be zero. Look at the smallest n so that the image, $\bar{y}$, of y in $\text{gr}(A)_n = \mathfrak{m}_0^n/\mathfrak{m}_0^{n+1}$ is not zero. By (∗), y can be written $y_0 + \eta$, where y_0 is a homogeneous form of degree n in the x_j's, and η lies in $\mathfrak{m}_0^{n+1}$. But then, y_0^2 + element of $\mathfrak{m}_0^{2n+1}$ is zero, i.e., $y^2 \in \mathfrak{m}_0^{2n+1}$, contradicting (∗). It follows that we have $y \in \bigcap_n \mathfrak{m}_0^n$. Now, if necessary, pass by base extension to $\bar{k}$, the algebraic closure of k. Then, by translation in the group, all the other maximal ideals, $\mathfrak{m}$, of A share the above property vis-à-vis y: $y^2 = 0 \Rightarrow y \in \bigcap_n \mathfrak{m}^n$. The Krull intersection theorem implies $y = 0$, as contended.

There remains only (∗) (which is standard commutative algebra once we have our corollary above). Consider the map

$$\delta_i: A = k \cdot 1 \oplus \mathfrak{m}_0 \to \mathfrak{m}_0/\mathfrak{m}_0^2 \to k$$

in which, on the right, we send x_i to $1 \in k$, all other x_j go to zero. Given $a \in A$, write $m^*(a) = \sum a_j \otimes b_j$, and form the k-derivation (cf. our corollary)

$$\Delta_i(a) = \sum_l a_l \delta_i(b_l).$$

It is easy to see that

$$\Delta_i(x_j) \equiv \delta_{ij} \quad \text{mod } \mathfrak{m}_0.$$

And now, if P is a form in the x_i, we get

$$\Delta_i P(x) = \sum_l \frac{\partial P}{\partial x_l} \Delta_i(x_l) \equiv \frac{\partial P}{\partial x_i} \quad \text{mod } \mathfrak{m}_0^{\deg P},$$

and so the Leibnitz rule yields

$$\Delta_t^{a_t} \ldots \Delta_1^{a_1}(x_1^{a_1} \ldots x_t^{a_t}) \equiv a_1! \ldots a_t! \quad \text{mod } \mathfrak{m}_0,$$

$$\Delta_t^{a_t} \ldots \Delta_1^{a_1} \text{(other monomials of same degree)} \equiv 0 \quad \text{mod } \mathfrak{m}_0.$$

We have enough linear functionals to separate the monomials; so, they are linearly independent. □

Here is a quick application of Cartier's result to the question of "lifting group schemes to characteristic 0 from characteristic $p > 0$". Suppose R is an integral domain of characteristic zero and **p** is a maximal ideal of R with $R/\mathbf{p}$ a field of characteristic $p > 0$. If G is a group scheme over $\mathrm{Spec}(R/\mathbf{p})$, we say G *lifts to* R (more properly to $S = \mathrm{Spec}\, R$) if and only if there is a *flat* group scheme, $\mathscr{G}$, over S whose base extension to $R/\mathbf{p}$ (i.e., fibre over **p**) is $R/\mathbf{p}$-isomorphic to G. In this case, $\mathscr{G}$ is a continuously varying family of group schemes over S whose fibre at **p** is given. Such a $\mathscr{G}$ is an example of a *deformation* of G; frequently, R is a local ring.

Now to the application: Let the characteristic of k be $p > 0$, then the group scheme α_p exists and its automorphism scheme is $\mathbb{G}_m$. Hence, we can make the semi-direct product

$$E(\alpha_p, \mu_p): 0 \to \alpha_p \to E(\alpha_p, \mu_p) \to \mu_p \to 0,$$

where μ_p acts on α_p as automorphisms according to its embedding in $\mathbb{G}_m$. In down-to-earth terms, $E(\alpha_p, \mu_p)$ represents the functor

$$E(\alpha_p, \mu_p)(L) = \left\{ \begin{pmatrix} a & b \\ 0 & 1 \end{pmatrix} \middle| a \in \mu_p(L), b \in \alpha_p(L), \text{matrix multiplication} \right\}$$

for each k-algebra L. The group scheme $E(\alpha_p, \mu_p)$ has order p^2; it is *non-commutative*. Now for the surprise: $E(\alpha_p, \mu_p)$ *admits no lifting to any ring of characteristic zero*. For otherwise, it could be lifted to a domain of characteristic zero, and we could form, $\mathscr{G}_0$, the generic fibre of the lifting, $\mathscr{G}$. By Cartier's theorem, $\mathscr{G}_0$ is étale, and by base extension to a separably closed field, it arises from an abstract group of order p^2. Consequently, $\mathscr{G}_0$ would be commutative, and it follows (as $\mathscr{G}_0$ is dense in $\mathscr{G}$) that $\mathscr{G}$ itself would be commutative. But then, $E(\alpha_p, \mu_p)$, a fibre of $\mathscr{G}$, would also be commutative, a contradiction.

Practically all the applications of finite group schemes (and their limits) to arithmetic come from commutative finite group schemes. We shall now examine these.

§4. Commutative Finite Group Schemes

Most of what we need has been done already, except for Cartier duality and some classification results. We suppose G is a finite group scheme over S, and we *assume such G are commutative unless explicit mention to the contrary is made*.

If $\mathscr{O}_G$ is the $\mathscr{O}_S$-algebra of G, then the $\mathscr{O}_S$-module

$$\mathscr{O}_G^D = \mathrm{Hom}_{\mathscr{O}_S\text{-modules}}(\mathscr{O}_G, \mathscr{O}_S)$$

is again the $\mathscr{O}_S$-algebra of a commutative group scheme, flat over S. Multiplication in $\mathscr{O}_G^D$ is $(m^*)^D$, the structure map is $(\varepsilon^*)^D$, etc. We set

$$G^D = \mathrm{Spec}(\mathscr{O}_G^D); \quad \text{so,} \quad \mathscr{O}_{G^D} = \mathscr{O}_G^D,$$

and call G^D the *Cartier dual* of G. Of course, we have

$$\#(G^D) = \#(G) \quad \text{and} \quad (G^D)^D \cong G.$$

Consider the "sheaf Hom"

$$\mathscr{H}om_S(A, B)(T) = \text{Hom}_T(A \times_S T, B \times_S T),$$

then it is easy to see that

$$G^D = \mathscr{H}om_{\text{Group Schemes}/S}(G, \mathbb{G}_m),$$

where the subscript "Group Schemes/S" refers to the subsheaf of group scheme homomorphisms. Clearly, G^D is the character group scheme of G as is usual. It is also easy to define the pairing $G \times G^D \to \mathbb{G}_m$ explicitly. For this, write $\mathbb{G}_m$ as Spec $\mathcal{O}_S[T, T^{-1}]$, and pick a trivializing open subset of S, say U, for the locally free sheaf $\mathcal{O}_G$. Let $e_1, \ldots, e_g$ be a basis for $\mathcal{O}_G|U$, and let $e_1^D, \ldots, e_g^D$ be the dual basis in $\mathcal{O}_{G^D}|U$. Then the map

$$\mathcal{O}_S[T, T^{-1}]|U \to \mathcal{O}_G|U \otimes_{\mathcal{O}_S|U} \mathcal{O}_{G^D}|U$$

is simply

(*) $$T \to \sum_{i=1}^{g} e_i \otimes e_i^D.$$

Note that under the canonical isomorphism

$$\mathcal{O}_{G^D} \otimes_{\mathcal{O}_S} \mathcal{O}_G \xrightarrow{\sim} \mathscr{H}om_{\mathcal{O}_S}(\mathcal{O}_G, \mathcal{O}_G),$$

the element $\sum e_i \otimes e_i^D$ goes over to the identity (module) map $\mathcal{O}_G \to \mathcal{O}_G$. Hence, the explicit map (*) really is canonical—it is independent of the choice of the e_i. Here are some examples.

(8) $\mathbb{Z}/r\mathbb{Z}$ is dual to μ_r.

(Hence, the dual of an étale group scheme may be connected and vice versa.) The general pairing is somewhat messy, but if $\mathcal{O}_S$ has characteristic $p > 0$ and if $r = p^t$, then it is rather nice. For this, recall that we have the exact sequences

$$0 \to \mu_{p^t} \to \mathbb{G}_m \xrightarrow{p^t} \mathbb{G}_m \to 0$$

and

$$0 \to \mathbb{Z}/p^t\mathbb{Z} \to \mathbb{W}_t \xrightarrow{\wp} \mathbb{W}_t \to 0,$$

where $\mathbb{W}_t$ is the Witt group scheme of length t over S; that is, $\mathbb{W}_t$ represents the functor

$$W_t(T) = \left\{ \langle \alpha_0, \ldots, \alpha_{t-1} \rangle \mid \alpha_j \in \Gamma(T, \mathcal{O}_T), \begin{array}{l} \text{and tuples are} \\ \text{added as Witt vectors} \end{array} \right\},$$

and where $\wp$ is the Artin–Schreier map:
$$\wp(\langle\alpha_0,\ldots,\alpha_{t-1}\rangle) = \langle\alpha_0^p,\ldots,\alpha_{t-1}^p\rangle - \langle\alpha_0,\ldots,\alpha_{t-1}\rangle.$$
(The case $t = 1$: $0 \to \mathbb{Z}/p\mathbb{Z} \to \mathbb{G}_a \xrightarrow{\wp} \mathbb{G}_a \to 0$ is familiar.)
First take the case $t = 1$. We have
$$\mathbb{Z}/p\mathbb{Z} = \operatorname{Spec} \mathcal{O}_S[X]/(X^p - X), \qquad \mu_p = \operatorname{Spec} \mathcal{O}_S[Y]/(Y^p - 1).$$
A basis for the algebra of $\mathbb{Z}/p\mathbb{Z}$ is $1, X, \ldots, X^{p-1}$; let the dual basis be $f_0, f_1, \ldots, f_{p-1}$. One can check that the f_i are *divided powers* of f_1; that is,
$$f_i = \frac{f_1^i}{i!}.$$
Let exp denote the "truncated exponential",
$$\exp(\xi) = 1 + \xi + \frac{\xi^2}{2!} + \cdots + \frac{\xi^{p-1}}{(p-1)!},$$
then it is easy to show that $Y = \exp(f_1)$. Hence, our prescription for the Cartier pairing shows that it is
$$T \to \exp(X \otimes \log Y).$$
In the general case ($t \geq 1$), we need the Artin–Hasse exponential. For this, we begin with the classical formula
$$e^{-s} = \prod_n (1 - s^n)^{\mu(n)/n}, \qquad \mu = \text{Möbius function}.$$
We set
$$F_p(s) = \prod_{(n,p)=1} (1 - s^n)^{\mu(n)/n},$$
and observe that $F_p(s)$ lies in $1 + s\mathbb{Z}_{(p)}[[s]]$. If
$$L(s) = -\sum_{r=0}^{\infty} \frac{s^{p^r}}{p^r},$$
then $F_p(s)$ is $\exp(L(s))$ (usual exponential). Now one finds that
$$F_p(as)F_p(bs) = \prod_{r \geq 0} F_p(\psi_r(a,b)s^{p^r});$$
so, we define the *Artin–Hasse exponential* by
$$E(\mathbf{a}; s) = E((a_0, a_1, \ldots); s) = \prod_{r \geq 0} F_p(a_r s^{p^r}),$$
where $\mathbf{a}$ is the Witt vector $(a_0, a_1, \ldots)$. It follows that
$$E(\mathbf{a}; s)E(\mathbf{b}; s) = E(\mathbf{a} + \mathbf{b}; s),$$
in which $\mathbf{a} + \mathbf{b}$ signifies addition of Witt vectors.

For the application to Cartier duality, we observe that $\mathbb{Z}/p^t\mathbb{Z}$ has

$\mathcal{O}_S$-algebra generated by $X_0, \ldots, X_{t-1}$, and a basis given by the monomials $X_0^{i_0} \ldots X_{t-1}^{i_{t-1}}$, where $0 \leq i_j \leq p - 1$. The dual basis possesses some special elements, namely the duals of the algebra generators $X_0, \ldots, X_{t-1}$, which we denote by $f_0, \ldots, f_{t-1}$. Once again Y is a truncated exponential,

$$Y = E((f_0, \ldots, f_{t-1}); 1) \quad \text{(truncated)};$$

the Cartier pairing is exactly

$$T \to E(\mathbf{X} \otimes \log \mathbf{Y}) = E((\mathbf{X}_i \otimes \mathbf{f}_i); 1) \quad \text{(truncated)}.$$

(9) In characteristic $p > 0$, again, let $f \colon \mathbb{W}_t \to \mathbb{W}_t$ be the *Frobenius map*, that is,

$$f((a_0, \ldots, a_{t-1})) = (a_0^p, \ldots, a_{t-1}^p).$$

Write β_t for the kernel of f on $\mathbb{W}_t$, then β_t is a finite group scheme of order p^t over S. The ring of β_t is

$$\mathcal{O}_S[X_0, \ldots, X_{t-1}]/(X_0^p, \ldots, X_{t-1}^p),$$

while that of α_{p^t} is $\mathcal{O}_S[Y]/(Y^{p^t})$. They are Cartier duals, the Cartier pairing is

$$T \to E(\mathbf{Y} \otimes (X_0, \ldots, X_{t-1}); 1),$$

and **Y** means the vector $(Y, 0, \ldots, 0)$. When $t = 1$, this gives the self-duality of α_p, via the pairing

$$T \to \exp(Y \otimes X) \quad \text{(truncated exp)}.$$

Slightly more generally, write

$$_s\beta_t = \ker f^s \quad \text{on } \mathbb{W}_t,$$

where $f^s(\langle a_0, \ldots, a_{t-1} \rangle)$ is just $\langle a_0^{p^s}, \ldots, a_{t-1}^{p^s} \rangle$. Then, $_1\beta_t$ is our old β_t and $_t\beta_1$ is our old α_{p^t}. These are finite group schemes over S, and the group schemes $_r\beta_s$ and $_s\beta_r$ are Cartier duals, via the Cartier pairing

$$T \to E((X_0, \ldots, X_{r-1}) \otimes (Y_0, \ldots, Y_{s-1}); 1).$$

(10) As a last example of Cartier duality, we sketch some results of Oort and Tate [12] on the classification of finite group schemes. We shall treat their work more fully below. Let $S = \operatorname{Spec} R$, where R is a complete, unequal characteristic local ring—let the residue field characteristic be $p > 0$. We introduce the group schemes G_a^c, with a, c in R, such that $ac = p$. The group scheme is given by

$$G_a^c = \operatorname{Spec} R[X]/(X^p - aX),$$

$$m^*(X) = 1 \otimes X + X \otimes 1 + c \sum_{i=1}^{p-1} \frac{X^i}{w_i} \otimes \frac{X^{p-i}}{w_{p-i}},$$

in which $w_1, \ldots, w_{p-1}$ are certain units of R. The elements X^i/w_i play the role of divided powers, indeed

$$w_i \equiv i! \mod \mathfrak{m}_R.$$

Now it turns out that G_a^c is isomorphic to G_b^d if and only if there is a unit, u, of R such that
$$b = u^{p-1}a \quad \text{and} \quad d = u^{1-p}c.$$
The Cartier dual of G_a^c is $G_{cw_{p-1}}^{a/w_{p-1}}$, and the Cartier pairing is
$$T \to 1 + \frac{1}{1-p} \sum_{i=1}^{p-1} \frac{(X \otimes Y)^i}{w_i} = \text{"generalized"} \exp(X \otimes Y).$$

In characteristic $p > 0$, and in certain applications, the Frobenius morphism (variants of which have appeared above) plays an important role. Here is how it comes about in general. There is a map $\phi_p \colon X \to X$ obtained by raising to the pth power in each stalk of $\mathcal{O}_X$. Suppose G is a finite group scheme over S, where S is a scheme of characteristic $p > 0$. Then there is a commutative diagram

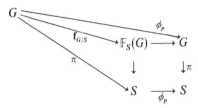

in which the outer square is *not* cartesian, but *the inner one is cartesian*. That is, $\mathbb{F}_S(G)$ is the product of G and S over S, as the diagram shows, and $\mathbf{f}_{G/S}$ is the morphism induced by π and ϕ_p (on G). The scheme $\mathbb{F}_S(G)$ is the *Frobenius group scheme* of G over S, and $\mathbf{f}_{G/S}$ is the *Frobenius morphism*. The scheme $\mathbb{F}_S(G)$ is a group scheme over S because it is a base extension, and the Frobenius morphism is a homomorphism of group schemes. Because the Frobenius operation is a base extension, it commutes with products and inverse limits. Also, observe that

$$\mathcal{O}_{\mathbb{F}_S(G)} = \mathcal{O}_G \otimes_{\mathcal{O}_S} \mathcal{O}_S, \quad \text{with} \quad \mathcal{O}_S \to \mathcal{O}_S \quad \text{via } p\text{th power}$$

and that $\mathbf{f}_{G/S}$ on the sheaf level is the map

$$\mathbf{f}(a \otimes \lambda) = a^p \lambda$$

taking $\mathcal{O}_G \otimes_{\mathcal{O}_S} \mathcal{O}_S$ to $\mathcal{O}_G$.

Suppose $S = \operatorname{Spec} R$, with R a complete local ring. For a finite group scheme over S, say G, the differentials $\Omega^1_{G/R}$ are given by $A \otimes \mathfrak{m}_0/\mathfrak{m}_0^2$. Here, $A = \Gamma(G, \mathcal{O}_G)$ and $\mathfrak{m}_0$ is $\ker \varepsilon^*$. Of course, $\mathfrak{m}_0/\mathfrak{m}_0^2$ is the cotangent space at the identity of G and can be identified with $\omega_{G/R}$—the space of invariant differentials of G (i.e., $\omega(uv) = \omega(u) + \omega(v)$). Write Λ_R for the ring of dual numbers over R, that is,

$$\Lambda_R = R[T]/(T^2).$$

We can consider the kernel of the map $G(\Lambda_R) \to G(R)$; it is easily seen that this kernel is the dual of $\mathfrak{m}_0/\mathfrak{m}_0^2$. These remarks lead immediately to the following proposition and its corollaries.

Proposition. *Let R be a complete, equicharacteristic local ring of characteristic $p > 0$ and let G be a finite, connected group scheme over R. Then, the kernel of $\mathbf{f}_{G/R}$ is never zero unless G is trivial. Hence, if R is Artinian, there is an integer $n > 0$ for which $G = \ker \mathbf{f}_{G/R}^n$. The smallest such n is the Frobenius height of G over R.*

PROOF. We examine the diagram

$$\begin{array}{ccccc} \text{Dual of} & (\mathfrak{m}_0/\mathfrak{m}_0^2) & \to & G(\Lambda_R) & \to & G(R) \\ & \downarrow \mathfrak{f} & & \downarrow \mathfrak{f} & & \downarrow \mathfrak{f} \\ \text{Dual of} & (\mathfrak{m}_0/\mathfrak{m}_0^2)_p & \to & \mathbb{F}_R(G)(\Lambda_R) & \to & \mathbb{F}_R(G)(R). \end{array}$$

As $\mathbf{f}(a \otimes \lambda) = a^p \lambda$ and $p \geq 2$, the leftmost vertical map is zero. If $\mathbf{f}$ were a monomorphism, we would find $\mathfrak{m}_0 = \mathfrak{m}_0^2$. So, Krull's theorem would show $\mathfrak{m}_0 = (0)$, i.e., $G = S$.

If R is Artinian, the sequence of kernels $\ker \mathbf{f}_{G/R}^n$ would eventually be stationary—so, it would have to stop at G. □

Corollary 1. *Every finite group scheme of prime order is automatically commutative.*

PROOF. Since multiplication in either order is continuous, the equalizer of these maps is closed; hence, it suffices to show every point of $G \times G$ lies in the equalizer. This means the problem is local. When R is a field, decompose G into connected and étale parts. Since G has order p, it is either one or the other. By faithfully flat-base extension, we may even assume R is algebraically closed. If G is étale, it has the form $\mathcal{H}$ for a group of order p; hence, G is commutative. If G is connected, it has Frobenius height one by the proposition. Hence, if $t \in \mathfrak{m}_0$, we find $t^p = 0$. Now the ring $\mathcal{O}_G$ is dual to the enveloping algebra of the Lie algebra of G (Lie $G = \ker(G(\Lambda_R) \to G(R))$), and so this enveloping algebra has rank p over R. But then, $\mathfrak{m}_0$ must be principal and, the enveloping algebra being generated by one element, is commutative. However, its multiplication is m^*; so G is commutative.

Finally, if R is just local, write $\bar{G}$ for $G \otimes_R R/\mathfrak{m}_R$, and observe that $\bar{G}$ is commutative. So, $\bar{G}^D$ is commutative, of rank p over $R/\mathfrak{m}_R$, and (by the above) its algebra, $\mathcal{O}_{\bar{G}^D}$, is generated by a single element. Lift this element to the module $\mathcal{O}_G^D$ over R, call the lifting t; we find $\mathcal{O}_G^D \supseteq R[t]$. Yet, $\mathcal{O}_G^D$ is free of rank p over R; so, Nakayama's lemma implies $\mathcal{O}_G^D$ is just $R[t]$. Thus, $\mathcal{O}_G^D$ is a commutative ring, which means G is a commutative group scheme. □

Corollary 2. *If R is a complete local ring with separably closed residue field (or, more generally, a strict Hensel local ring), and if the characteristic of the* residue field is $p > 0$, then a connected finite group scheme over R has order p^t, for some t.

PROOF. Let G be the connected group scheme over R, and write $\bar{G}$ for its fibre over the closed point ($\bar{G} = G \otimes_R R/\mathfrak{m}_R$). The scheme $\bar{G}$ must be connected, else we could lift a non-trivial idempotent of $\mathcal{O}_{G^{\text{et}}}$ from $R/\mathfrak{m}_R$ to R (by completeness). (Such an idempotent exists in $\mathcal{O}_{G^{\text{et}}}$ as $R/\mathfrak{m}_R$ is separably closed.) This would show G^{et} to be non-trivial, a contradiction. Since $\#(G) = \#(\bar{G})$, we are reduced to the case in which R is a separably closed field of characteristic $p > 0$. In this case, our proposition and induction on the Forbenius height of G, reduce us further to the case $G = \ker \mathbf{f}_{G/R}$. But, here, it is easy to see that, as an R-module, the algebra $\mathcal{O}_G$ is $R[t_1, \ldots, t_n]/(t_1^p, \ldots, t_n^p)$; the corollary is proved. □

Remark. Corollary 2 supplies the step missing in the proof of the Sylow type results of Section 3.

Corollary 3. *Let G be a finite group scheme of order m over a base scheme S. If m is invertible in $\mathcal{O}_S$, then G is étale over S.*

PROOF. According to [6], we may and do assume $S = \text{Spec } k$, where k is an algebraically closed field. Here, we use the decomposition into étale and connected parts, Lagrange's theorem, and Corollary 2 to conclude that G^0 is trivial. □

Oort and Tate [12] studied group schemes of prime order. It turns out that by assuming a mild restriction on the base scheme, S, a classification of these group schemes is achieved. The classification shows that the arithmetic in $\mathcal{O}_S$ is vitally connected to the number and complexity of the group schemes of order p over S.

We start by defining the basic ring, Λ_p, over which the classification is achieved. Let $\mathbb{Z}_p$ be the p-adic integers; there is a well-known multiplicative section of the residue class map $\mathbb{Z}_p \to \mathbb{Z}/p\mathbb{Z}$ (the Teichmüller representative). Call this section χ, and recall that it is given by

$$\chi(\bar{a}) = \lim_{n \to \infty} a^{p^n}, \quad \text{where } a \text{ lifts } \bar{a}.$$

It follows that the image of $\mathbb{Z}/p\mathbb{Z}^*$ in $\mathbb{Z}_p$ under χ consists of the $(p-1)$st roots of unity in $\mathbb{Z}_p$. Now we define Λ_p by

$$\Lambda_p = \mathbb{Z}\left[\chi(\mathbb{Z}/p\mathbb{Z}), \frac{1}{p(p-1)}\right] \cap \mathbb{Z}_p,$$

where the intersection is taken in $\mathbb{Q}_p$, the fraction field of $\mathbb{Z}_p$. In other words, Λ_p consists in adjoining to $\mathbb{Z}$ certain units from $\mathbb{Z}_p$, among them the ones occurring when we factor p in $\mathbb{Z}_p$ and use $\chi(\mathbb{Z}/p\mathbb{Z})$. For example, if $p = 2$, then Λ_p is $\mathbb{Z}$. But, if $p = 5$, then $\chi(\mathbb{Z}/5\mathbb{Z})$ is $\{0, i, i^2, i^3, 1\}$, and $\chi(2) = i$, so that $i \equiv 2$

(mod 5). Then $5 = (2 + i)(2 - i)$ in $\mathbb{Z}_5$ and $(2 + i)$ is a unit. Since $\frac{1}{2}$ lies in $\mathbb{Z}_p$, for odd p, we find

$$\Lambda_5 = \mathbb{Z}\left[i, \frac{1}{2(2+i)}\right].$$

Now let $S = \operatorname{Spec} \Lambda_p$, and let G be an S-group scheme of order p. Since G is commutative, the kernel of multiplication by p is a closed subgroup scheme of G. On the generic fibre of G over S, this kernel is the whole fibre; hence, the kernel of p is G, that is, G is killed by p. (In [12], one finds a much more general statement and better proof. In fact, the statement (proof by Deligne) is: *Every commutative group scheme of order m over any base is killed by m.*) Because G is killed by p, the group $(\mathbb{Z}/p\mathbb{Z})^*$ acts on G—this is the crucial observation.

We examine the group algebra $\mathcal{O}_S[\mathbb{Z}/p\mathbb{Z}^*]$ and its action on $\mathfrak{m}_0$ (the kernel of ε^* in $\mathcal{O}_G$). Write $[j]$ for the j-fold addition in G as an endomorphism, and consider the elements e_i of $\mathcal{O}_S[\mathbb{Z}/p\mathbb{Z}^*]$ given by

$$e_i = \frac{1}{(p-1)} \sum_{j \in \mathbb{Z}/p\mathbb{Z}^*} \chi^{-i}(j)[j].$$

It turns out that the e_i are orthogonal idempotents, and they split the augmentation ideal, $\mathfrak{m}_0$, into a direct sum of $(p - 1)$ ideals of $\mathcal{O}_G$

$$\mathfrak{m}_0 = e_1 \mathfrak{m}_0 \oplus \cdots \oplus e_{p-1} \mathfrak{m}_0.$$

The most important thing about this decomposition is that *it is an eigenspace decomposition*. That is, if we write I_i for $e_i \mathfrak{m}_0$, then *the ideal I_i is exactly the part of $\mathfrak{m}_0$ where the endomorphism $[m]$ acts like $\chi^i(m)$ for every $m \in \mathbb{Z}/p\mathbb{Z}^*$.* Thus,

$$f \in \Gamma(S, I_i) \Rightarrow [m](f) = \chi^i(m)f, \quad \text{all } m \in \mathbb{Z}/p\mathbb{Z}^*.$$

The ideal $\mathfrak{m}_0$ is locally free of rank $p - 1$; so, each I_i is locally free and Nakayama's lemma (reduction to a field) proves that each I_i has rank one. Moreover, one finds that I_i is $I_1^{\otimes i}$; so, we can summarize all this by

$$\begin{cases} \mathfrak{m}_0 = I_1 \oplus \cdots \oplus I_{p-1} = I_1 \oplus I_1^{\otimes 2} \oplus \cdots \oplus I_1^{\otimes p-1}, \\ \text{on } I_j, \text{ the elements of } \mathbb{Z}/p\mathbb{Z}^* \text{ act as endomorphisms like} \\ \quad \chi^j(\mathbb{Z}/p\mathbb{Z}^*), \text{ and} \\ I_j = I_1^{\otimes j} \text{ is } \mathcal{O}_S\text{-invertible.} \end{cases}$$

In the description of the Tate–Oort group schemes, certain units arise. These have already been mentioned when we discussed Cartier duality for these group schemes; they give rise to generalized divided powers and are defined by reference to μ_p. Let A be the affine ring of μ_p over Λ_p, so that A is $\Lambda_p[Z]/(Z^p - 1)$, and the kernel of ε^* on A is the principal ideal, $\mathfrak{m}$, generated by $Z - 1$. Of course, as A-module, we have the decomposition

$$\mathfrak{m} = \coprod_{j=1}^{p-1} A(Z^j - 1).$$

Now consider the elements

$$y_j = (p-1)e_j(1-Z) = \sum_{s \in \mathbb{Z}/p\mathbb{Z}^*} \chi^{-j}(s)(1-Z^s).$$

We find that $\mathfrak{m} = \coprod_{i=1}^{p-1} A y_i$, and thus the ideals I_i (for μ_p) are precisely the Ay_i. In terms of these y_i we can define our units w_i of Λ_p. They are given by

$$y^i = w_i y_i, \qquad w_1 = 1, \qquad y_1 = y.$$

One shows that $w_i \equiv i! \pmod{p}$; so, the y_i are generalized divided powers. Also, if we set $w_p = p w_{p-1}$, then $y^p = w_p y$. (The details of these arguments are in [12].)

To continue with the general construction, let X be a scheme over Λ_p, and look at the symmetric algebra on I_1 over $\mathcal{O}_X$

$$\operatorname{Symm}_{\mathcal{O}_X}(I_1) = \mathcal{O}_X \oplus I_1 \oplus I_1^{\otimes 2} \oplus \cdots.$$

The inclusion $I_1 \hookrightarrow \mathcal{O}_G$ gives a homomorphism $\operatorname{Symm}_{\mathcal{O}_X}(I_1) \to \mathcal{O}_G$. By what we have said above, this is surjective. Now consideration of the e_i above shows that under multiplication in $\mathcal{O}_G$ the $\mathcal{O}_X$-module $I_1^{\otimes p}$ goes to I_1; that is, there is a homomorphism

$$a \in \operatorname{Hom}_{\mathcal{O}_X}(I_1^{\otimes p}, I_1) = \operatorname{Hom}_{\mathcal{O}_X}(\mathcal{O}_X, I_1^{\otimes 1-p}).$$

Thus, a is a global section of $I_1^{\otimes 1-p}$, and we find that the kernel of the surjection $\operatorname{Symm}_{\mathcal{O}_X}(I_1) \to \mathcal{O}_G$ is generated by $(a-1) \otimes I_1^{\otimes p}$. We now apply all of this to the Cartier dual, G^D, of G. For the ideals, I_j, one can check that

$$(I^D)_j = (I_j)^D \cong I_j^{-1} \quad \text{(as invertible } \mathcal{O}_X\text{-modules)},$$

and we write a^D for the corresponding element of $\Gamma(X, I_1^{D \otimes (1-p)})$, i.e.,

$$a^D \in \Gamma(X, I_1^{\otimes (p-1)}).$$

Recall the Cartier pairing of example (10),

$$T \to 1 + \frac{1}{1-p} \sum_{i=1}^{p-1} \frac{Z^i}{w_i} \otimes \frac{Y^{p-i}}{w_{p-i}},$$

it sends T to a generating section of $I_1 \otimes I_1^D$; that is, the image of T gives an isomorphism of $I_1 \otimes I_1^D$ with $\mathcal{O}_X$. If we use this isomorphism, we find that $a \otimes a^D = w_p 1_{\mathcal{O}_X}$. Oort and Tate then use this to prove.

Theorem (Oort–Tate). *The correspondence*

$$G \mapsto (I_1^D, a, a^D)$$

is a bijection of the (isomorphism classes of) group schemes of order p over X (which lies over Λ_p) with the (isomorphism classes of) triples $(\mathcal{L}, \alpha, \beta)$ in which

(1) $\mathcal{L}$ *is an invertible $\mathcal{O}_X$-module;*
(2) $\alpha \in \Gamma(X, \mathcal{L}^{\otimes (p-1)})$ *and* $\beta \in \Gamma(X, \mathcal{L}^{\otimes (1-p)})$;
(3) $\alpha \otimes \beta = w_p 1_{\mathcal{O}_X}$.

The reconstruction of the functor which the group scheme determined by $(\mathscr{L}, \alpha, \beta)$ represents, and which we can denote $G_{\alpha\beta}^{\mathscr{L}}$, in terms of the data $\mathscr{L}$, α, β goes as follows: Let T be a scheme over X, then if $\zeta \in G_{\alpha\beta}^{\mathscr{L}}(T)$ we have the $\mathcal{O}_X$-homomorphism

$$\tilde{\zeta}: \mathcal{O}_G \to \mathcal{O}_T, \qquad G = G_{\alpha\beta}^{\mathscr{L}}.$$

Compose $\tilde{\zeta}$ with the inclusion $I_1 \subset \mathcal{O}_G$ and remember that $I_1 = \mathscr{L}^{-1}$, this gives

$$z: \mathscr{L}^{-1} \to \mathcal{O}_T, \quad \text{i.e.,} \quad z \in \Gamma(T, \mathscr{L} \otimes_{\mathcal{O}_X} \mathcal{O}_T).$$

The correspondence $\zeta \leftrightarrow z$ is bijective, so we obtain

$$G_{\alpha\beta}^{\mathscr{L}}(T) = \{z \in \Gamma(T, \mathscr{L} \otimes_{\mathcal{O}_X} \mathcal{O}_T) \mid z^{\otimes p} = \alpha \otimes z\},$$

and the group structure is given by

$$(z, y) \to z + y + \frac{\beta}{w_{p-1}} \otimes D_p(z \otimes 1, 1 \otimes y), \qquad \alpha\beta = pw_{p-1},$$

in which D_p is the polynomial

$$D_p(Z, Y) = \frac{w_{p-1}}{1-p} \sum_{i=1}^{p-1} \frac{Z^i}{w_i} \frac{Y^{p-i}}{w_{p-i}}.$$

(Note that $D_p(Z, Y) = (1/p)[(Z + Y)^p - Z^p - Y^p] \pmod{p}$.)

One further remark should be made. The completion of Λ_p is the p-adic integers, $\mathbb{Z}_p$; so, if R is a complete local ring with residue field characteristic $p > 0$, we can find a homomorphism $\Lambda_p \to R$. This means the Oort–Tate classification holds over R; moreover, line bundles over R are trivial so we are really facing the situation sketched in example (10). Over $\mathbb{Z}$, the Oort–Tate classification holds for group schemes of order 2, and μ_2 appears naturally in the form sketched at the very beginning.

Raynaud [14] extended this work of Oort and Tate. He considers a finite S-group scheme which is commutative and p-torsion (i.e., killed by p), and calls such S-group schemes "group schemes of type $(p, p, \ldots, p)$". Here is a rapid sketch of how he generalizes Oort–Tate to the new situation.

To get the correct endomorphisms (which one hopes will act on $\mathcal{O}_G$ to effect an eigenspace decomposition), choose a finite extension, F, of $\mathbb{Z}/p\mathbb{Z}$ and write $q = \#(F) = p^r$. There is an analog of the ring Λ_p, let us call it Λ_F, it arises as follows: Add the $(q-1)$st roots of unity to $\mathbb{Z}$, invert the integer $q - 1$, and adjoin to $\mathbb{Z}$ all primes of the field $\mathbb{Q}(\zeta_{q-1})$ lying above p except one of them (fixed in what follows). Here, ζ_{q-1} is a primitive $(q-1)$st root of unity. The choice of an omitted place above p, let us call it $\mathfrak{p}$, serves to delineate "good" characters. To see this, call a character *good* if the composed map

$$F \xrightarrow{\chi} \Lambda_F \xrightarrow[\text{at } \wp]{\text{red}} \kappa(\wp)$$

is a morphism of fields (i.e., additive). Once one good character, χ, is chosen all the others are $\chi_h = \chi^{p^h}$, with $0 \leq h \leq r-1$, and we have $\chi^{p^r} = \chi$. Of course, by choice $\chi_i^p = \chi_{i+1}$. The good characters form a p-basis for all the characters, for if ψ is *any* character $F \to \Lambda_F$, then

$$\psi = \prod_{i=0}^{r-1} \chi_i^{n_i}, \qquad 0 \leq n_i \leq p-1.$$

(In the Oort–Tate case, there is a unique good character—the Teichmüller character.) If another choice is made for the omitted place, the Galois group will move the old omitted place to the new one and move the old good characters to the new ones.

We now suppose G is slightly more special than just of type $(p, p, \ldots, p)$, namely, we assume G is an F vector space scheme (this means $G(T)$ is functorially an F vector space). We shall also assume G is an X-scheme and that the base, X, lies over Λ_F. Of course, these assumptions imply G has type $(p, p, \ldots, p)$, and that G^D also satisfies all these assumptions. For each character, χ, we form the element

$$e_\chi = \frac{1}{q-1} \sum_{\lambda \in F^*} \chi^{-1}(\lambda)[\lambda]$$

as in Oort–Tate. These e_χ are orthogonal idempotents, and they break $\mathfrak{m}_0$ ($= \ker \varepsilon^*$) into a direct sum of $\mathcal{O}_X$-modules

$$\mathfrak{m}_0 = \coprod_\chi \mathcal{L}_\chi, \quad \text{where} \quad \mathcal{L}_\chi = e_\chi \mathfrak{m}_0.$$

Further,

(a) each $\mathcal{L}_\chi$ is locally free, and
(b) $\mathcal{L}_\chi$ is the χ-eigenspace of $\mathfrak{m}_0$; that is, if $\lambda \in F^*$, then $[\lambda]$ acts on $\mathcal{L}_\chi$ via $\chi(\lambda)$.

Now one needs to assume the condition.

(RND) Each $\mathcal{L}_\chi$ is rank one (i.e., $\mathcal{O}_X$-invertible).

Condition (RND) is not really that restrictive. For one thing, it says $\#(G) = q$. For another, Raynaud proved that (RND) will hold if $\#(G) = q$ and if for at least one fibre either G or G^D is étale. (For example, if X has characteristic 0.) But, (RND) does not always hold: it fails for $\alpha_p \oplus \cdots \oplus \alpha_p$ (r times).

One also needs the analogs of the units w_i of Oort and Tate. For this, let $\chi_1, \ldots, \chi_n$ be characters, then write

$$w_{\chi_1 \ldots \chi_n} = (\chi_1 \cdots \chi_n)(n), \quad \text{i.e.,}$$

$$w_{\chi_1 \ldots \chi_n} = [n] \quad \text{on } \mathcal{L}_{\chi_1 \ldots \chi_n}.$$

If χ_i is a good character, set w_i equal to $\chi_i^p(p)$ ($= w_{\chi_i \ldots \chi_i}$). Raynaud proves w_i is *independent of i*, let us call it w. Now by studying Gauss sums, the special case $G = F^+$ (additive group of F), and the corresponding G^D, Raynaud proves:

(1) w_χ is invertible in Λ_F, for all characters χ;
(2) If $\chi = \prod_{i=0}^{r-1} \chi_i^{a_i}$, with χ_i the good characters and $0 \le a_i \le p - 1$, then $w_\chi \equiv a_0! \, a_1! \ldots a_r! \pmod{p}$;
(3) $w \equiv p! \pmod{p^2}$, $w = up$ with u a unit of Λ_F and $u \equiv -1 \pmod{p}$ (Wilson's theorem).

The classification proceeds as follows: Take the good characters $\chi_0, \chi_1, \ldots, \chi_{r-1}$ and form $\mathscr{J} = \mathscr{Q}_{\chi_0} \oplus \cdots \oplus \mathscr{Q}_{\chi_{r-1}}$. Then the symmetric algebra of $\mathscr{J}$ over $\mathcal{O}_X$ maps onto $\mathcal{O}_G$ provided G satisfies condition (RND). One analyzes its kernel in the same manner as Oort–Tate, and one finds the

Theorem (Raynaud). *The map* $G \mapsto (\mathscr{Q}_{\chi_0}, \ldots, \mathscr{Q}_{\chi_{r-1}}, c_i, d_i)$, *where* $c_i \colon \mathscr{Q}_{\chi_{i+1}} \to \mathscr{Q}_{\chi_i^p}$, *and* $d_i \colon \mathscr{Q}_{\chi_i^p} \to \mathscr{Q}_{\chi_{i+1}}$ *arise from the iterated comultiplication and multiplication in* $\mathcal{O}_G$, *is a bijection of the (isomorphism classes of) F-vector space group schemes satisfying* (RND) *over X with the (isomorphism classes of) systems formed of r invertible $\mathcal{O}_X$-modules $\mathscr{L}_0, \mathscr{L}_1, \ldots, \mathscr{L}_{r-1}$ and r pairs of $\mathcal{O}_X$-module maps* $\gamma_i \colon \mathscr{L}_{i+1} \to \mathscr{L}_i^p$, $\delta_i \colon \mathscr{L}_i^p \to \mathscr{L}_{i+1}$, *such that* $\delta_i \gamma_i = w 1_{\mathscr{L}_{i+1}}$ *for all i.*

If R is a complete local (more generally, Hensel) ring whose residue field contains a field with q elements, then R is a Λ_F-algebra; so, the above applies to Spec R. But here, locally free modules are free, and we get the much more down-to-earth assertion:

The F-vector space group schemes over R which satisfy (RND) *correspond to r pairs of elements of R*: $a_1, b_1, \ldots, a_r, b_r$ *with* $a_j b_j = w$ *for every j. Here* $\#(G) = q = p^r$, *and G is given by the equations*

$$X_i^p = b_i X_{i+1}, \qquad i = 1, 2, \ldots, r.$$

(The map m^* is given by a messy "universal" polynomial involving the a_j's as parameters.)

The F-vector space group schemes $G_{b_1 \ldots b_r}^{a_1 \ldots a_r}$ and $G_{d_1 \ldots d_r}^{c_1 \ldots c_r}$ (with obvious notations) are isomorphic if and only if there are units $u_1, \ldots, u_r$ of R with

$$u_{i+1} d_i = u_i^p b_i \quad \text{and} \quad u_{i+1} a_i = u_i^p c_i, \quad \text{all } i.$$

§5. Formal Groups

We intend no grand theory here, just what is needed for p-divisible groups (to be treated in the next section). To this end, R will always be a complete local ring (in particular, R could be a field or an Artinian local ring), and S will denote Spec R. For group schemes, we began with functors on the category of schemes; here, we restrict the category on which our functors are defined.

We look at schemes T over S which satisfy

(1) *T is finite over S (hence, affine); and*
(2) $\Gamma(T, \mathcal{O}_T)$ *is an R-module of finite length.*

(If R is a field or is Artinian local, then (1) implies (2).) To coin a word, let us call the schemes T satisfying (1) and (2) *very finite* over S. (*Warning: This is not standard terminology.*) The use of these very finite schemes is analogous to the use of C^∞-functions with compact support as test functions in the theory of distributions.

A *formal functor*, F, over S, is a cofunctor on the very finite schemes over S to sets, and a *formal group functor* over S is a formal functor so that each $F(T)$ has, in a functorial way, the structure of group.

Of course, every ordinary S-functor or S-group functor, say F, gives us a formal counterpart, call it $\hat{F}$, obtained by restricting F to the very finite S-schemes T. We call $\hat{F}$ the *formal completion of F*.

To treat questions of representability, let us call a commutative R-algebra, A, *profinite* if it is $\varprojlim A/\mathfrak{A}$ over a family of ideals $\mathfrak{A}$ such that $A/\mathfrak{A}$ is very finite over R.* Note that R is itself profinite, because $R = \varprojlim R/\mathfrak{m}_R^q$; and every R-algebra, A, *finite as R-module*, is also profinite. But, observe that the power series ring $R[[X_1, \ldots, X_n]]$, while *not finite over R, is still profinite*. To see this let

$$\mathscr{I} = (X_1, \ldots, X_n) + \mathfrak{m}_R R[[X_1, \ldots, X_n]],$$

then the power series ring is the projective limit of $R[[X_1, \ldots, X_n]]/\mathscr{I}^t$ as $t \to \infty$.

Each profinite R-algebra, A, defines a formal functor. We call this functor *the formal spectrum of A*, and denote it by Spf A. Here is the explicit definition:

(*) $$(\text{Spf } A)(T) = \text{Hom}_{R,\text{cont.}}(A, \Gamma(T, \mathcal{O}_T)),$$

where T is very finite over S and $\Gamma(T, \mathcal{O}_T)$ has the *discrete topology*.

Let us rewrite (*) solely in terms of $B = \Gamma(T, \mathcal{O}_T)$, so that it gives a (covariant) functor on very finite R-algebras:

(**) $$(\text{Spf } A)(B) = \text{Hom}_{R,\text{cont.}}(A, B) \quad \text{if B is very finite.}$$

The definition (**) immediately leads to an extension of Spf A as a functor on profinite R-algebras. Namely, suppose B is profinite, say $B = \varprojlim_\alpha B/\mathfrak{B}_\alpha$, then we set

$$(\text{Spf } A)(B) = \varprojlim_\alpha \text{Hom}_{R,\text{cont.}}(A, B/\mathfrak{B}_\alpha).$$

---

* We reject the barbarisms: "pro very finite", or "very profinite".

In this way, our formal functors are *really functors on profinite R-algebras*. Now we can make the obvious definition:

A *formal S-scheme* (respectively, a *formal S-group scheme* or (better) an *S-formal group*) is a representable functor on the category of profinite R-algebras to sets (resp. groups). That is, a formal S-scheme is just Spf A for some profinite R-algebra A. (If Spf A is a formal group, A will get more structure.)

If A is a finitely generated R-algebra, we have the ordinary representable functor Spec A. We can form the completion of Spec A, which we have denoted $\widehat{\text{Spec } A}$. It is easy to see that this completion is representable; that is, it is a formal S-scheme, Spf($\hat{A}$) for some profinite $\hat{A}$ over R. To do this, write a presentation for A

$$R[X_1, \ldots, X_m] \xrightarrow{\varphi} R[Y_1, \ldots, Y_n] \to A \to 0,$$

and observe that φ is given by m polynomials[†] in the Y's. If we replace the polynomial rings by power series rings, φ still is defined and we get a presentation for $\hat{A}$ ($=$ coker φ on power series)

$$R[[X_1, \ldots, X_m]] \xrightarrow{\varphi} R[[Y_1, \ldots, Y_n]] \to \hat{A} \to 0.$$

This $\hat{A}$ is the profinite R-algebra we seek to represent $\widehat{\text{Spec } A}$.

If Spf A is a formal group, then the group multiplication gives rise to a comultiplication

$$m^* \colon A \to A \mathbin{\hat{\otimes}_R} A,$$

where the caret over the tensor sign signifies the *complete tensor product*. If $A = \varprojlim A/\mathfrak{A}$, then $A \mathbin{\hat{\otimes}_R} A$ is just $\varprojlim (A/\mathfrak{A}) \otimes_R (A/\mathfrak{A})$. This operation yields a profinite R-algebra and is the coproduct in the category of profinite R-algebras.

One of the nice things about the formal category is the ease in deciding on representability (compare the ease in differentiating distributions). This is a result of Grothendieck [7] and goes as follows (we skip the proof):

Theorem (Grothendieck). *A functor from profinite R-algebras to sets (groups) is representable if and only if it is left-exact ($=$ commutes with fibred products and final elements).*

Let us now restrict attention to formal groups. We have

Proposition. *Every formal group, G, over $S = \text{Spec } R$ has a canonical exact sequence*

$$0 \to G^0 \to G \to G^{\text{et}} \to 0$$

[†] By linear change of variable in $R[X]$, we may assume no constant term.

in which G^0 is a connected normal formal subgroup of G and G^{et} is an étale formal group.

PROOF. Write $G = \text{Spf } A$, and $A = \varprojlim A/\mathfrak{A}$. Then, as in the proof of the corresponding exact sequence for ordinary group schemes, we have the R-algebras $(A/\mathfrak{A})^{et}$ and we pass these to the limit to get A^{et}. Of course, G^{et} is $\text{Spf}(A^{et})$, and it is easy to see that the group law on G gives one on G^{et} so that our surjection $G \to G^{et}$ is a homomorphism. Let the kernel be G^0. By the completeness of R, we can lift idempotents, and this shows that G^0 is connected. □

We say that G is *smooth* over S if G^0 is the formal spectrum of a power series ring over R. A smooth, connected formal group is called a *formal Lie group*; so, a formal Lie group is $\text{Spf}(R[[X_1, \ldots, X_n]])$ and the integer n is the *dimension* of the formal Lie group. The isomorphism

$$R[[X_1, \ldots, X_n]] \hat{\otimes}_R R[[Y_1, \ldots, Y_m]] \cong R[[S_1, \ldots, S_n, T_1, \ldots, T_m]],$$

in which $S_j = X_j \otimes 1$ and $T_j = 1 \otimes Y_j$, shows that to give

$$\text{Spf}(R[[X_1, \ldots, X_n]])$$

the structure of a formal Lie group we need to give n special power series in $2n$ variables. Suppose that $F_i(S_1, T_1, \ldots, S_n, T_n)$ for $1 \leq i \leq n$ are such given power series, and write

$$F(S, T) = (F_1(\ldots, S_i, T_i, \ldots), \ldots, F_n(\ldots, S_i, T_i, \ldots)).$$

Then, the rules we require of our power series result from translating the group axioms in these terms, they are:

(1) $X = F(X, 0) = F(0, X)$ (ε^*-law);
(2) $F(X, F(Y, Z)) = F(F(X, Y), Z)$ (associative law);
(3) $F(X, Y) = F(Y, X)$ (if the group is commutative).

The inverse law is automatic.

If $x_1, \ldots, x_n, y_1, \ldots, y_n$ are elements of some very finite R-algebra, B, and not units of B, and if x signifies $(x_1, \ldots, x_n)$, and similarly for y, then the group law on $(\text{Spf } A)(B)$ is given by

$$x * y = F(x, y) = x + y + \text{higher order terms in } x, y.$$

Recall that a complex Lie group is a group in the category of complex analytic manifolds. So, if $(X_1, \ldots, X_n)$ are coordinates near a point, then the group operations—being holomorphic—are given by complex power series in the coordinates of the points. The exponential map from the Lie algebra (= tangent space) shows that these series begin linearly. Thus, in the complex Lie group

$$xy = x + y + \text{higher terms in } x, y \quad (\text{converging in } \mathbb{C}).$$

What we have done is to replace the converging series by formal series, and this explains the name "formal Lie group".

If G is a smooth formal group, its dimension will be the dimension of its connected component. Étale formal groups are zero dimensional. This is standard terminology because the tangent space at the origin is the same for G and its connected component, and the vector space dimension of the tangent space is the dimension of G.

Returning to our basic definitions, we see that if A is a profinite ring $A = \varprojlim_\alpha A_\alpha$, instead of giving A we may as well give the projective mapping system $\{A_\alpha, \text{ maps } A_\beta \to A_\alpha\}$. Consequently, we obtain an inductive system of very finite R-schemes $X_\alpha = \text{Spec } A_\alpha$, and we can consider a formal scheme Spf A as an inductive system (X_α) of very finite R-schemes. As functors, some caution must be observed. If B is *very finite*, then

$$(\text{Spf } A)(B) = \text{Hom}_{R,\text{cont.}}\left(\varprojlim_\alpha A_\alpha, B\right) = \varinjlim_\alpha \text{Hom}_R(A_\alpha, B) = \varinjlim_\alpha X_\alpha(B).$$

That is, on *very finite* B, we have $(X_\alpha)(B) = \varinjlim_\alpha X_\alpha(B)$. But, if B is *profinite*, say $B = \varprojlim_\beta B_\beta$, then our definition of the functor Spf A was

$$(X_\alpha)(B) = (\text{Spf } A)(B) = \varprojlim_\beta \text{Hom}_{R,\text{cont.}}(A, B_\beta) = \varprojlim_\beta \varinjlim_\alpha X_\alpha(B_\beta),$$

and, *in general, we cannot interchange the two limits*. Thus, in general, $(X_\alpha)(B) \neq \varinjlim_\alpha X_\alpha(B)$, if B is merely profinite.

§6. p-Divisible Groups

A p-divisible group is a special kind of formal group. The definition can be formulated over any commutative ring R, or even over a scheme, but the essential case is where R is local (and, even more interesting, of residue characteristic $p > 0$).

We fix the prime p and a non-negative integer h.

Definition. A *p-divisible group over R of height h* is an inductive system (G_v, i_v) in which:

(1) G_v is a finite, commutative group scheme over R of order p^{hv}.
(2) For each v, we have the exact sequence

$$0 \to G_v \xrightarrow{i_v} G_{v+1} \xrightarrow{p^v} G_{v+1},$$

that is, i_v maps G_v (by closed immersion) to G_{v+1} and identifies it with the kernel of p^v on the latter group scheme.

The inductive system (G_v, i_v) gives a formal group; if $G_v = \text{Spec } A_v$ and we write $A = \varprojlim A_v$, then Spf A is the p-divisible group (G_v, i_v). Most often we work directly with the G_v and only occasionally with $G = \text{Spf } A$. Consider the diagram

GROUP SCHEMES, FORMAL GROUPS, AND p-DIVISIBLE GROUPS

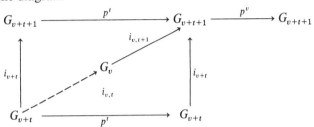

It shows that G_v is the kernel of p^v on G_{v+2} via the iterated injection $i_{v+1} \circ i_v$. This holds for all $t \geq 1$ on G_{v+t} as is clear; hence, G_v is the kernel of p^v on G and G is p-torsion (as it is $\varinjlim G_v$).

Next, the diagram

$$
\begin{array}{ccccc}
G_{v+t+1} & \xrightarrow{p^t} & G_{v+t+1} & \xrightarrow{p^v} & G_{v+t+1} \\
\uparrow{}_{i_{v+t}} & {}^{i_{v,t+1}}\nearrow & \uparrow{}_{i_{v+t}} & & \\
G_{v+t} & \xrightarrow{}_{i_{v,t}} G_v & & & \\
G_{v+t} & \xrightarrow{p^t} & G_{v+t} & &
\end{array}
$$

commutes. The map $G_{v+t} \to G_{v+t+1}$ via $i_{v+t} \circ p^t = p^t \circ i_{v+t}$ must factor through the kernel of p^v by (2) of the definition applied to G_{v+t}. But, this kernel is G_v as we have observed, so the dotted arrow (shown) exists, call it $j_{t,v}$. Then the lower triangle of our diagram

$$
\begin{array}{ccc}
 & G_v & \\
 \nearrow & & \searrow{}^{i_{v,t}} \\
G_{v+t} & \xrightarrow{p^t} & G_{v+t}
\end{array}
$$

commutes for every t. Now the kernel of $j_{t,v}$ is the kernel of p^t on G_{v+t}, that is, it is G_t. Since the order of G_{v+t} is the product of the orders of G_v and G_t by (1), we deduce that

$$0 \to G_t \xrightarrow{i_{t,v}} G_{v+t} \xrightarrow{j_{t,v}} G_v \to 0$$

is exact for every v and every t. It follows that the mapping $p: G \to G$ is an *isogeny*, meaning that it is onto and its kernel is a finite group scheme over S. We have therefore proved the

Proposition. *A p-divisible group over R is a p-torsion commutative formal group, G, over R for which $p: G \to G$ is an isogeny.*

Remarks. We recover G_v as the kernel of p^v on G, the height h appears as the exponent of the order of ker p, and $G = \varinjlim G_v$ because it is p-torsion.

EXAMPLES OF p-DIVISIBLE GROUPS

(1) If π is the fundamental group of S, and if T is a free $\mathbb{Z}_p$-module of rank h on which π operates continuously, then we shall set $T_v = T/p^v T$. The T_v form an inductive system of π-modules of order p^{vh} with associated étale

group schemes, $\mathcal{T}_v$, over S. We obtain the étale p-divisible group of height h, $\varinjlim \mathcal{T}_v$, over S, and each étale p-divisible group has this form (cf. example (7) of Section 3).

(2) Let $G_v = (\ker p^v \text{ on } \mathbb{G}_m) = \mu_{p^v}$. Then the μ_{p^v} form an inductive system, and we get the p-divisible group of $\mathbb{G}_m$: $\mathbb{G}_m(p) = \varinjlim \mu_{p^v}$. The height of $\mathbb{G}_m(p)$ is 1.

(3) (This is the most important—the motivating—example.) Let X be an abelian scheme of relative dimension d over S. If $_{p^v}X$ denotes the kernel of multiplication by p^v on X, then it is known that $_{p^v}X$ is a finite group scheme over S of order p^{2dv}, [11]. Consequently, the inductive system $(_{p^v}X, \text{incl.})$ gives a p-divisible group, $X(p)$, of height $2d$. We call $X(p)$ *the p-divisible group of the abelian scheme X*.

Since each p-divisible group fits into an exact sequence with kernel a connected and cokernel an étale p-divisible group, and since example (1) gives all étale p-divisible groups, we need to give a construction of connected p-divisible groups.

(4) (Connected p-divisible groups). Let Γ be an n-dimensional commutative formal Lie group over S. Assume the map $p\colon \Gamma \to \Gamma$ (multiplication by p, the residue characteristic of R) is an isogeny. If this is so, Γ is called divisible. Now repeat the process of examples (2) and (3) to make the p-divisible group $\Gamma(p) = \varinjlim (\ker p^n \text{ on } \Gamma)$. To actually see that the kernel of p^n on Γ has order p^{nh}, we use the flatness to base extend to the residue field which has characteristic $p > 0$. In characteristic p, the Frobenius morphism and its Cartier dual, the Verschiebung V, are related by the following diagram (in which G is a formal group)

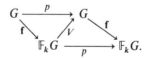

That is, $p = fV = Vf$. As Γ is connected, so is $\Gamma \otimes_R k$ because R is complete. Our diagram now shows that the kernel of p^n on Γ has order p^{nh}, all n; the height, h, of $\Gamma(p)$ is the exponent of the degree of the isogeny $p\colon \Gamma \to \Gamma$.

Example (4) turns out to be perfectly general. Indeed in Tate's article [19], one finds a sketch of the proof of the following characterization of connected p-divisible groups obtained by Serre and Tate.

Theorem. *The functor $\Gamma \rightsquigarrow \Gamma(p)$ is an equivalence of categories between the category of divisible, commutative formal Lie groups over R and the category of connected p-divisible groups over R.*

The essential content of this theorem (whose proof is omitted) is that every connected p-divisible group is smooth. Flatness reduces the proof of this to

GROUP SCHEMES, FORMAL GROUPS, AND p-DIVISIBLE GROUPS

the case of the special fibre, which lies over a field of characteristic $p > 0$. Here, the Frobenius–Verschiebung diagram above together with knowledge of the orders of the G_v allows one to give the proof.

There is the notion of Cartier duality for p-divisible groups. We have the diagram

$$\begin{array}{c} G_v \\ {}^{j_v = j_{1,v}}\swarrow \quad \searrow{}^{i_v = i_{v,1}} \\ G_{v+1} \xrightarrow{p} G_{v+1} \end{array}$$

for each v. Since each j_v is surjective, the system (G_v^D, j_v^D) forms a p-divisible group, called the *Cartier dual* of (G_v, i_v).

In the examples given above, the dual of $\mathbb{G}_m(p)$ is the étale p-divisible group $\mathbb{Q}_p/\mathbb{Z}_p = \varinjlim \mathbb{Z}/p^v\mathbb{Z}$. If X is an abelian scheme, then $\mathscr{H}\!om_{\text{groups}}(X, \mathbb{G}_m)$ is always trivial. But $\mathscr{E}\!xt^1_{\text{groups}}(X, \mathbb{G}_m)$ is an abelian scheme, call it X^D, and it is dual to X. (One should think of an extension $0 \to \mathbb{G}_m \to P \to X \to 0$ as defining a bundle over X with fibre $\mathbb{G}_m$. If one adds the zero section, there results a line bundle on X, i.e., an element of $\text{Pic}(X)$.) Now we consider the isogeny p^v on X,

$$0 \to {}_{p^v}X \to X \xrightarrow{p^v} X \to 0.$$

By using $\mathscr{H}\!om(-, \mathbb{G}_m)$, we get

$$0 \to \mathscr{H}\!om_{\text{groups}}({}_{p^v}X, \mathbb{G}_m) \to \mathscr{E}\!xt^1_{\text{groups}}(X, \mathbb{G}_m) \xrightarrow{p^v} \mathscr{E}\!xt^1_{\text{groups}}(X, \mathbb{G}_m) \to 0,$$

or (in terms of the dual abelian scheme, X^D)

$$0 \to ({}_{p^v}X)^D \to X^D \xrightarrow{p^v} X^D \to 0.$$

We deduce that $({}_{p^v}X)^D = {}_{p^v}(X^D)$; in other words that

$X^D(p)$ *is Cartier dual to* $X(p)$.

Since the connected component of a p-divisible group is a formal Lie group, the latter has a dimension and the dimension of our p-divisible group is just the dimension of its connected component. Moreover, it is clear that the height of a p-divisible group and its dual are the same. The dimension and height of p-divisible groups are connected through Cartier duality:

Proposition. *If G is a p-divisible group over R and G^D is its dual, then*

$$\dim G + \dim G^D = ht(G) = ht(G^D).$$

PROOF. The invariants do not change by restriction to the closed fibre; so, we may assume R is a field of characteristic $p > 0$. Then the Frobenius–Verschiebung diagram shows that

$$0 \to \ker \mathbf{f} \to \ker p \to \ker V \to 0$$

is exact. Since $\mathbf{f}|G^{\text{et}}$ is injective, $\ker \mathbf{f}$ is concentrated in G^0. Here, $\mathbf{f}$ operates by raising the variables to the pth power, therefore the order of ker $\mathbf{f}$ is $p^{\dim G}$. The map V is dual to $\mathbf{f}$; so, ker V is to ker $\mathbf{f}$ as G^D is to G. But then, the order of ker V is $p^{\dim G^D}$. By the exact sequence,

$$p^{\dim G} \cdot p^{\dim G^D} = \#(\ker p) = p^{ht(G)} \qquad \text{(definition of } ht(G)\text{).}$$

This finishes the proof. $\square$

In example (2), $\mathbb{G}_m(p)$ has height one and dimension one; its dual, $\mathbb{Q}_p/\mathbb{Z}_p$, has height one, dimension zero. In example (3), both $X(p)$ and $X^D(p)$ have height $2d$. The dimension of both $X(p)$ and $X^D(p)$ is d (the tangent space to the origin of X is d-dimensional and this tangent space is the Lie algebra for the formal Lie group $X(p)^0$). However, the height of $X(p)^0$ can be any integer n with $d \le n \le 2d$ depending on the p-torsion in the reduction of X. When the reduction of X mod $\mathfrak{m}_R$ is "ordinary", the height of $X(p)^0$ is d.

For applications to arithmetic, we need to study the points of G with values in a complete topologized ring. When we defined the points of Spf A with values in B as $\varprojlim (\text{Spf } A)(B/\mathfrak{B}_\beta)$, we really did not need that $B_\beta (= B/\mathfrak{B}_\beta)$ was very finite, what we needed was that B is topologized so that it is complete and so that B_β is discrete. Let us specialize R to be a complete discrete valuation ring whose residue field, k, has characteristic $p > 0$. If K is the fraction field of R, we will soon assume that K has characteristic zero, however, this is not yet necessary. Let L be the completion of an algebraic extension of K (with respect to an extension of the valuation of R), and write S for the ring of integers of L. Of course, $S = \varprojlim S/\mathfrak{m}_R^i S$, and so if $G = (G_v)$ is p-divisible over R, we get

$$G(S) = \varprojlim_i G(S/\mathfrak{m}_R^i S) = \varprojlim_i \varinjlim_v G_v(S/\mathfrak{m}_R^i S).$$

From the exact sequence

$$0 \to G_v \to G \xrightarrow{p^v} G \to 0$$

and the left-exactness of $\varprojlim$, we get the exact sequence

$$0 \to G_v(S) \to G(S) \xrightarrow{p^v} G(S).$$

Hence, the torsion subgroup of $G(S)$, denoted $G(S)_{\text{tors}}$, is

$$G(S)_{\text{tors}} = \varinjlim_v G_v(S) = \varinjlim_v \varprojlim_i G_v(S/\mathfrak{m}_R^i S).$$

This shows in clear detail that the two limiting operations cannot, in general, be interchanged. However, when G is étale, we find that $G_v(S/\mathfrak{m}_R^i) = G_v(S)$; so for G étale we get $G(S) = G(S)_{\text{tors}}$. Quite generally we have: *If $G = \text{Spf } A$, and S is complete with respect to its valuation topology, then*

$$G(S) = \text{Hom}_{R,\text{cont.}}(A, S),$$

where A has its adic-topology and S its valuation topology.

Look at the case in which G is connected. Then $A = R[[X_1, \ldots, X_n]]$, and the points of G in S (i.e., the elements of $G(S)$) are n-tuples $(\xi_1, \ldots, \xi_n)$, where each ξ_i lies in the maximal ideal of S. As S is complete, the power series defining the group multiplication $\xi *_\eta \eta$ converge in S, and $G(S)$ *is therefore an analytic group over L (p-adic analog of a complex Lie group)*. Observe as well that $G(S)$ is a $\mathbb{Z}_p$-module.

From now on, assume K has characteristic zero. Let $\bar{K}$ be the algebraic closure of K and write $\mathfrak{G}$ for the Galois group of $\bar{K}/K$.

Definition. If X is a commutative group scheme or a formal group over R, the *Tate module* of X, denoted $T_p(X)$, is

$$T_p(X) = \varprojlim_v (\ker(X(\bar{K}) \xrightarrow{p^v} X(\bar{K }))).$$

(Note that if $X = \varinjlim G_v = G$ is p-divisible, then $T_p(G) = \varprojlim_v G_v(\bar{K})$.) We can define the *Tate comodule* of the p-divisible group G (*not* standard nomenclature), denoted $\Phi_p(G)$, by

$$\Phi_p(G) = \varinjlim_v G_v(\bar{K}).$$

Remarks. (1) $T_p(G)$ and $\Phi_p(G)$ are $\mathbb{Z}_p$-modules.
(2) $\mathfrak{G}$ acts on both $T_p(G)$ and $\Phi_p(G)$ continuously.
(3) $T_p(G)$ and $\Phi_p(G)$ depend only upon the generic fibre, $G \otimes_R K$, of G over R.
(4) *The Tate module, $T_p(G)$, determines the generic fibre of G over R.* (To see this, note that in any case, $G \otimes_R K$ is étale as $\mathrm{char}(K) = 0$. Hence, $G \otimes_R K$ is determined by the Galois modules $(G_v \otimes_R K)(\bar{K})$, i.e., by $G_v(\bar{K})$. However, we have $T_p(G)/p^v T_p(G) = G_v(\bar{K})$, by definition of $T_p(G)$; so, our remark follows.)
(5) *The Tate comodule, $\Phi_p(G)$, determines the generic fibre of G over R.* (Similar argument to remark (4).)
(6) *As a $\mathbb{Z}_p$-module, $T_p(G)$ is $\mathbb{Z}_p^{ht(G)}$; and, similarly, $\Phi_p(G)$ is $(\mathbb{Q}_p/\mathbb{Z}_p)^{ht(G)}$.*
(7) *We have the canonical $\mathfrak{G}$-isomorphisms*

$$\Phi_p(G) \cong T_p(G) \otimes_{\mathbb{Z}_p} \mathbb{Q}_p/\mathbb{Z}_p,$$

$$T_p(G) \cong \mathrm{Hom}_{\mathbb{Z}_p}(\mathbb{Q}_p/\mathbb{Z}_p, \Phi_p(G)).$$

(8) *For the p-divisible group G, knowledge of any one of the Tate module, $T_p(G)$, the Tate comodule, $\Phi_p(G)$, the generic fibre over R, $G \otimes_R K$, implies knowledge of the other two.* (This is the conjunction of (4), (5), and (7).)

We wish to connect the Tate modules of G and G^D in order to get further information on each as well as more information on G and G^D from each of the Tate modules. For this, it turns out that we must introduce the true analog of $\mathbb{C}$ in the p-adic theory. We let C be the completion of the algebraic closure, $\bar{K}$, of K. The field C is algebraically closed and it is complete (but not locally compact); so, C is the analog of $\mathbb{C}$. Write R_C for the ring of integers in C.

Now for any completion, L, of an algebraic extension of K, with ring of integers S, we have
$$G_v(S) = G_v(L).$$
In particular, when $L = C$ and $S = R_C$, we get $G_v(R_C) = G_v(C)$. But, the comodule $\Phi_p(G)$ is the direct limit of the $G_v(R_C)$, and so we deduce
$$\Phi_p(G) = \text{torsion subgroup of } G(R_C) = \text{torsion subgroup of } G(C).$$
Consider the Cartier dual of G_v, it is $\mathcal{H}om_{\text{groups}}(G_v, \mathbb{G}_m)$. Consequently, by its very definition, we find
$$G_v^D(R_C) = \text{Hom}_{R_C\text{-groups}}(G_v \otimes_R R_C, \mathbb{G}_m),$$
(∗)
$$G_v^D(R_C) = \text{Hom}_{R_C\text{-groups}}(G_v \otimes_R R_C, \mu_{p^v}).$$

Remember that $G_v^D(\bar{K}) = G_v^D(R_C) = G_v^D(C)$, and observe that (∗) may also be written
$$G_v^D(R_C) = \text{Hom}_{R_C\text{-groups}}(G_v \otimes_R R_C, \mathbb{G}_m(p)).$$
Pass the latter equation to the projective limit over v; we get
(∗∗)
$$T_p(G^D) = \text{Hom}_{R_C\text{-groups}}(G \hat{\otimes}_R R_C, \mathbb{G}_m(p)).$$
Now (∗) yields the pairing
$$G_v^D(R_C) \times G_v(R_C) \to \mu_{p^v}(R_C),$$
which, when passed to the projective limit, yields in turn the duality pairing for Tate modules
(∗)′
$$T_p(G^D) \times T_p(G) \to T_p(\mathbb{G}_m) = T_p(\mathbb{G}_m(p)).$$
Pairing (∗)′ is a $\mathfrak{G}$-pairing, and if we write $\mathbb{Z}_p(1)$ for $T_p(\mathbb{G}_m)$, we get the $\mathfrak{G}$-pairing (for Tate modules)
(∗)″
$$T_p(G^D) \times T_p(G) \to \mathbb{Z}_p(1).$$
On the other hand, (∗∗) gives us the pairing
(∗∗∗)
$$T_p(G^D) \times G(R_C) \to \mathbb{G}_m(p)(R_C) = U_C,$$
where U_C is the group of units of R_C congruent to one modulo the maximal ideal. The Galois group, $\mathfrak{G}$, acts on (∗∗∗), and so we derive the homomorphism
$$G(R_C)^{\mathfrak{G}} \to \text{Hom}_{\mathfrak{G}}(T_p(G^D), U_C).$$
At this point, we need the fundamental fact—proved by Tate in his article [19]—that *the fixed points of C under its Galois action by $\mathfrak{G}$ are exactly K*, i.e.,
$$H^0(\mathfrak{G}, C) = K.$$
(This is done by Tate by analyzing ramification; independent proofs were

given separately by Ax and Dwork. We will omit these arguments.) Thus, our homomorphism becomes the fundamental homomorphism

$$\alpha_R \colon G(R) \to \operatorname{Hom}_{\mathfrak{G}}(T_p(G^D), U_C).$$

Note that if we were to succeed in proving α_R an isomorphism, then $T_p(G^D)$ would determine the R-points of G, and $(*)''$ would show that $T_p(G)$ determines the R-points of G. We aim to do this and more. In order to make sense of the arguments to follow, we make a small digression to discuss what is now known as the "Tate twist".

Let ζ be a primitive p^vth root of 1, then for each $s \in \mathfrak{G}$, the element $s\zeta$ is another p^vth root of 1, so that

$$s\zeta = \zeta^{\tau(s)}$$

for some element $\tau(s) \in (\mathbb{Z}/p^v\mathbb{Z})^*$. We pass to the limit over v and thereby get a character, τ, with values in the units of $\mathbb{Z}_p$. Thus, for every $\zeta \in \mathbb{Z}_p(1)$ $(= T_p(\mathbb{G}_m))$, we have

$$s\zeta = \zeta^{\tau(s)},$$

where $\tau \colon \mathfrak{G} \to (\mathbb{Z}_p)^*$ is a continuous map (character). The character τ is called the *cyclotomic character*. Now if M is a $\mathbb{Z}_p$-module and also a $\mathfrak{G}$-module, and if χ is any character of $\mathfrak{G}$ with values in $(\mathbb{Z}_p)^*$, then *we can twist the action of $\mathfrak{G}$ on M by χ*; that is, we can make a new $\mathfrak{G}$-action on M:

$$s * m = \chi(s) \cdot (sm), \qquad s \in \mathfrak{G}, \quad m \in M.$$

The module M with this new action is denoted $M(\chi)$ and is called M *twisted by χ*. The particular case of $\chi = \tau$ (the cyclotomic character) is called the *Tate twist*. Clearly,

$$M(\tau) = M \otimes_{\mathbb{Z}_p} \mathbb{Z}_p(1) \underset{(\mathrm{def})}{=} M(1).$$

Of course, for $n > 0$, we define

$$M(n) = M(\tau^n) = M \otimes_{\mathbb{Z}_p} \mathbb{Z}_p(1)^{\otimes n} = M \otimes_{\mathbb{Z}_p} \mathbb{Z}_p(n),$$

and if $n < 0$, we set

$$M(n) = M(\tau^n) = M \otimes_{\mathbb{Z}_p} \mathbb{Z}_p(-1)^{\otimes |n|} = \operatorname{Hom}_{\mathbb{Z}_p}(\mathbb{Z}_p(n), M).$$

Here,

$$\mathbb{Z}_p(-1) = \operatorname{Hom}_{\mathbb{Z}_p}(\mathbb{Z}_p(1), \mathbb{Z}_p).$$

Clearly, $M(0) = M$ and $\mathbb{Z}_p(0)$ is $\mathbb{Z}_p$ with trivial $\mathfrak{G}$-action. The same formalities hold for the Tate twist as hold for the Serre twist $(\otimes \mathcal{O}_X(1))$ of projective algebraic geometry. *Tensoring with $\mathbb{Z}_p(n)$ is called the n-fold Tate twist.* From now on, when we write $M(n)$ it will mean the n-fold Tate twist of M.

Now we return to the dualities involving the Tate modules of p-divisible groups. We have the duality pairing $(*)''$, and it shows that

$$T_p(G^D) \cong \operatorname{Hom}_{\mathbb{Z}_p, \mathrm{cont.}}(T_p(G)(-1), \mathbb{Z}_p),$$

that is,

$(*)'''$ $\quad T_p(G^D) \cong (T_p(G)(-1))^d \cong (T_p(G)^d)(1) \quad (d = \mathbb{Z}_p\text{-duality}).$

We shall need the vector spaces (over C)

$$W = \text{Hom}_{\mathbb{Z}_p}(T_p(G), C) \quad \text{and} \quad W^D = \text{Hom}_{\mathbb{Z}_p}(T_p(G^D), C).$$

By remark (6) above, these have dimension $h = ht(G)$ over C, and they are 𝔊-modules. Thus,

$$W^D = \text{Hom}_{\mathbb{Z}_p}(T_p(G^D), C) \cong \text{Hom}_{\mathbb{Z}_p}(T_p(G)^d(1), C)$$
$$\cong \text{Hom}_{\mathbb{Z}_p}(T_p(G)^d, C(-1))$$
$$\cong T_p(G) \otimes C(-1).$$

From this, we get the perfect duality of h-dimensional vector spaces over C

$(\dagger)$ $\quad\quad\quad\quad\quad\quad W \otimes_C W^D \to C(-1)$

and $(\dagger)$ is a 𝔊-pairing, as well.

The map α_R above is not sufficient, we need a differential form of it so that information about the tangent spaces may be gleaned. These spaces control the dimensions of G and G^D. To get this differential form of α_R, it will be necessary to *assume the residue field of R is perfect* and we do so from now on. Of course, this is no restriction in the applications to arithmetic.

The connected-étale sequence

$$0 \to G^0 \to G \to G^{et} \to 0$$

splits over the special fibre of R because the residue field is perfect. Since G is flat over R, this splitting (already true on the generic fibre) holds over R and we find

$$0 \to G^0 \to G \to G^{et} \to 0 \quad \text{is split exact.}$$

Thus, for each complete ring S,

$$0 \to G^0(S) \to G(S) \to G^{et}(S) \to 0$$

is exact. Now $p: G^0 \to G^0$ is an isogeny and $G^{et}(S) = G^{et}(S \otimes_R k)$. Because k is perfect, the maps

$$G^{et}_{v+1}(S \otimes_R k) \to G^{et}_v(S \otimes_R k) \quad \text{(mult. by } p\text{)},$$

are surjective for every v, and if we pass to R_C instead of S, the isogeny on G^0 gives the surjection $G^0(R_C) \to G^0(R_C)$. Hence the five lemma shows that

Lemma. *If the residue field of R is perfect, then the group $G(R_C)$ is divisible.*

The tangent space to G is the tangent space to its formal Lie group G^0; so, $t_G(L)$ means the points of the tangent space to G^0 (at its origin) with coordinates in L (where, L is a field over K). If $G^0 = \text{Spf } R[[X_1, \ldots, X_n]]$, then each $z \in t_G(L)$ is an R-linear derivation of the power series ring

GROUP SCHEMES, FORMAL GROUPS, AND p-DIVISIBLE GROUPS

$R[[X_1, \ldots, X_n]]$ into L. That is, z is R-linear and

$$z(fg) = f(0)z(g) + g(0)z(f).$$

This is the same as giving an R-linear map from I/I^2 to L, where $I = (X_1, \ldots, X_n)$.

Now we know from Section 3 that I/I^2 is the space of invariant one-forms on G^0, and it is easy to see that for each $\omega \in I/I^2$, there exists a unique power series $\Omega \in K[[X_1, \ldots, X_n]]$ with $\Omega(0) = 0$ and $d\Omega = \omega$. (Formal Poincaré Lemma.) Using this Ω for each ω, we can define the *logarithm map*

$$\log: G(S) \to t_G(L), \qquad L = \operatorname{Frac}(S).$$

Here is the definition:

$$\text{If} \quad x \in G(S), \quad \text{then} \quad \log(x)(\omega) = \Omega(x).$$

(There is another way to get the logarithm map, it goes as follows: For each $x \in G(S)$ and each $f \in I$, set

$$\log(x)(f) = \lim_{r \to \infty} \left(\frac{f(p^r x)}{p^r} \right),$$

and observe that $p^r x \in G^0(S)$ for $r \gg 0$ because $G^{\text{et}}(S)$ is torsion.)

The logarithm map is a $\mathbb{Z}_p$-homomorphism and is a local isomorphism. The kernel of log is $G(S)_{\text{tors}}$ (as is most easily seen from the second definition), and the logarithm will be surjective if $\operatorname{Frac}(S)$ ($= L$) is algebraically closed. This is true because the cokernel of log is torsion and $G(S)$ is divisible if L is algebraically closed—as in our lemma above. We can summarize these matters as follows:

$0 \to G(S)_{\text{tors}} \to G(S) \xrightarrow{\log} t_G(L)$ is exact,

$0 \to G(R_C)_{\text{tors}} \to G(R_C) \xrightarrow{\log} t_G(C) \to 0$ is exact,

$\log: G(S) \otimes_{\mathbb{Z}_p} \mathbb{Q}_p \to t_G(L)$ is an isomorphism.

If $G = \mathbb{G}_m(p)$, then $G(S)$ is the units of S congruent to 1 mod $\mathfrak{m}_S$ (the "principal units") and $t_G(L)$ is just L. The log is the usual p-adic logarithm. If $G = A(p)$ for an abelian scheme over R, then $G(S)$ is the subgroup of $A(S)$ consisting of those points of $A(S)$ which have finite p-power order after reduction mod $\mathfrak{m}_S$. The logarithm map was studied by E. Lutz for elliptic curves and A. Mattuck for abelian varieties. For further details on the analytic aspects of logarithms, exponentials, and tangent spaces, see [17].

We apply the logarithm map to the pairing (∗∗∗), and we get a new pairing

$$(\text{∗∗∗ log}): T_p(G^D) \times t_G(C) \to t_{\mathbb{G}_m(p)}(C) = C.$$

The Galois group, $\mathfrak{G}$, acts on this pairing; and, as $C^{\mathfrak{G}} = K$, we deduce the "differential of α_R", namely the map

$$d\alpha_R: t_G(K) \to \operatorname{Hom}_{\mathfrak{G}}(T_p(G^D), C).$$

We now have the following important theorem proved by Tate in his article on p-divisible groups.

Theorem (Tate). *If R is a complete, mixed characteristic, discrete valuation ring with perfect residue field of characteristic $p > 0$, and if G is a p-divisible group over R, then for an open subgroup $\mathfrak{H}$ of $\mathfrak{G}$ with fixed field L and ring of integers S, the two maps*

$$\alpha_S \colon G(S) \to \mathrm{Hom}_{\mathfrak{H}}(T_p(G^D), U_C),$$

$$d\alpha_S \colon t_G(L) \to \mathrm{Hom}_{\mathfrak{H}}(T_p(G^D), C)$$

are isomorphisms.

Remark. The source of the restriction to an open subgroup, $\mathfrak{H}$, of $\mathfrak{G}$ is a technical part of the proof to be sketched below. It essentially rests on Tate's method for analyzing the $\mathfrak{G}$-action on C by higher ramification.

PROOF. We have the commutative diagram

$$\begin{array}{ccccccccc}
0 & \to & \Phi_p(G) & \to & G(R_C) & \xrightarrow{\log} & t_G(C) & \to & 0 \\
& & \downarrow \alpha_0 & & \downarrow \alpha & & \downarrow d\alpha & & \\
0 & \to & \mathrm{Hom}(T_p(G^D), U_{C_{\mathrm{tois}}}) & \to & \mathrm{Hom}(T_p(G^D), U_C) & \to & \mathrm{Hom}(T_p(G^D), C) & \to & 0,
\end{array}$$

in which the middle and right-hand vertical arrows come from pairings (∗∗∗) and (∗∗∗ log) and the left-hand vertical arrow comes from the direct limit of perfect dualities on the finite layers. (See the discussion preceding pairing (∗)′.) Consequently, the left vertical arrow is an isomorphism. The kernel and cokernel of $d\alpha$ are linear; so the snake lemma implies the same for $\ker \alpha$ and $\mathrm{coker}\, \alpha$.

Now the trick is to examine α_R, the $\mathfrak{G}$-fixed part of α, and to show that it is injective. For suppose this is done, then our diagram shows $d\alpha$ is injective on the image of $\log|G(R)$. However, the latter image spans $t_G(K)$ and therefore $d\alpha_R$ is injective, too. But then the whole map $d\alpha$ is injective, because it factors as

$$t_G(C) \cong t_G(K) \otimes_K C \xrightarrow{d\alpha_R \otimes 1} \mathrm{Hom}_{\mathfrak{G}}(T_p(G^D), C) \otimes_K C \to \mathrm{Hom}(T_p(G^D), C)$$

and the right-hand map is injective. (For the latter, one must prove for a vector space over C, say Z, and a semi-linear $\mathfrak{G}$-action on Z, the map $Z^{\mathfrak{G}} \otimes_K C \to Z$ is injective. This is an argument involving a shortest linear dependence.)

Remember that the valuation on R is discrete. So, if x lies in $G^0(R)$ and its coordinates are in $\mathfrak{m}_R^i$, then px has all its coordinates in $\mathfrak{m}_R^{i+1}$ by the discreteness of the valuation. Thus $\bigcap p^r G^0(R) = (0)$. Yet, $\ker \alpha$ is a linear space and so it is uniquely divisible; thus, by what we have just said,

$$\ker \alpha \cap G^0(R) = (0).$$

Again, $\ker \alpha$ is a linear space; so, it is torsion free. But, $G(R)/G^0(R)$ is a torsion

GROUP SCHEMES, FORMAL GROUPS, AND p-DIVISIBLE GROUPS   71

group, and thus ker $\alpha \cap G(R) = \ker(\alpha_R)$ is zero. We have shown α_R and $d\alpha_R$ (and $d\alpha$) are injective.

The equation $C^{\mathfrak{G}} = K$ (and our diagram above) shows that

$$\operatorname{coker} \alpha_R \subseteq (\operatorname{coker} \alpha)^{\mathfrak{G}}, \quad \text{and}$$

$$\operatorname{coker} d\alpha_R \subseteq (\operatorname{coker} d\alpha)^{\mathfrak{G}}.$$

But, the snake lemma for our diagram says in particular that coker $\alpha \to$ coker $d\alpha$ is an isomorphism; hence,

$$\operatorname{coker} \alpha_R \to \operatorname{coker} d\alpha_R \quad \text{is injective.}$$

Consequently, the surjectivity of $d\alpha_R$ will imply the surjectivity of α_R. We are now free to concentrate on the K-linear map $d\alpha_R$ and argue via dimension. Recall the spaces W and W^D and their duality pairing (†) which is a $\mathfrak{G}$-pairing. Let δ, δ^D be the dimensions of $W^{\mathfrak{G}}$ and $(W^D)^{\mathfrak{G}}$ respectively. Now the injectivity of $d\alpha_R$ shows that

$$n = \dim G \leq \delta^D \quad \text{and} \quad n^D = \dim G^D \leq \delta \quad \text{(symmetry);}$$

we know that $n + n^D = h$, and we will show that

$$\delta + \delta^D \leq h.$$

This will complete the proof as then $n = \delta^D$ and $n^D = \delta$ which gives surjectivity.

We take fixed points of pairing (†), and get

(††) $\qquad\qquad W^{\mathfrak{G}} \otimes_K (W^D)^{\mathfrak{G}} \to C(-1)^{\mathfrak{G}}.$

Now for an open subgroup, $\mathfrak{H}$, of $\mathfrak{G}$, Tate showed that $C(-1)^{\mathfrak{H}} = (0)$, and $H^1(\mathfrak{H}, C(-1)) = (0)$. We may replace $\mathfrak{G}$ by $\mathfrak{H}$ through all the above; so, we may assume $\mathfrak{G} = \mathfrak{H}$. Equation (††) and our remarks imply that $W^{\mathfrak{G}} \otimes_K C$ and $(W^D)^{\mathfrak{G}} \otimes_K C$ are orthogonal subspaces of W and W^D, and so $\delta + \delta^D \leq h = \dim W = \dim W^D$. □

Three important corollaries issue from this theorem.

Corollary 1. *The Tate module $T_p(G)$ determines the dimension of G and the dimension of G^D.*

PROOF. $T_p(G)$ determines $G \otimes_R K$ (remark (4)), and $G^D \otimes_R K = (G \otimes_R K)^D$. Hence, $T_p(G)$ determines $G^D \otimes_R K$ and thereby (remark (7)) determines $T_p(G^D)$. Yet the theorem implies $t_{G^D}(L)$ is $\operatorname{Hom}_{\mathfrak{H}}(T_p(G), C)$ and $t_G(L)$ is $\operatorname{Hom}_{\mathfrak{H}}(T_p(G^D), C)$; that is, n^D and n are determined.

Corollary 2 (Hodge–Tate Decomposition). *The $\mathfrak{H}$-module $T_p(G) \otimes_K C$ is canonically isomorphic to the direct sum*

$$(t_G(L) \otimes_L C)(1) \oplus (t_{G^D}^*(L) \otimes_L C),$$

where $t_{G^D}^*(L)$ is the cotangent space to G^D at its origin.

(Here, $\mathfrak{H}$ is the open subgroup of $\mathfrak{G}$ for which $H^0(\mathfrak{H}, C(\pm 1))$ and $H^1(\mathfrak{H}, C(\pm 1))$ vanish, and L is the fixed field of $\mathfrak{H}$.)

PROOF. We shall assume $\mathfrak{G} = \mathfrak{H}$, and thus $L = K$, by passing everything up to the base field L if necessary. Now the map $d\alpha_R^D$ takes $t_{G^D}(K) \otimes_K C$ ($= t_{G^D}(C)$) injectively to a subspace of W and, similarly, $d\alpha_R$ takes $t_G(C)$ to a subspace of W^D. As, $H^0(\mathfrak{G}, C(-1))$ vanishes, these image subspaces are orthogonal complements in the pairing (†). There results an exact sequence of $\mathfrak{G}$-vector spaces

$$0 \to t_{G^D}(C) \xrightarrow{d\alpha_R^D} W \to t_G^*(C)(-1) \to 0.$$

The exact sequence splits as a sequence of $\mathfrak{G}$-modules because the obstruction lies in

$$H^1(\mathfrak{G}, \mathrm{Hom}(t_G^*(C)(-1), t_{G^D}(C))) = H^1(\mathfrak{G}, \mathrm{Hom}(C(-1), C))^{nn^D}$$
$$= H^1(\mathfrak{G}, C(1))^{nn^D} = (0).$$

The set of splittings is a torseur for

$$H^0(\mathfrak{G}, \mathrm{Hom}(t_G^*(C)(-1), t_{G^D}(C))) = H^0(\mathfrak{G}, C(1))^{nn^D} = (0);$$

and so there is a unique splitting. Thus we find the canonical decomposition

$$W = \mathrm{Hom}(T_p(G), C) \cong t_{G^D}(C) \oplus (t_G^*(C) \otimes C(-1)),$$

or

$$T_p(G)^d \otimes C \cong t_{G^D}(C) \oplus t_G^*(C)(-1).$$

Replace G by G^D, we get

$$T_p(G^D)^d \otimes C \cong t_G(C) \oplus t_{G^D}^*(C)(-1).$$

Remember that $T_p(G^D)^d$ is $T_p(G)(-1)$ and twist the last decomposition by $C(1)$; the corollary results. □

Corollary 3. *We have* $\Lambda^h(T_p(G) \otimes_K C) \cong C(n)$ *as an $\mathfrak{H}$-module.*

PROOF. The direct sum of $\mathfrak{H}$-modules

$$T_p(G) \otimes_K C = t_G(C)(1) \oplus t_{G^D}^*(C)$$

shows that

$$\Lambda^h(T_p(G) \otimes_K C) = \Lambda^n(t_G(C)(1)) \otimes \Lambda^{n^D} t_{G^D}^*(C).$$

This proves our corollary. □

Remarks. (1) The Tate module $T_p(X)$, for the case $X = A(p)$ in which A is an abelian scheme over R, represents the first homology group of A with coefficients in $\mathbb{Z}_p$. One sees this as $T_p(X)$ is obtained by passing the p^vth division points of A to the limit and these division points correspond to dividing the lattice (in the complex case) from which A arises. (See [11] and [16].) Therefore, the $\mathfrak{G}$-module $W = \text{Hom}(T_p(G), C)$ can be written as

$$W = \text{Hom}(H_1(A, \mathbb{Z}_p), C) = H^1(A, \mathbb{Q}_p) \otimes C.$$

(The cohomology group is the étale cohomology group [16] of A.) Now the tangent space to the dual abelian scheme, A^D, over R is known to be $H^1(A, \mathcal{O}_A) = H^1(A, \Omega_A^0)$ (sheaf cohomology), and the cotangent space (at the origin) of A itself is $H^0(A, \Omega_A^1)$. Consequently, the canonical decomposition proved in Corollary 2

$$W \cong t_{G^D}(C) \oplus t_G^*(C)(-1)$$

can be written

$$H^1(A, \mathbb{Q}_p) \otimes C \cong H^1(A, \Omega_A^0) \oplus H^0(A, \Omega_A^1)(-1).$$

This is visibly a "Hodge decomposition" in the p-adic cohomology, obtained after tensoring with the p-adic analog of the complex numbers, C; it explains why the decomposition of Corollary 2 is called the Hodge–Tate decomposition. It also shows the rich structure obtained from the Galois action by $\mathfrak{G}$.

Since the cohomology of an abelian variety is determined by the first cohomology group of the variety, the total cohomology of an abelian scheme over R has a Hodge–Tate decomposition. In [19], Tate asked whether projective, smooth schemes over a base Y admitted a Hodge–Tate decomposition in their p-adic cohomology after tensoring with C; that is, was the Hodge theorem valid in the non-archimedian case? Recently, Faltings proved that this is so [4]. Briefly and crudely, here is the idea of his proof: He constructs two cohomology theories, the p-adic one (tensored up to C) and the one modelled on a Hodge–Tate decomposition. To compare them, he embeds them both in one big cohomology theory and then he examines them on the projective spaces $\mathbb{P}^n$. He shows they agree on $\mathbb{P}^n$ for all n and concludes his proof from this.

(2) The meaning of Corollary 3 is that $\mathfrak{H}$ acts on $\Lambda^h T_p(G)$ via the character τ^n, where $n = \dim G$. This fact is used in the proof of the Šafarevič and Tate conjectures for abelian varieties over number fields by Faltings. Actually, it is true for the whole group $\mathfrak{G}$ as we shall see below by a different method due to Raynaud.

If G and G' are p-divisible groups, then $\text{Hom}(G, G')$ is a $\mathbb{Z}_p$-module. Any homomorphism $G \to G'$ will induce a corresponding homomorphism $T_p(G) \to T_p(G')$ and the latter will be a $\mathfrak{G}$-homomorphism (i.e., $\mathfrak{G}$-equivariant) because it is induced by a homomorphism defined over K. We therefore get a map

$$\theta: \text{Hom}(G, G') \to \text{Hom}_{\mathfrak{G}}(T_p(G), T_p(G')).$$

If G and G' come from abelian schemes A, A' over R so that $G = A(p)$ and $G' = A'(p)$, then there is an obvious map

$$\mathrm{Hom}(A, A') \otimes_{\mathbb{Z}} \mathbb{Z}_p \to \mathrm{Hom}(A(p), A'(p)),$$

and there results the commutative triangle

$$\begin{array}{c} \mathrm{Hom}_{\mathfrak{G}}(T_p G), T_p(G')) \\ {}^{T}\nearrow \qquad \nwarrow^{\theta} \\ \mathrm{Hom}(A, A') \otimes_{\mathbb{Z}} \mathbb{Z}_p \to \mathrm{Hom}(A(p), A'(p)) \end{array}$$

$(A(p) = G, A'(p) = G')$.

The Tate conjecture is the statement that if K is finitely generated over its prime field, the map T is an isomorphism. This was proved by Tate for the case of finite fields, by Zarhin for function fields over finite fields (equicharacteristic case), and is now proved by Faltings in the unequal characteristic case. Tate, in trying to prove his conjecture, developed the theory of p-divisible groups, and he was able to prove the analog of his conjecture in this theory, namely that the map θ is an isomorphism.

We shall give a sketch of this result which is the main theorem of [19].

Theorem (Tate). *If R is an integrally closed, noetherian domain with $\mathrm{Frac}(R) = K$ of characteristic zero, then for two p-divisible groups G, G' over R, the map*

$$\mathrm{Hom}_R(G, G') \xrightarrow[\text{gen'l fibre}]{\text{restr. to}} \mathrm{Hom}_K(G \otimes_R K, G' \otimes_R K)$$

is an isomorphism. Equivalently, the map

$$\theta: \mathrm{Hom}_R(G, G') \to \mathrm{Hom}_{\mathfrak{G}}(T_p(G), T_p(G'))$$

is an isomorphism.

We have the immediate corollary:

Corollary. *If $f: G \to G'$ is a homomorphism of p-divisible groups under the assumptions of the theorem, and if $f \otimes 1: G \otimes_R K \to G' \otimes_R K$ is an isomorphism, then f is an isomorphism.*

To prove the theorem, one first proves the corollary directly and deduces the theorem from it. This is what we now sketch.

First observe that $R = \bigcap R_P$, where P runs over the height one primes of R. Therefore, we achieve an immediate reduction to the case: R local, of dimension one, and normal, i.e., to the case that R is a discrete valuation ring and $\mathrm{Frac}(R)$ has characteristic zero. Furthermore, we may even assume R to be complete and that its residue field is algebraically closed. Now if the characteristic of k $(= R/\mathfrak{m}_R)$ is not p, then G and G' are étale and everything is trivial; so, we will assume the residue field of R has characteristic p. Thus, we are in exactly the situation of the previous pages.

Write $G = (G_v)$, $G' = (G'_v)$, and set $G_v = \mathrm{Spec}\, A_v$, and $G'_v = \mathrm{Spec}\, B_v$. We are given compatible homomorphisms $u_v: B_v \to A_v$ and our assumption is that $u_v \otimes 1: B_v \otimes_R K \to A_v \otimes_R K$ is an isomorphism for each v. It follows immediately that each map u_v is injective. If we prove that the discriminants of B_v and A_v are non-zero and equal, then u_v will be surjective for each v, as desired. Now we shall soon see that the discriminant of A_v over R is generated by $p^{nvp^{hv}}$, where $n = \dim G$, and $h = \operatorname{ht} G$. Since dimension and height are determined by the generic fibre (by previous results), the discriminant of B_v must be that of A_v.

To finish the corollary, we need to evaluate the discriminant of A_v—as above. Discriminants of étale algebras are the unit ideal; so, the connected-étale decomposition places us in the connected case. Here,

$$G = \operatorname{Spf} R[[X_1, \ldots, X_n]] = \operatorname{Spf} \mathscr{A}.$$

The map $\mathscr{A} \to \mathscr{A}$ via p^v (= addition p^v-times in the formal group), makes $\mathscr{A}$ a free module over itself of rank p^{vh}. Write $\mathscr{A}(p^v)$ for $\mathscr{A}$ when it is considered over itself via p^v. If $I = (X_1, \ldots, X_n)$, then $\mathscr{A}(p^v)/I\mathscr{A}(p^v)$ is just A_v, and therefore it suffices to prove the discriminant ideal of $\mathscr{A}(p^v)$ over $\mathscr{A}$ is generated by $p^{nvp^{hv}}$.

The map $w: \mathscr{A} \to \mathscr{A}(p^v)$ gives a corresponding map $dw: \Omega^1_{\mathscr{A}/R} \to \Omega^1_{\mathscr{A}(p^v)/R}$. The highest wedge of the latter map

$$\Lambda^n \, dw : \Lambda^n \Omega^1_{\mathscr{A}/R} \to \Lambda^n \Omega^1_{\mathscr{A}(p^v)/R},$$

takes $x \in \Lambda^n \Omega^1_{\mathscr{A}/R}$ to $ax \in \Lambda^n \Omega^1_{\mathscr{A}(p^v)/R}$ for some fixed $a \in \mathscr{A}(p^v)$. Now it is known that the discriminant of the $\mathscr{A}$-module $\mathscr{A}(p^v)$ is generated by $N_{\mathscr{A}(p^v)/\mathscr{A}}(a)$.*
But, we know from Section 3 that

$$\Omega^1_{\mathscr{A}/R} \cong \mathscr{A} \otimes_R \omega_{\mathscr{A}/R},$$

where $\omega_{\mathscr{A}/R}$ is the space of invariant differentials; and if $x_1, \ldots, x_n$ is a basis for $\omega_{\mathscr{A}/R}$, then $w(x_j) = p^v x_j$. Thus, $\Lambda^n \, dw(x) = p^{nv}x$; that is, $a = p^{nv}$. But then,

$$N_{\mathscr{A}(p^v)/\mathscr{A}}(a) = p^{nvp^{hv}},$$

as claimed.

Now that the corollary is proved, one proves the theorem from it as follows: The homomorphism f on the generic fibres: $G \otimes_R K \to G' \otimes_R K$ gives a homomorphism $T_p(G) \to T_p(G')$. The graph, M, of the latter homomorphism is a $\mathbb{Z}_p$-direct summand of $T_p(G \times G')$. It can be shown (and this we omit), that such a summand arises from a homomorphism $\varphi: \Gamma \to G \times G'$ of p-divisible groups so that $T_p(\varphi): T_p(\Gamma) \to M$ is an isomorphism. Admitting this, we see that $\operatorname{pr}_1 \circ \varphi$ maps Γ to G, and on Tate modules (or, what is the same, generic fibres) is an isomorphism. Our corollary implies $\operatorname{pr}_1 \circ \varphi$ is an isomorphism of p-divisible groups. But then, the generic fibre homomorphism f extends across R by

$$G \xrightarrow{(\operatorname{pr}_1 \circ \phi)^{-1}} \Gamma \xrightarrow{\phi} G \times G' \xrightarrow{\operatorname{pr}_2} G'.$$

As uniqueness is clear, the proof is finished. □

---

* This depends on the existence of a trace map

$$\Lambda^n \Omega^1_{\mathscr{A}(p^v)/R} \to \Lambda^n \Omega^1_{\mathscr{A}/R}.$$

§7. Applications of Groups of Type $(p, p, \ldots, p)$ to p-Divisible Groups

Recall that in the proof of the lemma preceding the Sylow theorem of Section 3, the existence of a subgroup was proved (by the second method) by taking the scheme-theoretic closure of the subgroup on the generic fibre in the ambient scheme. This closure operation yielded a group scheme from a group scheme, it commuted with products, and the resulting closure was flat over the base.

Let R be a complete discrete valuation ring of mixed characteristic. Then there may be many ways of extending a finite commutative group scheme from K ($= \text{Frac}(R)$) to R as a finite R-group scheme (remember: flatness is assumed), but Raynaud [14] showed that if extension is possible at all there will be a maximal and a minimal way to extend. Moreover, Cartier duality will interchange these extremal extensions.

Let v be the valuation of R, and write $v(p) = e$; the number e is the ramification index of R over $\mathbb{Z}_p$. Raynaud [14] showed the following facts about extending group schemes from K to R:

(1) If $e < p - 1$ (so that R is certainly tamely ramified at worst over $\mathbb{Z}_p$), then each K-group scheme of p-power order can be extended in at most one way (if extension is possible at all) to a finite R-group scheme.

(2) Under the hypotheses of (1), if G, G' are p-power order R-group schemes, then the map

$$\text{Hom}_{R\text{-groups}}(G, G') \to \text{Hom}_{K\text{-groups}}(G \otimes_R K, G' \otimes_R K)$$

is an isomorphism. If u_K is extended by $u \in \text{Hom}(G, G')$, then ker u and coker u are flat over R, and the map

$$\text{Ext}_{R\text{-groups}}(G, G') \to \text{Ext}_{K\text{-groups}}(G \otimes_R K, G' \otimes_R K)$$

is injective.

(3) Assume the residue field of R is algebraically closed and $e \leq p - 1$. Then each p-power order R-group scheme possesses a composition series whose factors are F-vector space schemes for varying finite fields F. (This permits an analysis of these p-power group schemes by using Raynaud's results sketched in Section 4.)

(4) Each finite K-group scheme killed by a power of p has a composition series whose factors are F-vector space schemes. If G is an F-vector space scheme over K, then there is a condition on the representation of $\mathfrak{G}$ ($= \text{Gal}(\bar{K}/K)$) in F^* which is necessary and sufficient in order that G extend to R. The condition is satisfied if $e \geq p - 1$.

Tate's main theorem on p-divisible groups (proving the Tate conjecture for them) characterizes the p-divisible group $G = (G_v)$ by its generic fibre: $G \otimes_R K = (G_v \otimes_R K)$. But, when can one extend a p-divisible group from K to all of R? Using the ideas of scheme-theoretic closure, Raynaud proves.

Proposition. *Let* (G_v) *be a p-divisible group over* K ($=$ Frac R, *where* R *is a complete discrete valuation ring of mixed characteristic) and suppose each* G_v *can be separately extended to a finite R-group scheme. Then there exists a unique p-divisible group* (Γ_v) *over* R *whose generic fibre is* (G_v).

This proposition and result (4) above give the argument skipped at the end of the proof of Tate's main theorem on *p*-divisible groups.

Lastly, there is the matter of the Galois action of $\mathfrak{G}$ on the highest wedge of the Tate module of a *p*-divisible group. Corollary 3 in Section 6 asserts this action is via $\tau^{\dim G}$ for a suitable open subgroup, $\mathfrak{H}$, of $\mathfrak{G}$. Actually, this is true for $\mathfrak{G}$ as well; here is a sketch of Raynaud's proof.

Theorem (Raynaud–Tate). *If* R *is a mixed characteristic complete discrete valuation ring with fraction field* K, *and if* G *is a p-divisible group over* R, *then the* $\mathfrak{G}$ ($= \mathrm{Gal}(\bar{K}/K)$)*-action on* $\Lambda^h(T_p(G) \otimes_K C)$ *is given by* τ^n, *where* $h = \mathrm{ht}(G)$ *and* $n = \dim G$.

PROOF. (Raynaud). Call a *p*-divisible group, *ordinary*, if its connected component of identity is the dual of an étale *p*-divisible group. Now the character τ acts trivially if and only if $T_p(G)$ is unramified; so, for ordinary *p*-divisible groups, our theorem is clear. Next, by decomposing the special fibre of G, call it $\bar{G}$, we may assume $\bar{G}$ is connected.

There is a vast machinery concerning the deformations of formal Lie groups over fields; this machinery shows that there is a formal power series ring $\mathcal{O} = R[[X_1, \ldots, X_t]]$ so that $V = \mathrm{Spec}\,\mathcal{O}$ plays the role of a versal parameter space for the deformations of $\bar{G}$. Moreover, there is a *p*-divisible group, $\mathscr{G}$, lying over V which is the versal *p*-divisible group [10]. By some results of Cartier, Raynaud shows that $\bar{G}$ may be deformed to an ordinary *p*-divisible group equicharacteristically, i.e., over the generic point of $k[[X_1, \ldots, X_t]]$—where $k = R/\mathfrak{m}_R$. Because $(V, \mathscr{G})$ is the versal pair, this means that $\bar{\mathscr{G}} = \mathscr{G} \otimes_R k$ is ordinary at the generic point of $\bar{V} = V \otimes_R k$. Dimension and height are preserved by deformation. Write x for the generic point of $\bar{V}$, then $\bar{\mathscr{G}}_x$ is an ordinary *p*-divisible group of height h and dimension n. Here is a picture of V, $\bar{V}$, x, R, and k.

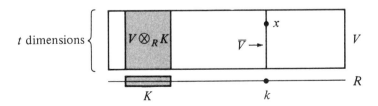

By passing to the strict Henselization of V, we may assume k is separably closed. As $\bar{\mathscr{G}}_x$ is ordinary, and as the étale fundamental group of V is trivial, we see that $\bar{\mathscr{G}}_x$ has

$$0 \to \mathbb{G}_m(p)^n \to \bar{\mathscr{G}}_x \to (\mathbb{Q}_p/\mathbb{Z}_p)^{h-n} \to 0$$

for its connected-étale decomposition. But then, the Galois module $\Lambda^h(T_p(\mathscr{G}))(-n)$ is unramified at x. It is also unramified at the points of $V \otimes_R K$ which are of characteristic zero. Hence, $\Lambda^h(T_p(\mathscr{G}))(-n)$ is unramified at *all* points of codimension ≤ 1 of V. The theorem of purity of the branch locus implies that $\Lambda^h(T_p(\mathscr{G}))(-n)$ *is unramified over all of V*. But, some point of V corresponds to our original p-divisible group G by versality; hence $\Lambda^h(T_p(G))(-n)$ is unramified over R, which is what we wanted to prove. □

REFERENCES

[1] Artin, M. *Grothendieck Topologies*. Seminar Notes. Harvard University: Cambridge, MA, 1962.

[2] Artin, M. and Grothendieck, A. *Séminaire Géométrie Algébrique*, 3. Lecture Notes in Mathematics, 151, 152, 153. Springer-Verlag: New York, 1970.

[3] Cartier, P. *Colloque sur la Théorie des Groupes Algébriques*, Bruxelles, 1962.

[4] Faltings, G. Hodge–Tate decompositions in the étale cohomology of schemes, (To appear.)

[5] Grothendieck, A. and Dieudonné, J. Éléments de géométrie algébrique, IV, No. 2. *Publ. Math. I.H.E.S.*, **24** (1965).

[6] Grothendieck, A. and Dieudonné, J. Éléments de géométrie algébrique, IV, No. 4. *Publ. Math. I.H.E.S.*, **32** (1967).

[7] Grothendieck, A. Technique de descente et théorèmes d'éxistence en géométrie algébrique, II. *Séminaire Bourbaki*, Éxposé 195, 1960.

[8] Grothendieck, A. Technique de descente et théorèmes d'éxistence en géométrie algébrique, III. *Séminaire Bourbaki*, Éxposé 212, 1961.

[9] Grothendieck, A. Technique de descente et théorèmes d'éxistence en géométrie algébrique, IV. *Séminaire Bourbaki*, Éxposé 221, 1961.

[10] Grothendieck, A. *Séminaire Géométrie Algébrique*, 1. Lecture Notes in Mathematics, 224. Springer-Verlag: New York, 1971.

[11] Milne, J. Abelian varieties over fields, this volume.

[12] Oort, F. and Tate, J. Group schemes of finite order. *Ann. Sci. École Norm. Sup.*, 3 (1970), 1–21.

[13] Raynaud, M. Passage au quotient par une rélation d'équivalence plate. *Proceedings of a Conference on Local Fields*, Driebergen, 1966. Springer-Verlag: Berlin, Heidelberg, New York, 1967, pp. 78–85.

[14] Raynaud, M. Schémas en groupes de type $(p, p, \ldots, p)$, *Bull. Soc. Math. Fr.*, **102** (1974), 241–280.

[15] Ribet, K. Recent results of J. M. Fontaine on p-divisible groups over $\mathbb{Z}$. Talk at the Storrs conference.

[16] Rosen, M. Abelian varieties over the complex numbers, this volume.

[17] Serre, J.-P. *Lie Algebras and Lie Groups*. Benjamin: New York, 1965.

[18] Tate, J. Arithmetic of elliptic curves. *Invent. Math.*, **23** (1974), 179–206.

[19] Tate, J. p-Divisible groups. *Proceedings of a Conference on Local Fields*, Driebergen, 1966. Springer-Verlag: Berlin, Heidelberg, New York, 1967, pp. 158–183.

[20] Waterhouse, W. *Introduction to Affine Group Schemes*. Graduate Texts in Mathematics, 66. Springer-Verlag: New York, 1979.

[21] Fontaine, J. M. Il n'y a pas de variété abélienne sur $\mathbb{Z}$, *Invent. Math.*, **81** (1985), 515–538.

CHAPTER IV

Abelian Varieties over $\mathbb{C}$

MICHAEL ROSEN

(Notes by F. O. McGuinness, Fordham University)

§0. Introduction

These lecture notes present, in outline, the theory of abelian varieties over the complex numbers. They focus mainly on the analytic side of the subject. In the first section we prove some basic results on complex tori. The second section is devoted to a discussion of isogenies. The third section (the longest) describes the necessary and sufficient conditions that a complex torus must satisfy in order to be isomorphic to an abelian variety. In the fourth section we describe the construction of the dual abelian variety and the concluding two sections discuss polarizations and the moduli space of principally polarized abelian varieties. Proofs for the most part are omitted or only sketched. Details can be found in [SW] or [L-A] (see the list of references at the end of this chapter). For the algebraic–geometric study of abelian varieties over arbitrary fields, the reader is referred to [M-AV] and to the articles of J. S. Milne in this volume.

The author would like to extend a special note of thanks to F. O. McGuinness who reworked the original sketchy notes into a coherent manuscript and made a number of very useful improvements, additions, and clarifications.

§1. Complex Tori

An *abelian variety* A is a complete and connected algebraic group defined over the field of complex numbers. Thus A comes equipped with a multiplication $m: A \times A \to A$ and an inverse map $i: A \to A$ which are morphisms of varieties and satisfy the usual group axioms. The complex points $A(\mathbb{C})$ is then a connected, compact, complex Lie group. We will begin by considering properties of such objects. Let T be an arbitrary connected, compact, complex Lie group. Then:

(1) T is a commutative group

To see why this is so, let V denote the tangent space to T at the identity element e. Consider the adjoint representation of T on V:

$$\text{Ad}: T \to \text{Aut}_{\mathbb{C}}(V).$$

(Ad(t) is the differential of the conjugation map $u \to tut^{-1}$ on T.) The coordinate functions with respect to a basis of V are holomorphic on the compact complex manifold T and so must be constants. Thus $\text{Ad}(t) = \text{Ad}(e) = \text{Id}$ for all $t \in T$. It is now easy to check that the exponential map, $\exp: V \to T$, maps V onto a subgroup of the center of T. Since T is connected, $\exp(V)$ generates T, and so T is commutative.

(2) T is a complex torus

A more refined analysis shows $\exp$ is a surjective homomorphism from V to T with kernel Λ a discrete subgroup. Recall that a discrete subgroup of a real vector space with compact quotient is called a lattice. Thus, $T \approx V/\Lambda$, is a complex torus. See [M-AV] for the proof.

From now on we will write the group law on T additively and denote the identity element by 0.

(3) Holomorphic 1-forms

The representation of T as a complex torus can be achieved in another way. Let Ω be the vector space of holomorphic 1-forms on T. Define $H_1(T, \mathbb{Z}) \to \Omega^* = \text{Hom}_{\mathbb{C}}(\Omega, \mathbb{C})$ by $\gamma \to (\omega \mapsto \int_\gamma \omega)$ where $\int_\gamma \omega$ is the integral of ω around the integral 1-cycle γ.

This map is injective and the image Λ is a lattice in Ω^*. Now define $T \to \Omega^*/\Lambda$ by $p \mapsto (\omega \to \int_0^p \omega)$. Note that $\int_0^p \omega$ is well-defined modulo Λ. This yields an isomorphism

$$T \approx \Omega^*/\Lambda.$$

(4) Mappings between complex tori

Suppose T_1 and T_2 are complex tori and $\phi: T_1 \to T_2$ is a holomorphic map. If $\phi(0) = 0$ then ϕ is a homomorphism. This is implied by (3) above since ϕ induces a linear map $\Omega_1^* \to \Omega_2^*$ which takes Λ_1 to Λ_2. In general, ϕ is a homomorphism followed by a translation. If we write $T_j = V_j/\Lambda_j$ then every holomorphic homomorphism from T_1 to T_2 is induced by a $\mathbb{C}$ linear map from V_1 to V_2 such that $\phi(\Lambda_1) \subseteq \Lambda_2$. We continue to call this map ϕ. This yields two faithful representations:

$$\rho_{\mathbb{C}}: \text{Hom}(T_1, T_2) \to \text{Hom}_{\mathbb{C}}(V_1, V_2),$$

$$\rho_{\mathbb{Z}}: \text{Hom}(T_1, T_2) \to \text{Hom}_{\mathbb{Z}}(\Lambda_1, \Lambda_2),$$

called the complex representation and the rational representation respectively. The fact that $\rho_{\mathbb{Z}}$ is faithful shows immediately that $\mathrm{Hom}(T_1, T_2)$ is a finitely generated, torsion-free, abelian group of rank $\leq 4(\dim T_1)(\dim T_2)$.

The case $T_1 = T_2 = T$ is of particular interest. $\mathrm{Hom}(T, T) = \mathrm{End}(T)$ is a ring which we will discuss further below. Here we present another way of looking at endomorphisms of complex tori. Suppose Λ is a lattice in an even-dimensional real vector space V. The real torus V/Λ will be a complex torus if V has the structure of a complex vector space. A complex structure on V is given by an $\mathbb{R}$-linear map $J: V \to V$ such that $J^2 = -\mathrm{Id}$ (set $iv = Jv$). Any $\mathbb{Z}$-linear map $\phi: \Lambda \to \Lambda$ defines an $\mathbb{R}$-endomorphism of $V = \mathbb{R} \otimes_{\mathbb{Z}} \Lambda$ and therefore an endomorphism of V/Λ. The map ϕ is an endomorphism of the *complex* torus V/Λ if and only if $\phi \circ J = J \circ \phi$. Thus, $\mathrm{End}(T) = \{\phi \in \mathrm{End}_{\mathbb{Z}}(\Lambda) | \phi \circ J = J \circ \phi\}$. Continuing in this direction leads to the definition of the Hodge group, a certain $\mathbb{Q}$-algebraic subgroup of $\mathrm{Gl}(V)$ whose complex points contain J. However, we will not pursue this.

(5) The image and kernel of a morphism.

If $\phi: T_1 \to T_2$ is a morphism of complex tori then $\mathrm{im}\,\phi$ is a subtorus of T_2 while $\ker \phi$ is a closed subgroup of T_1 whose connected component is a subtorus of finite index in $\ker \phi$. Both these facts are easily established.

§2. Isogenies of Complex Tori

A morphism $\phi: T_1 \to T_2$ is an *isogeny* if it is a surjective homomorphism with finite kernel. The order of the kernel is called the *degree of ϕ*, $\deg(\phi)$.

EXAMPLE. Let δ be the identity map on T, a complex torus. Let $m > 0$ be an integer. The map $m\delta: T \to T$ is an isogeny of degree m^{2d} where $d = \dim T$. To see this, write $T = V/\Lambda$. Then $\ker(m\delta) = (1/m)\Lambda/\Lambda \approx \Lambda/m\Lambda \approx (\mathbb{Z}/m\mathbb{Z})^{2d}$.

If $\phi_1: T_1 \to T_2$, and $\phi_2: T_2 \to T_3$ are isogenies, then so is $\phi_2 \circ \phi_1$ and degrees multiply: $\deg(\phi_2 \circ \phi_1) = \deg(\phi_1)\deg(\phi_2)$.

We say that T_1 and T_2 are *isogenous*, $T_1 \sim T_2$, if there is an isogeny $\phi: T_1 \to T_2$. The next proposition shows that isogeny is an equivalence relation.

Proposition. *Let $d = \dim T_1 = \dim T_2$. If $\phi: T_1 \to T_2$ is an isogeny of degree m there is a unique isogeny $\psi: T_2 \to T_1$ of degree m^{2d-1} such that $\psi \circ \phi = m\delta_1$ and $\phi \circ \psi = m\delta_2$. ψ is called the* dual isogeny *to ϕ.*

PROOF. Since $\ker \phi \subseteq \ker(m\delta_1)$, a map ψ exists which makes the following diagram commutative:

One checks that ψ is the desired isogeny. Uniqueness is straightforward.

Since $\psi \circ \phi = m\delta_1$ one sees $(\phi \circ \psi - m\delta_2) \circ \phi = 0$. It follows that $\phi \circ \psi = m\delta_2$ because ϕ is onto. Finally, taking the degree of both sides of $\psi \circ \phi = m\delta_1$ yields $\deg \psi = m^{2d-1}$. □

Let the dual isogeny ψ be denoted by $\tilde{\phi}$. Then $\tilde{\tilde{\phi}} = m^{2d-2}\phi$ and $\widetilde{\phi_2 \circ \phi_1} = \tilde{\phi}_1 \circ \tilde{\phi}_2$.

We now begin our study of End(T). We define $\text{End}_0(T) = \text{End}(T) \otimes_{\mathbb{Z}} \mathbb{Q}$. $\text{End}_0(T)$ is a finite-dimensional $\mathbb{Q}$-algebra and End(T) can be considered as an order in it. We note that $\phi \in \text{End}(T)$ is an isogeny if and only if it is invertible in $\text{End}_0(T)$.

A complex torus T is called *simple* if it contains no proper complex subtorus. We say T is of *semisimple type* (this is not standard terminology) if it is isogenous to a product of simple complex tori. If T is simple then the usual Schur's lemma argument shows that $\text{End}_0(T)$ is a division algebra.

Proposition. *If T is of semisimple type, then $\text{End}_0(T)$ is a semisimple $\mathbb{Q}$-algebra.*

PROOF. Write $T \sim T_1^{n_1} \times T_2^{n_2} \times \cdots \times T_m^{n_m}$ where the T_j are simple and pairwise non-isogenous. Then, $\text{End}_0(T) \approx M_{n_1}(D_1) \otimes \cdots \otimes M_{n_m}(D_m)$ where $D_j = \text{End}_0(T_j)$ is a finite-dimensional division algebra over $\mathbb{Q}$. □

To study $\text{End}_0(T)$ further we recall the complex representation

$$\rho_{\mathbb{C}}: \text{End}_0(T) \to \text{End}_{\mathbb{C}} V$$

and the rational representation $\rho_{\mathbb{Q}} = \rho_{\mathbb{Z}} \otimes \mathbb{Q}$

$$\rho_{\mathbb{Q}}: \text{End}_0(T) \to \text{End}_{\mathbb{Q}}(\Lambda \otimes_{\mathbb{Z}} \mathbb{Q}).$$

Proposition. $\rho_{\mathbb{Q}} \otimes \mathbb{C} \approx \rho_{\mathbb{C}} \otimes \bar{\rho}_{\mathbb{C}}$.

See [SW, Lemma 39, p. 70], for the (simple) proof.

Let $\phi \in \text{End}(T)$. We define the *characteristic polynomial*, char(ϕ, x) of ϕ to be $\det(\rho_{\mathbb{Z}}(\phi) - xI)$. Note that char($\phi, x$) $\in \mathbb{Z}[x]$ is a monic polynomial of degree $2d$ where $d = \dim T$. We can easily extend this definition to $\phi \in \text{End}_0(T)$.

Lemma. *Let $\psi \in \text{End}(T)$. Then $\det(\rho_{\mathbb{Z}}(\psi)) = \deg(\psi)$ (if ψ is not an isogeny we define $\deg(\psi) = 0$).*

PROOF. $\rho_{\mathbb{Z}}(\psi): \Lambda \to \Lambda$ is 1-1 if and only if $\det(\rho_{\mathbb{Z}}(\psi)) \neq 0$. Thus, if ψ is an isogeny

$$\det(\rho_{\mathbb{Z}}(\psi)) = [\Lambda : \rho_{\mathbb{Z}}(\psi)\Lambda] = [\rho_{\mathbb{Z}}(\psi)^{-1}\Lambda : \Lambda] = \deg(\psi). \quad \square$$

Using the lemma we can give an intrinsic characterization of char(ϕ, x).

Proposition. *For all but finitely many integers n, $\phi - n\delta$ is an isogeny. The* *characteristic polynomial* char(ϕ, x) *is the unique polynomial such that for all* $n \in \mathbb{Z}$, char$(\phi, n) = \deg(\phi - n\delta)$.

PROOF. $\phi - n\delta$ is an isogeny if and only if $\det(\rho_{\mathbb{Z}}(\phi) - n\delta) \neq 0$. Thus if n is not a root of char(ϕ, x), $\phi - n\delta$ is an isogeny. In this case, by the lemma, char$(\phi, n) = \det(\rho_{\mathbb{Z}}(\phi) - n\delta) = \deg(\phi - n\delta)$. □

We note that this proposition makes sense in characteristic p and can be used to define the characteristic polynomial in the abstract theory.

We conclude this section by making a few remarks about the l-*adic representations*. For $n \in \mathbb{Z}$ define $T[n] = \ker(n\delta)$. As we have seen, $T[n]$ is isomorphic to $(\mathbb{Z}/n\mathbb{Z})^{2d}$. If $\phi \in \text{End}(T)$ then $\phi(T[n]) \subseteq T[n]$. For a prime number l consider the inverse system $\{T[l^m] | m \geq 1\}$ where $l\delta: T[l^{m+1}] \to T[l^m]$ are the transition maps. An endomorphism ϕ induces a map of this inverse system and thus acts on proj lim $T[l^m] = T_l(T)$, the l-*adic Tate module*. Let $V_l(T) = T_l(T) \otimes_{\mathbb{Z}_l} \mathbb{Q}_l$. Then we have a representation

$$\rho_{\mathbb{Q}_l}: \text{End}_0(T) \to \text{End}_{\mathbb{Q}_l}(V_l(T)).$$

It is easy to check that $T_l(T) \approx \Lambda \otimes \mathbb{Z}_l$ both as a $\mathbb{Z}_l$ and as an End(T) module. Thus the l-adic representations are all equivalent to the rational representation. In working over $\mathbb{C}$ they provide no new information. However, when working with abelian varieties over arbitrary fields the l-adic representations can always be defined whereas an analogue of the rational representation need not exist.

§3. Abelian Varieties

We will be using some standard terminology from the theory of complex analytic manifolds. We assume known the definitions of holomorphic and meromorphic functions on such manifolds, as well as the definitions of divisors, positive divisors, etc. See [SW, §3], for a concise discussion. Another good reference is [SHAF, Chap. VIII].

Let $\mathcal{M}(T)$ be the field of meromorphic functions on the complex torus T. Since T is compact the only holomorphic functions are constants. How big is $\mathcal{M}(T)$? How can one construct elements of $\mathcal{M}(T)$? We quote a general theorem of Siegel. See [SHAF] for a proof.

Theorem. *Let M be a compact, connected, complex manifold of dimension d. Then $\mathcal{M}(M)$ has transcendence degree over $\mathbb{C}$ at most d. If d is attained then $\mathcal{M}(M)$ is a finitely generated field over $\mathbb{C}$.*

If $M = X(\mathbb{C})$, the complex points on a non-singular algebraic variety X, then $\mathcal{M}(M) \approx \mathbb{C}(X)$, the field of rational functions on X. Thus, in this case,

$\mathcal{M}(M)$ is a finitely generated field of transcendence degree $d = \dim X$. This shows that if the complex torus T is an abelian variety $\mathcal{M}(T)$ is "big." Later on we will give an example of a torus T with $\mathcal{M}(T) = \mathbb{C}$. This never happens when T has dimension 1. Let $E = \mathbb{C}/\Lambda$ be a one-dimensional complex torus. Then $\mathcal{M}(E)$ is generated by the Weierstrass elliptic function $\mathscr{P}(z, \Lambda)$ and its derivative $\mathscr{P}'(z, \Lambda)$. These functions are connected by the well-known equation

$$\mathscr{P}'(z, \Lambda)^2 = 4\mathscr{P}(z, \Lambda)^3 - g_2\mathscr{P}(z, \Lambda) - g_3,$$

where g_2 and g_3 are constants satisfying $g_2^3 - 27g_3^2 \neq 0$. The map

$$E \to [1, \mathscr{P}(z, \Lambda), \mathscr{P}'(z, \Lambda)]$$

extends to an imbedding of E into $\mathbb{P}^2$ as a non-singular cubic plane curve. Thus one-dimensional complex tori are one-dimensional abelian varieties or elliptic curves.

In higher dimensions the situation is more complicated. There are non-trivial conditions on a complex torus in order that it correspond to an abelian variety. To explain these conditions we need to review some linear algebra.

Suppose V is a finite-dimensional complex vector space. A map

$$H: V \times V \to \mathbb{C}$$

is a *Hermitian form on V* if:

(i) for fixed $v \in V$,

$$u \mapsto H(u, v)$$

is a linear map $V \to \mathbb{C}$;

(ii) for fixed $u \in V$,

$$v \mapsto H(u, v)$$

is an *antilinear* map $V \to \mathbb{C}$;

(iii) $H(u, v) = \overline{H(v, u)}$ for all u, v in V.

(Of course, (i) + (iii) $\Rightarrow$ (ii)).

If H is a Hermitian form, then we will always write S for the real part of H and E for the imaginary part.

Thus $H(u, v) = S(u, v) + iE(u, v)$, $u, v \in V$, and $S, E: V \times V \to \mathbb{R}$ are real bilinear. We also see that:

$$S(u, v) = E(iu, v),$$

$$S(iu, iv) = S(u, v), \qquad E(iu, iv) = E(u, v),$$

$$S \text{ is symmetric} \qquad (S(u, v) = S(v, u)),$$

$$E \text{ is antisymmetric} \qquad (E(u, v) = -E(v, u)).$$

Conversely, if E is a real, antisymmetric bilinear form on V satisfying

$E(iu, iv) = E(u, v)$, then $H(u, v) = E(iu, v) + iE(u, v)$ is a Hermitian form. The set of Hermitian forms on V form a group under pointwise addition and subtraction.

Definition. Suppose $T = V/\Lambda$ is a complex torus. A *Riemann form* on T is a Hermitian form H on V such that $E = \operatorname{Im} H$ is integer valued on Λ, i.e. $E(\lambda_1, \lambda_2) \in \mathbb{Z}$, for all $\lambda_1, \lambda_2 \in \Lambda$. If $H(u, u) \geq 0$ for all $u \in V$ we say H is a *positive Riemann form*. If H is positive definite, i.e. $H(u, u) > 0$ for all $u \in V$, $u \neq 0$, we say H is a *non-degenerate Riemann form* on T.

One sometimes calls H a *Hermitian Riemann form* on T, and $E = \operatorname{Im} H$ an *alternating Riemann form* on T.

Theorem A. *A complex torus T is the manifold of complex points on an abelian variety if and only if T possesses a non-degenerate Riemann form.*

The proof will be sketched later. The idea is to construct *theta functions* using the non-degenerate Riemann form on T and use these to construct a projective embedding of T.

We will now discuss some naturally occurring Riemann forms. If $\dim T = 1$ we have $T = \mathbb{C}/\Lambda$ where $\Lambda = \mathbb{Z}\lambda_1 + \mathbb{Z}\lambda_2$ with $\operatorname{Im}(\lambda_1/\lambda_2) > 0$. Regard $\mathbb{C}$ as a two-dimensional vector space over $\mathbb{R}$, and define $E(z, w)$ by the equation $z \wedge w = E(z, w)\lambda_1 \wedge \lambda_2$. Then $E(z, w)$ is a Riemann form on T. $E(iz, iw) = E(z, w)$ follows from the fact that multiplication by i is area preserving. Every other Riemann form on T is an integral multiple of E. Thus, the Riemann form does not usually occur explicitly in the theory of elliptic functions.

Here is another class of complex tori for which it is possible to explicitly write down a Riemann form. Suppose K is a CM field, i.e. a totally imaginary quadratic extension of a totally real number field, K^+. Examples are provided by imaginary quadratic number fields, and cyclotomic fields. Set $[K : \mathbb{Q}] = 2d$, and let $\Phi = \{\phi_1, \ldots, \phi_d\}$ be a subset of distinct complex imbeddings $K \subset \mathbb{C}$ such that if $\phi \in \Phi$, $\bar{\phi} \notin \Phi$, where $\bar{\phi}$ is the complex conjugate embedding. Φ provides an isomorphism, which we continue to call Φ, of $K \otimes_\mathbb{Q} \mathbb{R}$ with $\mathbb{C}^d$, which takes $\alpha \otimes 1$ to $(\phi_1(\alpha), \phi_2(\alpha), \ldots, \phi_d(\alpha))$. Let $\mathscr{A}$ be an integral ideal in K. It can be shown that $\Phi(\mathscr{A})$ is a lattice in $\mathbb{C}^d$. Set $A = \mathbb{C}^d/\Phi(\mathscr{A})$. We proceed to find a Riemann form on A. A simple calculation shows we can find an algebraic integer $\xi \in K$ such that $K = K^+(\xi)$, $-\xi^2 \in K^+$ and is totally positive, and $\operatorname{Im} \phi_j(\xi) > 0$ for $j = 1, \ldots, d$. For $z, w \in \mathbb{C}^d$ define $E(z, w) = \sum_{j=1}^{d} \phi_j(\xi)(\bar{z}_j w_j - z_j \bar{w}_j)$. $E(z, w)$ is $\mathbb{R}$-bilinear, anti-symmetric, and $E(iz, w)$ is symmetric and positive definite. A calculation shows that for α, $\beta \in K$, $E(\Phi(\alpha), \Phi(\beta)) = t(\xi\tilde{\alpha}\beta)$ where $\alpha \to \tilde{\alpha}$ is the non-trivial automorphism of K/K^+ and t is the trace from K to $\mathbb{Q}$. Thus, $E(z, w)$ takes integral values on $\Phi(\mathscr{A})$, and is a non-degenerate Riemann form on A. The ring of integers $\mathcal{O}_K$ of K imbeds in $\operatorname{End}(\mathscr{A})$ via the map which associates to ω the diagonal matrix whose iith coefficient is $\phi_i(\omega)$. Thus, $K \subset \operatorname{End}_0(A)$. In this situation, A is said to admit complex multiplication by K and the corresponding abelian variety is said to be of CM type (K, Φ). The assumption that $\mathscr{A}$ is an integral ideal is unnecessarily restrictive. It suffices to assume that $\mathscr{A}$ is a $\mathbb{Z}$-lattice in K. For this and much more on abelian varieties of CM type see [L–CM] and [SHIM].

Returning to the general theory we make the following convenient definition. A complex torus is an *abelian manifold* if it possesses a non-degenerate Riemann form.

Restricting the Riemann form shows that a subtorus of an abelian manifold is again an abelian manifold. One can show that a quotient of an abelian manifold is also an abelian manifold. This is a corollary of the following important result.

Theorem (Poincaré Reducibility Theorem). *Suppose A is an abelian manifold and $A_1 \subset A$ an abelian submanifold. Then there is an abelian submanifold A_2 such that $A_1 \cap A_2$ is finite and A is isogenous to $A_1 \times A_2$.*

PROOF. *Sketch.* Write $A = V/\Lambda$ with Riemann form H. Then $A_1 = V_1/\Lambda_1$ where $V_1 \subseteq V$ is a complex subspace and $\Lambda_1 = V_1 \cap \Lambda$. Set $V_2 = V_1^\perp$, the orthogonal complement of V_1 with respect to H, and set $\Lambda_2 = V_2 \cap \Lambda$. It can be shown that Λ_2 is a lattice in V_2 and so $A_2 = V_2/\Lambda_2$ is an abelian submanifold of A. Moreover, $\Lambda_1 + \Lambda_2$ is of finite index in Λ. The map $A_1 \times A_2 \to A$ given by $(a_1, a_2) \to a_1 + a_2$ is an isogeny. See [SW, Theorem 34, Cor. 3] or [L–A, p. 117] for more details. □

Corollary. *An abelian manifold A is of semisimple type and so $\mathrm{End}_0(A)$ is a semisimple $\mathbb{Q}$-algebra (see the second proposition in Section 2).*

We will now discuss some analytic results which will lead to the introduction of theta functions. We will explain Poincaré's basic result (Theorem B) that every periodic divisor is generated by a theta function, and then the important theorem of Frobenius (Theorem C) which computes the dimension of a certain vector space of theta functions. Then, finally, we will be in a position to state the Lefschetz Embedding Theorem (Theorem D) of which Theorem A is an immediate consequence.

Suppose V is a d-dimensional vector space over $\mathbb{C}$ and Λ is a lattice in V. Put $T = V/\Lambda$ and let $\pi\colon V \to T$ be the projection map. A function f on V is *periodic* with respect to Λ if $f(z + \lambda) = f(z)$ for all $z \in V$, $\lambda \in \Lambda$. Such a function gives rise to a function on T, and conversely if g is a function on T then $f = g \circ \pi$ is a periodic function on V. A *Cartier divisor* D on V is given by a family $\{(U_\alpha, f_\alpha)\}$ where the U_α form an open covering of V, f_α is meromorphic on U_α, not identically zero, and f_α/f_β is holomorphic on $U_\alpha \cap U_\beta$ for all α, β. The divisor is called *positive* if the function f_α are holomorphic. If $a \in V$ the *translate* of D by a, D_a, is given by $\{(U_\alpha + a, f_\alpha(z - a)\}$. If $D_\lambda = D$ for all $\lambda \in \Lambda$ we say D is a *periodic divisor*. Note that the divisor of a periodic meromorphic function is a periodic divisor. Let $\mathscr{D}(V)$ and $\mathscr{D}(T)$ be the group of divisors on V and T respectively. Then π induces a homomorphism

$$\pi^*: \mathscr{D}(T) \to \mathscr{D}(V).$$

The image of π^* consists of the periodic divisors.

Divisors on V are easier to analyze than those on T. For example, we have the following facts:

(1) Suppose $g \in \mathscr{M}(V)$ has trivial divisor. Then g is a nowhere vanishing holomorphic function and we can write

$$g(z) = \mathbf{e}(h(z)),$$

where $h(z)$ is holomorphic and $\mathbf{e}(z) = \exp(2\pi i z)$, $i = \sqrt{-1}$.

(2) (Cousin's Theorem). Every divisor on V is principal, i.e. we can set all $f_\alpha = f$, a single function meromorphic on V. In cohomological terms this says $H^1(V, \mathcal{O}^*) = (0)$ where $\mathcal{O}^*$ is the sheaf of nowhere vanishing holomorphic functions on V.

Suppose D' is a divisor on T and $\pi^*(D') = D$ is the corresponding periodic divisor on V. By Cousin's theorem, $D = (f)$. Since $D_\lambda = D$ for all $\lambda \in \Lambda$, we see $f(z + \lambda) = U_\lambda(z) f(z)$ for all $\lambda \in \Lambda$ where $U_\lambda(z)$ is a nowhere vanishing holomorphic function. Thus, by (1), $U_\lambda(z) = \mathbf{e}(h_\lambda(z))$ where $h_\lambda(z)$ is holomorphic. Setting $\lambda = \lambda_1 + \lambda_2$ we find the following consistency condition

$$h_{\lambda_1 + \lambda_2}(z) \equiv h_{\lambda_1}(z + \lambda_2) + h_{\lambda_2}(z) \mod \mathbb{Z}.$$

We wish to choose $h_\lambda(z)$ to be as simple as possible. The simplest choice leading to a fruitful theory is

$$h_\lambda(z) = L(z, \lambda) + J(\lambda)$$

where $L(z, \lambda)$ is linear in z, and $J(\lambda)$ is a constant.

Definition. Let $L: V \times \Lambda \to \mathbb{C}$ and $J: \Lambda \to \mathbb{C}$ be maps with $L(z, \lambda)$ linear in z for all $\lambda \in \Lambda$. A holomorphic (resp. meromorphic) *theta function* for Λ of type (L, J) is a holomorphic (resp. meromorphic) function θ on V such that $\theta(z + \lambda) = \mathbf{e}(L(z, \lambda) + J(\lambda))\theta(z)$ for all $z \in V$, $\lambda \in \Lambda$.

Theorem B (Poincaré). *For every divisor D' on T, the periodic divisor $\pi^*(D') = D$ is the divisor of a meromorphic theta function, $\pi^*(D') = (\theta)$. If D' is a positive (holomorphic) divisor, then θ is a holomorphic theta function.*

Note that Theorem B is a sharp form of Cousin's theorem in the special case of periodic divisors. The early proofs were quite complicated. The proof usually quoted today is due to A. Weil ("Théorémes fondamentaux de la théorie des fonctions thêta", *Seminar Bourbaki*, 1948/49).

A question which naturally arises is to what extent θ is determined by D'. An easy exercise shows that a theta function has trivial divisor if and only if $\theta(z) = \mathbf{e}(q(z) + l(z) + c)$ where $q(z)$ is a quadratic form, $l(z)$ is linear, and c is a constant. We call such theta functions *trivial theta functions*. Then:

$$\mathscr{D}(T) \approx \frac{\text{group of theta functions}}{\text{trivial theta functions}}.$$

Suppose θ is a theta funcction of type (L, J). We now show how to associate a Hermitian form to θ.

The consistency conditions explained earlier impose the following restrictions on L and J:

(a) $J(\lambda + \mu) - J(\lambda) - J(\mu) \equiv L(\lambda, \mu) \mod \mathbb{Z}$;
(b) $L(\lambda, \mu) \equiv L(\mu, \lambda) \mod \mathbb{Z}$;
(c) $L(z, \lambda + \mu) = L(z, \lambda) + L(z, \mu)$.

Note that (b) follows from (a). Condition (c) implies that $L(z, \lambda)$ can be extended to an $\mathbb{R}$-bilinear function on V. Define $E(z, w) = L(z, w) - L(w, z)$. Then, E is an anti-symmetric $\mathbb{R}$-bilinear function on $V \times V$ which assumes integer values on $\Lambda \times \Lambda$, by (b). This last condition implies E is real valued on $V \times V$. Moreover, we have the following result.

Lemma. $E(iz, iw) = E(z, w)$.

PROOF. $E(iz, iw) = L(iz, iw) - L(iw, iz) = i(L(z, iw) - L(w, iz))$ and $E(z, w) = L(z, w) - L(w, z) = -i(L(iz, w) - L(iw, z))$. Thus $E(iz, iw) - E(z, w) = i(E(iz, w) - E(iw, z))$ must be zero since it is in $\mathbb{R} \cap i\mathbb{R}$. □

Define $H(z, w) = E(iz, w) + iE(z, w)$. Then H is a Riemann form on $T = V/\Lambda$, called *the Riemann form associated to* θ.

Suppose $\theta(z) = \mathbf{e}(q(z) + l(z) + c)$ is a trivial theta function. Let $B(z, w) = q(z + w) - q(z) - q(w)$. $B(z, w)$ is $\mathbb{C}$-bilinear and symmetric. A short calculation shows that $\theta(z + \lambda) = \mathbf{e}(B(z, \lambda) + q(\lambda) + l(\lambda))\theta(z)$. Thus $E(z, \lambda) = B(z, \lambda) - B(\lambda, z) = 0$ and the Riemann form associated with a trivial theta function is zero. Thus there is a homomorphism from the divisor group $\mathscr{D}(T)$ to the group of Hermitian forms on T given by:

$$D' \to \pi^*(D') = (\theta) \to H.$$

We can refine this further.

Proposition. *Suppose $D = \pi^*(D')$ is a positive divisor. If $D = (\theta)$, then θ is an entire function and the corresponding Riemann form H is positive, i.e. $H(z, z) \geq 0$ for all $z \in V$.*

See [SW, p. 31, Lemma 31] for the proof. If H is positive definite we say θ is *non-degenerate* and that the corresponding divisor on T is *ample*. That ample divisors in this sense are ample in the sense of algebraic geometry will be shown later.

Since we can multiply a given theta function by a trivial theta function without changing the corresponding divisor it is natural to look for a normal form.

Proposition. *Let θ be a theta function and H the associated Riemann form. Then there is a theta function $\tilde{\theta}$, unique up to a multiplicative constant, such that $\tilde{\theta}/\theta$ is a trivial theta function, and*

$$\tilde{\theta}(z + \lambda) = \mathbf{e}\left(\frac{1}{2i}H\left(z + \frac{\lambda}{2}, \lambda\right) + K(\lambda)\right)\tilde{\theta}(z),$$

where $K(\lambda)$ is real valued and

$$K(\lambda_1 + \lambda_2) - K(\lambda_1) - K(\lambda_2) \equiv \tfrac{1}{2}E(\lambda_1, \lambda_2) \mod \mathbb{Z}.$$

$\tilde{\theta}$ is called the *normalized theta function* associated to θ. Set $\psi(\lambda) = \mathbf{e}(K(\lambda))$. Then ψ satisfies $\psi(\lambda_1 + \lambda_2) = \psi(\lambda_1)\psi(\lambda_2)\mathbf{e}(\tfrac{1}{2}E(\lambda_1, \lambda_2))$. ψ is called *the associated quadratic character* of θ. Note that $|\psi(\lambda)| = 1$.

If $\text{Th}(L, J)$ denotes the vector space of theta functions of type (L, J) and $\text{Th}_{\text{norm}}(H, \psi)$ the space of normalized theta functions with associated Riemann form H and quadratic character ψ, then one can find a trivial theta function θ_0 such that multiplication by θ_0 gives an isomorphism

$$\text{Th}(L, J) \approx \text{Th}_{\text{norm}}(H, \psi).$$

See [L–A, Chap. VI, §2] for this and the proof of the above proposition.

At this point it is easy to show that every $f \in \mathcal{M}(T)$ can be represented as a quotient of holomorphic theta functions of the same type. One can write $(f) = D_0 - D_\infty$ where D_0 and D_∞ are positive divisors. $\pi^*(D_\infty) = (\theta_\infty)$ by Poincaré's theorem, Theorem B. Now, $\pi^*(f)\theta_\infty$ has divisor $\pi^*(D_0)$ and so is a holomorphic theta function, θ_0, of the same type as θ_∞. Thus $\pi^*f = \theta_0/\theta_\infty$ as asserted. Conversely, the quotient of two theta functions of the same type is a periodic meromorphic function. This leads to the problem of constructing all holomorphic theta functions of a given type. This is accomplished by a theorem of Frobenius. Before stating this theorem it is necessary to review the definition of the Pfaffian of an alternating form.

Let x_{ij}, $1 \leq i < j \leq 2d$ be $d(2d - 1)$ elements algebraically independent over $\mathbb{Q}$. Set $x_{ji} = -x_{ij}$ and consider the antisymmetric matrix $X = (x_{ij})$. There is a unique polynomial $\text{Pf}(x)$ of degree d such that $\det X = \text{Pf}(x)^2$ and $\text{Pf}(x)$ takes the value 1 on

$$\begin{pmatrix} 0 & I_d \\ -I_d & 0 \end{pmatrix}.$$

If $G = (g_{ij})$ is an antisymmetric $2d \times 2d$ matrix with coefficients in any ring we define $\text{Pf}(G)$ to be the value of $\text{Pf}(x)$ when g_{ij} is substituted for x_{ij}. $\text{Pf}(G)$ is called the *Pfaffian* of G. Let Λ be a free $\mathbb{Z}$-module of rank $2d$ and E an alternating form on Λ. If $\{\lambda_1, \lambda_2, \ldots, \lambda_{2d}\}$ is a basis of Λ, set $\text{Pf}(E) = \text{Pf}(E(\lambda_i, \lambda_j))$. This is well defined up to sign.

Lemma. *Let Λ be a free $\mathbb{Z}$-module of rank $2d$ and E a non-degenerate alternating form on Λ. Then there is a basis $\{\lambda_1, \lambda_2, \ldots, \lambda_{2d}\}$ of Λ such that $E(\lambda_i, \lambda_j) = 0$ for $1 \leq i, j \leq d$, $E(\lambda_{d+i}, \lambda_{d+j}) = 0$ for $1 \leq i, j \leq d$, and $E(\lambda_i, \lambda_{d+j}) = e_i \delta_{ij}$ for $1 \leq i, j \leq d$ where $e_1 | e_2 | \ldots | e_d$ are positive integers. Finally, $\mathrm{Pf}(E) = e_1 e_2 \ldots e_d$.*

This lemma is due to Frobenius. A basis with the given properties is called a *symplectic basis* for Λ. If $\mathscr{E}$ is the diagonal matrix with diagonal entries $e_1, e_2, \ldots, e_d$ then $(E(\lambda_i, \lambda_j))$ has the form

$$\begin{pmatrix} 0 & \mathscr{E} \\ -\mathscr{E} & 0 \end{pmatrix}.$$

Theorem C (Frobenius). *Suppose (L, J) is a type, H the associated Riemann form, $E = \mathrm{Im}\, H$. Assume H is positive definite. The vector space of holomorphic theta functions of type (L, J) over $\mathbb{C}$ has dimension $\mathrm{Pf}(E)$.*

PROOF. Sketch. Choose a symplectic basis for Λ with respect to E. It is easy to check that $\lambda_1, \lambda_2, \ldots, \lambda_d$ are a basis for V over $\mathbb{C}$. Let $z_1, z_2, \ldots, z_d$ be the coordinate functions on V with respect to this basis. By multiplying by a suitable trivial theta function, the space that we are examining is isomorphic to the space of holomorphic θ on V satisfying the equations

$$\theta(z + \lambda_i) = \theta(z), \qquad 1 \leq i \leq d,$$

$$\theta(z + \lambda_{d+i}) = \mathbf{e}(e_i z + c_i)\theta(z), \qquad 1 \leq i \leq d,$$

for some fixed constants $c_1, c_2, \ldots, c_d$. The first set of these equations show that we can expand θ as a Fourier series

$$\theta(z) = \sum_{n \in \mathbb{Z}^d} a(n)\mathbf{e}(n \cdot z).$$

The second set of equations imposes recurrence relations on the set of coefficients $a(n)$ which show that all the $a(n)$ can be expressed in terms of those $a(n)$ with $0 \leq n_i \leq e_i - 1$ where $n = (n_1, n_2, \ldots, n_d)$. This gives an upper bound of $e_1 e_2 \ldots e_d = \mathrm{Pf}(E)$ on the dimension of the given space of theta functions. To get equality one must show that for a collection of $a(n)$ satisfying the recurrence relations the corresponding formal Fourier series is in fact a holomorphic function. We omit the proof but note that here the assumption that H is positive definite comes into play. □

Theorems B and C allow us to answer all the questions previously raised about complex tori. We first discuss a form of the Riemann–Roch theorem.

Let D be a positive divisor on T and define, as usual

$$\mathscr{L}(D) = \{f \in \mathscr{M}(T) | (f) + D \geq 0\}.$$

There is a holomorphic theta function θ_0 such that $\pi^*(D) = (\theta_0)$. Define

$\mathscr{L}(\theta_0)$ to be the space of all holomorphic theta functions with the same type as θ_0. Then $\theta \to \theta/\theta_0$ gives an isomorphism of $\mathscr{L}(\theta_0)$ with $\mathscr{L}(D)$. Theorem C gives the dimension of this space.

Theorem. *Suppose $D_0, D_1, \ldots, D_m$ are positive divisors on T and that D_0 is ample. Then there is a polynomial P of degree d such that*

$$\dim_{\mathbb{C}} \mathscr{L}\left(\sum_{j=0}^{m} r_j D_j \right) = P(r_0, r_1, \ldots, r_m)$$

whenever $r_j \geq 0$ for all j and $r_0 > 0$.

PROOF. Set $\pi^*(D_j) = (\theta_j)$ and let H_j be the Hermitian form corresponding to θ_j. Then $\sum_{j=0}^{m} r_j H_j$ corresponds to $\prod_{j=0}^{m} \theta_j^{r_j}$ and is positive definite if $r_j \geq 0$ for $1 \leq j \leq m$ and $r_0 > 0$. Then we have

$$\dim_{\mathbb{C}} \mathscr{L}\left(\sum_{j=0}^{m} r_j D_j \right) = \dim_{\mathbb{C}} \mathscr{L}(\theta_0^{r_0} \theta_1^{r_1} \ldots \theta_m^{r_m})$$
$$= \mathrm{Pf}(r_0 E_0 + r_1 E_1 + \cdots + r_m E_m)$$

which is a polynomial in the r_j of the type described in the theorem. □

Corollary. *If D is an ample divisor on T, then $\dim_{\mathbb{C}} \mathscr{L}(rD) = r^d \dim_{\mathbb{C}} \mathscr{L}(D)$ for $r > 0$.*

Suppose T is a complex torus of dimension d which possesses an ample divisor. The above theorem can be used to prove Siegel's theorem (see the beginning of this section) for T. We prove the first part as follows. Suppose $f_1, f_2, \ldots, f_m \in \mathscr{M}(T)$ with $m > d$, where $d = \dim T$. There exists an ample divisor D such that $(f_j) + D \geq 0$ for $1 \leq j \leq m$. Set $\pi^*(D) = (\theta_0)$ and $\theta_j = f_j \theta_0$. The θ_j are holomorphic theta functions of the same type as θ_0. For $r = \sum_{j=1}^{m} r_j$ there are $\binom{m+r}{r}$ monomials $\theta_0^{r_0} \theta_1^{r_1} \ldots \theta_m^{r_m}$ as the r_j vary over non-negative integers. These are all in $\mathscr{L}(\theta_0^r)$ which has dimension $r^d \mathrm{Pf}(E)$ where E is the alternating form corresponding to θ_0. Since $m > d$ we have $\binom{m+r}{m} > r^d \mathrm{Pf}(E)$ for large r. The corresponding monomials are then linearly dependent and this gives an algebraic relation among the f_j. Thus the transcendence degree of $\mathscr{M}(T)$ over $\mathbb{C}$ is $\leq d$. See [L–A, Chap. VI, §6] for the proof that when equality holds, $\mathscr{M}(T)$ is finitely generated over $\mathbb{C}$.

Let A be an abelian manifold. Recall that this means A is a complex torus with a non-degenerate Riemann form H, i.e. H is positive definite. Frobenius' theorem, Theorem C, implies the existence of a theta function θ on A with H its corresponding Hermitian form. Let D be the divisor on A induced by θ.

Theorem D (Lefschetz Embedding Theorem). *Let A be an abelian manifold and D the divisor on A constructed in the above remarks. Then $\mathscr{L}(3D)$ considered as a linear system on A gives a projective embedding $A \to \mathbb{P}^N(\mathbb{C})$.*

See [L-A] or [SW] for the proof. The map is obtained as follows. $\mathscr{L}(3D) \approx \mathscr{L}(\theta^3)$. Let $\theta_0, \theta_1, \theta_2, \ldots, \theta_N$ be a basis of $\mathscr{L}(\theta^3)$. Then $t \in A$ goes to $[\theta_0(t), \theta_1(t), \ldots, \theta_N(t)] \in \mathbb{P}^N(\mathbb{C})$. Since all the θ_j are of the same type, the map is well defined. One must show it is defined everywhere, is 1-1, and that the image is a non-singular subvariety of $\mathbb{P}^N(\mathbb{C})$.

Let $E = \operatorname{Im} H$ be the alternating Riemann form corresponding to H. Then $N = \dim \mathscr{L}(\theta^3) - 1 = 3^d \operatorname{Pf}(E) - 1$ where $d = \dim_\mathbb{C} A$. One can also show that the degree of the embedding is $d! \, 3^d \operatorname{Pf}(E)$.

Suppose $d = 1$. Then $N = 2$ and the degree of the embedding is 3 if we use the Riemann form constructed earlier on $\mathbb{C}/\Lambda = A$; A embeds as a non-singular plane cubic. Thus, the Lefschetz Embedding Theorem can be considered as a vast generalization of the work of Weierstrass on elliptic functions. We remark in passing that the theta function that arises in this context is the Weierstrass σ-function, $\sigma(z)$, whose induced divisor on $\mathbb{C}/\Lambda$ is just the zero element.

Theorem D is, of course, a very explicit form of Theorem A. Conversely, if A is an abelian variety over $\mathbb{C}$ then A has a projective embedding. The pull-back of a hyperplane section is an ample divisor D on A and if $\pi^*(D) = (\theta)$ then the Hermitian form corresponding to θ is a non-degenerate Riemann form on $A(\mathbb{C})$, i.e. $A(\mathbb{C})$ is an abelian manifold. Thus the existence of a non-degenerate Riemann form is a necessary and sufficient condition for a complex torus to be the manifold of complex points on an abelian variety!

§4. The Néron–Severi Group and the Picard Group

In this section we assume $A = V/\Lambda$ is an abelian manifold. We define some groups of divisors on A.

$\mathscr{D}$ = group of all divisors on A.

$\mathscr{D}_a$ = group of divisors on A whose corresponding Riemann form is 0.

$\mathscr{D}_l$ = group of principal divisors.

The divisors in $\mathscr{D}_a$ are said to be *algebraically equivalent to zero*, those in $\mathscr{D}_l$ are said to be *linearly equivalent to zero*. We have $\mathscr{D}_l \subseteq \mathscr{D}_a \subseteq \mathscr{D}$. Define the *Néron–Severi group* to be $\operatorname{NS}(A) = \mathscr{D}/\mathscr{D}_a$, the *Picard Group* to be $\operatorname{Pic}(A) = \mathscr{D}/\mathscr{D}_l$, and $\operatorname{Pic}^0(A) = \mathscr{D}_a/\mathscr{D}_l$. Then we have the exact sequence

$$(0) \to \operatorname{Pic}^0(A) \to \operatorname{Pic}(A) \to \operatorname{NS}(A) \to (0).$$

Proposition. $\operatorname{NS}(A)$ *is a torsion free finitely generated abelian group (and so a free abelian group) of* rank $\leq d(2d - 1)$ *where* $d = \dim A$.

PROOF. The Riemann form H associated to a divisor is completely determined by $E = \operatorname{Im} H$ restricted to Λ. These form a group isomorphic to a subgroup of the $2d \times 2d$ antisymmetric matrices with integer coefficients. The latter group is free abelian of rank $d(2d - 1)$. Thus $\operatorname{NS}(A)$ injects into a free abelian group of rank $d(2d - 1)$. □

When $d = 1$, $\operatorname{NS}(A) \approx \mathbb{Z}$, the isomorphism being given by $D \to \deg(D)$.

We will next show that $\operatorname{Pic}^0(A)$ can be given the structure of an abelian manifold, $\hat{A}$, *the dual abelian manifold of A*.

Suppose $[D]$ is the class in $\operatorname{Pic}^0(A)$ of a divisor D. Then D corresponds to a normalized theta function θ which only depends on the class of D. Since the Riemann form associated to θ is trivial we have $\theta(z + \lambda) = \mathbf{e}(K(\lambda))\theta(z)$ where $K(\lambda) \in \mathbb{R}$ and satisfies $K(\lambda_1 + \lambda_2) \equiv K(\lambda_1) + K(\lambda_2) \bmod \mathbb{Z}$. Then $\chi_D(\lambda) = \mathbf{e}(K(\lambda))$ is a character of Λ. Note that χ_D is the trivial character if and only if $\theta \in \mathcal{M}(A)$. Denoting the Pontryagin dual of Λ by $\hat{\Lambda}$ we get a monomorphism $\operatorname{Pic}^0(A) \hookrightarrow \hat{\Lambda}$ by $[D] \to \chi_D$. Since $\Lambda \approx \mathbb{Z}^{2d}$ we have $\hat{\Lambda} \approx (\mathbb{R}/\mathbb{Z})^{2d}$ is a real torus of dimension $2d$. To give $\operatorname{Pic}^0(A)$ the structure of an abelian manifold we will show $\operatorname{Pic}^0(A) \hookrightarrow \hat{\Lambda}$ is an isomorphism, that $\hat{\Lambda}$ has the structure of a *complex* torus, and finally that the resulting complex torus has a nondegenerate Riemann form.

Suppose X is an ample divisor on A, θ the corresponding theta function, H the corresponding Riemann form, and $E = \operatorname{Im} H$. For $t \in V$ we set $\theta_t(z) = \theta(z - t)$. The divisor corresponding to θ_t is X_t, the translation of X by t (actually, by the image of t in A). A calculation shows that the normalized theta function associated to θ_t/θ has multiplier $\mathbf{e}(-E(t, \lambda))$. In the first place this shows $X_t - X$ is algebraically equivalent to zero. Secondly, since E is non-degenerate, every character of $\hat{\Lambda}$ is of the form $\lambda \to \mathbf{e}(-E(t, \lambda))$ for suitable $t \in V$. This proves

Proposition. $\operatorname{Pic}^0(A) \to \hat{\Lambda}$ *given by* $D \to \chi_D$ *is an isomorphism. Moreover, every* $D \in \mathcal{D}_a$ *is linearly equivalent to $X_t - X$ for suitable $t \in A$.*

Corollary. *If X is an ample divisor on A, the map $\phi_X \colon A \to \operatorname{Pic}^0(A)$ given by $\phi_X(t) = [X_t - X]$ is surjective with finite kernel of order $\det(E) = \operatorname{Pf}(E)^2$.*

PROOF OF COROLLARY. The surjectivity is given by the theorem. The kernel is precisely $\{t \in V \mid E(t, \lambda) \in \mathbb{Z}, \text{ for all } \lambda \in \Lambda\}/\Lambda$. It is straightforward to see this is a finite group of order $\operatorname{Pf}(E)^2$ (use a symplectic basis for Λ). □

To put the structure of a complex torus on $\operatorname{Pic}^0(A) \approx \hat{\Lambda}$ we consider the space V^* of antilinear functionals on V. Explicitly,

$$V^* = \{f \in \operatorname{Hom}_\mathbb{R}(V, \mathbb{C}) \mid f(\alpha t) = \overline{\alpha} f(t), \alpha \in \mathbb{C}, t \in V\}.$$

V^* is a complex vector space of the same dimension as V. We have a nondegenerate $\mathbb{R}$-bilinear pairing $\langle \ , \ \rangle \colon V^* \times V \to \mathbb{R}$ given by $\langle \xi, t \rangle = \operatorname{Im} \xi(t)$.

Define Λ^* by $\Lambda^* = \{\xi \in V^* | \langle \xi, \lambda \rangle \in \mathbb{Z}, \text{ for all } \lambda \in \Lambda\}$. It is not hard to see that Λ^* is a lattice in V^*. The following lemma follows from the definitions and the non-degeneracy of the pairing $\langle \xi, \lambda \rangle$.

Lemma. *For $\xi \in V^*$ define $\chi_\xi(\lambda) = \mathbf{e}(-\langle \xi, \lambda \rangle)$. Then $\xi \to \chi_\xi$ gives an isomorphism from V^*/Λ^* to $\hat{\Lambda}$.*

This gives $\hat{\Lambda}$ the structure of a complex torus.

If H is the given Riemann form on A, then $t \to H(t, \cdot)$ is an isomorphism of V with V^* as complex vector spaces. Under this isomorphism one checks that Λ goes to Λ^*. Thus we have an epimorphism $\phi_H: V/\Lambda \to V^*/\Lambda^*$. The various maps we have defined are tied together by the following commutative diagram

$$A = V/\Lambda \xrightarrow{\phi_H} V^*/\Lambda^*$$
$$\phi_X \downarrow \qquad \approx \downarrow \xi \to \chi_\xi$$
$$\mathrm{Pic}^0(A) \xrightarrow{\approx} \hat{\Lambda}$$

It remains to exhibit a Riemann form on V^*/Λ^*. The map $V \to V^*$ (which we also denote by ϕ_H) given by $t \to H(t, \cdot)$ is an isomorphism as we have already pointed out. Define

$$H^*(\xi, \eta) = H(\phi_H^{-1}(\xi), \phi_H^{-1}(\eta)).$$

H^* is certainly a Hermitian form on V^* but $E^* = \mathrm{Im}\, H^*$ need not be integer valued on Λ^*. Using the above commutative diagram we see the kernel of $\phi_H: V/\Lambda \to V^*/\Lambda^*$ is finite implying $\phi_H^{-1}(\Lambda^*)/\Lambda$ is finite. Thus an appropriate integer multiple of H^* is a Riemann form on V^*/Λ^*. By "transport of structure," $\mathrm{Pic}^0(A)$ becomes an abelian manifold, $\hat{A}$ called the *dual abelian manifold* of A.

The association $A \leftrightarrow \hat{A}$ is a genuine duality. If $\rho: A \to B$ is a morphism of abelian manifolds, then $\hat{\rho}: \hat{B} \to \hat{A}$ is the morphism of abelian manifolds induced by pulling back divisors. One can show A is canonically isomorphic to $\hat{\hat{A}}$, etc.

For every ample divisor X on A, $\phi_X: A \to \hat{A}$ is an isogeny with kernel of order $\mathrm{Pf}(E)^2$ where E is the alternating Riemann form corresponding to X.

Before leaving the topic of dual abelian manifolds we briefly discuss the *Rosati involution*. If X is an ample divisor on A the isogeny $\phi_X \in \mathrm{Hom}(A, \hat{A})$ is an isomorphism in $\mathrm{Hom}_0(A, \hat{A})$. Let $\phi_X^{-1} \in \mathrm{Hom}_0(\hat{A}, A)$ be its inverse. If $\rho: A \to A$ let $\hat{\rho}: \hat{A} \to \hat{A}$ be the dual morphism. Then $\rho \to \hat{\rho}$ extends to a map from $\mathrm{End}_0(A) \to \mathrm{End}_0(\hat{A})$. For $\rho \in \mathrm{End}_0(A)$ define $\rho' = \phi_X^{-1} \circ \hat{\rho} \circ \phi_X$. The map $\rho \to \rho'$ is an involution, i.e. $(\rho_1 \circ \rho_2)' = \rho_2' \circ \rho_1'$ and $\rho'' = \rho$, called the Rosati involution on $\mathrm{End}_0(A)$ (we suppress the dependence on X in the notation). Let tr denote the trace map on the semisimple $\mathbb{Q}$-algebra $\mathrm{End}_0(A)$. Then $\mathrm{tr}(\rho'\rho) > 0$ for all $\rho \neq 0$. All this follows from

Proposition. *Let X be an ample divisor on A, and H the corresponding Riemann form. Then $H(\rho'z, w) = H(z, \rho w)$ for all $z, w \in V$, i.e. ρ' is the adjoint of ρ with respect to H.*

The proof, which is not hard, follows from carefully unwinding the definitions.

In the case of abelian varieties over finite fields, the existence and positivity of the Rosati involution can be used to prove the Riemann hypothesis for the associated zeta function (see [M-AV, Chap. IV, §21] for this). Another application is to classify the endomorphism rings of abelian manifolds. This theory was developed in the 1930s by A. A. Albert and others under the rubric of "Riemann matrices." For example, if $D = \text{End}_0(A)$ is a division algebra, then the center of D is either totally real or else a CM field. [M-AV, Chap. IV, §21], gives the main results.

§5. Polarizations and Polarized Abelian Manifolds

For many purposes, it is natural to consider not just an abelian manifold A, but A together with the choice of a non-degenerate Riemann form. Roughly speaking, a polarized abelian manifold is a pair (A, H) where H is a non-degenerate Riemann form. We actually use a slightly different definition: say two Riemann forms H_1, H_2 are *equivalent* if there exists $n_1, n_2 \in \mathbb{N}$ such that $n_1 H_1 = n_2 H_2$. Then a *polarized abelian manifold* is an abelian manifold A together with an equivalence class of Riemann forms on A that contains a non-degenerate Riemann form. Such an equivalence class is called a (homogeneous) *polarization* of A. We use the notation $(A, \tilde{H})$ for a polarized abelian manifold, where $\tilde{H}$ is the equivalence class of the Riemann form H.

Note that a non-degenerate Riemann form on A corresponds to an algebraic equivalence class of a non-degenerate positive divisor. Such a divisor is ample on A, and gives rise to a projective embedding of A. Then a polarized abelian manifold $(A, \tilde{H})$ corresponds to giving the abelian manifold A together with an equivalence class of projective embeddings of A.

A morphism of polarized abelian manifolds $\phi: (A_1, \tilde{H}_1) \to (A_2, \tilde{H}_2)$ is a morphism $\phi: A_1 \to A_2$ such that $\phi^* H_2 \in \tilde{H}_1$.

A justification for the introduction of the notion of polarization is given by the following result.

Theorem. *The automorphism group of a polarized abelian manifold is finite.*

PROOF. Let $(A, \tilde{H})$ be a polarized abelian manifold, where H is a non-degenerate Riemann form. Suppose $\sigma \in \text{Aut}(A, \tilde{H})$. Then $H(\sigma x, \sigma y) = H(x, y)$ (and not just an integer multiple of H). Here we have lifted σ to a linear map on V, the universal cover of A. Thus σ belongs to the compact group of linear maps preserving H. On the other hand, σ is determined by its restriction to the lattice Λ, on which it preserves the $\mathbb{Z}$-valued alternating form $E = \text{Im } H$. So σ belongs to a discrete group also. But the intersection of a compact set with a discrete set is finite! □

A special role is played by the *principally-polarized abelian manifolds*. These are the polarized abelian manifolds $(A, \tilde{H})$ for which there is $H \in \tilde{H}$ with $\text{Pf}(\text{Im } H) = 1$. Thus with respect to a symplectic basis of Λ, $E = \text{Im } H$ is given by the matrix

$$\begin{pmatrix} 0 & I \\ -I & 0 \end{pmatrix}.$$

Examples of principally polarized (p.p.) abelian manifolds are given by the *Jacobian varieties* of non-singular algebraic curves (or Riemann surfaces). If Γ is a non-singular algebraic curve of genus $g > 0$, defined over $\mathbb{C}$, then one can show that $\text{Pic}^0(\Gamma) = \{\text{divisors of degree 0 on } \Gamma\}/\{\text{linear equivalence}\}$ has the structure of an abelian variety $J = \text{Jac}(\Gamma)$, the Jacobian of Γ. (The Abel–Jacobi theorem, proved in [L–A, Chap. IV] shows that $\text{Pic}^0(\Gamma)$ has the structure of a complex torus, while Riemann's relations, proved in [L–A, Chap. IV, §4], imply that $\text{Pic}^0(\Gamma)$ is an abelian manifold). Fix $p_0 \in \Gamma$. Then the map $\alpha: \Gamma \to J$ given by $p \to [p - p_0]$ induces a map $\alpha^{(g-1)}: \Gamma^{(g-1)} \to J$ given by $\alpha^{(g-1)}(p_1, p_2, \ldots, p_{g-1}) = \sum_{j=1}^{g-1} \alpha(p_j)$. Here $\Gamma^{(g-1)}$ is the symmetric product of Γ with itself $g - 1$ times. The image of $\Gamma^{(g-1)}$ in J is well determined up to translation and is a non-degenerate divisor, the theta divisor θ. It can be shown that θ determines a principal polarization on J.

Torelli's theorem (see [G–H, p. 359] for a proof) says that the pair (J, θ) determines Γ up to isomorphism. More precisely, if $(J, \theta) \approx (J', \theta')$, where $J' = \text{Jac}(\Gamma')$, then $\Gamma \approx \Gamma'$.

The following proposition is often useful.

Proposition. *Every polarized abelian manifold is isogenous to a principally polarized abelian manifold.*

PROOF. Let $(A, \tilde{H})$ be a polarized abelian manifold of dimension d. As usual, $A = V/\Lambda$ and $E = \text{Im } H$ is integer valued on Λ. Let $\{\lambda_1, \lambda_2, \ldots, \lambda_{2d}\}$ be a symplectic basis for Λ. In particular, $E(\lambda_j, \lambda_{d+j}) = e_j$, $1 \leq j \leq d$, for some integers e_j. Define a new lattice

$$\Lambda' = \sum_{j=1}^{d} \frac{1}{e_j} \lambda_j \mathbb{Z} + \sum_{j=1}^{d} \lambda_{d+j} \mathbb{Z}.$$

Then E as an alternating form on Λ', is integer valued, and has determinant 1. Let $A' = V/\Lambda'$. Then, $A = V/\Lambda \to V/\Lambda' = A'$ is an isogeny and A' is principally polarized (by E). □

If $(A, \tilde{H})$ is a polarized abelian manifold, then $\phi_H: A \to \hat{A}$, the dual abelian manifold, is an isogeny with kernel of order det $E = \operatorname{Pf}(E)^2$. In particular, if A is principally polarized, $A \approx \hat{A}$. It follows that Jacobians are self-dual.

§6. The Space of Principally Polarized Abelian Manifolds

For $d \geq 1$, let $\mathscr{A}_d$ be the set of isomorphism classes of principally polarized abelian manifolds. We will indicate how $\mathscr{A}_d$ can be given the structure of a complex analytic space.

Consider the case $d = 1$. Every abelian manifold of dimension 1 is principally polarized, so the polarization is irrelevant. Let $\mathscr{H} = \{\tau \in \mathbb{C} | \operatorname{Im} \tau > 0\}$, the Poincaré upper half plane. One has $\mathscr{A}_1 \approx \operatorname{Sl}_2(\mathbb{Z})\backslash \mathscr{H}$ given by

$$\mathbb{C}/\mathbb{Z}\omega_1 + \mathbb{Z}\omega_2 \to \tau = \omega_1/\omega_2$$

where $\operatorname{Im}(\omega_1/\omega_2) > 0$. The action of $\operatorname{Sl}_2(\mathbb{Z})$ on $\mathscr{H}$ is given by

$$\begin{pmatrix} a & b \\ c & d \end{pmatrix}(\tau) = \frac{a\tau + b}{c\tau + d}.$$

This material, which is fairly familiar, will be seen as a special case of our further considerations.

If $A = V/\Lambda$ is a principally polarized abelian variety we choose a $\mathbb{C}$-basis for V and a symplectic basis for Λ. One then sees that A has a concrete representation as $(\mathbb{C}^d/\langle \omega_1, \ldots, \omega_{2d}\rangle, H)$ where $\omega_1, \ldots, \omega_{2d}$ are $\mathbb{R}$-linearly independent column vectors and H is a Riemann form whose imaginary part E has matrix $(E(\omega_i, \omega_j)) = J$ where $J = \begin{pmatrix} 0 & I \\ -I & 0 \end{pmatrix}$. The $d \times 2d$ complex matrix $\Omega = (\omega_1, \omega_2, \ldots, \omega_{2d})$ is called the *period matrix*.

Two questions arise.

(1) What conditions on Ω express the condition that the alternating form E on $\langle \omega_1, \omega_2, \ldots, \omega_{2d}\rangle$ given by J is the imaginary part of a non-degenerate Riemann form?
(2) When do two period matrices Ω and Ω' correspond to isomorphic principally polarized abelian manifolds?

Some calculations (see [L-A, Chap. VIII, §1]) show that Ω determines an abelian manifold if and only if the following two conditions hold.

(RI) $\Omega J \Omega^t = 0$ ($\Omega^t =$ transpose of Ω),

(RII) $2i(\bar{\Omega}J^{-1}\Omega^t)^{-1} > 0$ (>0 means positive definite).

These conditions are known as Riemann's relations. (RI) is equivalent to the condition $E(iz, iw) = E(z, w)$, and (RII) is equivalent to the condition that H be positive definite. In fact, the matrix of H with respect to the standard basis of $\mathbb{C}^d$ is $2i(\bar{\Omega}J^{-1}\Omega^t)^{-1}$.

If we write $\Omega = (\Omega_1, \Omega_2)$ with Ω_1, Ω_2 complex $d \times d$ matrices, the Riemann relations take the form

(RI') $$\Omega_2 \Omega_1^t - \Omega_1 \Omega_2^t = 0,$$
(RII') $$2i(\Omega_2 \bar{\Omega}_1^t - \Omega_1 \bar{\Omega}_2^t) > 0.$$

From these relations it is easy to show that both Ω_1 and Ω_2 are invertible.

Let $\mathscr{R}$ be the set of complex matrices $\Omega = (\Omega_1 \Omega_2)$ satisfying (RI') and (RII'). $\mathrm{Gl}_d(\mathbb{C})$ acts on $\mathscr{R}$ by multiplication on the left and $\mathrm{Sp}_{2d}(\mathbb{Z}) = \{M \in \mathrm{Gl}_{2d}(\mathbb{Z}) | MJM^t = J\}$ acts on $\mathscr{R}$ by multiplication on the right. One sees that

$$\mathscr{A}_d \approx \mathrm{Gl}_d(\mathbb{C}) \backslash \mathscr{R} / \mathrm{Sp}_{2d}(\mathbb{Z}).$$

The action of $\mathrm{Gl}_d(\mathbb{C})$ gives an isomorphism of abelian varieties, while the action of $\mathrm{Sp}_{2d}(\mathbb{Z})$ corresponds to a change of symplectic basis.

Thus $\Omega = (\Omega_1 \Omega_2) \sim (\tau, I)$ where $\tau = \Omega_2^{-1} \Omega_1$. The conditions (RI') and (RII') assert that τ is symmetric and Im τ is positive definite.

Definition. Let $\mathscr{H}_d$ be the space of $d \times d$ complex matrices τ which are symmetric and Im τ is positive definite. $\mathscr{H}_d$ is called *the Siegel upper half space*. It is a complex manifold of dimension $d(d + 1)/2$.

When is $(\tau, I) \sim (\tau', I)$? Write $M = \begin{pmatrix} A & B \\ C & D \end{pmatrix}$. Then, $(\tau, I)M = (\tau A + C, \tau B + D) \sim ((\tau B + D)^{-1}(\tau A + C), I)$. Thus, we must have

$$\tau' = (\tau B + D)^{-1}(\tau A + C)$$

for some $\begin{pmatrix} A & B \\ C & D \end{pmatrix}$ in $\mathrm{Sp}_{2d}(\mathbb{Z})$.

To put things in somewhat more familiar form we note that $\mathrm{Sp}_{2d}(\mathbb{Z})$ is invariant under transpose and that $\tau^t = \tau$. Thus, we let $\mathrm{Sp}_{2d}(\mathbb{Z})$ act on $\mathscr{H}_d$ by

$$\begin{pmatrix} A & B \\ C & D \end{pmatrix} : \tau \to (A\tau + B)(C\tau + D)^{-1}.$$

The final conclusion is

$$\boxed{\mathscr{A}_d \approx \mathrm{Sp}_{2d}(\mathbb{Z}) \backslash \mathscr{H}_d.}$$

By analyzing this more carefully we see $\mathscr{A}_d$ is parametrized by a $d(d + 1)/2$ complex analytic space. In fact, this space can be given the structure of a normal quasi-projective variety.

When $d = 1$ we recover the familiar $\mathscr{A}_1 \approx \mathrm{Sl}_2(\mathbb{Z}) \backslash \mathscr{H}$ which, via the j-function, is isomorphic to the complex plane.

If one investigates in a similar manner the "space" of complex tori of dimension d it turns out to depend on d^2 complex parameters. Thus one suspects that when $d > 1$ there are complex tori which are not abelian varieties since

$$d^2 - \frac{d(d+1)}{2} = \frac{d(d-1)}{2} \geq 1 \quad \text{when } d \geq 2.$$

This is indeed the case.

As an example, let $d = 2$ and set

$$\Omega = \begin{pmatrix} \alpha + i & \beta & 1 & 0 \\ \gamma & \delta + i & 0 & 1 \end{pmatrix},$$

where $\alpha, \beta, \gamma,$ and δ are real and algebraically independent over $\mathbb{Q}$. Let T be the torus $\mathbb{C}^2/\langle\Omega\rangle$, where $\langle\Omega\rangle$ is the lattice generated by the columns of Ω. We will show $\mathcal{M}(T) = \mathbb{C}$. Suppose $f \in \mathcal{M}(T)$ is not a constant. Then $(f) = D_0 - D_\infty$ where D_∞ is a non-zero positive divisor on T. As shown in [SHAF, pp. 354–356], D_∞ corresponds to an integral, non-zero, antisymmetric matrix A such that $\Omega A \Omega^t = 0$. Writing this out shows there is a non-trivial linear relation with coefficients in $\mathbb{Z}$ between $1, \beta, \gamma, \alpha + i, \delta + i$, and $\alpha\delta - \beta\gamma + i(\alpha + \delta)$ and so between $1, \beta, \gamma, \alpha, \delta, \alpha\delta - \beta\gamma$. This contradicts the algebraic independence of $\alpha, \beta, \gamma,$ and δ.

Here is a heuristic argument that the generic abelian variety is simple. If A is a principally polarized abelian variety of dimension d which is not simple then A is isogenous to $B \times C$ where $\dim B = b$ and $\dim C = c$ with $1 \leq b, c < d$, and $b + c = d$. The modulus of B is in $\mathcal{A}_b$, a $[b(b+1)/2]$-dimensional space, and the modulus of C is in $\mathcal{A}_c$, a $[c(c+1)/2]$-dimensional space. Thus the moduli of A with a factor of dimension b lie in a subspace of $\mathcal{A}_d$ of dimension

$$\frac{b(b+1)}{2} + \frac{(d-b)(d-b+1)}{2}.$$

The maximum of these dimensions for $1 \leq b < d$ is easily seen to be that given by $b = [d/2]$, namely $[(d^2 + 2d + 1)/4]$ (here $[x]$ means the greatest integer $\leq x$). But,

$$\left[\frac{d^2 + 2d + 1}{4}\right] < \frac{d(d+1)}{2} \quad \text{for all } d > 1.$$

Finally, we show the generic abelian variety A has $\mathbb{Z}$ for its endomorphism ring. Let $\tau \in \mathcal{A}_d$ and A be the abelian variety corresponding to (τ, I). Then multiplication by $g \in M_d(\mathbb{C})$ gives an element of $\text{End}(A)$ if and only if there is a matrix $M \in M_{2d}(\mathbb{Z})$ such that $g(\tau, I) = (\tau, I)M$. Let $M = \begin{pmatrix} A & B \\ C & D \end{pmatrix}$. We must have $(g\tau, g) = (\tau A + C, \tau B + D)$ so $g = \tau B + D$ and $g\tau = \tau A + C$. Thus, $\tau B\tau + D\tau - \tau A - C = 0$. Let $\tau = (t_{ij})$. If $B = C = 0$ and $A = D = nI$ for $n \in \mathbb{Z}$ there is no condition imposed on τ. Otherwise the t_{ij} must satisfy certain non-trivial quadratic polynomials with coefficients in $\mathbb{Z}$. In general, this cannot happen. For example, suppose t_{ij} with $i \leq j$ are algebraically independent over $\mathbb{Q}$. Then a simple calculation shows that $\tau B\tau + D\tau - \tau A - C = 0$ can only happen when $C = B = 0$ and $A = D = nI$ for some $n \in \mathbb{Z}$. For the corresponding abelian variety the endomorphism ring is precisely $\mathbb{Z}$. Of course, when $d = 1$ we need only require that $t = t_{11}$ not be quadratic over $\mathbb{Q}$.

ANNOTATED REFERENCES

[G–H] Griffiths, P. and Harris, J. *Principles of Algebraic Geometry*. Wiley: New York, 1978.
Chapter 2 (Riemann Surfaces and Algebraic Curves) contains a discussion of Jacobians and abelian varieties from the complex analytic point of view.

[I] Igusa, J. *Theta Functions*. Springer-Verlag: New York, 1972.
This book contains an extensive treatment of both the analytic and algebraic aspects of the theory of theta functions. Chapter IV discusses the equations defining abelian varieties under the projective embedding obtained from theta functions.

[L–A] Lang, S. *Introduction to Algebraic and Abelian Functions*, 2nd edn. Springer-Verlag: New York, 1982.
Chapters VI, VII, VIII, X are relevant for the content of these lectures. Chapter IX constructs abelian manifolds whose algebra of endomorphisms contains a quaternion algebra.

[L–CM] Lang, S. *Complex Multiplication*. Springer-Verlag: New York, 1983.
A complete exposition of the theory of complex multiplication of abelian varieties.

[M–AV] Mumford, D. *Abelian Varieties*, Oxford University Press: Oxford, 1970. (2nd edn 1974).
Chapter I treats the analytic theory using line bundles instead of divisors. Most of the book gives a general treatment of abelian varieties from the point of view of schemes.

[M–CJ] Mumford, D. *Curves and Their Jacobians*. University of Michigan Press: Ann Arbor, 1975.
An excellent survey of algebraic curves and their Jacobians.

[M–T] Mumford, D. *Tata Lectures on Theta*, I. Birkhäuser-Verlag: Basle, 1983.
A more down-to-earth treatment of theta functions than [I].

[R] Robert, A. Introduction aux Variétés Abéliennes Complexes, *Enseign. Math.*, **28** (1982), 91–137.
A self-contained discussion of the criterion for a complex torus to be projectively embeddable.

[SHAF] Shafarevich, I. R. *Basic Algebraic Geometry*, Springer-Verlag: New York, 1977.
Part 3 contains an excellent treatment of complex analytic manifolds.

[SHIM] Shimura, G., and Taniyama, Y. *Complex Multiplication of Abelian Varieties*, Mathematical Society of Japan: Tokyo, 1961.
The original treatment of the subject of the title, due to Shimura, Taniyama and Weil.

[SI] Siegel, C. L. *Topics in Complex Function Theory*, Vol. III. Wiley: New York, 1972.

An exposition of the classical approach to Riemann surfaces and Abelian integrals.
[SW] Swinnerton-Dyer, H. P. F. *Analytic Theory of Abelian Varieties*. Cambridge University Press: Cambridge, 1974.
A treatment of abelian manifolds close to these lectures.
[WE] Weil, A. *Introduction à l'Étude des Variétés Kählériennes*. Hermann: Paris, 1971.
Weil's approach to the analytic theory, based on Hodge theory.

CHAPTER V

Abelian Varieties

J. S. MILNE

This chapter reviews the theory of abelian varieties emphasizing those points of particular interest to arithmetic geometers. In the main it follows Mumford's book [16] except that most results are stated relative to an arbitrary base field, some additional results are proved, and étale cohomology is included. Many proofs have had to be omitted or only sketched. The reader is assumed to be familiar with [10, Chaps. II, III] and (for a few sections that can be skipped) some étale cohomology. The last section of Chapter VII, "Jacobian Varieties", contains bibliographic notes for both chapters.

Conventions

The algebraic closure of a field k is denoted by $\bar{k}$ and its separable closure by k_s. For a scheme V over k and a k-algebra R, V_R denotes $V \times_{\text{spec}(k)} \text{spec}(R)$, and $V(R)$ denotes $\text{Mor}_k(\text{spec}(R), V)$. By a scheme over k, we shall always mean a scheme of finite type over k.

A variety V over k is a separated scheme of finite type over k such that $V_{\bar{k}}$ is integral (that is, reduced and irreducible). It is nonsingular if $V_{\bar{k}}$ is regular. Note that with these definitions, if V is a variety (and is nonsingular) then V_K is integral (and is regular) for all fields $K \supset k$, and a product of (nonsingular) varieties is a (nonsingular) variety; moreover, $V(k_s)$ is nonempty. A k-rational point of V is often identified with a closed point v of V such that $k(v) = k$.

All statements are relative to a fixed group field: if V and W are varieties over k, then a sheaf or divisor on V, or a morphism $V \to W$, is automatically meant to be a defined over k (not over some "universal domain" as in the pre-scheme days).

Divisor means Cartier divisor, except that because most of our varieties are nonsingular we can usually think of them as Weil divisors. If $\pi: W \to V$ is a map and D is a divisor on V with local equation f near v, then π^*D (or

$\pi^{-1}D$) is the divisor on W with local equation $f \circ \pi$ near $\pi^{-1}(v)$. The invertible sheaf defined by D is denoted by $\mathscr{L}(D)$.

The tangent space to V at v is denoted by $T_v(V)$. Canonical isomorphisms are often denoted by $=$. The two projection maps $p: V \times W \to V$ and $q: V \times W \to W$ are always so denoted. The kernel of multiplication by n, $X \to X$, is denoted by X_n. An equivalence class containing x is often denoted by $[x]$.

§1. Definitions

A *group variety* over k is a variety V over k together with morphisms

$$m: V \times V \to V \quad \text{(multiplication)},$$
$$\text{inv}: V \to V \quad \text{(inverse)},$$

and an element $\varepsilon \in V(k)$ such that the structure on $V(\bar{k})$ defined by m and inv is that of a group with identity element ε.

Such a quadruple $(V, m, \text{inv}, \varepsilon)$ is a group in the category of varieties over k, i.e., the diagrams [22, §2] commute. (To see this, note that two morphisms with domain a variety W are equal if they become equal over $\bar{k}$, and that $W(\bar{k})$ is dense in $W_{\bar{k}}$.) Thus, for every k-algebra R, $V(R)$ acquires a group structure, and these group structures depend functorially on R.

For $a \in V(\bar{k})$, the projection map $p: V_{\bar{k}} \times V_{\bar{k}} \to V_{\bar{k}}$ induces an isomorphism $V_{\bar{k}} \times \{a\} \xrightarrow{\approx} V_{\bar{k}}$, and we define t_a to be the composite

$$V_{\bar{k}} \approx V_{\bar{k}} \times \{a\} \subset V_{\bar{k}} \times V_{\bar{k}} \xrightarrow{m} V_{\bar{k}}.$$

On points t_a is the translation map $P \mapsto m(P, a)$. Similarly, for any point $a \in V$, there is a translation map $t_a: V_{k(a)} \to V_{k(a)}$. In particular, if $a \in V(k)$, then t_a maps V into V.

A group variety is automatically nonsingular: as does any variety, it contains a nonempty, nonsingular open subvariety U, and the translates of $U_{\bar{k}}$ cover $V_{\bar{k}}$.

A complete group variety is called an *abelian variety*. As we shall see, they are projective and (fortunately) commutative. Their group laws will be written additively.

An affine group variety is called a *linear algebraic group*. Each such variety can be realized as a closed subgroup of GL_n for some n [24, 3.4].

§2. Rigidity

Theorem 2.1 (Rigidity Theorem). *Let $f: V \times W \to U$ be a morphism of varieties over k. If V is complete and*

$$f(V \times \{w_0\}) = \{u_0\} = f(\{v_0\} \times W)$$

for some $u_0 \in U(k)$, $v_0 \in V(k)$, $w_0 \in W(k)$, then $f(V \times W) = \{u_0\}$.

PROOF. Let U_0 be an open affine neighborhood of u_0. The projection map $q: V \times W \to W$ is closed (this is what it means for V to be complete), and so the set $Z = q(f^{-1}(U - U_0))$ is closed in W. Note that a closed point w of W lies outside Z if and only if $f(V \times \{w\}) \subset U_0$. In particular, $w_0 \in W - Z$ and so $W - Z$ is a dense open subset of W. As $V \times \{w\}$ is complete and U_0 is affine, $f(V \times \{w\})$ must be a point whenever w is a closed point of $W - Z$, [14, p. 104]; in fact, $f(V \times \{w\}) = f(\{v_0\} \times \{w\}) = \{u_0\}$. Thus f is constant on the dense subset $V \times (W - Z)$ of $V \times W$, and so is constant. □

Corollary 2.2. *Every morphism $f: A \to B$ of abelian varieties is the composite of a homomorphism $h: A \to B$ with a translation t_a, $a = -f(0) \in B(k)$.*

PROOF. After replacing f with $t_a \circ f$, $a = -f(0)$, we can assume that $f(0) = 0$. Define $\varphi: A \times A \to B$ to be $f \circ m_A - m_B \circ (f \times f)$, so that on points $\varphi(a, a') = f(a + a') - f(a) - f(a')$. Then $\varphi(A \times \{0\}) = 0 = \varphi(\{0\} \times A)$, and so the theorem shows that $\varphi = 0$ on $A \times A$. Thus $f \circ m_A = m_B \circ (f \times f)$, which is what we mean by f being a homomorphism. □

Remark 2.3. The corollary shows that the group structure on A is uniquely determined by the choice of a zero element.

Corollary 2.4. *The group law on an abelian variety A is commutative.*

PROOF. Commutative groups are distinguished by the fact that the map taking an element to its inverse is a homomorphism. The preceding corollary shows that inv: $A \to A$ is a homomorphism. □

Corollary 2.5. *Let V and W be complete varieties over k with rational points $v_0 \in V(k)$, $w_0 \in W(k)$, and let A be an abelian variety. Then a morphism $h: V \times W \to A$ such that $h(v_0, w_0) = 0$ can be written uniquely as $h = f \circ p + g \circ q$ with $f: V \to A$ and $g: W \to A$ morphisms such that $f(v_0) = 0$, $g(w_0) = 0$.*

PROOF. Define f to be $V = V \times \{w_0\} \xrightarrow{h} A$ and g to be $W = \{v_0\} \times W \xrightarrow{h} A$, so that $k \stackrel{\text{df}}{=} h - (f \circ p + g \circ q)$ is the map such that on points $k(v, w) = h(v, w) - h(v, w_0) - h(v_0, w)$. Then

$$k(V \times \{w_0\}) = 0 = k(\{v_0\} \times W),$$

and so the theorem shows that $k = 0$. □

§3. Rational Maps into Abelian Varieties

We improve some of the results in the last section.

Recall [10, I, 4] that a *rational map* $f: V \dashrightarrow W$ of varieties is an equivalence class of pairs (U, f_U) with U a dense open subset of V and f_U a morphism

$U \to W$; two pairs (U, f_U) and $(U', f_{U'})$ are equivalent if f_U and $f_{U'}$ agree on $U \cap U'$. There is a largest open subset U of V such that f defines a morphism $U \to W$, and f is said to be *defined* at the points of U.

Theorem 3.1. *A rational map $f: V \dashrightarrow A$ from a nonsingular variety to an abelian variety is defined on the whole of V.*

PROOF. Combine the next two lemmas. □

Lemma 3.2. *A rational map $f: V \dashrightarrow W$ from a normal variety to a complete variety is defined on an open subset U of V whose complement $V - U$ has codimension ≥ 2*

PROOF. Let $f_U: U \to W$ be a representative of f, and let v be a point of $V - U$ of codimension 1 in V (that is, whose closure $\{\bar{v}\}$ has codimension 1). Then $\mathcal{O}_{V,v}$ is a discrete valuation ring (because V is normal) whose field of fractions is $k(V)$, and the valuative criterion of properness [10, II, 4.7] shows that the map $\text{spec}(k(V)) \to W$ defined by f extends to a map $\text{spec}(\mathcal{O}_{V,v}) \to W$. This implies that f has a representative defined on a neighborhood of v, and so the set on which f is defined contains all points of codimension ≤ 1. This proves the lemma. □

Lemma 3.3. *Let $f: V \dashrightarrow G$ be a rational map from a nonsingular variety to a group variety. Then either f is defined on all of V or the points where it is not defined form a closed subset of pure codimension 1 in V.*

PROOF. See [2, 1.3]. □

Theorem 3.4. *Let $f: V \times W \to A$ be a morphism from a product of nonsingular varieties into an abelian variety. If*
$$f(V \times \{w_0\}) = \{a_0\} = f(\{v_0\} \times W)$$
for some $a_0 \in A(K)$, $v_0 \in V(k)$, and $w_0 \in W(k)$, then $f(V \times W) = \{a_0\}$.

PROOF. We can assume k to be algebraically closed. Consider first the case that V has dimension 1. Then V can be embedded in a nonsingular complete curve $\bar{V}$, and (3.1) shows that f extends to a map $\bar{f}: \bar{V} \times W \to A$. Now (2.1) shows that $\bar{f}$ is constant.

In the general case, let C be an irreducible curve on V passing through v_0 and nonsingular at v_0, and let $\tilde{C} \to C$ be the normalization of C. Then f defines a morphism $\tilde{C} \times W \to A$ which the preceding argument shows to be constant. Therefore $f(C \times W) = \{a_0\}$, and the next lemma completes the proof. □

Lemma 3.5. *Let V be an integral scheme of finite type over a field k, and assume*

V is nonsingular at a point $v_0 \in V(k)$; then the union of the integral one-dimensional subschemes passing through v_0 and nonsingular at v_0 is dense in V.

PROOF. By induction it suffices to show that the union of the integral subschemes of codimension 1 passing through v_0 and smooth at v_0 is dense in V. We can assume that V is affine and v_0 is the origin. For H a hyperplane passing through v_0 but not containing $T_{v_0}(V)$, $V \cap H$ is smooth at v_0. Let V_H be the component of $V \cap H$ passing through v_0, regarded as an integral subscheme of V and let Z be a closed subset of V containing all V_H. Regard Z as a reduced subscheme of V, and let $C_{v_0}(Z)$ be the tangent cone to Z at v_0 [14, III.3]. Clearly $T_{v_0}(V) \cap H = T_{v_0}(V_H) = C_{v_0}(V_H) \subset C_{v_0}(Z) \subset C_{v_0}(V) = T_{v_0}(V)$, and it follows that $C_{v_0}(Z) = T_{v_0}(V)$. As $\dim C_{v_0}(Z) = \dim(Z)$ (see [14, III.3, p. 320]), this implies that $Z = V$. □

Corollary 3.6. *Every rational map $f\colon G \dashrightarrow A$ from a group variety to an abelian variety is the composite of a homomorphism $h\colon G \to A$ with a translation.*

PROOF. Theorem 3.1 shows that f is a morphism. The rest of the proof is the same as that of (2.2). □

Remark 3.7. The corollary shows that A is determined by $k(A)$ up to the choice of a zero element. In particular, if A and B are abelian varieties and $k(A)$ is isomorphic to $k(B)$, then A is isomorphic to B (as an abelian variety).

Corollary 3.8. *Every rational map $f\colon \mathbb{P}^1 \dashrightarrow A$ is constant.*

PROOF. The variety $\mathbb{P}^1 - \{\infty\}$ becomes a group variety under addition, and $\mathbb{P}^1 - \{0, \infty\}$ becomes a group variety under multiplication. Therefore the last corollary shows that there exist $a, b \in A(k)$ such that

$$f(x+y) = f(x) + f(y) + a, \quad \text{all} \quad x, y \in \bar{k} = \mathbb{P}^1(\bar{k}) - \{\infty\},$$
$$f(xy) = f(x) + f(y) + b, \quad \text{all} \quad x, y \in \bar{k}^\times = \mathbb{P}^1(\bar{k}) - \{0, \infty\}.$$

This is clearly impossible unless f is constant. □

Recall that a variety V of dimension d is *unirational* if there is an embedding of $\bar{k}(V)$ into a purely transcendental extension $\bar{k}(X_1, \ldots, X_d)$ of $\bar{k}$. Such an embedding corresponds to a rational map $\mathbb{P}^d_{\bar{k}} \dashrightarrow V_{\bar{k}}$ whose image is dense in $V_{\bar{k}}$.

Corollary 3.9. *Every rational map from a unirational variety to an abelian variety is constant.*

PROOF. We can suppose k to be algebraically closed. By assumption there is a rational map $\mathbb{A}^d \dashrightarrow V$ with dense image, and the composite of this with a rational map $f\colon V \dashrightarrow A$ extends to a morphism $\bar f\colon \mathbb{P}^1 \times \cdots \times \mathbb{P}^1 \to A$. According to (2.5), $\bar f(x_1,\ldots,x_n) = \sum f_i(x_i)$ for some morphisms $f_i\colon \mathbb{P}^1 \to A$, and (3.8) shows that each f_i is constant. □

§4. Review of the Cohomology of Schemes

In order to prove some of the theorems concering abelian varieties, we shall need to make use of results from the cohomology of coherent sheaves. The first of these is Grothendieck's relative version of the theorem asserting that the cohomology groups of coherent sheaves on complete varieties are finite dimensional.

Theorem 4.1. *If $f\colon V \to T$ is a proper morphism of Noetherian schemes and $\mathscr{F}$ is a coherent $\mathcal{O}_V$-module, then the higher direct image sheaves $R^r f_* \mathscr{F}$ are coherent $\mathcal{O}_T$-modules for all $r \geq 0$.*

PROOF. When f is projective, this is proved in [10, III, 8.8]. Chow's lemma [10, II, Ex. 4.10] allows one to extend the result to the general case [9, III.3.2.1]. □

The second result describes how the dimensions of the cohomology groups of the members of a flat family of coherent sheaves vary.

Theorem 4.2. *Let $f\colon V \to T$ be a proper flat morphism of Noetherian schemes, and let $\mathscr{F}$ be a locally free $\mathcal{O}_V$-module of finite rank. For each t in T, write V_t for the fibre of V over t and $\mathscr{F}_t$ for the inverse image of $\mathscr{F}$ on V_t.*

(a) *The formation of the higher direct images of $\mathscr{F}$ commutes with flat base change. In particular, if $T = \operatorname{spec}(R)$ is affine and R' is a flat R-algebra, then $H^r(V', \mathscr{F}') = H^r(V, \mathscr{F}) \otimes_R R'$, where $V' = V \times_{\operatorname{spec}(R)} \operatorname{spec}(R')$ and $\mathscr{F}'$ is the inverse image of $\mathscr{F}$ on V'.*

(b) *The function $t \mapsto \chi(\mathscr{F}_t) \stackrel{\mathrm{df}}{=} \sum (-1)^r \dim_{k(t)} H^r(V_t, \mathscr{F}_t)$ is locally constant on T.*

(c) *For each r, the function $t \mapsto \dim_{k(t)} H^r(V_t, \mathscr{F}_t)$ is upper semicontinuous (that is, it jumps on closed subsets).*

(d) *If T is integral and $\dim_{k(t)} H^r(V_t, \mathscr{F}_t)$ is equal to a constant s for all t in T, then $R^r f_* \mathscr{F}$ is a locally free $\mathcal{O}_T$-module and the natural maps $R^r f_* \mathscr{F} \otimes_{\mathcal{O}_T} k(t) \to H^r(V_t, \mathscr{F}_t)$ are isomorphisms.*

(e) *If $H^1(V_t, \mathscr{F}_t) = 0$ for all t in T, then $R^1 f_* \mathscr{F} = 0$, $f_* \mathscr{F}$ is locally free, and the formation of $f_* \mathscr{F}$ commutes with base change.*

PROOF. (a) The statement is local on the base, and so it suffices to prove it for the particular case in which we have given an explicit statement. In

[16, §5, p. 46], a complex $K^{\cdot}$ of R-modules is constructed such that for all R-algebras R', $H^r(V', \mathscr{F}') = H^r(K^{\cdot} \otimes_R R')$. In our case, R' is flat over R, and so $H^r(K^{\cdot} \otimes_R R') = H^r(K^{\cdot}) \otimes_R R'$, which equals $H^r(V, \mathscr{F}) \otimes_R R'$.

(b), (c), (d). These are proved in [16, §5].

(e). The hypothesis implies that $R^1 f_* \mathscr{F} = 0$ ([10, III, 12.11a]), and it follows that $f_* \mathscr{F} \otimes_{\mathcal{O}_T} k(t) \to H^0(V_t, \mathscr{F}_t)$ is surjective for all t ([10, III, 12.11b]) and so is an isomorphism. Now this last reference (applied with $i = 0$) shows that $f_* \mathscr{F}$ is locally free. □

§5. The Seesaw Principle

We shall frequently need to consider the following situation: V is a variety over k, T is a scheme of finite type over k, and $\mathscr{L}$ is an invertible sheaf on $V \times T$. For $t \in T$, $\mathscr{L}_t$ will then always denote the invertible sheaf $(1 \times \iota)^* \mathscr{L}$ on $V_t = V_{k(t)} = (V \times T) \times_T t$, where ι is the inclusion of $t = \mathrm{spec}(k(t))$ into T. There is the diagram

$$(V \times T, \mathscr{L}) \leftarrow (V_t, \mathscr{L}_t).$$
$$\downarrow \qquad\qquad \downarrow$$
$$T \xleftarrow{\iota} t$$

It is often useful to regard $\mathscr{L}$ as defining a family of invertible sheaves on V parametrized by T.

Theorem 5.1. *Let V be a complete variety and T an integral scheme of finite type over k, and let $\mathscr{L}$ and $\mathscr{M}$ be invertible sheaves on $V \times T$. If $\mathscr{L}_t \approx \mathscr{M}_t$ for all $t \in T$, then there exists an invertible sheaf $\mathscr{N}$ on T such that $\mathscr{L} \approx \mathscr{M} \otimes q^* \mathscr{N}$.*

PROOF. By assumption, $(\mathscr{L} \otimes \mathscr{M}^{-1})_t$ is trivial for all $t \in T$, and so $H^0(V_t, (\mathscr{L} \otimes \mathscr{M}^{-1})_t) \approx H^0(V_t, \mathcal{O}_{V_t}) = k(t)$. Therefore (4.2d) shows that the sheaf $\mathscr{N} = q_*(\mathscr{L} \otimes \mathscr{M}^{-1})$ is invertible. Consider the natural map $q^* \mathscr{N} = q^* q_*(\mathscr{L} \otimes \mathscr{M}^{-1}) \xrightarrow{\alpha} \mathscr{L} \otimes \mathscr{M}^{-1}$. As $(\mathscr{L} \otimes \mathscr{M}^{-1})_t \approx \mathcal{O}_{V_t}$, the restriction of α to the fibre V_t is isomorphic to the natural map $\alpha_t \colon \mathcal{O}_{V_t} \otimes \Gamma(V_t, \mathcal{O}_{V_t}) \to \mathcal{O}_{V_t}$, which is an isomorphism. Now Nakayama's lemma implies that α is surjective, and because both $q^* \mathscr{N}$ and $\mathscr{L} \otimes \mathscr{M}^{-1}$ are invertible sheaves, it follows that α is an isomorphism (if R is a local ring, then a surjective R-linear map $R \to R$ is an isomorphism because it must send 1 to a unit). □

Corollary 5.2 (Seesaw Principle). *Suppose in addition to the hypotheses of the theorem that $\mathscr{L}_v \approx \mathscr{M}_v$ for at least one $v \in V(k)$. Then $\mathscr{L} \approx \mathscr{M}$.*

PROOF. The theorem shows that $\mathscr{L} \approx \mathscr{M} \otimes q^* \mathscr{N}$ for some $\mathscr{N}$ on T. On pulling back by $T = \{v\} \times T \subset V \times T$, we obtain an isomorphism

$\mathscr{L}_v \approx \mathscr{M}_v \otimes q^*\mathscr{N}_v$. As $\mathscr{L}_v \approx \mathscr{M}_v$ and $(q^*\mathscr{N})_v = \mathscr{N}$, this shows that $\mathscr{N}$ is trivial. □

The next result shows that the condition $\mathscr{L}_t \approx \mathscr{M}_t$ of the theorem needs only to be checked for t in some dense subset of T (for example, it needs only to be checked for t the generic point of T).

Theorem 5.3. *Let V be a complete variety, and let $\mathscr{L}$ be an invertible sheaf on $V \times T$. Then $\{t \in T \mid \mathscr{L}_t \text{ is trivial}\}$ is closed in T.*

Lemma 5.4. *An invertible sheaf $\mathscr{L}$ on a complete variety is trivial if and only if both it and its dual $\mathscr{L}^{-1}$ have nonzero global sections.*

PROOF. The sections define nonzero homomorphisms $s_1: \mathcal{O}_V \to \mathscr{L}$ and $s_2: \mathcal{O}_V \to \mathscr{L}^{-1}$. The dual of s_2 is a homomorphism $s_2^\vee: \mathscr{L} \to \mathcal{O}_V$, and $s_2^\vee \circ s_1$, being nonzero, is an isomorphism (note that $\mathrm{Hom}(\mathcal{O}_V, \mathcal{O}_V) = H^0(V, \mathcal{O}_V) = k$). Because $\mathscr{L}$ is an invertible sheaf, this implies that s_1 is also an isomorphism. □

PROOF OF (5.3). The lemma identifies the set of t for which $\mathscr{L}_t$ is trivial with the set of t for which both $\dim H^0(V_t, \mathscr{L}_t) > 0$ and $\dim H^0(V_t, \mathscr{L}_t^{-1}) > 0$. Part (c) of (4.2) shows that this set is closed. □

Remark 5.5. Let V, T, and $\mathscr{L}$ be as at the start of the section with V complete. We shall say that $\mathscr{L}$ defines a *trivial family* of sheaves on T if $\mathscr{L} \approx q^*\mathscr{N}$ for some invertible sheaf $\mathscr{N}$ on T. According to (5.1), in the case that T is integral, $\mathscr{L}$ defines a trivial family if and only if each $\mathscr{L}_t$ is trivial. Returning to the general situation, let Z be the closed subset of T determined by (5.3). Clearly Z has the following property: A morphism $f: T' \to T$ from an integral scheme to T factors through Z if and only if $(1 \times f)^*\mathscr{L}$ defines a trivial family on V. This result can be significantly strengthened: there exists a unique closed subscheme Z of T (not necessarily reduced) such that a morphism $f: T' \to T$ (with T' not necessarily integral) factors through the inclusion morphism $Z \hookrightarrow T$ if and only if $(1 \times f)^*\mathscr{L}$ defines a trivial family on V. See [16, §10, p. 89].

§6. The Theorems of the Cube and the Square

Theorem 6.1 (Theorem of the Cube). *Let U, V, W be complete varieties over k with base points $u_0 \in U(k)$, $v_0 \in V(k)$, $w_0 \in W(k)$. An invertible sheaf $\mathscr{L}$ on $U \times V \times W$ is trivial if its restrictions to $\{u_0\} \times V \times W$, $U \times \{v_0\} \times W$, and $U \times V \times \{w_0\}$ are all trivial.*

PROOF. Because $\mathscr{L} | U \times V \times \{w_0\}$ is trivial, the seesaw principle shows that it suffices to prove that $\mathscr{L} | z \times W$ is trivial for a dense set of z in $U \times V$. Next one shows that U can be taken to be a complete curve ((3.5) accomplishes this reduction when u_0 is nonsingular). This case is proved in [16, §6, pp. 57–58] when k is algebraically closed, and the next lemma shows that we may assume that. □

Lemma 6.2. *Let $\mathscr{L}$ be an invertible sheaf on a complete variety V over a field k; if $\mathscr{L}$ becomes trivial on $V_{\bar{k}}$ then it is trivial on V.*

PROOF. The triviality of $\mathscr{L}$ on $V_{\bar{k}}$ implies that both $H^0(V_{\bar{k}}, \mathscr{L})$ and $H^0(V_{\bar{k}}, \mathscr{L}^{-1})$ are nonzero. As $H^0(V_{\bar{k}}, \mathscr{L}^{\pm 1}) = H^0(V, \mathscr{L}^{\pm 1}) \otimes_k \bar{k}$ (see (4.2a)), Lemma 5.4 shows that $\mathscr{L}$ is trivial. □

Remark 6.3. At least in the case that k is algebraically closed, it is not necessary to assume in (6.1) that W is complete [16, §6, p. 55], nor even that it is a variety [16, §10, p. 91].

Corollary 6.4. *Let A be an abelian variety, and let $p_i: A \times A \times A \to A$ be the projection onto the ith factor; let $p_{ij} = p_i + p_j$ and $p_{ijk} = p_i + p_j + p_k$. For any invertible sheaf $\mathscr{L}$ on A, the sheaf*

$$p_{123}^*\mathscr{L} \otimes p_{12}^*\mathscr{L}^{-1} \otimes p_{23}^*\mathscr{L}^{-1} \otimes p_{13}^*\mathscr{L}^{-1} \otimes p_1^*\mathscr{L} \otimes p_2^*\mathscr{L} \otimes p_3^*\mathscr{L}$$

on $A \times A \times A$ is trivial.

PROOF. The restriction of the sheaf to $\{0\} \times A \times A (= A \times A)$ is

$$m^*\mathscr{L} \otimes p^*\mathscr{L}^{-1} \otimes m^*\mathscr{L}^{-1} \otimes q^*\mathscr{L}^{-1} \otimes \mathscr{O}_{A \times A} \otimes p^*\mathscr{L} \otimes q^*\mathscr{L},$$

which is trivial. Similarly its restrictions to $A \times \{0\} \times A$ and $A \times A \times \{0\}$ are trivial, which implies that it is trivial on $A \times A \times A$. □

Corollary 6.5. *Let f, g, h be morphisms from a variety V to an abelian variety A. For any invertible sheaf $\mathscr{L}$ on A, the sheaf*

$$(f + g + h)^*\mathscr{L} \otimes (f + g)^*\mathscr{L}^{-1} \otimes (g + h)^*\mathscr{L}^{-1} \otimes (f + h)^*\mathscr{L}^{-1}$$

$$\otimes f^*\mathscr{L} \otimes g^*\mathscr{L} \otimes h^*\mathscr{L}$$

on V is trivial.

PROOF. The sheaf in question is the inverse image of the sheaf in (6.4) by $(f, g, h): V \to A \times A \times A$. □

Corollary 6.6. *Consider the map $n_A: A \to A$ equal to multiplication by n. For all invertible sheaves $\mathscr{L}$ on A,*

$$n_A^*\mathscr{L} \approx \mathscr{L}^{(n^2+n)/2} \otimes (-1)^* \mathscr{L}^{(n^2-n)/2}.$$

In particular,

$$n_A^*\mathscr{L} \approx \mathscr{L}^{n^2} \quad \text{if } \mathscr{L} \text{ is symmetric (i.e., } \mathscr{L} \approx (-1)_A^*\mathscr{L})$$

$$n_A^*\mathscr{L} \approx \mathscr{L}^n \quad \text{if } \mathscr{L} \text{ is antisymmetric (i.e., } \mathscr{L}^{-1} \approx (-1)_A^*\mathscr{L}).$$

PROOF. On applying the last corollary to the maps $n_A, 1_A, (-1)_A: A \to A$ we find that $(n+1)_A^* \mathscr{L}^{-1} \otimes n_A^* \mathscr{L}^2 \otimes (n-1)_A^* \mathscr{L}^{-1} \approx \mathscr{L}^{-1} \otimes (-1)^* \mathscr{L}^{-1}$. This fact can be used to prove the corollary by induction, starting from the easy cases $n = 0, 1, -1$. □

Theorem 6.7 (Theorem of the Square). *For all invertible sheaves $\mathscr{L}$ on an abelian variety A and points $a, b \in A(k)$,*

$$t_{a+b}^* \mathscr{L} \otimes \mathscr{L} \approx t_a^* \mathscr{L} \otimes t_b^* \mathscr{L}.$$

PROOF. Apply (6.5) with f the identity map on A and g and h the constant maps with images a and b. □

Remark 6.8. When tensored with $\mathscr{L}^{-2}$, the isomorphism in (6.7) becomes

$$t_{a+b}^* \mathscr{L} \otimes \mathscr{L}^{-1} \approx (t_a^* \mathscr{L} \otimes \mathscr{L}^{-1}) \otimes (t_b^* \mathscr{L} \otimes \mathscr{L}^{-1}).$$

Thus the map $\varphi_{\mathscr{L}}$,

$$a \mapsto t_a^* \mathscr{L} \otimes \mathscr{L}^{-1}: A(k) \to \mathrm{Pic}(A),$$

is a homomorphism. Therefore, if $\sum_{i=1}^n a_i = 0$ in $A(k)$, then

$$t_{a_1}^* \mathscr{L} \otimes t_{a_2}^* \mathscr{L} \otimes \cdots \otimes t_{a_n}^* \mathscr{L} \approx \mathscr{L}^n.$$

Remark 6.9. We write $\sim$ for linear equivalence of divisors, so that $D \sim D'$ if and only if $\mathscr{L}(D) \approx \mathscr{L}(D')$. Also, we write D_a for the translate $t_a D = D + a$ of D. Note that $t_a^* \mathscr{L}(D) = \mathscr{L}(t_a^{-1} D) = \mathscr{L}(D_{-a})$. The isomorphisms in (6.7) and (6.8) become the relations:

$$D_{a+b} + D \sim D_a + D_b, \qquad a, b \in A(k),$$

$$\sum_{i=1}^n D_{a_i} \sim nD, \qquad \text{if } \sum a_i = 0 \text{ in } A(k).$$

§7. Abelian Varieties Are Projective

For D a divisor on a variety V we write

$$L(D) = \{f \in k(V) | (f) + D \geq 0\} \cup \{0\} = H^0(V, \mathscr{L}(D)),$$

$$|D| = \{(f) + D | f \in L(D)\} = \text{the complete linear system containing } D.$$

A projective embedding of an elliptic curve can be constructed as follows: let $D = P_0$, where P_0 is the zero element of A, and choose a suitable basis 1, x, y of $L(3D)$; then the map $A \to \mathbb{P}^2$ defined by $\{1, x, y\}$ identifies A with the cubic projective curve

$$Y^2 Z + a_1 XYZ + a_3 YZ^2 = X^3 + a_2 X^2 Z + a_4 XZ^2 + a_6 Z^3.$$

(See [10, IV, 4.6].) This argument can be extended to every abelian variety.

Theorem 7.1. *Every abelian variety is projective.*

PROOF. We first prove this under the assumption that the abelian variety A is defined over an algebraically closed field.

Recall [10, II, 7.8.2] that a variety is projective if it has a very ample linear system, and that a linear system $\mathfrak{d}$ is very ample if:

(a) it separates points (for any pair a, b of distinct closed points on the variety, there is a D in $\mathfrak{d}$ such that $a \in D$ but $b \notin D$); and
(b) it separates tangent vectors (for any closed point a and tangent vector t to the variety at a, there exists a $D \in \mathfrak{d}$ such that $a \in D$ but $t \notin T_a(D)$).

The first step of the proof is to show that there exists a linear system that separates 0 from the other points of A and separates tangent vectors at 0. More precisely, we show that there exists a finite set $\{Z_i\}$ of prime divisors on A such that:

(a) $\bigcap Z_i = \{0\}$; and
(b) for any $t \in T_0(A)$ there exists a Z_i such that $t_i \notin T_0(Z_i)$.

The second step is to show that if $D = \sum Z_i$, then $|3D|$ is very ample.

The existence of the set $\{Z_i\}$ is an immediate consequence of the observations:

(i) for any closed point $a \neq 0$ of A, there is a prime divisor Z such that $0 \in Z$, $a \notin Z$;
(ii) for any $t \in T_0(A)$, there is a prime divisor Z passing through 0 such that $t \notin T_0(Z)$.

The proof of (ii) is obvious: choose an open affine neighborhood U of 0, let Z_0 be an irreducible component of $A \cap H$ where H is any hyperplane through 0 not containing t, and take Z to be the closure of Z_0. The proof of (i) will be equally obvious once we have shown that 0 and a are contained in a single open affine subset of A. Let U again be an open affine neighborhood of 0, and let $U + a$ be its translate by a. Choose a closed point u of $U \cap (U + a)$. Then both u and $u + a$ lie in $U + a$, and so $U + a - u$ is an open affine neighborhood of both 0 and a.

Now let D be the divisor $\sum Z_i$ where $(Z_i)_{1 \leq i \leq n}$ satisfies (a) and (b). For any family $(a_i, b_i)_{1 \leq i \leq n}$ of closed points of A, the theorem of the square (6.9) shows that

$$\sum_i (Z_{i, a_i} + Z_{i, b_i} + Z_{i, -a_i - b_i}) \sim \sum_i 3Z_i = 3D.$$

Let a and b be distinct closed points of A. By (a), for some i, say $i = 1$, Z_i does not contain $b - a$. Choose $a_1 = a$. Then Z_{1, a_1} passes through a but not b. The sets

$$\{b_1 | Z_{1, b_1} \text{ passes through } b\},$$

$$\{b_1 | Z_{1, -a_1 - b_1} \text{ passes through } b\},$$

are proper closed subsets of A. Therefore, it is possible to choose a b_1 that lies in neither. Similarly a_i and b_i for $i \geq 2$ can be chosen so that none of the Z_{i,a_i}, Z_{i,b_i}, or $Z_{i,-a_i-b_i}$ passes through b. Then a is in the support of $\sum(Z_{i,a_i} + Z_{i,b_i} + Z_{i,-a_i-b_i})$ but b is not, which shows that $|3D|$ separates points. The proof that it separates tangents is similar.

The final step is to show that if $A_{\bar{k}}$ is projective, then so also is A_k. Let D be an ample divisor on $A_{\bar{k}}$; then D is defined over a finite extension of k, and the following statements explain how to construct from D an ample divisor on A.

(a) Let D be a divisor on A; if $|D_{\bar{k}}|$ is very ample, then so also is $|D|$. (The map $A_{\bar{k}} \hookrightarrow \mathbb{P}^n$ defined by $|D_{\bar{k}}|$ is obtained by base change from that defined by $|D|$.)
(b) If $|D_1|$ and $|D_2|$ are ample, then so also is $|D_1 + D_2|$. (See [10, II, Ex. 7.5].)
(c) If D is a divisor on $A_{k'}$, where k' is a finite Galois extension of k with Galois group G, then $\sum \sigma D$, $\sigma \in G$, arises from a divisor on A. (This is obvious.)
(d) If D is a divisor on $A_{k'}$, where k' is a finite purely inseparable extension of k such that $k'^{p^m} \subset k$, then $p^m D$ arises from a divisor on A. (Regard D as the Cartier divisor defined by a family of pairs (f_i, U_i'), $f_i \in k'(A)$, and let U_i be the image of U_i' in A; then $k'(A)^{p^m} \subset k(A)$, and so the pairs $(f_i^{p^m}, U_i)$ define a divisor on A whose inverse image on $A_{k'}$ is $p^m D$.) □

Corollary 7.2. *Every abelian variety has a symmetric ample invertible sheaf.*

PROOF. According to the theorem, it has an ample invertible sheaf $\mathscr{L}$. As multiplication by -1 is an isomorphism, $(-1)^*\mathscr{L}$ is ample, and therefore $\mathscr{L} \otimes (-1)^*\mathscr{L}$ is ample [10, II, Ex. 7.5] and symmetric. □

Remark 7.3. If $\mathscr{L}$ is an ample invertible sheaf on A, then by definition $\mathscr{L}^n$ is very ample for some n. It is an important theorem that in fact $\mathscr{L}^3$ will be very ample (see [16, §17, p. 163]). The three is needed, as in the above proof, so that one can apply the theorem of the square.

§8. Isogenies

Let $f: A \to B$ be a homomorphism of abelian varieties. The kernel N of f in the sense of [22, §2] is a closed subgroup scheme of A of finite type over k. When k has characteristic zero, N is reduced [22, §3], and so its identity component N^0 is an abelian variety (possibly zero); in general, N will be an extension of a finite group scheme by an abelian variety. If f is surjective and has finite kernel then it is called an *isogeny*.

Proposition 8.1. *For a homomorphism $f: A \to B$ of abelian varieties, the following statements are equivalent:*

(a) f is an isogeny;
(b) $\dim A = \dim B$ and f is surjective;
(c) $\dim A = \dim B$ and $\operatorname{Ker}(f)$ is a finite group scheme;
(d) f is finite, flat, and surjective.

PROOF. As $f(A)$ is closed in B, the equivalence of the first three statements follows from the theorem on the dimension of fibres of morphisms; see [14, I.8].

Clearly (d) implies (a), and so assume (a). Because f is a homomorphism, the translation map t_b can be used to show that the (scheme-theoretic) fibre $f^{-1}(b)$ is isomorphic to $f^{-1}(0)_{k(b)}$. Therefore f is quasi-finite. It is also projective ([10, II, Ex. 4.9]), and this shows that it is finite ([10, III, Ex. 11.2]). The sheaf $f_* \mathcal{O}_A$ is a coherent $\mathcal{O}_B$-module, and $\dim_{k(b)}(f_* \mathcal{O}_A \otimes k(b)) = \dim_k(f_* \mathcal{O}_A \otimes k(0))$ is independent of b, and so (4.2d) shows that $f_* \mathcal{O}_A$ is locally free. □

The *degree* of an isogeny $f: A \to B$ is defined to be the order of the kernel of f (as a finite group scheme); equivalently, it is the rank of $f_* \mathcal{O}_A$ as a locally free $\mathcal{O}_B$-module. Clearly, $\deg(g \circ f) = \deg(g)\deg(f)$. Let $n = \deg(f)$; then $\operatorname{Ker}(f) \subset \operatorname{Ker}(n_A)$ and so n_A factors as $n_A = g \circ f$ with g an isogeny $B \to A$.

For an integer n we write n_A, or simply n, for the morphism $a \mapsto na: A \to A$.

Theorem 8.2. *Let A be an abelian variety of dimension g, and let $n > 0$ be an integer. Then $n_A: A \to A$ is an isogeny of degree n^{2g}; it is étale if and only if the characteristic of k does not divide n.*

PROOF. From (7.2) we know there is an ample symmetric invertible sheaf $\mathcal{L}$ on A, and according to (6.6) $n_A^* \mathcal{L} \approx \mathcal{L}^{n^2}$. The restriction of an ample invertible sheaf to a closed subscheme is again ample, and so the restriction of $n_A^* \mathcal{L}$ to $\operatorname{Ker}(n_A)$ is both trivial and ample. This is impossible unless $\operatorname{Ker}(n_A)$ has dimension zero. We have shown that n_A is an isogeny.

In proving that n_A has degree n^{2g} we shall use some elementary intersection theory from [21, IV.1]. Clearly we may assume k is algebraically closed.

Let V be a smooth projective variety of dimension g. If $D_1, \ldots, D_g$ are effective divisors on V such that $\bigcap D_i$ has dimension zero, then their intersection number is defined by the equations

$$(D_1, \ldots, D_g) = \sum_v (D_1, \ldots, D_g)_v \quad \text{(sum over } v \in \bigcap D_i\text{)},$$

$$(D_1, \ldots, D_g)_v = \dim_k(\mathcal{O}_{V,v}/(f_{1,v}, \ldots, f_{g,v})),$$

where $f_{i,v}$ is a local equation for D_i near v. The definition is extended by linearity to noneffective divisors whose components intersect properly. Then one checks that $(D_1, \ldots, D_g)$ is unchanged if each D_i is replaced by a linearly equivalent divisor and shows that this can be used to extend the definition to all g-tuples of divisors (loc. cit.). In particular $(D^g) = (D, D, \ldots)$ is defined.

Lemma 8.3. *Let V and W be smooth projective varieties of dimension g, and let*

$f: W \to V$ be a finite flat map of degree d. Then for any divisors $D_1, \ldots, D_g$ on V

$$(f^*D_1, \ldots, f^*D_g) = d(D_1, \ldots, D_g).$$

PROOF. It suffices to prove the equality in the case that the D_i are effective and $\bigcap D_i$ is finite. Let $v \in \bigcap D_i$. Then $(f_*\mathcal{O}_W) \otimes_{\mathcal{O}_V} \mathcal{O}_{V,v} = \prod_{f(w)=v} \mathcal{O}_{W,w}$, which is therefore a free $\mathcal{O}_{V,v}$-module of rank d. If $f_{i,v}$ is a local equation for D_i near v, then $f_{i,v} \circ f$ is a local equation for f^*D_i near each of the points in $f^{-1}(v)$. Therefore

$$\sum_{f(w)=v} (f^*D_1, \ldots, f^*D_g)_w = \sum_{f(w)=v} \dim_k(\mathcal{O}_{W,w}/(f_{1,v} \circ f, \ldots, f_{g,v} \circ f))$$

$$= \dim_k((\prod_{f(w)=v} \mathcal{O}_{W,w}) \otimes_{\mathcal{O}_{V,v}} (\mathcal{O}_{V,v}/(f_{1,v}, \ldots, f_{g,v})))$$

$$= d(D_1, \ldots, D_g)_v. \qquad \square$$

We apply this theory to a divisor D on A such that D is linearly equivalent to $(-1)^*D$ (i.e., such that $\mathscr{L}(D)$ is symmetric). Let $d = \deg(n_A)$. Then (8.3) shows that $((n_A^*D)^g) = d(D^g)$, but (6.6) shows that n_A^*D is linearly equivalent to $n^2 D$ and therefore that $((n_A^*D))^g = ((n^2 D)^g) = n^{2g}(D^g)$. These equalities imply $d = n^{2g}$ provided we can find a D for which $(D^g) \neq 0$. Choose D to be very ample (see (7.2)), and let $A \subset \mathbb{P}^N$ be the embedding defined by $|D|$. Then for any hyperplane sections $H_1, \ldots, H_g$ of A in $\mathbb{P}^N$, $(D^g) = (H_1, \ldots, H_g)$, and this is obviously positive.

It remains to prove the second assertion of the theorem. For a homomorphism $f: A \to B$, let $(df)_0: T_0(A) \to T_0(B)$ be the map on tangent spaces defined by f. It is neither surprising nor difficult to show that $d(f + g)_0 = (df)_0 + (dg)_0$ (cf. [16, §4, p. 42]). Therefore $(dn_A)_0$ is multiplication by n on the k-vector-space $T_0(A)$, and so $(dn_A)_0$ is an isomorphism (and n_A is étale at zero) if and only if the characteristic of k does not divide n. By using the translation maps, one shows that a homomorphism is étale at zero if and only if it is étale at all points. $\qquad \square$

Remark 8.4. If k is separably algebraically closed and n is not divisible by its characteristic, then the theorem says that the kernel $A_n(k)$ of $n: A(k) \to A(k)$ has n^{2g} elements. As this is also true for all n' dividing n, it follows that $A_n(k)$ is a free $\mathbb{Z}/n\mathbb{Z}$-module of rank $2g$. Therefore for all primes $l \neq \mathrm{char}(k)$, $T_l A \stackrel{\mathrm{df}}{=} \varprojlim A_{l^n}(k)$ is a free $\mathbb{Z}_l$-module of rank $2g$. Note that an element $a = (a_n)$ of $T_l A$ is a sequence $a_1, a_2, a_3, \ldots$ of elements of $A(k)$ such that $la_1 = 0$ and $la_n = a_{n-1}$ for all n.

When k is not separably algebraically closed then we define $T_l A = T_l A_{k_s}$. In this case there is a continuous action of $\mathrm{Gal}(k_s/k)$ on $T_l A$.

Remark 8.5. Assume that k is algebraically closed of characteristic $p \neq 0$. Then $A_p \stackrel{\mathrm{df}}{=} \mathrm{Ker}(p_A)$ is a finite group scheme of order p^{2g} killed by p. Therefore (see [22]) $A_p \approx (\mathbb{Z}/p\mathbb{Z})^r \times \mu_p^s \times \alpha_p^t$ for some r, s, t such that $r + s + t = 2g$. It is known that $r = s$ and $r \leq g$ (the inequality is a consequence of the fact that

$(dp_A)_0 = 0$). All values of r, s, and t are possible subject to these constraints. The case $r = g$ is the "general" case. For example when $g = 1$, then $r = 0$ only for supersingular elliptic curves and there are only finitely many of these over a given k [16, §22, p. 216].

§9. The Dual Abelian Variety: Definition

Let $\mathscr{L}$ be an invertible sheaf on A. Recall (6.8) that the map
$$\varphi_\mathscr{L}: A(k) \to \mathrm{Pic}(A), \qquad a \mapsto t_a^* \mathscr{L} \otimes \mathscr{L}^{-1}$$
is a homomorphism. Define
$$K_\mathscr{L} = \{a \in A \mid \text{the restriction of } m^*\mathscr{L} \otimes q^*\mathscr{L}^{-1} \text{ to } \{a\} \times A \text{ is trivial}\}.$$

According to (5.3), $K_\mathscr{L}$ is a closed subset of A, and we regard it as a reduced subscheme of A. For a in $A(k)$, the maps
$$A = \{a\} \times A \hookrightarrow A \times A \underset{q}{\overset{m}{\rightrightarrows}} A$$
$$\text{send} \quad P \longmapsto (a, P) \begin{matrix} \mapsto a + P \\ \mapsto P \end{matrix},$$
and so $m^*\mathscr{L} \otimes q^*\mathscr{L}^{-1}|_{\{a\} \times A}$ can be identified with $t_a^*\mathscr{L} \otimes \mathscr{L}^{-1}$ on A. Thus
$$K_\mathscr{L}(k) = \{a \in A(k) \mid t_a^*\mathscr{L} \approx \mathscr{L}\}.$$
Note that (6.2) implies that the definition of $K_\mathscr{L}$ commutes with a change of the base field.

Proposition 9.1. *Let $\mathscr{L}$ be an invertible sheaf such that $H^0(A, \mathscr{L}) \neq 0$. Then $\mathscr{L}$ is ample if and only if $K_\mathscr{L}$ has dimension zero, i.e., if and only if $t_a^*\mathscr{L} \approx \mathscr{L}$ on $A_{\bar{k}}$ for only a finite set of $a \in A(\bar{k})$.*

PROOF. Let s be a nonzero global section of $\mathscr{L}$, and let D be its divisor of zeros. Then D is effective and $\mathscr{L} = \mathscr{L}(D)$, and so the result [16, §6, p. 60] applies. □

We shall be more concerned in this section with the $\mathscr{L}$ of opposite type.

Proposition 9.2. *For $\mathscr{L}$ an invertible sheaf on A, the following conditions are equivalent:*

(a) $K_\mathscr{L} = A$;
(b) $t_a^*\mathscr{L} \approx \mathscr{L}$ on $A_{\bar{k}}$ for all $a \in A(\bar{k})$;
(c) $m^*\mathscr{L} \approx p^*\mathscr{L} \otimes q^*\mathscr{L}$.

PROOF. The equivalence of (a) and (b) follows from the remarks in the first paragraph of this section. Clearly (c) implies that for all $a \in A$,

$m^*\mathscr{L} \otimes q^*\mathscr{L}^{-1}|_{\{a\} \times A} \approx p^*\mathscr{L}|_{\{a\} \times A}$, which is trivial. Thus (c) implies (a), and the converse follows easily from the seesaw principle (5.2) because $m^*\mathscr{L} \otimes q^*\mathscr{L}^{-1}|_{\{a\} \times A}$ and $p^*\mathscr{L}|_{\{a\} \times A}$ are both trivial for all $a \in A$ and $m^*\mathscr{L} \otimes q^*\mathscr{L}^{-1}|_{A \times \{0\}} = \mathscr{L} = p^*\mathscr{L}|_{A \times \{0\}}$. □

Define $\text{Pic}^0(A)$ to be the group of isomorphism classes of invertible sheaves on A satisfying the conditions of (9.2). Note that if f and g are maps from some k-scheme S into A and $\mathscr{L} \in \text{Pic}^0(A)$, then

$$(f+g)^*\mathscr{L} \approx (f,g)^*m^*\mathscr{L} \stackrel{(c)}{\approx} (f,g)^*(p^*\mathscr{L} \otimes q^*\mathscr{L}) \approx f^*\mathscr{L} \otimes g^*\mathscr{L}.$$

From this it follows that $n^*\mathscr{L} \approx \mathscr{L}^n$ all $n \in \mathbb{Z}$, $\mathscr{L} \in \text{Pic}^0(A)$.

Remark 9.3. An invertible sheaf $\mathscr{L}$ lies in $\text{Pic}^0 A$ if and only if it occurs in an algebraic family containing a trivial sheaf, i.e., there exists a connected variety T and an invertible sheaf $\mathscr{M}$ on $A \times T$ such that, for some $t_0, t_1 \in T(k)$, $\mathscr{M}_{t_0}$ is trivial and $\mathscr{M}_{t_1} \approx \mathscr{L}$. The sufficiency of the condition can be proved directly using the theorem of the cube [16, §8, (vi)]; the necessity follows from the existence of the dual abelian variety (see below).

Roughly speaking, the dual (or Picard) variety $A^\vee$ of A is an abelian variety over k such that $A^\vee(\bar{k}) = \text{Pic}^0(A_{\bar{k}})$; moreover, there is to be an invertible sheaf (the Poincaré sheaf) $\mathscr{P}$ on $A \times A^\vee$ such that for all $a \in A^\vee(\bar{k})$, the inverse image of $\mathscr{P}$ on $A \times \{a\} = A_{\bar{k}}$ represents a as an element of $\text{Pic}^0(A_{\bar{k}})$. One usually normalizes $\mathscr{P}$ so that $\mathscr{P}|_{\{0\} \times A^\vee}$ is trivial.

The precise definition is as follows: an abelian variety $A^\vee$ is the *dual abelian variety* of A and an invertible sheaf $\mathscr{P}$ on $A \times A^\vee$ is the *Poincaré sheaf* if:

(a) $\mathscr{P}|_{\{0\} \times A^\vee}$ is trivial and $\mathscr{P}|_{A \times \{a\}}$ lies in $\text{Pic}^0(A_{k(a)})$ for all $a \in A^\vee$; and
(b) for every k-scheme T and invertible sheaf $\mathscr{L}$ on $A \times T$ such that $\mathscr{L}|_{\{0\} \times T}$ is trivial and $\mathscr{L}|_{A \times \{t\}}$ lies in $\text{Pic}^0(A_{k(t)})$ for $t \in T$, there is a unique morphism $f: T \to A^\vee$ such that $(1 \times f)^*\mathscr{P} \approx \mathscr{L}$.

Remark 9.4. (a) Clearly the pair $(A^\vee, \mathscr{P})$ is uniquely determined up to a unique isomorphism by these conditions.

(b) On applying condition (b) with $T = \text{spec } K$, K a field, one finds that $A^\vee(K) = \text{Pic}^0(A_K)$. In particular $A^\vee(\bar{k}) = \text{Pic}^0(A_{\bar{k}})$, and every element of $\text{Pic}^0(A_{\bar{k}})$ is represented exactly once in the family $(\mathscr{P}_a)_{a \in A^\vee(\bar{k})}$. The map $f: T \to A^\vee$ in condition (b) sends $t \in T(\bar{k})$ to the unique $a \in A^\vee(\bar{k})$ such that $\mathscr{L}_t \approx \mathscr{P}_a$.

(c) By using the description of tangent vectors in terms of maps from the dual numbers to $A^\vee$ [10, II, Ex. 2.8], one can show easily that there is a canonical isomorphism $T_0(A^\vee) \xrightarrow{\approx} H^1(A, \mathcal{O}_A)$; in particular, $\dim A^\vee = \dim A$. In the case that $k = \mathbb{C}$, there is an isomorphism $H^1(A, \mathcal{O}_A) \xrightarrow{\approx} H^1(A^{an}, \mathcal{O}_{A^{an}})$ (cohomology relative to the complex topology), and one shows that $\exp: T_0(A^\vee) \to A(\mathbb{C})$ induces an isomorphism $H^1(A^{an}, \mathcal{O}_{A^{an}})/H^1(A^{an}, \mathbb{Z}) \xrightarrow{\approx} A(\mathbb{C})$.

One expects of course that $A^{\vee\vee} = A$. Mumford [16] gives an elegant proof of this.

Proposition 9.5. *Let $\mathscr{P}$ be an invertible sheaf on the product $A \times B$ of two abelian varieties of the same dimension, and assume that the restrictions of $\mathscr{P}$ to $A \times \{0\}$ and $\{0\} \times B$ are both trivial. Then B is the dual of A and $\mathscr{P}$ is the Poincaré sheaf if and only if $\chi(A \times B, \mathscr{P}) = \pm 1$.*

PROOF. [16, §13, p. 131]. □

Note that the second condition is symmetric between A and B; therefore if $(B, \mathscr{P})$ is the dual of A, then $(A, s^*\mathscr{P})$ is the dual of B, where $s: B \times A \to A \times B$ is the morphism switching the factors.

§10. The Dual Abelian Variety: Construction

We can include only a brief sketch—for the details, see [16, §8, §§10–12].

Proposition 10.1. *Let $\mathscr{L}$ be an invertible sheaf on A; then the image of $\varphi_{\mathscr{L}}: A(k) \to \mathrm{Pic}(A)$ is contained in $\mathrm{Pic}^0(A)$; if $\mathscr{L}$ is ample and k is algebraically closed, then $\varphi_{\mathscr{L}}$ maps onto $\mathrm{Pic}^0(A)$.*

PROOF. Let $b \in A(k)$; in order to show that $\varphi_{\mathscr{L}}(b)$ is in $\mathrm{Pic}^0(A)$, we have to check that $t_a^*(\varphi_{\mathscr{L}}(b)) = \varphi_{\mathscr{L}}(b)$ for all $a \in A(\bar{k})$. But

$$t_a^*(\varphi_{\mathscr{L}}(b)) = t_a^*(t_b^*\mathscr{L} \otimes \mathscr{L}^{-1}) = t_{a+b}^*\mathscr{L} \otimes (t_a^*\mathscr{L})^{-1},$$

which the theorem of the square (6.7) shows to be isomorphic to

$$t_b^*\mathscr{L} \otimes \mathscr{L}^{-1} = \varphi_{\mathscr{L}}(b).$$

This shows that $\varphi_{\mathscr{L}}$ maps into $\mathrm{Pic}^0(A)$, and for the proof that it maps onto, we refer the reader to [16, §8, p. 77]. □

Let $\mathscr{L}$ be an invertible sheaf on A, and consider

$$\mathscr{L}^* = m^*\mathscr{L} \otimes p^*\mathscr{L}^{-1} \otimes q^*\mathscr{L}^{-1}$$

on $A \times A$. Then $\mathscr{L}^*|_{\{0\} \times A} = \mathscr{L} \otimes \mathscr{L}^{-1}$, which is trivial, and for a in $A(\bar{k})$, $\mathscr{L}^*|_{A \times \{a\}} = t_a^*\mathscr{L} \otimes \mathscr{L}^{-1} = \varphi_{\mathscr{L}}(a)$, which, as we have just seen, lies in $\mathrm{Pic}^0(A_{\bar{k}})$. Therefore, if $\mathscr{L}$ is ample, then $\mathscr{L}^*$ defines a family of sheaves on A parametrized by A such that each element of $\mathrm{Pic}^0(A_{\bar{k}})$ is represented by $\mathscr{L}_a^*$ for a (nonzero) finite number of a in $A(\bar{k})$. Consequently, if $(A^{\vee}, \mathscr{P})$ exists, then there is a unique isogeny $\varphi: A \to A^{\vee}$ such that $(1 \times \varphi)^*\mathscr{P} = \mathscr{L}^*$. Moreover $\varphi = \varphi_{\mathscr{L}}$, and the fibres of $A(\bar{k}) \to A^{\vee}(\bar{k})$ are the equivalence classes for the relation "$a \sim a'$ if and only if $\mathscr{L}_a \approx \mathscr{L}_{a'}$".

In characteristic zero, we even know what the kernel of φ as a finite subgroup scheme of A must be because it is determined by its underlying set: it equals $K_\mathscr{L}$ with its unique reduced subscheme structure. Therefore, in this case we define $A^\vee$ to be the quotient $A/K_\mathscr{L}$ (see [16, §7, p. 66 or §12, p. 111] for the construction of quotients). The action of $K_\mathscr{L}$ on the second factor of $A \times A$ lifts to an action on $\mathscr{L}^*$ over $A \times A$, and on forming the quotient we obtain a sheaf $\mathscr{P}$ on $A \times A^\vee$ such that $(1 \times \varphi_\mathscr{L})^*\mathscr{P} = \mathscr{L}^*$.

Assume further that k is algebraically closed. It easy to check that the pair $(A^\vee, \mathscr{P})$ just constructed has the correct universal property for families of sheaves $\mathscr{M}$ parametrized by normal k-schemes. Let $\mathscr{M}$ on $A \times T$ be such a family, and let $\mathscr{F}$ be the invertible sheaf $q_{12}^*\mathscr{M} \otimes q_{13}^*\mathscr{P}^{-1}$ on $A \times T \times A^\vee$, where q_{ij} is the projection onto the (i,j)th factor. Then $\mathscr{F}|_{A \times (t,b)} \approx \mathscr{M}_t \otimes \mathscr{P}_b^{-1}$, and so if we let Γ denote the closed subset of $T \times A^\vee$ of points (t, b) such $\mathscr{F}|_{A \times (t,b)}$ is trivial, then $\Gamma(k)$ is the graph of a map $T(k) \to A^\vee(k)$ sending a point t to the unique point b such that $\mathscr{P}_b \approx \mathscr{F}_t$. Regard Γ as a closed reduced subscheme of $T \times A^\vee$. Then the projection $\Gamma \to T$ has separable degree 1 because it induces a bijection on points (see [21, II, 5]). As k has characteristic zero, it must in fact have degree 1, and now the original form of Zariski's Main Theorem [14, III.9, p. 413] shows that $\Gamma \to T$ is an isomorphism. The morphism $f: T \approx \Gamma \xrightarrow{q} A^\vee$ has the property that $(1 \times f)^*\mathscr{P} = \mathscr{M}$, as required.

When k has nonzero characteristic, then $A^\vee$ is still the quotient of A by a subgroup $\mathscr{K}_\mathscr{L}$ having support $K_\mathscr{L}$, but $\mathscr{K}_\mathscr{L}$ need not be reduced. Instead one defines $\mathscr{K}_\mathscr{L}$ to be the maximal *subscheme* of A such that the restriction of $m^*\mathscr{L} \otimes q^*\mathscr{L}^{-1}$ to $\mathscr{K}_\mathscr{L} \times A$ defines a trivial family on A (see 5.5), and takes $A^\vee = A/\mathscr{K}_\mathscr{L}$. The proof that this has the correct universal property is similar to the above, but involves much more.

§11. The Dual Exact Sequence

Let $f: A \to B$ be a homomorphism of abelian varieties, and let $\mathscr{P}_B$ be the Poincaré sheaf on $B \times B^\vee$. The invertible sheaf $(f \times 1)^*\mathscr{P}_B$ on $A \times B^\vee$ gives rise to a homomorphism $f^\vee: B^\vee \to A^\vee$ such that $(1 \times f^\vee)^*\mathscr{P}_A \approx (f \times 1)^*\mathscr{P}_B$. On points f is simply the map $\mathrm{Pic}^0(B) \to \mathrm{Pic}^0(A)$ sending the isomorphism class of an invertible sheaf on B to it inverse image on A.

Theorem 11.1. *If $f: A \to B$ is an isogeny with kernel N, then $f^\vee: B^\vee \to A^\vee$ is an isogeny with kernel $N^\vee$, the Cartier dual of N. In other words, the exact sequence*

$$0 \to N \to A \to B \to 0$$

gives rise to a dual exact sequence

$$0 \to N^\vee \to B^\vee \to A^\vee \to 0.$$

PROOF. See [16, §15, p. 143]. □

There is another approach to this theorem which offers a different insight. Let $\mathscr{L}$ be an invertible sheaf on A whose class is in $\text{Pic}^0(A)$, and let L be the line bundle associated with $\mathscr{L}$. The isomorphism $p^*\mathscr{L} \otimes q^*\mathscr{L} \to m^*\mathscr{L}$ of (9.2) gives rise to a map $m_L: L \times L \to L$ lying over $m: A \times A \to A$. The absence of nonconstant regular functions on A forces numerous compatibility properties of m_L, which are summarized by the following statement.

Proposition 11.2. *Let $G(\mathscr{L})$ denote L with the zero section removed; then, for some k-rational point e of $G(\mathscr{L})$, m_L defines on $G(\mathscr{L})$ the structure of a commutative group variety with identity element e relative to which $G(\mathscr{L})$ is an extension of A by $\mathbb{G}_m$.*

Thus $\mathscr{L}$ gives rise to an exact sequence

$$E(\mathscr{L}): 0 \to \mathbb{G}_m \to G(\mathscr{L}) \to A \to 0.$$

The commutative group varieties over k form an abelian category, and so it is possible to define $\text{Ext}^1_k(A, \mathbb{G}_m)$ to be the group of classes of extensions of A by $\mathbb{G}_m$ in this category. We have:

Proposition 11.3. *The map $\mathscr{L} \mapsto E(\mathscr{L})$ defines an isomorphism*

$$\text{Pic}^0(A) \to \text{Ext}^1_k(A, \mathbb{G}_m).$$

Proofs of these results can be found in [20, VII, §3]. They show that the sequence

$$0 \to N^{\vee}(k) \to B^{\vee}(k) \to A^{\vee}(k)$$

can be identified with the sequence of Exts

$$0 \to \text{Hom}_k(N, \mathbb{G}_m) \to \text{Ext}^1_k(B, \mathbb{G}_m) \to \text{Ext}^1_k(A, \mathbb{G}_m).$$

(The reason for the zero at the left of the second sequence is that $\text{Hom}_k(A, \mathbb{G}_m) = 0$.)

The isomorphism in (11.3) extends to any base [17, III.18]. This means that if we let $\mathscr{E}xt^r$ denote Ext in the category of sheaves on the flat site over $\text{spec}(k)$ (see [13, III.1.5(e)]), then $A^{\vee}$ can be identified with the sheaf $\mathscr{E}xt^1(A, \mathbb{G}_m)$, and the exact sequence

$$0 \to N^{\vee} \to B^{\vee} \to A^{\vee} \to 0$$

can be identified with

$$0 \to \mathscr{H}om(N, \mathbb{G}_m) \to \mathscr{E}xt^1(B, \mathbb{G}_m) \to \mathscr{E}xt^1(A, \mathbb{G}_m) \to 0.$$

§12. Endomorphisms

The main result in this section is that $\text{End}^0(A) \stackrel{\text{df}}{=} \text{End}(A) \otimes \mathbb{Q}$ is a finite-dimensional semisimple algebra over $\mathbb{Q}$. As in the classical case, the semisimplicity follows from the existence of approximate complements for abelian subvarieties. If W is a subspace of a vector space V, one way of constructing a complement W' for W is to choose a nondegenerate bilinear form on V and take $W' = W^\perp$; equivalently, choose an isomorphism $V \to \check{V}$ and take W' to be the kernel of $V \to \check{V} \to \check{W}$. The same method works for abelian varieties.

Proposition 12.1. *Let B be an abelian subvariety of A; then there is an abelian variety $B' \subset A$ such that $B \cap B'$ is finite and $B + B' = A$, i.e., such that $B \times B' \to A$ is an isogeny.*

PROOF. Choose an ample sheaf $\mathscr{L}$ on A and define B' to be the reduced subscheme of the zero component of the kernel of $A \xrightarrow{\phi_\mathscr{L}} A^\vee \to B^\vee$; this is an abelian variety. From the theorem on the dimension of fibres of morphisms, $\dim B' \geq \dim A - \dim B$. The restriction of the morphism $A \to B^\vee$ to B is $\phi_{\mathscr{L}|B}: B \to B^\vee$, which has finite kernel because $\mathscr{L}|B$ is ample. Therefore $B \cap B'$ is finite, and so $B \times B' \to A$ is an isogeny. □

Define an abelian variety to be *simple* if it has no proper nonzero abelian subvarieties. Then, as in the classical case, each abelian variety A is isogenous to a product $\prod A_i^{r_i}$ of powers of nonisogenous simple abelian varieties A_i; the r_i are uniquely determined and the A_i are uniquely determined up to isogeny. Each $\mathrm{End}^0(A_i)$ is a skew field, $\mathrm{End}^0(A_i^{r_i})$ is equal to the matrix algebra $M_{r_i}(\mathrm{End}^0(A_i))$, and $\mathrm{End}^0(A) = \prod \mathrm{End}^0(A_i^{r_i})$.

Lemma 12.2. *For any prime $l \neq \mathrm{char}(k)$, the natural map*

$$\mathrm{Hom}(A, B) \to \mathrm{Hom}_{\mathbb{Z}_l}(T_l A, T_l B)$$

is injective; in particular, $\mathrm{Hom}(A, B)$ is torsion free.

PROOF. Let $\varphi: A \to B$ be a homomorphism such that $T_l \varphi = 0$; then $\varphi(A_{l^n}(\bar{k})) = 0$ for all n. For any simple abelian subvariety A' of A, this implies that the kernel of $\varphi|A'$ is not finite and therefore must equal the whole of A'. It follows that $\varphi = 0$. □

A function $f: V \to K$ on a vector space V over a field K is said to be a (*homogeneous*) *polynomial function* of degree d if for every finite linearly independent set $\{e_1, \ldots, e_n\}$ of elements of V, $f(x_1 e_1 + \cdots + x_n e_n)$ is a (homogenous) polynomial function of degree d in the x_i with coefficients in K.

Lemma 12.3. *Assume K is infinite, and let $f: V \to K$ be a function such that $f(xv + w)$ is a polynomial in x with coefficients in K, for all v, w in V; then f is a polynomial function.*

PROOF. We show by induction on n that, for every subset $\{v_1, \ldots, v_n, w\}$ of V, $f(x_1 v_1 + \cdots + x_n v_n + w)$ is a polynomial in the x_i. For $n = 1$, this is true by hypothesis; assume it for $n - 1$. The original hypothesis applied with $v = v_n$ shows that

$$f(x_1 v_1 + \cdots + x_n v_n + w) = a_0(x_1, \ldots, x_{n-1}) + \cdots + a_d(x_1, \ldots, x_{n-1}) x_n^d$$

for some d, with the a_i functions $k^{n-1} \to k$. Choose distinct elements $c_0, \ldots, c_d$ of K; on solving the system of linear equations

$$f(x_1 v_1 + \cdots + x_{n-1} v_{n-1} + c_j v_n + w) = \sum a_i(x_1, \ldots, x_{n-1}) c_j^i,$$

$$j = 0, 1, \ldots, d,$$

for a_i, we obtain an expression for a_i as a linear combination of the terms $f(x_1 v_1 + \cdots + x_{n-1} v_{n-1} + c_j v_n + w)$, which the induction assumption says are polynomials in $x_1, \ldots, x_{n-1}$. □

Let A be an abelian variety of dimension g over k. For $\varphi \in \text{End}(A)$, we define deg φ to be the degree of φ in the sense of Section 8 if φ is an isogeny and otherwise we set deg $\varphi = 0$. As $\deg(n\varphi) = \deg n_A \deg \varphi = n^{2g} \deg \varphi$, we can extend this notion to all of $\text{End}^0(A)$ by setting $\deg \varphi = n^{-2g} \deg(n\varphi)$ if $n\varphi \in \text{End}(A)$.

Proposition 12.4. *The function $\varphi \mapsto \deg \varphi : \text{End}^0(A) \to \mathbb{Q}$ is a homogeneous polynomial function of degree $2g$ on $\text{End}^0(A)$.*

PROOF. As $\deg(n\varphi) = n^{2g} \deg \varphi$, the lemma shows that it suffices to prove that $\deg(n\varphi + \psi)$ is a polynomial of degree $\leq 2g$ in n for $n \in \mathbb{Z}$ and fixed φ, $\psi \in \text{End}(A)$. Let D be a very ample divisor on A, and let $D_n = (n\varphi + \psi)^*D$. Then (see (8.3)), $\deg(n\varphi + \psi)(D^g) = (D_n^g)$, where $g = \dim A$, and so it suffices to prove that (D_n^g) is a polynomial of degree $\leq 2g$ in n. Corollary (6.5) applied to the maps $n\varphi + \psi, \varphi, \varphi : A \to A$ and the sheaf $\mathscr{L} = \mathscr{L}(D)$ shows that

$$D_{n+2} - 2D_{n+1} - (2\varphi)^*D + D_n + 2(\varphi^*D) \sim 0,$$

i.e., $\quad D_{n+2} - 2D_{n+1} + D_n = D'$, where $D' = 2(\varphi^*D) - (2\varphi)^*D$.

An induction argument now shows that

$$D_n = \frac{n(n-1)}{2} D' + nD_1 - (n-1)D_0$$

and so

$$(D_n^g) = \left(\frac{n(n-1)}{2}\right)^g (D'^g) + \cdots$$

is a polynomial in n of degree $\leq 2g$. □

Theorem 12.5. *For any abelian varieties A and B, $\text{Hom}(A, B)$ is a free $\mathbb{Z}$-module of finite rank $\leq 4 \dim A \dim B$; for each prime $l \neq \text{char}(k)$, the natural map*

$$\text{Hom}(A, B) \otimes \mathbb{Z}_l \to \text{Hom}(T_l A, T_l B)$$

is injective with torsion-free cokernel.

PROOF. Clearly it suffices to prove the second statement.

Lemma 12.6. *Let $\varphi \in \text{Hom}(A, B)$; if φ is divisible by l^n in $\text{Hom}(T_l A, T_l B)$, then it is divisible by l^n in $\text{Hom}(A, B)$.*

PROOF. The hypothesis implies that φ is zero on $A_{l^n}(\bar{k})$. As A_{l^n} is an étale subgroup scheme of A, this means that φ is zero on A_{l^n} and therefore factors as $\varphi = \varphi' \circ l^n$:

$$0 \to A_{l^n} \to A \xrightarrow{l^n} A \to 0$$
$$\searrow_\varphi \quad \downarrow \varphi'$$
$$B \qquad \square$$

Lemma 12.7. *If A is simple, then $\text{End}(A) \otimes \mathbb{Z}_l \to \text{End}(T_l A)$ is injective.*

PROOF. We have to show that if $e_1, \ldots, e_r$ are linearly independent over $\mathbb{Z}$ in $\text{End}(A)$, then $T_l(e_1), \ldots, T_l(e_r)$ are linearly independent over $\mathbb{Z}_l$ in $\text{End}(T_l A)$. Let P be the polynomial function on $\text{End}^0(A)$ such that $P(\varphi) = \deg(\varphi)$ for all φ. Note that every nonzero element φ of $\text{End}(A)$ is an isogeny, and therefore $P(\varphi)$ is a positive integer. Let M be the $\mathbb{Z}$-submodule of $\text{End}^0(A)$ generated by the e_i. The map $P: \mathbb{Q}M \to \mathbb{Q}$ is continuous for the real topology, and so $U = \{v | P(v) < 1\}$ is an open neighborhood of 0. As $(\mathbb{Q}M \cap \text{End } A) \cap U = 0$, we see that $\mathbb{Q}M \cap \text{End}(A)$ is discrete in $\mathbb{Q}M$, and therefore is a finitely generated $\mathbb{Z}$-module. It follows that:

(∗)   there exists an integer N such that $N(\mathbb{Q}M \cap \text{End } A) \subset M$.

Suppose that $T_l(e_1), \ldots, T_l(e_r)$ are linearly dependent, so that there exist $a_i \in \mathbb{Z}_l$, not all divisible by l, such that $\sum a_i T_l(e_i) = 0$. Choose integers n_i close to the a_i for the l-adic topology. Then $T_l(\sum n_i e_i) = \sum n_i T_l(e_i)$ is divisible by a high power of l in $\text{End}(T_l A)$, and so $\sum n_i e_i$ is divisible by a high power of l in $\text{End}(A)$. This contradicts (∗) when the power is sufficiently great, because then, for some m, $(N/l^m) \sum n_i e_i$ will lie in $N(\mathbb{Q}M \cap \text{End } A)$ but not M. $\square$

We are now ready to prove (12.5). Because $\text{Hom}(A, B)$ and $\text{Hom}(T_l A, T_l B)$ are direct summands of $\text{End}(A \times B)$ and $\text{End}(T_l(A \times B))$, it suffices to prove (12.5) in the case that $A = B$. Lemma 12.7 shows that $\text{End}^0(A)$ is finite dimensional over $\mathbb{Q}$ if A is simple, and this implies that it is finite dimensional for all A. It follows that $\text{End}(A)$ is finitely generated over $\mathbb{Z}$ because it is obviously torsion-free. Clearly now condition (∗) holds, and so the same argument as above shows that $\text{End}(A) \otimes \mathbb{Z}_l \to \text{End}(T_l A)$ is injective. Lemma 12.6 shows that its cokernel is torsion-free. $\square$

Define the *Néron–Severi group* $\text{NS}(A)$ of an abelian variety to be the quotient group $\text{Pic}(A)/\text{Pic}^0(A)$. Clearly $\mathscr{L} \mapsto \varphi_{\mathscr{L}}$ defines an injection $\text{NS}(A) \hookrightarrow \text{Hom}(A, A^\vee)$, and so (12.5) has the following consequence.

Corollary 12.8. *The Néron–Severi group of an abelian variety is a free $\mathbb{Z}$-module of finite rank.*

Proposition 12.4 shows that, for each α in $\operatorname{End}^0(A)$, there is a polynomial $P_\alpha(X) \in \mathbb{Q}[X]$ of degree $2g$ such that, for all rational numbers r, $P_\alpha(r) = \deg(\alpha - r_A)$. Let $\alpha \in \operatorname{End}(A)$, and let D be an ample symmetric divisor on A; then the calculation in the proof of (12.4) shows that

$$P_\alpha(-n) = \deg(\alpha + n) = (D_n^g)/(D^g),$$

where $D_n = (n(n-1)/2)D' + n(\alpha + 1_A)^*D - (n-1)\alpha^*D$, with

$$D' = 2D - 2_A^*D \sim 2D.$$

In particular, we see that P_α is monic and that it has integer coefficients when $\alpha \in \operatorname{End}(A)$. We call P_α the *characteristic polynomial* of α and we define the *trace* of α by the equation

$$P_\alpha(X) = X^{2g} - \operatorname{Tr}(\alpha)X^{2g-1} + \cdots + \deg(\alpha).$$

Proposition 12.9. *For all $l \neq \operatorname{char}(k)$, $P_\alpha(X)$ is the characteristic polynomial of α acting on $T_l A \otimes \mathbb{Q}_l$; hence the trace and degree of α are the trace and determinant of α acting on $T_l A \otimes \mathbb{Q}_l$.*

PROOF. We need two elementary lemmas.

Lemma 12.10. *Let $P(X) = \prod(X - a_i)$ and $Q(X) = \prod(X - b_i)$ be monic polynomials of the same degree with coefficients in $\mathbb{Q}_l$; if $|\prod F(a_i)|_l = |\prod F(b_i)|_l$ for all $F \in \mathbb{Z}[T]$, then $P = Q$.*

PROOF. See [12, VII, 1, Lemma 1]. □

Lemma 12.11. *Let E be an algebra over a field K, and let $\delta: E \to K$ be a polynomial function on E (regarded as a vector space over K) such that $\delta(\alpha\beta) = \delta(\alpha)\delta(\beta)$ for all $\alpha, \beta \in E$. Let $\alpha \in E$, and let $P = \prod(X - a_i)$ be the polynomial such that $P(x) = \delta(\alpha - x)$. Then $\delta(F(\alpha)) = \pm\prod F(a_i)$ for any $F \in K[T]$.*

PROOF. After extending K, we may assume that the roots $b_1, b_2, \ldots$ of F and of P lie in K; then

$$\delta(F(\alpha)) = \delta\left(\prod_j (\alpha - b_j)\right) = \prod_j \delta(\alpha - b_j) = \prod_j P(b_j) = \prod_{i,j} (b_j - a_i)$$
$$= \pm\prod_i F(a_i). \quad \square$$

We now prove (12.9). Clearly we may assume $k = k_s$. For any $\beta \in \operatorname{End}(A)$

$$|\deg(\beta)|_l = |\#(\operatorname{Ker}(\beta))|_l = \#(\operatorname{Ker}(\beta)(l))^{-1}$$
$$= \#(\operatorname{Coker}(T_l\beta))^{-1} = |\det(T_l\beta)|_l.$$

Consider $\alpha \in \operatorname{End}(A)$, and let $a_1, a_2, \ldots$ be the roots of P_α. Then for any polynomial $F \in \mathbb{Z}[T]$.

$$|\prod F(a_i)|_l = |\deg F(\alpha)|_l \qquad \text{by (12.11)}$$
$$= |\det T_l(F(\alpha))|_l$$
$$= |\prod F(b_i)|_l, \qquad \text{by (12.11)}$$

where the b_i are the eigenvalues of $T_l\beta$. By Lemma 12.10, this proves the proposition. □

Let D be a simple algebra of finite degree over $\mathbb{Q}$, and let K be the centre of D. The reduced trace and reduced norm of D over K satisfy

$$\operatorname{Tr}_{D/K}(\alpha) = [D:K]^{1/2}\operatorname{Trd}_{D/K}(\alpha), \quad \mathrm{N}_{D/K}(\alpha) = \operatorname{Nrd}_{D/K}(\alpha)^{[D:K]^{1/2}}, \quad \alpha \in D.$$

We shall always set $\operatorname{Trd} = \operatorname{Tr}_{K/\mathbb{Q}} \circ \operatorname{Trd}_{D/K}$ and $\operatorname{Nrd} = \mathrm{N}_{K/\mathbb{Q}} \circ \operatorname{Nrd}_{D/K}$. Let $V_1, \ldots, V_f$, $f = [K:\mathbb{Q}]$, be the nonisomorphic simple representations of D over $\overline{\mathbb{Q}}$; each has degree d where $d^2 = [D:K]$. The representation $V = \bigoplus V_i$ is called the reduced representation of D. For any α in D, $\operatorname{Trd}(\alpha) = \operatorname{Tr}(\alpha|V)$ and $\operatorname{Nrd}(\alpha) = \operatorname{Det}(\alpha|V)$.

Proposition 12.12. *Let D be a simple subalgebra of $\operatorname{End}^0(A)$ (this means D and $\operatorname{End}^0(A)$ have the same identity element), and let d, f, K, and V be as above. Then $2g/fd$ is an integer, and $\mathbb{Q}_l \otimes T_l A$ is a direct sum of $2g/fd$ copies of $\mathbb{Q}_l \otimes_{\mathbb{Q}} V$; consequently $\operatorname{Tr}(\alpha) = (2g/fd) \operatorname{Trd}(\alpha)$ and $\deg(\alpha) = \operatorname{Nrd}(\alpha)^{2g/fd}$ for all α in D.*

PROOF. Assume $\mathbb{Q}_l \otimes T_l V$ becomes isomorphic to $\bigoplus m_i V_i$ over $\overline{\mathbb{Q}}_l$, $m_i \geq 0$, and let σ_i be the embedding of K into $\overline{\mathbb{Q}}$ corresponding to V_i. Then, for any α in K, the characteristic polynomial of α on V_i is $(X - \sigma_i\alpha)^d$, and so $P_\alpha(X) = \prod (X - \sigma_i\alpha)^{dm_i}$. As $P_\alpha(X)$ has coefficients in $\mathbb{Q}$, it follows easily that the m_i must be equal. □

Remark 12.13. The group $\operatorname{NS}(A)$ is a functor of A. Direct calculations show that t_a acts as the identity on $\operatorname{NS}(A)$ for all a in $A(k)$ (because $\varphi_{t_a^*\mathscr{L}} = \varphi_{\mathscr{L}}$) and n acts as n^2 (because -1 acts as 1, and so $n^*\mathscr{L} = \mathscr{L}^{n^2}$ in $\operatorname{NS}(A)$ by (6.6)).

§13. Polarizations and the Cohomology of Invertible Sheaves

For many purposes the correct higher dimensional analogue of an elliptic curve is not an abelian variety but a polarized abelian variety.

A *polarization* λ on an abelian variety A is an isogeny $\lambda: A \to A^\vee$ such that $\lambda_{\overline{k}} = \varphi_{\mathscr{L}}$ for some ample invertible sheaf $\mathscr{L}$ on $A_{\overline{k}}$. The *degree* of a polarization is its degree as an isogeny. An abelian variety together with a polarization is called a *polarized abelian variety*; there is an obvious notion of a morphism of polarized abelian varieties. If λ has degree 1, then (A, λ) is said to belong to the *principal family* and λ is said to be a *principal polarization*.

Example 13.1. If A has dimension 1, then $\mathrm{NS}(A) = \mathbb{Z}$. For each integer d, there is a unique polarization of degree d^2; it is $\varphi_\mathscr{L}$ where $\mathscr{L} = \mathscr{L}(D)$ for D any effective divisor of degree d.

Remark 13.2. If λ is a polarization, there need not exist an $\mathscr{L}$ on A such that $\lambda = \varphi_\mathscr{L}$. Suppose, for example, that k is perfect and $G = \mathrm{Gal}(\bar{k}/k)$. By assumption, there is an $\mathscr{L}$ on $A_{\bar{k}}$ such that $\varphi_\mathscr{L} = \lambda_{\bar{k}}$. As $\lambda_{\bar{k}}$ is fixed by the action of G on $\mathrm{Hom}(A_{\bar{k}}, A_{\bar{k}}^\vee)$, the class $[\mathscr{L}]$ of $\mathscr{L}$ in $\mathrm{NS}(A_{\bar{k}})$ will also be fixed by G. Unfortunately this does not imply that $[\mathscr{L}]$ lifts to an element of $\mathrm{Pic}(A)$: there is a sequence of Galois cohomology groups

$$0 \to A^\vee(k) \to \mathrm{Pic}(A) \to \mathrm{NS}(A_{\bar{k}})^G \to H^1(G, A^\vee(\bar{k}))$$

and the obstruction in $H^1(G, A^\vee(\bar{k}))$ may be nonzero. However, if k is finite, an easy lemma [16, §21, p. 205] shows that $H^1(G, A^\vee(\bar{k})) = 0$ and therefore $\lambda = \varphi_\mathscr{L}$ for some $\mathscr{L}$ in $\mathrm{Pic}(A)$.

There is an important formula for the degree of a polarization, which it is convenient to state as part of a more general theorem.

Theorem 13.3. *Let $\mathscr{L}$ be an invertible sheaf on A, and write*

$$\chi(\mathscr{L}) = \sum (-1)^i \dim_k H^i(A, \mathscr{L}).$$

(a) *The degree of $\varphi_\mathscr{L}$ is $\chi(\mathscr{L})^2$.*
(b) *(Riemann–Roch). If $\mathscr{L} = \mathscr{L}(D)$, then $\chi(\mathscr{L}) = (D^g)/g!$.*
(c) *If $\dim K_\mathscr{L} = 0$, then there is exactly one integer r for which $H^r(A, \mathscr{L})$ is nonzero.*

PROOF. Combine [16, §16, p. 150] with (4.2a). □

Exercise 13.4. Verify (13.3) for elliptic curves using only the results in [10, IV].

Remark 13.5. The definition of polarization we have adopted is the one that is most useful for moduli questions. It differs from Weil's original notion (see [12, p. 193], [19, §5]).

§14. A Finiteness Theorem

Theorem 14.1. *Let k be a finite field, and let g and d be positive integers. Up to isomorphism, there are only finitely many abelian varieties A over k of dimension g possessing a polarization of degree d^2.*

PROOF. First assume dim $A = 1$. Then A automatically has a polarization of degree 1, defined by $\mathscr{L} = \mathscr{L}(P)$ for any $P \in A(k)$. The linear system $|3P|$ defines an embedding $A \hookrightarrow \mathbb{P}^2$, and the image is a cubic curve in $\mathbb{P}^2$. The cubic curve is determined by a polynomial of degree 3 in three variables. As there are only finitely many such polynomials with coefficients in k, we have shown that there are only finitely many isomorphism classes of A's.

The proof in the general case is essentially the same. By (13.2) we know there exists an ample invertible sheaf $\mathscr{L}$ on A such that $\varphi_{\mathscr{L}}$ is a polarization of degree d^2. Let $\mathscr{L} = \mathscr{L}(D)$; then, by (13.3), $\chi(\mathscr{L}) = d$ and $(D^g) = \chi(\mathscr{L})g! = d(g!)$. As $\mathscr{L}^3 = \mathscr{L}(3D)$, $\chi(\mathscr{L}^3) = ((3D)^g)/g! = 3^g d$. Moreover $\mathscr{L}^3$ is very ample (see (7.3)); in particular $H^0(A, \mathscr{L}^3) \neq 0$, and so (13.3c) shows that dim $H^0(A, \mathscr{L}^3) = \chi(\mathscr{L}^3) = 3^g d$. The linear system $|3D|$ therefore gives an embedding $A \hookrightarrow \mathbb{P}^{3^g d - 1}$.

Recall [21, I.6] that if V is a smooth variety of dimension g in $\mathbb{P}^N$, then the *degree* of V is $(D_1, \ldots, D_g)$ where $D_1, \ldots, D_g$ are hyperplane sections of V. Moreover, there is a polynomial, called the *Cayley* or *Chow form* of V,

$$F_V(a_0^{(0)}, \ldots, a_N^{(0)}; \ldots; a_0^{(g)}, \ldots, a_N^{(g)})$$

associated with V, which is a polynomial separately homogeneous of degree deg V in each of $g + 1$ sets of $N + 1$ variables. If we regard each set of variables $a_0^{(i)}, \ldots, a_N^{(i)}$ as defining a hyperplane,

$$H^{(i)}: a_0^{(i)} X_0 + \cdots + a_N^{(i)} X_N = 0,$$

then F_V is defined by the condition:

$$F_V(H^{(0)}, \ldots, H^{(g)}) = 0 \Leftrightarrow A \cap H^{(0)} \cap \cdots \cap H^{(g)} \text{ is nonempty.}$$

A theorem states that F_V uniquely determines V.

Returning to the proof of (14.1), we see that the degree of A in $\mathbb{P}^{3^g d - 1}$ is $((3D)^g) = 3^g d(g!)$. It is therefore determined by a polynomial F_A of degree $3^g d(g!)$ in each of $g + 1$ sets of $3^g d$ variables with coefficients in k. There are only finitely many such polynomials.

Remark 14.2. Of course, Theorem 14.1 is trivial if one assumes the existence of moduli varieties. However, everything used in the above proof (and much more) is required for the construction of moduli varieties.

Remark 14.3. The assumption that A has a polarization of a given degree plays a crucial role in the above proof. Nevertheless, we shall see in (18.9) below that it can be removed from the statement of the theorem.

§15. The Étale Cohomology of an Abelian Variety

The usual cohomology groups $H^r(A(\mathbb{C}), \mathbb{Z})$ of an abelian variety are described by the statements:

(a) A representation of $A(\mathbb{C})$ as a quotient $A(\mathbb{C}) = \mathbb{C}^g/L$ determines an isomorphism $H^1(A(\mathbb{C}), \mathbb{Z}) \xrightarrow{\approx} \text{Hom}(L, \mathbb{Z})$.

(b) The cup-product pairings define isomorphisms
$$\Lambda^r H^1(A(\mathbb{C}), \mathbb{Z}) \xrightarrow{\approx} H^r(A(\mathbb{C}), \mathbb{Z}) \quad \text{for all } r.$$

To prove (a), note that $\mathbb{C}^g$ is the universal covering space of $A(\mathbb{C})$, and that L is its group of covering transformations. Therefore, $\pi_1(A(\mathbb{C}), 0) = L$, and for any pointed manifold (M, m), $H^1(M, \mathbb{Z}) = \text{Hom}(\pi_1(M, m), \mathbb{Z})$. Statement (b) can be proved by observing that $A(\mathbb{C})$ is homeomorphic to a product of $2g$ circles and using the Künneth formula (see [16, §1, p. 3]), or by using the same argument as that given below for the étale topology.

Theorem 15.1. *Let A be an abelian variety of dimension g over an algebraically closed field k, and let l be a prime different from $\text{char}(k)$.*

(a) *There is a canonical isomorphism $H^1(A_{\text{et}}, \mathbb{Z}_l) \xrightarrow{\approx} \text{Hom}_{\mathbb{Z}_l}(T_l A, \mathbb{Z}_l)$.*

(b) *The cup-product pairings define isomorphisms*
$$\Lambda^r H^1(A_{\text{et}}, \mathbb{Z}_l) \xrightarrow{\approx} H^r(A_{\text{et}}, \mathbb{Z}_l) \quad \text{for all } r.$$

In particular, $H^r(A_{\text{et}}, \mathbb{Z}_l)$ is a free $\mathbb{Z}_l$-module of rank $\binom{2g}{r}$.

PROOF. If $\pi_1^{\text{et}}(A, 0)$ now denotes the étale fundamental group, then $H^1(A, \mathbb{Z}_l) = \text{Hom}_{\text{conts}}(\pi_1^{\text{et}}(A, 0), \mathbb{Z}_l)$. For each n, $l_A^n: A \to A$ is a finite étale covering of A with group of covering transformations $\text{Ker}(l_A^n) = A_{l^n}(k)$. By definition $\pi_1^{\text{et}}(A, 0)$ classifies such coverings, and therefore there is a canonical epimorphism $\pi_1^{\text{et}}(A, 0) \twoheadrightarrow A_{l^n}(k)$ (see [13, I.5]). On passing to the inverse limit, we get an epimorphism $\pi_1^{\text{et}}(A, 0) \twoheadrightarrow T_l A$, and consequently an injection $\text{Hom}_{\mathbb{Z}_l}(T_l A, \mathbb{Z}_l) \hookrightarrow H^1(A, \mathbb{Z}_l)$.

To proceed further we need to work with other coefficient groups. Let R be $\mathbb{Z}_l$, $\mathbb{F}_l$, or $\mathbb{Q}_l$, and write $H^*(A)$ for $\bigoplus_{r \geq 0} H^r(A_{\text{et}}, R)$. The cup-product pairing makes this into a graded, associative, anticommutative algebra. There is a canonical map $H^*(A) \otimes H^*(A) \to H^*(A \times A)$, which the Künneth formula shows to be an isomorphism when R is a field. In this case, the addition map $m: A \times A \to A$ defines a map
$$m^*: H^*(A) \to H^*(A \times A) = H^*(A) \otimes H^*(A).$$

Moreover, the map $a \mapsto (a, 0): A \to A \times A$ identifies $H^*(A)$ with the direct summand $H^*(A) \otimes H^0(A)$ of $H^*(A) \otimes H^*(A)$. As $m \circ (a \mapsto (a, 0)) = \text{id}$, the projection of $H^*(A) \otimes H^*(A)$ onto $H^*(A) \otimes H^0(A)$ sends $m^*(x)$ to $x \otimes 1$. As the same remark applies to $a \mapsto (0, a)$, this shows that
$$m^*(x) = x \otimes 1 + 1 \otimes x + \sum x_i \otimes y_i, \quad \deg(x_i), \deg(y_i) > 0.$$

Lemma 15.2. *Let H^* be a graded, associative, anticommutative algebra over a perfect field K. Assume that there is map $m^*: H^* \to H^* \otimes H^*$ satisfying the above identity. If $H^0 = K$ and $H^r = 0$ for all r greater than some integer d, then*

$\dim(H^1) \leq d$, and when equality holds, H^* is isomorphic to the exterior algebra on H^1.

PROOF. A fundamental structure theorem for Hopf algebras [3, Theorem 6.1] shows that H^* is equal to the associative algebra generated by certain elements x_i subject only to the relations imposed by the anticommutativity of H^* and the nilpotence of each x_i. The product of the x_i has degree $\sum \deg(x_i)$, from which it follows that $\sum \deg(x_i) \leq d$. In particular, the number of x_i of degree 1 is $\leq d$; as this number is equal to the dimension of H^1, this shows that its dimension is $\leq d$. When equality holds, all the x_i must have degree 1; moreover their squares must all be zero because otherwise there would be a nonzero element $x_1 x_2 \ldots x_i^2 \ldots x_d$ of degree $d + 1$. Hence H^* is identified with the exterior algebra on H^1. $\square$

When R is $\mathbb{Q}_l$ or $\mathbb{F}_l$, the conditions of the lemma are fulfilled with $d = 2g$ [13, VI, 1.1]. Therefore $H^1(A, \mathbb{Q}_l)$ has dimension $\leq 2g$. But $H^1(A, \mathbb{Q}_l) = H^1(A, \mathbb{Z}_l) \otimes \mathbb{Q}_l$, and so the earlier calculation shows that $H^1(A, \mathbb{Q}_l)$ has dimension $2g$. The lemma now shows that $H^r(A, \mathbb{Q}_l) = \Lambda^r(H^1(A, \mathbb{Q}_l))$, and, in particular, that its dimension is $\binom{2g}{r}$. This implies that $H^r(A, \mathbb{Z}_l)$ has rank $\binom{2g}{r}$. The exact sequence [13, V, 1.11]

$$\cdots \to H^r(A, \mathbb{Z}_l) \xrightarrow{l} H^r(A, \mathbb{Z}_l) \to H^r(A, \mathbb{F}_l) \to H^{r+1}(A, \mathbb{Z}_l) \xrightarrow{l} H^{r+1}(A, \mathbb{Z}_l) \to \cdots$$

now shows that $\dim(H^1(A, \mathbb{F}_l)) \geq 2g$, and so the lemma implies that this dimension equals $2g$ and that $\dim(H^r(A, \mathbb{F}_l)) = \binom{2g}{r}$. On looking at the exact sequence again, we see that $H^r(A, \mathbb{Z}_l)$ must be torsion-free for all r. Consequently, $\Lambda^r H^1(A, \mathbb{Z}_l) \to H^r(A, \mathbb{Z}_l)$ is injective because it becomes so when tensored with $\mathbb{Q}_l$, and it is surjective because it becomes so when tensored with $\mathbb{F}_l$. This completes the proof.

Remark 15.3. In the course of the above proof, we have shown that the maximal abelian l-quotient of $\pi_1^{\text{et}}(A, 0)$ is isomorphic to $T_l A$. In fact, it is known that $\pi_1^{\text{et}}(A, 0) = TA$, where $TA = \varprojlim A_n(k)$. In order to prove this one has to show that all finite étale coverings of A are isogenies. This is accomplished by the following theorem ([14, §18, p. 167]): *Let A be an abelian variety over an algebraically closed field, and let $f: B \to A$ be a finite étale covering with B connected; then it is possible to define on B the structure of an abelian variety relative to which f is an isogeny.*

Remark 15.4. We have shown that the following three algebras are isomorphic:

(i) $H^*(A, \mathbb{Z}_l)$ with its cup-product structure;
(ii) $\Lambda^* H^1(A, \mathbb{Z}_l)$ with its wedge-product structure;
(iii) the dual of $\Lambda^* T_l A$ with its wedge-product structure.

If we denote the pairing

$$T_l A \times H^1(A, \mathbb{Z}_l) \to \mathbb{Z}_l$$

by $\langle \cdot | \cdot \rangle$, then the pairing
$$\Lambda^r T_l A \times H^r(A, \mathbb{Z}_l) \to \mathbb{Z}_l$$
is determined by
$$(a_1 \wedge \cdots \wedge a_r, b_1 \cup \cdots \cup b_r) = \det(\langle a_i | b_j \rangle).$$
See [5, §8].

Remark 15.5. Theorem 15.1 is still true if k is only separably closed (see [13, II, 3.17]). If A is defined over a field k, then the isomorphism
$$\Lambda^* \operatorname{Hom}(T_l, \mathbb{Z}_l) \to H^*(A_{k_s}, \mathbb{Z}_l)$$
is compatible with the natural actions of $\operatorname{Gal}(k_s/k)$.

§16. Pairings

As we discussed in Section 11, if M and N denote the kernels of an isogeny f and its dual $f^\vee$, then there is a canonical pairing $M \times N \to \mathbb{G}_m$ which identifies each group scheme with the Cartier dual of the other. In the case that f is multiplication by m, $m_A: A \to A$, then $f^\vee$ is $m_{A^\vee}: A^\vee \to A^\vee$, and so the general theory gives a pairing $\bar{e}_m: A_m \times A_m^\vee \to \mathbb{G}_m$. If we assume further that m is not divisible by the characteristic of k, then this can be identified with a nondegenerate pairing of $\operatorname{Gal}(\bar{k}/k)$-modules
$$\bar{e}_m: A_m(\bar{k}) \times A_m^\vee(\bar{k}) \to \bar{k}^\times.$$
This pairing has a very explicit description. Let $a \in A_m(\bar{k})$ and let $a' \in A_m^\vee(\bar{k}) \subset \operatorname{Pic}^0(A_{\bar{k}})$. If a' is represented by the divisor D on $A_{\bar{k}}$, then $m_A^{-1} D$ is linearly equivalent to mD (see the paragraph following (9.2)), which is linearly equivalent to zero. Therefore there are functions f and g on $A_{\bar{k}}$ such that $mD = (f)$ and $m_A^{-1} D = (g)$. Since the divisor
$$(f \circ m_A) = m_A^{-1}((f)) = m_A^{-1}(mD) = m(m_A^{-1} D) = (g^m),$$
we see that $g^m/f \circ m_A$ is a constant function c on $A_{\bar{k}}$. In particular,
$$g(x+a)^m = cf(mx + ma) = cf(mx) = g(x)^m.$$
Therefore $g/g \circ t_a$ is a function on $A_{\bar{k}}$ whose mth power is one. This means that it is an mth root of 1 in $\bar{k}(A)$ and can be identified with an element of $\bar{k}$. It is shown in [16, §20, p. 184] that $\bar{e}_m(a, a') = g/g \circ t_a$.

Lemma 16.1. *Let m and n be integers not divisible by the characteristic of k. Then for all $a \in A_{mn}(\bar{k})$ and $a' \in A_{mn}^\vee(\bar{k})$,*
$$\bar{e}_{mn}(a, a')^n = \bar{e}_m(na, na').$$

PROOF. Let D represent a', and let $(mn)_A^{-1}(D) = (g)$ and $m_A^{-1}(nD) = (g')$. Then
$$(g' \circ n_A) = n_A^{-1}((g')) = n_A^{-1}(m_A^{-1}(nD)) = n(mn)_A^{-1}(D) = (g^n),$$

and so $g^n = c(g' \circ n_A)$ for some constant function c. Therefore
$$(g(x)/g(x+a))^n = g'(nx)/g'(nx + na),$$
and this equals $\bar{e}_m(na, na')$ for all x. □

Regard $\bar{e}_m$ as taking values in $\mu_m = \{\zeta \in \bar{k} | \zeta^m = 1\}$, and let $\mathbb{Z}_l(1) = \varprojlim \mu_{l^n}$ for l a prime not equal to the characteristic of k. (Warning: We sometimes write $\mathbb{Z}_l(1)$ additively and sometimes multiplicatively.) The lemma allows us to define a pairing $e_l: T_l A \times T_l A^\vee \to \mathbb{Z}_l(1)$ by the rule
$$e_l((a_n), (a'_n)) = (\bar{e}_{l^n}(a_n, a'_n)).$$
For a homomorphism $\lambda: A \to A^\vee$, we define pairings
$$\bar{e}_m^\lambda: A_m \times A_m \to \mu_m, \quad (a, a') \mapsto \bar{e}_m(a, \lambda a'),$$
$$e_l^\lambda: T_l A \times T_l A \to \mathbb{Z}_l(1), \quad (a, a') \mapsto e_l(a, \lambda a').$$
If $\lambda = \varphi_\mathscr{L}$, $\mathscr{L} \in \text{Pic}(A)$, then we write $\bar{e}_m^\mathscr{L}$ and $e_l^\mathscr{L}$ for $\bar{e}_m^\lambda$ and e_l^λ.

Lemma 16.2. *There are the following formulas: for a homomorphism $f: A \to B$,*
(a) $\bar{e}_m(a, f^\vee(b)) = \bar{e}_m(f(a), b)$, $a \in A_m$, $b \in B_m$;
(b) $e_l(a, f^\vee(b)) = e_l(f(a), b)$, $a \in T_l A$, $b \in T_l B$;
(c) $e_l^{f^\vee \circ \lambda \circ f}(a, a') = e_l^\lambda(f(a), f(a'))$, $a, a' \in T_l A$, $\lambda \in \text{Hom}(B, B^\vee)$;
(d) $e_l^{f^*\mathscr{L}}(a, a') = e_l^\mathscr{L}(f(a), f(a'))$, $a, a' \in T_l A$, $\mathscr{L} \in \text{Pic}(B)$.
Moreover,
(e) $\mathscr{L} \mapsto e_l^\mathscr{L}$ *is a homomorphism* $\text{Pic}(A) \to \text{Hom}(\Lambda^2 T_l A, \mathbb{Z}_l(1))$.

PROOF. Let a and b be as in (a); let the divisor D on B represent b, and let $m_B^{-1} D = (g)$. Then $\bar{e}_m(f(a), b) = g(x)/g(x + f(a))$ for all x. On the other hand, $f^{-1}D$ represents $f^\vee(b)$ on A, and $m_A^{-1}f^{-1}D = f^{-1}m_B^{-1}D = (g \circ f)$, and so $\bar{e}_m(a, f^\vee(b)) = g(f(x))/g(f(x) + f(a))$. This proves (a), and (b) and (c) follow immediately. Formula (d) follows from (c) because
$$\varphi_{f^*\mathscr{L}}(a) = t_a^* f^* \mathscr{L} \otimes f^* \mathscr{L}^{-1} = f^* t_{fa}^* \mathscr{L} \otimes f^* \mathscr{L}^{-1} = f^*(\varphi_\mathscr{L}(fa))$$
$$= f^\vee \circ \varphi_\mathscr{L} \circ f(a),$$
which shows that $\varphi_{f^*\mathscr{L}} = f^\vee \circ \varphi_\mathscr{L} \circ f$. Finally, (e) follows from the fact that $\varphi_{\mathscr{L} \otimes \mathscr{L}'} = \varphi_\mathscr{L} + \varphi_{\mathscr{L}'}$.

Example 16.3. Let A be an abelian variety over $\mathbb{C}$. The exact sequence of sheaves on $A(\mathbb{C})$ (here $\mathcal{O}_A$ denotes the sheaf of holomorphic functions on $A(\mathbb{C})$)
$$0 \to \mathbb{Z} \to \mathcal{O}_A \xrightarrow{e^{2\pi i(\cdot)}} \mathcal{O}_A^\times \to 0$$
gives rise to an exact sequence
$$H^1(A(\mathbb{C}), \mathbb{Z}) \to H^1(A(\mathbb{C}), \mathcal{O}) \to H^1(A(\mathbb{C}), \mathcal{O}^\times) \to H^2(A(\mathbb{C}), \mathbb{Z}) \to H^2(A(\mathbb{C}), \mathcal{O}).$$

As $H^1(A(\mathbb{C}), \mathcal{O}^\times) = \text{Pic}(A)$ and $H^1(A(\mathbb{C}), \mathcal{O})/H^1(A(\mathbb{C}), \mathbb{Z}) = A^\vee(\mathbb{C})$ (see (9.4c)), we can extract from this an exact sequence
$$0 \to \text{NS}(A) \to H^2(A(\mathbb{C}), \mathbb{Z}) \to H^2(A(\mathbb{C}), \mathcal{O}_A).$$

Let $\lambda \in \text{NS}(A)$, and let E^λ be its image in $H^2(A(\mathbb{C}), \mathbb{Z})$. Then (see Section 15) E^λ can be regarded as a skew-symmetric form on $H_1(A(\mathbb{C}), \mathbb{Z})$. It is a non-degenerate Riemann form if and only if λ is ample. As was explained above, λ induces a pairing e_l^λ, and it is shown in [16, §24, p. 237] that the diagram

$$\begin{array}{ccccc} E^\lambda: & H_1(A, \mathbb{Z}) \times H_1(A, \mathbb{Z}) & \to & \mathbb{Z} \\ & \downarrow \quad\quad\quad\quad \downarrow & & \downarrow \\ e_l^\lambda: & T_l A \quad\quad \times \quad\quad T_l A & \to & \mathbb{Z}_l(1) \end{array}$$

commutes with a minus sign if the maps $H^1(A(\mathbb{C}), \mathbb{Z}) \to T_l A$ are taken to be the obvious ones and $\mathbb{Z} \to \mathbb{Z}_l(1)$ is taken to be $m \mapsto \zeta^m$, $\zeta = (\ldots, e^{2\pi i/l^n}, \ldots)$; in other words, $e_l^\lambda(a, a') = \zeta^{-E^\lambda(a, a')}$.

In the remainder of this section, we shall show how étale cohomology can be used to give short proofs (except for the characteristic k part) of some important results concerning polarizations. Proofs not using étale cohomology can be found in [16, §§20, 23].

The family of exact sequences of sheaves
$$0 \to \mu_{l^n} \to \mathbb{G}_m \xrightarrow{l^n} \mathbb{G}_m \to 0,$$
$l \neq \text{char}(k)$, $n \geq 1$, plays the same role for the étale topology that the exponential sequence in (16.3) plays for the complex topology. As $\text{Pic}(A) = H^1(A, \mathbb{G}_m)$ (étale cohomology), these sequences give rise to cohomology sequences
$$0 \to \text{Pic}(A_{\bar k})/l^n \text{Pic}(A_{\bar k}) \to H^2(A_{\bar k}, \mu_{l^n}) \to H^2(A_{\bar k}, \mathbb{G}_m)_{l^n} \to 0.$$
Note that $\text{Pic}^0(A_{\bar k}) = A^\vee(\bar k)$ is divisible, and so $\text{Pic}(A_{\bar k})/l^n \text{Pic}(A_{\bar k}) = \text{NS}(A_{\bar k})/l^n \text{NS}(A_{\bar k})$. On passing to the inverse limit over these sequences, we get an exact sequence
$$0 \to \text{NS}(A_{\bar k}) \otimes \mathbb{Z}_l \to H^2(A_{\bar k}, \mathbb{Z}_l(1)) \to T_l H^2(A_{\bar k}, \mathbb{G}_m) \to 0,$$
where $T_l M$ for any group M is $\varprojlim M_{l^n}$. Note that $T_l M$ is always torsion-free. As in the above example, an element λ of $\text{NS}(A_{\bar k})$ defines a skew-symmetric pairing $E_l^\lambda: T_l A \times T_l A \to \mathbb{Z}_l(1)$, and one can show as in the previous case that $E_l^\lambda = -e_l^\lambda$ (in fact, this provides a convenient alternative definition of e_l^λ in the case that λ arises from an element of $\text{NS}(A_{\bar k})$).

We now assume that k is algebraically closed.

Theorem 16.4. *Let $f: A \to B$ be an isogeny of degree prime to the characteristic of k, and let $\lambda \in \text{NS}(A)$. Then $\lambda = f^*(\lambda')$ for some $\lambda' \in \text{NS}(B)$ if and only if, for* all l dividing $\deg(f)$, there exists an e_l in $\mathrm{Hom}(\Lambda^2 T_l B, \mathbb{Z}_l(1))$ such that $e_l^\lambda(a, a') = e_l(f(a), f(a'))$ all $a, a' \in T_l A$.

PROOF. The necessity is obvious from (16.2c). For the converse, consider for each $l \neq \mathrm{char}(k)$ the commutative diagram

$$\begin{array}{ccccccc} 0 & \to & \mathrm{NS}(A) \otimes \mathbb{Z}_l & \to & H^2(A, \mathbb{Z}_l(1)) & \to & T_l(H^2(A, \mathbb{G}_m)) \\ & & \uparrow & & \uparrow & & \uparrow \\ 0 & \to & \mathrm{NS}(B) \otimes \mathbb{Z}_l & \to & H^2(B, \mathbb{Z}_l(1)) & \to & T_l(H^2(B, \mathbb{G}_m)). \end{array}$$

The right-hand vertical arrow is injective because there exists an isogeny $f': B \to A$ such that $f \circ f'$ is multiplication by $\deg(f)$ on B (see Section 8) and $T_l(H^2(B, \mathbb{G}_m))$ is torsion-free. A diagram chase now shows that λ is in the image of $\mathrm{NS}(B) \otimes \mathbb{Z}_l \to \mathrm{NS}(A) \otimes \mathbb{Z}_l$ for all l dividing $\deg(f)$, and the existence of f' shows that it is in the image for all remaining primes. This implies that it is in the image of $\mathrm{NS}(B) \to \mathrm{NS}(A)$ because $\mathrm{NS}(A)$ is a finitely generated $\mathbb{Z}$-module. □

Corollary 16.5. *Assume $l \neq \mathrm{char}(k)$. An element λ of $\mathrm{NS}(A)$ is divisible by l^n if and only if e_l^λ is divisible by l^n in $\mathrm{Hom}(\Lambda^2 T_l A, \mathbb{Z}_l(1))$.*

PROOF. Apply the proposition to $l_A^n: A \to A$. □

Proposition 16.6. *Assume $\mathrm{char}(k) \neq 2, l$. A homomorphism $\lambda: A \to A^\vee$ is of the form $\varphi_{\mathscr{L}}$ for some $\mathscr{L} \in \mathrm{Pic}(A)$ if and only if e_l^λ is skew-symmetric.*

PROOF. If λ is in the subgroup $\mathrm{NS}(A)$ of $\mathrm{Hom}(A, A^\vee)$, we already know that e_l^λ is skew-symmetric. Conversely, suppose e_l^λ is skew-symmetric, and let $\mathscr{L}$ be the pull-back of the Poincaré sheaf $\mathscr{P}$ by $(1, \lambda): A \to A \times A^\vee$. For all $a, a' \in T_l A$,

$$\begin{aligned} e_l(a, \varphi_{\mathscr{L}} a') = e_l^{\mathscr{L}}(a, a') &= e_l^{\mathscr{P}}((a, \lambda a), (a', \lambda a')) \quad \text{(by 16.2d)} \\ &= e_l(a, \lambda a') - e_l(a', \lambda a) \quad \text{(see the next lemma)} \\ &= e_l^\lambda(a, a') - e_l^\lambda(a', a) \\ &= 2 e_l^\lambda(a, a') \quad \text{(because e_l^λ is skew-symmetric)} \\ &= e_l(a, 2\lambda a'). \end{aligned}$$

As e_l is nondegenerate, this shows that $2\lambda = \varphi_{\mathscr{L}}$, and (16.5) shows that $\mathscr{L}$ is divisible by 2 in $\mathrm{NS}(A)$. □

Lemma 16.7. *Let $\mathscr{P}$ be the Poincaré sheaf on $A \times A^\vee$. Then*

$$e_l^{\mathscr{P}}((a, b), (a', b')) = e_l(a, b') - e_l(a', b)$$

for $a, a' \in T_l A$ and $b, b' \in T_l A^\vee$.

PROOF. Because $\mathbb{Z}_l(1)$ is torsion-free, it suffices to prove the identity for b and b' in a subgroup of finite index in $T_l A^\vee$. Therefore we can assume that $b = \lambda c$ and $b' = \lambda c'$ for some polarization $\lambda = \varphi_{\mathscr{L}}$ of A and elements c and c' of $T_l A$. From Section 10 we know that $(1 \times \lambda)^* \mathscr{P} = m^* \mathscr{L} \otimes p^* \mathscr{L}^{-1} \otimes q^* \mathscr{L}^{-1}$, and so

$$e_l^{\mathscr{P}}((a, b), (a', b')) = e_l^{(1 \times \lambda)^* \mathscr{P}}((a, c), (a', c'))$$
$$= e_l^{\mathscr{L}}(a + c, a' + c') - e_l^{\mathscr{L}}(a, a') - e_l^{\mathscr{L}}(c, c')$$
$$= e_l^{\mathscr{L}}(a, c') - e_l^{\mathscr{L}}(a', c)$$
$$= e_l(a, b') - e_l(a', b). \qquad \square$$

For a polarization $\lambda: A \to A^\vee$, define

$$e^\lambda: \mathrm{Ker}(\lambda) \times \mathrm{Ker}(\lambda) \to \mu_m$$

as follows: suppose m kills $\mathrm{Ker}(\lambda)$, and let a and a' be in $\mathrm{Ker}(\lambda)$; choose a b such that $mb = a'$, and let $e^\lambda(a, a') = \bar{e}_m(a, \lambda b)$; this makes sense because $m(\lambda b) = \lambda(mb) = 0$. Also it is independent of the choice of b and m because if $mnb' = a'$ and $nc = a$, then

$$\bar{e}_{mn}(a, \lambda b') = \bar{e}_{mn}(c, \lambda b')^n = \bar{e}_m(a, \lambda n b') \qquad \text{(by 16.1)}$$

and so

$$\bar{e}_{mn}(a, \lambda b')/\bar{e}_m(a, \lambda b) = \bar{e}_m(a, \lambda(nb' - b)) = \bar{e}_m^\lambda(a, nb' - b)$$
$$= \bar{e}_m^\lambda(nb' - b, a)^{-1}$$
$$= 1 \quad \text{as } \lambda a = 0.$$

Let $a = (a_n)$ and $a' = (a'_n)$ be in $T_l A$. If $\lambda a_m = 0 = \lambda a'_m$ for some m, then

$$e^\lambda(a_m, a'_m) = \bar{e}_{l^m}(a_m, \lambda a'_{2m}) = \bar{e}_{l^{2m}}(a_{2m}, \lambda a'_{3m})^{l^m} = \bar{e}_{l^{2m}}^\lambda(a_{2m}, a'_{2m}).$$

Note that this implies that e^λ is skew-symmetric.

Proposition 16.8. *Let $f: A \to B$ be an isogeny of degree prime to $\mathrm{char}(k)$, and let $\lambda: A \to A^\vee$ be a polarization of A. Then $\lambda = f^*(\lambda')$ for some polarization λ' on B if and only if $\mathrm{Ker}(f) \subset \mathrm{Ker}(\lambda)$ and e^λ is trivial on $\mathrm{Ker}(f) \times \mathrm{Ker}(f)$.*

PROOF. We will assume the second condition and construct an e_l in $\mathrm{Hom}(\Lambda^2 T_l B, \mathbb{Z}_l(1))$ such that $e_l^\lambda(a, a') = e_l(fa, fa')$ for all a, a' in $T_l A$; then (16.4) will show the existence of λ'. Let $b, b' \in T_l B$; for some m there will exist $a, a' \in T_l A$ such that $l^m b = f(a)$ and $l^m b' = f(a')$. If we write $a = (a_n)$ and $a' = (a'_n)$, then these equations imply that $f(a_m) = 0 = f(a'_m)$, and therefore that a_m and a'_m are in $\mathrm{Ker}(\lambda)$ and that $e^\lambda(a_m, a'_m) = 0$. The calculation preceding the statement of the proposition now shows that $\bar{e}_{l^{2m}}^\lambda(a_{2m}, a'_{2m}) = 0$ and therefore that $e_l^\lambda(a, a')$ is divisible by l^{2m}. We can therefore define $e_l(b, b') = l^{-2m} e_l^\lambda(a, a')$. This proves the sufficiency of the second condition, and the necessity is easy. $\qquad \square$

Remark 16.9. The degrees of λ and λ' are related by $\deg(\lambda) = \deg(\lambda') \cdot \deg(f)^2$, because $\lambda = f^\vee \circ \lambda' \circ f$.

Corollary 16.10. *Let A be an abelian variety having a polarization of degree prime to* $\operatorname{char}(k)$. *Then A is isogenous to a principally polarized abelian variety.*

PROOF. Let λ be a polarization of A, and let l be a prime dividing the degree of λ. Choose a subgroup N of $\operatorname{Ker}(\lambda)$ of order l, and let $B = A/N$. As e^λ is skew-symmetric, it must be zero on $N \times N$, and so the last proposition implies that B has a polarization of degree $\deg(\lambda)/l^2$. □

Corollary 16.11. *Let λ be a polarization of A, and assume that $\operatorname{Ker}(\lambda) \subset A_m$ with m prime to $\operatorname{char}(k)$. If there exists an element α of $\operatorname{End}(A)$ such that $\alpha(\operatorname{Ker}(\lambda)) \subset \operatorname{Ker}(\lambda)$ and $\alpha^\vee \circ \lambda \circ \alpha = -\lambda$ on A_{m^2}, then $A \times A^\vee$ is principally polarized.*

PROOF. Let $N = \{(a, \alpha a) \mid a \in \operatorname{Ker}(\lambda)\} \subset A \times A$. Then $N \subset \operatorname{Ker}(\lambda \times \lambda)$, and for $(a, \alpha a)$ and $(a', \alpha a')$ in N

$$e^{\lambda \times \lambda}((a, \alpha a), (a', \alpha a')) = e^\lambda(a, a') + e^\lambda(\alpha a, \alpha a')$$
$$= \bar{e}_m(a, \lambda b) + \bar{e}_m(a, \alpha^\vee \circ \lambda \circ \alpha(b)) \quad \text{where} \quad mb = a'$$
$$= \bar{e}_m(a, \lambda b) + \bar{e}_m(a, -\lambda b)$$
$$= 0.$$

Therefore, (16.8) applied to $A \times A \to (A \times A)/N$ and the polarization $\lambda \times \lambda$ on $A \times A$ shows that $(A \times A)/N$ is principally polarized. The kernel of $(a, a') \mapsto (a, \alpha a + a'): A \times A \to (A \times A)/N$ is $\operatorname{Ker}(\lambda) \times \{0\}$, and so the map induces an isomorphism $A^\vee \times A \to (A \times A)/N$. □

Remark 16.12 (Zarhin's Trick). Let A and λ be as in the statement of the corollary. Then there always exists an α satisfying the conditions for (A^4, λ^4) and therefore $(A \times A^\vee)^4$ is principally polarized. To see this choose integers a, b, c, d such that $a^2 + b^2 + c^2 + d^2 \equiv -1 \pmod{m^2}$, and let

$$\alpha = \begin{bmatrix} a & -b & -c & -d \\ b & a & d & -c \\ c & -d & a & b \\ d & c & -b & a \end{bmatrix} \in M_4(\mathbb{Z}) \subset \operatorname{End}(A^4).$$

Clearly $\alpha(\operatorname{Ker}(\lambda^4)) \subset \operatorname{Ker}(\lambda^4)$. Moreover $\alpha^\vee$ can be identified with the transpose $\alpha^{\operatorname{tr}}$ of α (as a matrix), and so

$$\alpha^\vee \circ \lambda^4 \circ \alpha = \alpha^{\operatorname{tr}} \circ \lambda^4 \circ \alpha = \lambda^4 \circ \alpha^{\operatorname{tr}} \circ \alpha.$$

But $\alpha^{\operatorname{tr}} \circ \alpha = (a^2 + b^2 + c^2 + d^2)I_4$.

Remark 16.13. In [16, §§20, 23] there is a different and much more profound treatment of the above theory using finite group schemes. In particular, it is possible to remove the restrictions on l or a degree being prime to the characteristic in the results (16.4) through (16.12).

Remark 16.14. Some of the above results extend to fields that are not algebraically closed. For example, if A is an abelian variety over a perfect field, then (16.5) implies immediately that a polarization λ of A can be written as l^m times a polarization if and only if e_l^λ is divisible by l^m; similarly (16.11) implies that the same result holds over a perfect field. On the other hand (16.10) seems to be false unless one allows a field extension (roughly speaking, it is necessary to divide out by half the kernel of the polarization λ, which need not be rational over k).

§17. The Rosati Involution

Fix a polarization λ on A. As λ is an isogeny $A \to A^\vee$, it has an inverse in $\mathrm{Hom}^0(A^\vee, A) \stackrel{\mathrm{df}}{=} \mathrm{Hom}(A^\vee, A) \otimes \mathbb{Q}$. The *Rosati involution* on $\mathrm{End}^0(A)$ corresponding to λ is

$$\alpha \mapsto \alpha^\dagger = \lambda^{-1} \circ \alpha^\vee \circ \lambda.$$

This has the following obvious properties:

$$(\alpha + \beta)^\dagger = \alpha^\dagger + \beta^\dagger, \quad (\alpha\beta)^\dagger = \beta^\dagger \alpha^\dagger, \quad a^\dagger = a \quad \text{for } a \in \mathbb{Q}.$$

For any $a, a' \in T_l A \otimes \mathbb{Q}, l \neq \mathrm{char}(k)$,

$$e_l^\lambda(\alpha a, a') = e_l(\alpha a, \lambda a') = e_l(a, \alpha^\vee \circ \lambda a') = e_l^\lambda(a, \alpha^\dagger a'),$$

from which it follows that $\alpha^{\dagger\dagger} = \alpha$.

Remark 17.1. The second condition on α in (16.11) can now be stated as $\alpha^\dagger \circ \alpha = -1$ on A_{m^2} (provided $\alpha^\dagger$ lies in $\mathrm{End}(A)$).

Proposition 17.2. *Assume that k is algebraically closed. Then the map*

$$\mathscr{L} \mapsto \lambda^{-1} \circ \varphi_\mathscr{L}, \mathrm{NS}(A) \otimes \mathbb{Q} \to \mathrm{End}^0(A),$$

identifies $\mathrm{NS}(A) \otimes \mathbb{Q}$ *with the subset of* $\mathrm{End}^0(A)$ *of elements fixed by* $\dagger$.

PROOF. Let $\alpha \in \mathrm{End}^0(A)$, and let l be an odd prime $\neq \mathrm{char}(k)$. According to (16.6), $\lambda \circ \alpha$ is of the form $\varphi_\mathscr{L}$ if and only if $e_l^{\lambda \circ \alpha}(a, a') = -e_l^{\lambda \circ \alpha}(a', a)$ for all $a, a' \in T_l A \otimes \mathbb{Q}$. But

$$e_l^{\lambda \circ \alpha}(a, a') = e_l^\lambda(a, \alpha a') = -e_l^\lambda(\alpha a', a) = -e_l(a', \alpha^\vee \circ \lambda(a)),$$

and so this is equivalent to $\lambda \circ \alpha = \alpha^\vee \circ \lambda$, that is, to $\alpha = \alpha^\dagger$. $\square$

Theorem 17.3. *The bilinear form*
$$(\alpha, \beta) \mapsto \mathrm{Tr}(\alpha \circ \beta^\dagger) \colon \mathrm{End}^0(A) \times \mathrm{End}^0(A) \to \mathbb{Q}$$
is positive definite. More precisely, if $\lambda = \varphi_{\mathcal{L}(D)}$, *then*
$$\mathrm{Tr}(\alpha \circ \alpha^\dagger) = \frac{2g}{(D^g)}(D^{g-1} \cdot \alpha^*(D)).$$

PROOF. As D is ample and $\alpha^*(D)$ is effective, the intersection number $(D^{g-1} \cdot \alpha^*(D))$ is positive. Thus the second statement implies the first. Clearly it suffices to prove it with k algebraically closed.

Lemma 17.4. *Let A be an abelian variety over an algebraically closed field, and let $\mathbb{Z}_l(g) = \mathbb{Z}_l(1)^{\otimes g}$. Then there is a canonical generator ε of $\mathrm{Hom}(\Lambda^{2g}(T_l A), \mathbb{Z}_l(g))$ with the following property: if $D_1, \ldots, D_g$ are divisors on A and $e_i = e_l^{\mathcal{L}(D_i)} \in \mathrm{Hom}(\Lambda^2 T_l A, \mathbb{Z}_l(1))$, then $e_1 \wedge \cdots \wedge e_g$ is the multiple $(D_1, D_2, \ldots, D_g)\varepsilon$ of ε.*

PROOF. See [16, §20, Theorem 3, p. 190]. (From the point of view of étale cohomology, ε corresponds to the canonical generator of $H^{2g}(A, \mathbb{Z}_l(g))$, which is equal to the cohomology class of any point on A. If c_i is the class of D_i in $H^2(A, \mathbb{Z}_l(1))$, then the compatibility of intersection products with cup-products shows that $(D_1, \ldots, D_g)\varepsilon = c_1 \cup \cdots \cup c_g$. Consequently, the lemma follows from (15.4).) □

PROOF OF (17.3). From the lemma, we find that
$$e_l^\lambda \wedge \cdots \wedge e_l^\lambda = (D^g)\varepsilon,$$
$$e_l^\lambda \wedge \cdots \wedge e_l^\lambda \wedge e_l^{\alpha^*(\lambda)} = (D^{g-1} \cdot \alpha^*(D))\varepsilon.$$

It suffices therefore to show that, for some basis $a_1, \ldots, a_{2g}$ of $T_l A \otimes \mathbb{Q}$,
$$\frac{\langle a_1 \wedge \cdots \wedge a_{2g} | e_l^\lambda \wedge \cdots \wedge e_l^\lambda \wedge e_l^{\alpha^*(\lambda)} \rangle}{\langle a_1 \wedge \cdots \wedge a_{2g} | e_l^\lambda \wedge \cdots \wedge e_l^\lambda \rangle} = \frac{1}{2g} \mathrm{Tr}(\alpha \circ \alpha^\dagger).$$

(See (15.4).) Choose the basis $a_1, a_2, \ldots, a_{2g}$ so that
$$e_l^\lambda(a_{2i-1}, a_{2i}) = 1 = -e_l^\lambda(a_{2i}, a_{2i-1}), \qquad i = 1, 2, \ldots, g,$$
$$e_l^\lambda(a_i, a_j) = 0, \qquad \text{otherwise.}$$

Let $f_1, \ldots, f_{2g}$ be the dual basis; then for $j \neq j'$,
$$\langle a_i \wedge a_{i'} | f_j \wedge f_{j'} \rangle = \begin{vmatrix} f_j(a_i) & f_{j'}(a_i) \\ f_j(a_{i'}) & f_{j'}(a_{i'}) \end{vmatrix} = \begin{cases} 1 & \text{if } i = j, i' = j', \\ -1 & \text{if } i = j', i' = j, \\ 0 & \text{otherwise.} \end{cases}$$

Therefore $e_l^\lambda = \sum_{i=1}^g f_{2i-1} \wedge f_{2i}$, and so $e_l^\lambda \wedge \cdots \wedge e_l^\lambda = g!(f_1 \wedge \cdots \wedge f_{2g})$.

Thus
$$\langle a_1 \wedge \cdots \wedge a_{2g} | e_l^\lambda \wedge \cdots \wedge e_l^\lambda \rangle = \langle a_1 \wedge \cdots \wedge a_{2g} | g!(f_1 \wedge \cdots \wedge f_{2g}) \rangle = g!.$$

Similarly,
$$\langle a_1 \wedge \cdots \wedge a_{2g} | e_l^\lambda \wedge \cdots \wedge e_l^\lambda \wedge e_l^{\alpha^*(\mathscr{L})} \rangle$$

$$= (g-1)! \sum_{i=1}^{g} e_l^\lambda(\alpha a_{2i-1}, \alpha a_{2i})$$

$$= \frac{(g-1)!}{2} \sum (e_l^\lambda(a_{2i-1}, \alpha^\dagger \alpha a_{2i}) + e_l^\lambda(\alpha^\dagger \alpha a_{2i-1}, a_{2i}))$$

$$= \frac{g!}{2g} \mathrm{Tr}(\alpha^\dagger \alpha),$$

which completes the proof. □

Proposition 17.5. *Let λ be a polarization of the abelian variety A.*

(a) *The automorphism group of (A, λ) is finite.*
(b) *For any integer $n \geq 3$, an automorphism of (A, λ) acting as the identity on $A_n(\bar{k})$ is equal to the identity.*

PROOF. Let α be an automorphism of A. In order for α to be an automorphism of (A, λ), we must have $\lambda = \alpha^\vee \circ \lambda \circ \alpha$, and therefore $\alpha^\dagger \alpha = 1$, where $\dagger$ is the Rosati involution defined by λ. Consequently,

$$\alpha \in \mathrm{End}(A) \cap \{\alpha \in \mathrm{End}(A) \otimes \mathbb{R} | \mathrm{Tr}(\alpha^\dagger \alpha) = 2g\},$$

and the first of these sets is discrete in $\mathrm{End}(A) \otimes \mathbb{R}$, while the second is compact. This proves (a).

Assume further that α acts as the identity on A_n. Then $\alpha - 1$ is zero on A_n, and so it is of the form $n\beta$ with $\beta \in \mathrm{End}(A)$ (see (12.6)). The eigenvalues of α and β are algebraic integers, and those of α are roots of 1 because it has finite order. The next lemma shows that the eigenvalues of α equal 1.

Lemma 17.6. *If ζ is a root of 1 such that for some algebraic integer γ and rational integer $n \geq 3$, $\zeta = 1 + n\gamma$, then $\zeta = 1$.*

PROOF. If $\zeta \neq 1$, then after raising it to a power, we may assume that it is a primitive pth root of 1 for some prime p. Then $N_{\mathbb{Q}(\zeta)/\mathbb{Q}}(1 - \zeta) = p$, and so the equation $1 - \zeta = -n\gamma$ implies $p = \pm n^{p-1} N(\gamma)$. This is impossible because p is prime. □

We have shown that α is unipotent and therefore that $\alpha - 1 = n\beta$ is nilpotent. Suppose that $\beta \neq 0$. Then $\beta' = \beta^\dagger \beta \neq 0$, because $\mathrm{Tr}(\beta^\dagger \beta) > 0$. As $\beta' = \beta'^\dagger$, this implies that $\mathrm{Tr}(\beta'^2) > 0$ and so $\beta'^2 \neq 0$. Similarly, $\beta'^4 \neq 0$, and so on, which contradicts the nilpotence of β. □

Remark 17.7. Let (A, λ) and (A', λ') be polarized abelian varieties over a field k, and assume that A and A' have all their points of order n rational over k for some $n \geq 3$. Then any isomorphism $\alpha: (A, \lambda) \to (A', \lambda')$ defined over the separable closure k_s of k is automatically defined over k because, for all $\sigma \in \text{Gal}(k_s/k)$, $\alpha^{-1} \circ \sigma\alpha$ is an automorphism of (A, λ) fixing the points of order n and therefore is the identity map.

Remark 17.8. On combining the results in Section 12 with (17.3), we see that the endomorphism algebra $\text{End}^0(A)$ of a simple abelian variety A is a skew field together with an involution † such that $\text{Tr}(\alpha \circ \alpha^\dagger) > 0$ for all nonzero α.

§18. Two More Finiteness Theorems

The first theorem shows that an abelian variety can be endowed with a polarization of a fixed degree d in only a finite number of essentially different ways. The second shows that an abelian variety has only finitely many non-isomorphic direct factors.

Theorem 18.1. *Let A be an abelian variety over a field k, and let d be an integer; then there exist only finitely many isomorphism classes of polarized abelian varieties (A, λ) with λ of degree d.*

Fix a polarization λ_0 of A, and let † be the Rosati involution on $\text{End}^0(A)$ defined by λ_0. The map $\lambda \mapsto \lambda_0^{-1} \circ \lambda$ identifies the set of polarizations of A with a subset of the set $\text{End}^0(A)^\dagger$ of elements of $\text{End}^0(A)$ fixed by †. As $\text{NS}(A_{\bar{k}})$ is a finitely generated abelian group, there exists an N such that all the $\lambda_0^{-1} \circ \lambda$ are contained in a lattice $L = N^{-1}\text{End}(A)^\dagger$ in $\text{End}^0(A)^\dagger$. Note that L is stable under the action
$$\alpha \mapsto u^\dagger \alpha u, \quad u \in \text{End}(A)^\times, \quad \alpha \in \text{End}^0(A)$$
of $\text{End}(A)^\times$ on $\text{End}^0(A)$.

Let λ be a polarization of A, and let $u \in \text{End}(A)^\times$. Then u defines an isomorphism $(A, u^\vee \circ \lambda \circ u) \xrightarrow{\approx} (A, \lambda)$, and $\lambda_0^{-1} \circ (u^\vee \circ \lambda \circ u) = u^\dagger \circ (\lambda_0^{-1} \circ \lambda) \circ u$. Thus to each isomorphism class of polarized abelian varieties (A, λ), we can associate an orbit of $\text{End}(A)^\times$ in L. Recall (12.12) that the map $\alpha \mapsto \deg(\alpha)$ is a positive power of the reduced norm on each simple factor of $\text{End}^0(A)$, and so Nrd is bounded on the set of elements of L with degree d. These remarks show that the theorem is a consequence of the following result on algebras.

Proposition 18.2. *Let E be a finite-dimensional semisimple algebra over $\mathbb{Q}$ with an involution †, and let R be an order in E. Let L be a lattice in $E^\dagger$ that is stable under the action $e \mapsto u^\dagger e u$ of $R^\times$ on E. Then for any integer d, $\{v \in L \mid \text{Nrd}(v) \leq d\}$ is the union of a finite number of orbits.*

This proposition will be proved using a general result from the reduction theory of arithmetic subgroups.

Theorem 18.3. *Let G be a reductive group over $\mathbb{Q}$, and let Γ be an arithmetic subgroup of G; let $G \to \mathrm{Gl}(V)$ be a representation of G over $\mathbb{Q}$, and let L be a lattice in V that is stable under Γ. If X is a closed orbit of G in V, then $L \cap X$ is the union of a finite number of orbits of Γ.*

PROOF. See [4, 9.11]. □

Remark 18.4. (a) An algebraic group G is *reductive* if its identity component is an extension of a semisimple group by a torus. A subgroup Γ of $G(\mathbb{Q})$ is *arithmetic* if it is commensurable with $G(\mathbb{Z})$ for some $\mathbb{Z}$-structure on G.

(b) The following example may give the reader some idea of the nature of the above theorem. Let $G = \mathrm{SL}_n$, and let $\Gamma = \mathrm{SL}_n(\mathbb{Z})$. Then G acts in a natural way on the space V of quadratic forms in n variables with rational coefficients, and Γ preserves the lattice L of such forms with integer coefficients. Let q be a quadratic form with nonzero discriminant d. By the orbit X of q we mean the image $G \cdot q$ of G under the map of algebraic varieties $g \mapsto g \cdot q \colon G \to V$. The theory of quadratic forms shows that $X(\bar{\mathbb{Q}})$ is equal to the set of all quadratic forms (with coefficients in $\bar{\mathbb{Q}}$) of discriminant d. Clearly this is closed, and so the theorem shows that $X \cap L$ contains only finitely many $\mathrm{SL}_n(\mathbb{Z})$-orbits: the quadratic forms with integer coefficients and discriminant d fall into a finite number of proper equivalence classes.

We shall apply (18.3) with G a reductive group such that

$$G(\mathbb{Q}) = \{e \in E \mid \mathrm{Nrd}(e) = \pm 1\},$$

$\Gamma = R^\times$, $V = E^\dagger$, and $L \subset V$ the lattice in (18.2). In order to prove (18.2), we shall show

(a) there exists a reductive group G over $\mathbb{Q}$ with $G(\mathbb{Q})$ as described and having Γ as an arithmetic subgroup;
(b) the orbits of G on V are all closed;
(c) for any rational number d, $V_d \stackrel{\mathrm{df}}{=} \{v \in V \mid \mathrm{Nrd}(v) = d\}$ is the union of a finite number of orbits of G.

Then (18.3) will show $L \cap V_d$ comprises only finitely many Γ-orbits, as is asserted by (18.2).

To prove (a), embed E into some matrix algebra $M_n(\mathbb{Q})$. Then the condition that $\mathrm{Nrd}(e) = \pm 1$ can be expressed as a polynomial equation in the matrix coefficients of e, and this polynomial equation defines a linear algebraic group G over $\mathbb{Q}$ such that $G(S) = \{e \in E \otimes S \mid \mathrm{Nrd}(e) = \pm 1\}$ for all $\mathbb{Q}$-algebras S. Over $\bar{\mathbb{Q}}$, E is isomorphic to a product of matrix algebras $\prod M_{n_i}(\bar{\mathbb{Q}})$; consequently, $G(\bar{\mathbb{Q}}) = \{(e_i) \in \prod \mathrm{GL}_{n_i}(\bar{\mathbb{Q}}) \mid \prod \det(e_i) = \pm 1\}$. From this it is clear that the identity component of G is an extension of $\prod \mathrm{PGL}_{n_i}$

by a torus, and so G is reductive. It is easy to see that Γ is an arithmetic subgroup of $G(\mathbb{Q})$.

To prove (b), we need the following lemma from the theory of algebras with involution.

Lemma 18.5. *Let E be a semisimple algebra over an algebraically closed field K of characteristic zero, and let $\dagger$ be an involution of E fixing the elements of K. Then every element e of E such that $e^\dagger = e$ can be written $e = ca^\dagger a$ where c is in the centre of E and $\mathrm{Nrd}(a) = 1$.*

PROOF. Lacking a good proof, we make use of the classification of pairs $(E, \dagger)$. Each pair is a direct sum of pairs of the following types:

(A_n) $E = M_n(K) \times M_n(K)$ and $(e_1, e_2)^\dagger = (e_2^{\mathrm{tr}}, e_1^{\mathrm{tr}})$;
(B_n) E is the matrix algebra $M_n(K)$ and $e^\dagger = e^{\mathrm{tr}}$;
(C_n) $E = M_{2n}(K)$ and $e^\dagger = J^{-1} e^{\mathrm{tr}} J$ with J an invertible alternating matrix.

(See, for example, [25].) In the cases (B_n) and (C_n), the lemma follows from elementary linear algebra; in the case (A_n), $e = (e', e'^{\mathrm{tr}})$, and we can take $c = d(I_n, I_n)$ and $a = (e'/d, I_n)$, where $d = \det(e')^{1/n}$. □

From the lemma, we see that if G_e is the isotropy group at $e \in V$, then there is an isomorphism $g \mapsto ag: G_e \to G_1$ defined over $\bar{\mathbb{Q}}$. In particular, all isotropy groups have the same dimension, and therefore all orbits of G in V have the same dimension. This implies that they are all closed, because every orbit of minimal dimension is closed (see, for example, [11, 8.3]).

It remains to prove (c). Let $v, v' \in V_d \otimes \mathbb{C}$, and write $v = ca^\dagger a$, $v' = c'a'^\dagger a'$ with c, c' and a, a' as in the lemma. Clearly v and v' are in the same orbit if and only if c and c' are. Note that c and c' lie in $V_d \otimes \mathbb{C}$. Let Z be the subalgebra of the centre of $E \otimes \mathbb{C}$ of elements fixed by $\dagger$. Then c and c' are in Z, and they lie in the same orbit of G if $c/c' \in Z^2$. But Z is a finite product of copies of $\mathbb{R}$ and $\mathbb{C}$, and so $Z^\times / Z^{\times 2}$ is finite. □

Corollary 18.6. *Let k be a finite field, and let g and d be positive integers. Up to isomorphism, there are only finitely many polarized abelian varieties (A, λ) over k with $\dim A = g$ and $\deg \lambda = d^2$.*

PROOF. From (14.1) we know that there are only finitely many possible A's, and (18.1) shows that for each A there are only finitely many λ's. □

We come now to the second main result of this section. An abelian variety A' is said to be a *direct factor* of an abelian variety A if $A \approx A' \times A''$ for some abelian variety A''.

Theorem 18.7. *Up to isomorphism, an abelian variety A has only finitely many direct factors.*

PROOF. To each direct factor A' of A, there corresponds an element e of End(A) defined by $A \xrightarrow{\sim} A' \times A'' \xrightarrow{p} A' \to A' \times A'' \xrightarrow{\sim} A$. Moreover $e^2 = e$, and A' is determined by e because it equals the kernel of $1 - e$. If $e' = ueu^{-1}$ with u in End$(A)^\times$, then $u(1 - e)u^{-1} = 1 - e'$, and so e and e' correspond to isomorphic direct factors. These remarks show that the theorem is a consequence of the next lemma. □

Lemma 18.8. *Let E be a semisimple algebra of finite dimension over $\mathbb{Q}$, and let R be an order in E. Then $R^\times$, acting on the set of idempotents of R by inner automorphisms, has only finitely many orbits.*

PROOF. Apply (18.3) with G the algebraic group such that $G(\mathbb{Q}) = E^\times$; take Γ to be the arithmetic group $R^\times$, V to be E with G acting by inner automorphisms, and L to be R. Then the idempotents in E form a finite set of orbits under G, and each of these orbits is closed. In proving these statements we may replace $\mathbb{Q}$ by $\bar{\mathbb{Q}}$ and assume E to be a matrix algebra. Then each idempotent is conjugate to one of the form $e = \text{diag}(1, \ldots, 1, 0, \ldots, 0)$, and the stabilizer G_e of e is a parabolic subgroup of G and so G/G_e is a projective variety (see [11, 21.3]) which implies that its image Ge in V is closed. □

Corollary 18.9. *Let k be a finite field; for each integer g, there exist only finitely many isomorphism classes of abelian varieties of dimension g over k.*

PROOF. Let A be an abelian variety of dimension g over k. From (16.12) we know that $(A \times A^\vee)^4$ has a principal polarization, and according to (14.1), the abelian varieties of dimension $8g$ over k having principal polarizations form only finitely many isomorphism classes. The result therefore follows from (18.7). □

§19. The Zeta Function of an Abelian Variety

Throughout this section, A will be an abelian variety over a finite field k with q elements, and k_m will be the unique subfield of $\bar{k}$ with q^m elements. Thus the elements of k_m are the solutions of $c^{q^m} = c$. We write N_m for the order of $A(k_m)$.

Theorem 19.1. *There are algebraic integers $a_1, \ldots, a_{2g}$ such that:*

(a) *the polynomial $P(X) = \prod(X - a_i)$ has coefficients in $\mathbb{Z}$;*
(b) $N_m = \prod(1 - a_i^m)$ *for all $m \geq 1$; and*
(c) *(Riemann hypothesis) $|a_i| = q^{1/2}$.*

In particular, $|N_m - q^{mg}| \leq 2gq^{m(g-1/2)} + (2^{2g} - 2g - 1)q^{m(g-1)}$.

The proof will use the Frobenius morphism. For a variety V over k, this is defined to be the morphism $\pi_V: V \to V$ which is the identity map on the underlying topological space of V and is the map $f \mapsto f^q$ on $\mathcal{O}_V$. For example, if $V = \mathbb{P}^n = \operatorname{Proj}(k[X_0, \ldots, X_n])$, then π_V is defined by the homomorphism of rings

$$X_i \mapsto X_i^q: k[X_0, \ldots, X_n] \to k[X_0, \ldots, X_n]$$

and induces the map on points

$$(x_0 : \ldots : x_n) \mapsto (x_0^q : \ldots : x_n^q): \mathbb{P}^n(\bar{k}) \to \mathbb{P}^n(\bar{k}).$$

For any map $\varphi: W \to V$, it is obvious that $\varphi \circ \pi_W = \pi_V \circ \varphi$. Therefore, if $A \subset \mathbb{P}^n$ is a projective embedding of A, then π_A induces the map $(x_0 : \ldots : x_n) \mapsto (x_0^q : \ldots : x_n^q)$ on $A(\bar{k})$. In particular, we see that the kernel of $1 - \pi_A^m: A(\bar{k}) \to A(\bar{k})$ is $A(k_m)$. Note that π_A maps zero to zero, and therefore (see (2.2)) is a homomorphism. Clearly π always defines the zero map on tangent spaces (look at its action on the cotangent space), and so $d(1 - \pi_A^m)_0: T_0(A) \to T_0(A)$ is the identity map. Therefore, $1 - \pi_A^m$ is étale, and the order N_m of its kernel in $A(\bar{k})$ is equal to its degree. Let P be the characteristic polynomial of π_A. It is a monic polynomial of degree $2g$ with integer coefficients, and if we let $a_1, \ldots, a_{2g}$ be its roots, then (12.9) shows that $\prod(X - a_i^m)$ is the characteristic polynomial of π_A^m. Consequently,

$$N_m = \deg(\pi_A^m - 1) = \prod(1 - a_i^m).$$

This proves (a) and (b) of the theorem with the added information that P is the characteristic polynomial of π_A. Part (c) follows from the next two lemmas.

Lemma 19.2. *Let $\dagger$ be the Rosati involution on $\operatorname{End}^0(A)$ defined by a polarization of A; then $\pi_A^\dagger \circ \pi_A = q_A$.*

PROOF. As was noted in (13.2), the polarization will be defined by an ample sheaf $\mathscr{L}$ on A. We have to show that $\pi_A^\vee \circ \varphi_{\mathscr{L}} \circ \pi_A = q\varphi_{\mathscr{L}}$. It follows from the definition of π_A that $\pi_A^* \mathscr{L} \approx \mathscr{L}^q$. Therefore, for all $a \in A(\bar{k})$,

$$\pi_A^\vee \circ \varphi_{\mathscr{L}} \circ \pi_A(a) = \pi_A^*(t_{\pi a}^* \mathscr{L} \otimes \mathscr{L}^{-1}) = t_a^*(\pi_A^* \mathscr{L}) \otimes (\pi_A^* \mathscr{L})^{-1} = q\varphi_{\mathscr{L}}(a),$$

as required. $\square$

Lemma 19.3. *Let α be an element of $\operatorname{End}^0(A)$ such that $\alpha^\dagger \circ \alpha$ is an integer r; for any root a of P_α, $|a|^2 = r$.*

PROOF. Note that $\mathbb{Q}(\alpha)$ is stable under $\dagger$. The argument terminating the proof of (17.5) shows that $\mathbb{Q}(\alpha)$ contains no nilpotent elements, and therefore is a product of fields. The tensor product $\mathbb{Q}(\alpha) \otimes \mathbb{R}$ is a product of copies of $\mathbb{R}$ and $\mathbb{C}$. Moreover $\dagger$ extends to an $\mathbb{R}$-linear involution of $\mathbb{Q}(\alpha) \otimes \mathbb{R}$, and $\operatorname{Tr}(\beta^\dagger \beta) \geq 0$ for all $\beta \neq 0$, with inequality holding on a dense subset. It follows easily that each factor K of $\mathbb{Q}(\alpha) \otimes \mathbb{R}$ is stable under $\dagger$ and that $\dagger$ is the identity map if K is real, and is complex conjugation if K is complex. Thus, for each homomorphism ι of $\mathbb{Q}(\alpha)$ into $\mathbb{C}$, $\iota(\alpha^\dagger)$ is the complex conjugate of $\iota\alpha$. The hypothesis of the theorem therefore states that $|\iota\alpha|^2 = r$, which, in essence, is also the conclusion. □

The zeta function of a variety V over k is defined to be the formal power series $Z(V, t) = \exp(\sum N_m t^m/m)$.

Corollary 19.4. *Let* $P_r(t) = \prod(1 - a_{i,r}t)$, *where the* $a_{i,r}$ *run through the products* $a_{i_1} a_{i_2} \ldots a_{i_r}$, $0 < i_1 < \cdots < i_r \leq 2g$, a_i *a root of* $P(t)$.

Then
$$Z(A, t) = \frac{P_1(t) \ldots P_{2g-1}(t)}{[P_0(t) \ldots P_{2g}(t)]}.$$

PROOF. Take the logarithm of each side, and use the identity
$$-\log(1-t) = 1 + t + t^2/2 + t^3/3 + \cdots. \quad \square$$

Remark 19.5. (a) The polynomial $P_r(t)$ is the characteristic polynomial of π acting on $\Lambda^r T_l A$.

(b) Let $\zeta(V, s) = Z(V, q^{-s})$; then (19.1c) implies that the zeros of $\zeta(V, s)$ lie on the lines $\mathrm{Re}(s) = 1/2, 3/2, \ldots, (2g-1)/2$ and the poles on the lines $\mathrm{Re}(s) = 0, 1, \ldots, 2g$.

Remark 19.6. The isomorphism $\Lambda^r T_l A \approx H^r(A_{\mathrm{et}}, \mathbb{Q}_l)^\vee$ and the above results show that
$$N_m = \sum (-1)^r \mathrm{Tr}(\pi | H^r(A_{\mathrm{et}}, \mathbb{Q}_l))$$
and that
$$Z(A, t) = \prod \det(1 - \pi t | H^r(A_{\mathrm{et}}, \mathbb{Q}_l))^{(-1)^r}.$$

§20. Abelian Schemes

Let S be a scheme; a group scheme $\pi: \mathscr{A} \to S$ over S is an *abelian scheme* if π is proper and smooth and the geometric fibres of π are connected. The second condition means that, for all maps $\bar{s} \to S$ with $\bar{s}$ the spectrum of an algebraically closed field, the pull-back $\mathscr{A}_{\bar{s}}$ of $\mathscr{A}$ to $\bar{s}$ is connected. In the presence of the first condition, it is equivalent to the fibres of π being abelian varieties. Thus an abelian scheme over S can be thought of as a continuous family of abelian varieties parametrized by S.

Many results concerning abelian varieties extend to abelian schemes.

Proposition 20.1 (Rigidity Lemma). *Let S be a connected scheme, and let $\pi: \mathscr{V} \to S$ be a proper flat map whose fibres are varieties; let $\pi': \mathscr{V}' \to S$ be a* second S-scheme, and let $f: \mathscr{V} \to \mathscr{V}'$ be a morphism of S-schemes. If for some point s of S, the image of $\mathscr{V}_s$ in $\mathscr{V}'_s$ is a single point, then f factors through S (that is, there exists a map $f': S \to \mathscr{V}'$ such that $f = f' \circ \pi$).

PROOF. See [15, 6.1]. □

Corollary 20.2. (a) *Every morphism of abelian schemes carrying the zero section into the zero section is a homomorphism.*
(b) *The group structure on an abelian scheme is uniquely determined by the choice of a zero section.*
(c) *An abelian scheme is commutative.*

PROOF. (a) Apply the proposition to the map $\varphi: \mathscr{A} \times \mathscr{A} \to \mathscr{B}$ defined as in the proof of (2.2).
(b) This follows immediately from (a).
(c) The map $a \mapsto a^{-1}$ is a homomorphism. □

Our next result shows that an abelian variety cannot contain a nonconstant algebraic family of subvarieties.

Proposition 20.3. *Let A be an abelian variety over a field k, and let S be a k-scheme such that $S(k) \neq \emptyset$. For any injective homomorphism $f: \mathscr{B} \hookrightarrow A \times S$ of abelian schemes over S, there is an abelian subvariety B of A (defined over k) such that $f(\mathscr{B}) = B \times S$.*

PROOF. Let $s \in S(k)$, and let $B = \mathscr{B}_s$. Then f_s identifies B with a subvariety of A. The map $h: \mathscr{B} \xrightarrow{f} A \times S \twoheadrightarrow (A/B) \times S$ has fibre $B_s \to A \to A/B_s$ over s, which is zero, and so (20.1) shows that $h = 0$. It follows that $f(\mathscr{B}) = B \times S$. □

Recall that a finitely generated extension K of a field k is *regular* if it is linearly disjoint from $\bar{k}$.

Corollary 20.4. *Let K be a regular extension of a field k.*

(a) *Let A be an abelian variety over k. Then every abelian subvariety of A_K is defined over k.*
(b) *If A and B are abelian varieties over k, then every homomorphism $\alpha: A_K \to B_K$ is defined over k.*

PROOF. (a) There exists a variety V over k such that $k(V) = K$. After V has been replaced by an open subvariety, we can assume that B extends to an abelian scheme over V (cf. (20.9) below). If V has a k-rational point, then the proposition shows that B is defined over k. In any case, there exists a finite Galois extension k' of k and an abelian subvariety B' of $A_{k'}$ such that $B'_{Kk'} = B_{Kk'}$ as subvarieties of $A_{Kk'}$. The equality uniquely determines B' as a subvariety of $A_{k'}$. As σB has the same property for any $\sigma \in \text{Gal}(k'/k)$, we must have $\sigma B = B$, and this shows that B is defined over k.

(b) Part (a) shows that the graph of α is defined over k. □

Theorem 20.5. *Let K/k be a regular extension of fields, and let A be an abelian variety over K. Then there exists an abelian variety B over k and a homomorphism $f: B_K \to A$ with finite kernel having the following universal property: for any abelian variety B' and homomorphism $f': B'_K \to A$ with finite kernel, there exists a unique homomorphism $\varphi: B' \to B$ such that $f' = f \circ \varphi_K$.*

PROOF. Consider the collection of pairs (B, f) with B an abelian variety over k and f a homomorphism $B_K \to A$ with finite kernel, and let A^* be the abelian subvariety of A generated by the images the f. Consider two pairs (B_1, f_1) and (B_2, f_2). Then the identity component C of the kernel of $(f_1, f_2): (B_1 \times B_2)_K \to A$ is an abelian subvariety of $B_1 \times B_2$, which (20.4) shows to be defined over k. The map $(B_1 \times B_2/C)_K \to A$ has finite kernel and image the subvariety of A generated by $f_1(B_1)$ and $f_2(B_2)$. It is now clear that there is a pair (B, f) such that the image of f is A^*. Divide B by the largest subgroup scheme N of $\text{Ker}(f)$ to be defined over k. Then it is not difficult to see that the pair $(B/N, f)$ has the correct universal property (given $f': B'_K \to A$, note that for a suitable C contained in the kernel of $(B/N)_K \times B'_K \to A$, the map $b \mapsto (b, 0): B/N \to (B/N) \times B'/C$ is an isomorphism). □

Remark 20.6. The pair (B, f) is obviously uniquely determined up to a unique isomorphism by the condition of the theorem; it is called the *K/k-trace* of A. (For more details on the K/k-trace and the reverse concept, the K/k-image, see [12, VIII].)

Proposition 20.7. *Let $\mathcal{A}$ be an abelian scheme of relative dimension g over S, and let $n_\mathcal{A}$ be multiplication by n on $\mathcal{A}$. Then $n_\mathcal{A}$ is flat, surjective, and finite, and its kernel $\mathcal{A}_n$ is a finite flat group scheme over S of order n^{2g}. Moreover $n_\mathcal{A}$ (and therefore its kernel) is étale if and only if n is not divisible by any of the characteristics of the residue fields of S.*

PROOF. The map $n_\mathcal{A}$ is flat because $\mathcal{A}$ is flat over S and multiplication by n is flat on each fibre of $\mathcal{A}$ over S (see Section 8). (For the criterion of flatness used here, see [7, IV, 5.9] or [6, III, 5.4, Prop. 2.3].) Moreover $n_\mathcal{A}$ is proper [10, II, 4.8e] with finite fibres, and hence is finite (see, for example, [13, I, 1.10]). It follows that $\mathcal{A}_n$ is flat and finite, and (8.2) shows that it has order n^{2g}. The remaining statement also follows from (8.2). □

Corollary 20.8. *Let S be an connected normal scheme, and let A be an abelian variety over the field of rational functions k of S. Assume that A extends to an abelian scheme over S, and let n be an integer which is prime to the characteristics of the residue fields of S. Then for any point $P \in A(k)$, the normaliza-*

tion of S in $k(n^{-1}P)$ is étale over S. (By $k(n^{-1}P)$ we mean the field generated over k by the coordinates of the points Q such that $nQ = P$.)

PROOF. The hypotheses imply that $\mathscr{A}_n$ is étale over S. Let k' be the composite of the fields of rational functions of the components of $\mathscr{A}_n$, and let k'' be the Galois closure of k'. Then the normalization of S in k'' is étale over S and $A_n(k'')$ has n^{2g} elements. We may replace k with k'' and so assume A has all its points of order n rational in k. The point P extends (by the valuative criterion of properness) to a section s of $\mathscr{A}$ over S. The pull-back of the covering $n_\mathscr{A}: \mathscr{A} \to \mathscr{A}$ to S by means of the section s is a finite étale covering $S' \to S$, and s lifts to a section in $\mathscr{A}(S')$. Let S_0 be any connected component of S'; then the field K of rational functions of S contains $k(n^{-1}P)$, and S_0 is the normalization of S in K. □

Remark 20.9. Let S be an integral Noetherian scheme, and let A be an abelian variety over its field of rational functions K. Choose a projective embedding $A \hookrightarrow \mathbb{P}^n$ and let $\mathscr{A}$ be the closure of A in $\mathbb{P}^n_S$. Then $\pi: \mathscr{A} \to S$ is projective, and its generic fibre is a smooth variety. As $\mathcal{O}_S \to \pi_*\mathcal{O}_\mathscr{A}$ is an isomorphism at the generic point and $\mathcal{O}_S$ and $\pi_*\mathcal{O}_\mathscr{A}$ are coherent, there will be an open subset over which it is an isomorphism and therefore over which π has connected fibres [10, III, 11.5]. The existence of a section implies the fibres will be geometrically connected there. Also there will be an open subset over which $\mathscr{A}$ is smooth [10, III, Ex. 10.2], and an open subset where the group structure extends. These remarks show that there is an open subset U of S such that $\mathscr{A}$ extends to an abelian scheme over U.

When S is locally the spectrum of a Dedekind domain, we can be more precise. Then the projective embedding of A determines a unique extension of A to a flat projective scheme $\pi: \mathscr{A} \to S$ (see [10, III, 9.8]). The R-module $\pi_*\mathcal{O}_\mathscr{A}$ is finitely-generated (because π is proper) and torsion-free (because π is flat). It is therefore a projective R-module, and its rank is one because its tensor product with K is $\Gamma(A, \mathcal{O}_A) = K$. Now, as before, the geometric fibres of $\mathscr{A}$ are connected. We conclude: the choice of a projective embedding defines a flat projective extension $\mathscr{A}$ of A to S; $\mathscr{A}$ will be an abelian scheme over an open set U of S.

It is clear from looking at the example of an elliptic curve, that the extended scheme $\mathscr{A}$ over S depends on the choice of the projective embedding of A, but [2, 1.4] shows that its restriction to U does not. The purpose of the theory of Néron models is to replace $\mathscr{A}$ by a "minimal" (nonproper) extension which is unique.

Using the above results, it is possible to give a short proof of a weak form of the Mordell–Weil theorem.

Theorem 20.10. *Let A be an abelian variety over a number field k, and let n be integer such that all points of A of order n are rational over k. Then $A(k)/nA(k)$ is a finite group.*

PROOF. Let $a \in A(k)$, and let $b \in A(\bar{k})$ be such that $nb = a$. For σ in the Galois group of $\bar{k}$ over k, define $\varphi_a(\sigma)$ to be $\sigma b - b$. Then $a \mapsto \varphi_a$ defines an injection $A(k)/nA(k) \hookrightarrow \mathrm{Hom}(G, A_n(k))$.

Let spec(R) be an open subset of the spectrum of the ring of integers of k such that A extends to an abelian scheme $\mathscr{A}$ over spec(R) and n is invertible in R. Let k' be the maximal abelian extension of k of exponent n unramified outside the finite set of primes not corresponding to prime ideals of R. Then (20.8) shows that φ_a factors through the group Gal(k'/k) for all a. This proves the theorem because k' is a finite extension of k. □

Remark 20.11. Using the theory of heights, one can show that for an abelian variety over a number field k, $A(k)/nA(k)$ finite implies $A(k)$ is finitely generated (see [23]). As the hypothesis of (20.10) always holds after a finite extension of k, this proves the Mordell–Weil theorem: for any abelian variety A over a number field k, $A(k)$ is finitely generated.

Remark 20.12. Let A and B be polarized abelian varieties over a number field k, and assume that they both have good reduction outside a given finite set of primes S; let l be an odd prime. If A and B are isomorphic over $\bar{k}$ (as polarized abelian varieties), then they are isomorphic over an extension k' of k unramified outside S and l and of degree $\leq$ (order of $\mathrm{Gl}_{2g}(\mathbb{F}_l))^2$. (Because the l-torsion points of A and B are rational over such a k', and we can apply (17.7).)

In contrast to abelian varieties, abelian schemes are not always projective, even if the base scheme is the spectrum of an integral local ring of dimension one or an Artinian ring (see [18, XII]). If $\mathscr{A}$ is projective over S, then the dual abelian scheme $\mathscr{A}^\vee$ is known to exist (see [8]); if $\mathscr{A}$ is not projective then $\mathscr{A}^\vee$ exists only as an algebraic space (see [1]). In either case, a *polarization* of $\mathscr{A}$ is defined to be a homomorphism $\lambda: \mathscr{A} \to \mathscr{A}^\vee$ such that, for all geometric points $\bar{s}$ of the base scheme S, $\lambda_{\bar{s}}$ is of the form $\varphi_{\mathscr{L}}$ for some ample invertible sheaf $\mathscr{L}$ on $\mathscr{A}_{\bar{s}}$. Alternatively, λ is a polarization if $\lambda_s: \mathscr{A}_s \to \mathscr{A}_s^\vee$ is a polarization of abelian varieties for all $s \in S$. If S is connected, then the degree of λ_s is independent of s and is called the *degree* of λ.

For a field k and fixed integers g and d, let $\mathscr{F}_{g,d}$ be the functor associating with each k-scheme of finite type the set of isomorphism classes of polarized abelian schemes of dimension g and which have a polarization of degree d^2.

Theorem 20.13. *There exists a variety $M_{g,d}$ over k and a natural transformation $i: \mathscr{F}_{g,d} \to M_{g,d}$ such that:*

(a) $i(K): \mathscr{F}_{g,d}(K) \to M_{g,d}(K)$ *is a bijection for any algebraically closed field containing k;*

(b) *for any variety N over k and natural transformation $j: \mathscr{F}_{g,d} \to N$, there is a unique morphism $\varphi: M_{g,d} \to N$ such that $\varphi \circ i = j$.*

PROOF. This one of the main results of [15]. □

The variety $M_{g,d}$ is uniquely determined up to a unique isomorphism by the conditions of (20.13); it is the (coarse) *moduli variety* for polarized abelian varieties of dimension g and degree d^2. By introducing level structures, one can define a functor that is representable by a fine moduli variety—see the article by C.-L. Chai in these proceedings.

REFERENCES

[1] Artin, M. Algebraization of formal moduli I, in *Global Analysis*. Princeton University Press: Princeton, NJ, 1969, pp. 21–71.
[2] Artin, M. Néron models, this volume, pp. 213–230.
[3] Borel, A. Sur la cohomologie des espaces fibrés principaux et des espace homogènes de groupes de Lie compacts. *Ann. Math.*, **64** (1953), 115–207.
[4] Borel, A. *Introduction aux Groupes Arithmétiques*. Hermann: Paris, 1958.
[5] Bourbaki, N. *Algèbre Multilinéaire*. Hermann: Paris, 1958.
[6] Bourbaki, N. *Algèbre Commutative*. Hermann: Paris, 1961, 1964, 1965.
[7] Grothendieck, A. *Revêtements Étales et Groupe Fondamental* (SGA1, 1960–61). Lecture Notes in Mathematics, 224. Springer-Verlag: Heidelberg, 1971.
[8] Grothendieck, A.: Technique de descente et théorèmes d'existence en géométrie algébrique V. Les schémas de Picard: Théorèmes d'existence. *Séminaire Bourbaki*, Éxposé 232, 1961/62.
[9] Grothendieck, A. (with Dieudonné, J.). Eléments de géométrie algébrique. *Publ. Math. I.H.E.S.*, **4, 8, 11, 17, 20, 24, 28, 32** (1960–67).
[10] Hartshorne, R. *Algebraic Geometry*. Springer Verlag: Hcidelberg, 1977.
[11] Humphreys, J. *Linear Algebraic Groups*. Springer-Verlag: Heidelberg, 1975.
[12] Lang, S. *Abelian Varieties*. Interscience: New York, 1959.
[13] Milne, J. *Étale Cohomology*. Princeton University Press: Princeton, NJ, 1980.
[14] Mumford, D. *Introduction to Algebraic Geometry*. Lecture Notes, Harvard University: Cambridge, MA, 1967.
[15] Mumford D. *Geometric Invariant Theory*. Springer-Verlag: Heidelberg, 1965.
[16] Mumford D. *Abelian Varieties*. Oxford University Press: Oxford, 1970.
[17] Oort, F. *Commutative Group Schemes*. Lecture Notes in Mathematics. Springer-Verlag: Heidelberg, 1966.
[18] Raynaud, M. *Faisceaux Amples sur les Schémas en Groupes et les Espace Homogènes*. Lecture Notes in Mathematics, 119. Springer-Verlag: Heidelberg, 1970.
[19] Rosen, M. (notes by F. McGuinness). Abelian varieties over $\mathbb{C}$, this volume, pp. 79–101.
[20] Serre, J.-P. *Groupes Algèbriques et Corps de Classes*. Hermann: Paris, 1959.
[21] Shafarevich, I. *Basic Algebraic Geometry*. Springer-Verlag: Heidelberg, 1974.
[22] Shatz, S. Group schemes, formal groups, and p-divisible groups, this volume, pp. 29–78.
[23] Silverman, J. The theory of height functions, this volume, pp. 151–166.
[24] Waterhouse, W. *Introduction to Affine Group Schemes*. Springer-Verlag: Heidelberg, 1979.
[25] Weil, A. Algebras with involution and classical groups. *J. Indian Math. Soc.*, **24** (1960), 589–623.

CHAPTER VI

The Theory of Height Functions

JOSEPH H. SILVERMAN

The Classical Theory of Heights

§1. Absolute Values

The following notations and normalizations will be used throughout this chapter:

$K/\mathbb{Q}$ a number field.
M_K the set of absolute values on K extending the usual absolute values on $\mathbb{Q}$. (That is, the p-adic absolute values are normalized so that $|p|_p = 1/p$.)

$\|\cdot\|_v = |\cdot|_v^{[K_v:\mathbb{Q}_v]}$.

§2. Height on Projective Space

The height of a point $P = [x_0, \ldots, x_n]$ in $\mathbb{P}^n(K)$ is a measure of the "arithmetic complexity" of the point.

Definition. The *height of P (relative to K)* is defined by the formula

$$H_K(P) = \prod_{v \in M_K} \max\{\|x_0\|_v, \ldots, \|x_n\|_v\}.$$

Remarks. (1) The height of P is well defined (independent of the choice of homogeneous coordinates for P). This is easily checked using the product formula.

(2) For a finite extension L/K, we have the formula
$$H_L(P) = H_K(P)^{[L:K]}.$$
This is checked using the formula $\sum [L_w : K_v] = [L:K]$, where the sum is over the places $w \in M_L$ lying over a given $v \in M_K$.

(3) Note that $H_K(P) \geq 1$ for all points P, since we can always choose homogeneous coordinates for P with some $x_i = 1$.

Using remark (2) above, we can define a height function which is independent of the field of definition.

Definition. Let $P \in \mathbb{P}^n(\bar{\mathbb{Q}})$. The *absolute height of P* is defined by
$$H(P) = H_K(P)^{1/[K:\mathbb{Q}]},$$
where K is any number field with $P \in \mathbb{P}^n(K)$. It is also often convenient to deal with the *logarithmic height*
$$h(P) = \log H(P).$$

EXAMPLE. Let $P \in \mathbb{P}^n(\mathbb{Q})$. Write $P = [x_0, \ldots, x_n]$ with $x_i \in \mathbb{Z}$ and $\gcd(x_0, \ldots, x_n) = 1$. Then
$$H(P) = \max\{|x_0|, \ldots, |x_n|\}.$$

Theorem 2.1 (Finiteness Theorem). *Let C and d be constants. Then*
$$\{P \in \mathbb{P}^n(\bar{\mathbb{Q}}) : H(P) \leq C \text{ and } [\mathbb{Q}(P) : \mathbb{Q}] \leq d\}$$
is a finite set.

PROOF. From the above example, the theorem is clear if we restrict to points in $\mathbb{P}^n(\mathbb{Q})$. We will reduce the general theorem to this case.

Choose homogeneous coordinates $P = [x_0, \ldots, x_n]$ with some $x_i = 1$. Then, since $H([x_0, \ldots, x_n]) \geq H([1, x_i])$ for any i, we are reduced to the case that $P = [1, x] \in \mathbb{P}^1(\bar{\mathbb{Q}})$ and $[\mathbb{Q}(x) : \mathbb{Q}] = d$. Let $x^{(1)}, \ldots, x^{(d)}$ be the conjugates of x over $\mathbb{Q}$, and let $1 = s_0, \ldots, s_d$ be the elementary symmetric polynomials in $x^{(1)}, \ldots, x^{(d)}$. Then each s_j is in $\mathbb{Q}$, and x is a root of the polynomial
$$F(T) = \sum_{j=0}^{d} (-1)^j s_j T^{d-j} = \prod_{i=1}^{d} T - x^{(i)} \in \mathbb{Q}[T].$$

Now using the triangle inequality, one easily checks that $H(s_j) \leq c_j H(x)^j$ for certain constants c_j which do not depend on x. Hence applying the above example to the point $[s_0, \ldots, s_d] \in \mathbb{P}^d(\mathbb{Q})$, we see that there are only finitely many possibilities for the polynomial $F(T)$, and so only finitely many possibilities for x. □

§3. Heights on Projective Varieties

For this section, V will denote a smooth projective variety defined over $\bar{\mathbb{Q}}$, and all morphisms, divisors, etc., will be assumed to be defined over $\bar{\mathbb{Q}}$. We will also use V to denote the geometric points of V (i.e. $V(\bar{\mathbb{Q}})$). In order to define a height function on V, we take a map from V into projective space and use the height function from the previous section.

Definition. Let $F: V \to \mathbb{P}^n$ be a morphism. The *(logarithmic) height on V relative to F* is defined by

$$h_F: V \to \mathbb{R}, \qquad h_F(P) = h(F(P)).$$

As usual, such a map $F: V \to \mathbb{P}^n$ is associated to an invertible sheaf (or line bundle) on V, namely the pull-back of the twisting sheaf, $F^*\mathcal{O}_\mathbb{P}(1)$. Naturally, many different maps will give rise to the same sheaf. Of crucial importance is the next result, which says that the corresponding height functions are essentially the same.

Theorem 3.1. *Let $\mathscr{L}$ be a sheaf without basepoints on V, and let*

$$F: V \to \mathbb{P}^n \quad \text{and} \quad G: V \to \mathbb{P}^m$$

be two maps of V which are associated to $\mathscr{L}$. Then

$$h_F = h_G + O(1) \quad \text{on } V.$$

Remarks. (1) The last statement means that the quantity $|h_F(P) - h_G(P)|$ is bounded as P ranges over V.

(2) The condition that a map F be associated to $\mathscr{L}$ can also be phrased in terms of divisors. Thus choose a divisor E on V so that $\mathscr{L} \approx \mathcal{O}_V(E)$, and let H be a hyperplane in $\mathbb{P}^n$ not containing $F(V)$. Then F is associated to $\mathscr{L}$ if and only if $F^*(H)$ is linearly equivalent to E.

PROOF. Let E be a divisor in the linear system of $\mathscr{L}$. (That is, $E \geq 0$ and $\mathscr{L} \approx \mathcal{O}_V(E)$.) Then on the complement of E, we can write $F = [f_0, \ldots, f_n]$ and $G = [g_0, \ldots, g_m]$ with rational functions f_i and g_j such that

$$(f_i) = D_i - E \quad \text{and} \quad (g_j) = D'_j - E \quad \text{for divisors } D_i, D'_j \geq 0.$$

(Cf. [Har, II.7.8.1].)

Further, the fact that F has no basepoints means that the D_i's have no point in common. Let K be a common field of definition for V, $f_0, \ldots, f_n$, and $g_0, \ldots, g_n$.

Now pick any j, and look at the ideal $\mathscr{I} = (f_0/g_j, \ldots, f_n/g_j)$ in the ring $\mathscr{R} = K[f_0/g_j, \ldots, f_n/g_j]$. Since $(f_i/g_j) = D_i - D'_j$ and the D_i's have no point in common, it follows that $\mathscr{I}$ is the unit ideal. [Suppose not. Then there would be a maximal ideal $\mathcal{M}$ of $\mathcal{R}$ containing $\mathcal{J}$. Since Spec($\mathcal{R}$) is isomorphic to an open subset of V containing the complement of D_j', $\mathcal{M}$ would correspond to a point P of V not in D_j' such that $(f_i/g_j)(P) = 0$ for all i. But then P would lie in the support of every D_i, yielding a contradiction.] Hence we can find a polynomial $\phi_j(T_0, \ldots, T_n) \in K[T_0, \ldots, T_n]$ having no constant term such that

$$\phi_j(f_0/g_j, \ldots, f_n/g_j) = 1.$$

Taking the v-adic absolute value and using the triangle inequality, one easily finds a constant $C_1 = C_1(v, F, G, \phi_j) > 0$ such that for all P in the complement of D_j',

$$\max\{|f_0/g_j(P)|_v, \ldots, |f_n/g_j(P)|_v\} \geq C_1.$$

Further, we may take $C_1 = 1$ for all but finitely many v, independent of P. (Note that it may be necessary to extend K so that $P \in V(K)$.)

Next multiply through by $|g_j(P)|_v$. Then the inequality also holds for $g_j(P) = 0$, so taking the maximum over j yields

$$\max\{|f_0(P)|_v, \ldots, |f_n(P)|_v\} \geq C_2 \max\{|g_0(P)|_v, \ldots, |g_m(P)|_v\}$$

for a constant $C_2 = C_2(v, F, G) > 0$, where P ranges over the complement of E and $C_2 = 1$ for all but finitely many v. Now raise to the $[K_v : \mathbb{Q}_v]$ power, multiply over all $v \in M_K$, and take the $[K : \mathbb{Q}]$th root. This gives

$$H(F(P)) \geq C_3 H(G(P)),$$

with $C_3 = C_3(F, G) > 0$, as P ranges over the complement of E in V. Next, since $\mathscr{L}$ has no basepoints, we can choose finitely many divisors $E_1, \ldots, E_r$ in the linear system for $\mathscr{L}$ so that the E_i's have trivial intersection. In this way we obtain the above inequality on all of V. Taking logarithms gives one of the desired bounds, and the other follows by symmetry. □

Definition. The group of *functions* mod O(1) on V, denoted $\mathscr{H}(V)$, is defined by

$$\mathscr{H}(V) = \{\text{functions } h: V \to \mathbb{R}\}/\{\text{bounded functions}\}.$$

Definition. Let $\mathscr{L}$ be a sheaf without basepoints on V. The *height function associated to* $\mathscr{L}$ is the class of functions $h_\mathscr{L} \in \mathscr{H}(V)$ obtained by taking the height function h_F for any map F associated to $\mathscr{L}$. (From Theorem 3.1, $h_\mathscr{L}$ is well-defined.)

Proposition 3.2. *Let $\mathscr{L}$ and $\mathscr{M}$ be basepoint-free sheaves on V. Then*

$$h_{\mathscr{L} \otimes \mathscr{M}} = h_\mathscr{L} + h_\mathscr{M} + O(1).$$

PROOF. Let $F = [f_0, \ldots, f_n]$ and $G = [g_0, \ldots, g_m]$ be maps associated to $\mathscr{L}$ and $\mathscr{M}$ respectively. Then the map

$$[\ldots, f_i g_j, \ldots]_{0 \leq i \leq n, 0 \leq j \leq m}: V \to \mathbb{P}^{nm+n+m}$$

is associated with $\mathscr{L} \otimes \mathscr{M}$. (This is just the composition of the map $F \times G\colon V \times V \to \mathbb{P}^n \times \mathbb{P}^m$ with the Segre embedding $\mathbb{P}^n \times \mathbb{P}^m \to \mathbb{P}^{nm+n+m}$.) Now the desired result follows immediately from

$$\max\{\ldots, |f_i g_j|_v, \ldots\} = \max\{\ldots, |f_i|_v, \ldots\} \max\{\ldots, |g_j|_v, \ldots\}. \qquad \square$$

Definition. Let $\mathscr{L} \in \mathrm{Pic}(V)$ be any invertible sheaf. Choose sheaves $\mathscr{L}_1$ and $\mathscr{L}_2$ without basepoints such that $\mathscr{L} = \mathscr{L}_1 \otimes \mathscr{L}_2^{-1}$. Then the *height function on V associated to $\mathscr{L}$* is the function defined by

$$h_\mathscr{L} = h_{\mathscr{L}_1} - h_{\mathscr{L}_2} \in \mathscr{H}(V).$$

Note that from Theorem 3.1 and Proposition 3.2, this is independent of the choice of the $\mathscr{L}_i$'s. The fact that any invertible sheaf is the difference of basepoint-free sheaves follows from standard results in algebraic geometry (such as [Har, II.5.17].)

Theorem 3.3 (Height Machine). (a) *There exists a unique homomorphism*

$$\mathrm{Pic}(V) \to \mathscr{H}(V),$$

$$\mathscr{L} \to h_\mathscr{L},$$

with the property that if $\mathscr{L}$ has no basepoints and $F\colon V \to \mathbb{P}^n$ is a morphism associated to $\mathscr{L}$, then

$$h_\mathscr{L} = h_F + O(1).$$

(b) *If $f\colon V \to W$ is a morphism of smooth varieties, and $\mathscr{L}$ is an invertible sheaf on W, then*

$$h_{f^*\mathscr{L}} = h_\mathscr{L} \circ f + O(1).$$

[*That is, the homomorphism in* (a) *is functorial with respect to morphisms of smooth varieties.*]

PROOF. (a) This follows immediately from Theorem 3.1 and Proposition 3.2.
(b) If $\mathscr{L}$ has no basepoints, and $F\colon W \to \mathbb{P}^n$ is associated to $\mathscr{L}$, then $F \circ f$ is associated to $f^*\mathscr{L}$. $\qquad \square$

Corollary 3.4 (Finiteness). *If $\mathscr{L}$ is an ample sheaf on V, then for all constants C and d, the set*

$$\{P \in V(\overline{\mathbb{Q}})\colon h_\mathscr{L} \leq C \text{ and } [\mathbb{Q}(P)\colon \mathbb{Q}] \leq d\}$$

is finite.

PROOF. Choose some integer m so that $\mathscr{L}^{\otimes m}$ is very ample, and let $F\colon V \to \mathbb{P}^n$ be an embedding associated to $\mathscr{L}^{\otimes m}$. Then for P in this set,

$$C \geq h_\mathscr{L}(P) = (1/m) h_{\mathscr{L}^{\otimes m}}(P) + O(1) = h(F(P)) + O(1).$$

This reduces the problem to the usual height on $\mathbb{P}^n$, to which we can apply Theorem 2.1. □

There are two other basic properties of height functions which we should mention, but we will refer the reader to the literature for their proofs.

Proposition 3.5 (Positivity). *Let $\mathscr{L}$ be an invertible sheaf on V whose base locus B is not all of V. (That is, there is a rational map $F: V \to \mathbb{P}^n$ associated to $\mathscr{L}$ which is a morphism on the complement of B.) Then*

$$h_{\mathscr{L}} \geq O(1) \quad \text{on the complement of } B.$$

PROOF. [La, Chap. 4, Prop. 5.2]. □

Proposition 3.6 (Quasi-equivalence). *Let $\mathscr{L}$ and $\mathscr{M}$ be algebraically equivalent sheaves on V, and assume that $\mathscr{L}$ is ample. Then for all $\varepsilon > 0$,*

$$(1 - \varepsilon)h_{\mathscr{L}} - O(1) \leq h_{\mathscr{M}} \leq (1 + \varepsilon)h_{\mathscr{L}} + O(1).$$

(Here the $O(1)$ constants will depend on ε. Notice in particular that

$$h_{\mathscr{M}}(P)/h_{\mathscr{L}}(P) \to 1 \quad \text{as} \quad h_{\mathscr{L}}(P) \to \infty.)$$

PROOF. [La, Chap. 4, Prop. 5.3]. □

§4. Heights on Abelian Varieties

Using the height machine (Theorem 3.3), any relation between sheaves gives a corresponding relation between height functions. On an abelian variety, the theorem of the cube gives such a relation. This leads to the fundamental fact that on an abelian variety, a height function behaves essentially quadratically with respect to the group law.

Theorem 4.1. *Let $A/\bar{\mathbb{Q}}$ be an abelian variety, and let $\mathscr{L}$ be an invertible sheaf on A. Then for all $P, Q, R \in A$,*

$$h_{\mathscr{L}}(P + Q + R) - h_{\mathscr{L}}(P + Q) - h_{\mathscr{L}}(P + R) - h_{\mathscr{L}}(Q + R)$$
$$+ h_{\mathscr{L}}(P) + h_{\mathscr{L}}(Q) + h_{\mathscr{L}}(R) = O(1).$$

(As usual, the $O(1)$ constant depends on A and $\mathscr{L}$, but is independent of P, Q, R. This says that up to a bounded amount, $h_{\mathscr{L}}$ is a quadratic function.)

PROOF. Define maps

$$\pi_1, \pi_2, \pi_3, \pi_{12}, \pi_{13}, \pi_{23}, \pi_{123}: A \times A \times A \to A$$

by projecting onto the indicated components and then adding. For example,

$$\pi_1(P, Q, R) = P \quad \text{and} \quad \pi_{23}(P, Q, R) = Q + R.$$

Then the theorem of the cube ([Mum, Chap. 2, §6, Cor. 2]) says that

$$\pi_{123}^*\mathscr{L} \otimes \pi_{12}^*\mathscr{L}^{-1} \otimes \pi_{13}^*\mathscr{L}^{-1} \otimes \pi_{23}^*\mathscr{L}^{-1} \otimes \pi_1^*\mathscr{L} \otimes \pi_2^*\mathscr{L} \otimes \pi_3^*\mathscr{L}$$

is isomorphic to the trivial sheaf. Hence using the height machine (Theorem 3.3), we obtain (note $h_{\pi^*\mathscr{L}}(P, Q, R) = h_{\mathscr{L}}(\pi(P, Q, R)) + O(1)$ from Theorem 3.3b))

$$h_{\mathscr{L}}(\pi_{123}(P, Q, R)) - h_{\mathscr{L}}(\pi_{12}(P, Q, R)) - h_{\mathscr{L}}(\pi_{13}(P, Q, R)) - h_{\mathscr{L}}(\pi_{23}(P, Q, R))$$
$$+ h_{\mathscr{L}}(\pi_1(P, Q, R)) + h_{\mathscr{L}}(\pi_2(P, Q, R)) + h_{\mathscr{L}}(\pi_3(P, Q, R)) = O(1).$$

This is exactly the desired relation. □

Corollary 4.2. (a) *Let n be an integer, and let* $[n]: A \to A$ *be the multiplication-by-n map. Then*

$$h_{\mathscr{L}} \circ [n] = ((n^2 + n)/2)h_{\mathscr{L}} + ((n^2 - n)/2)h_{\mathscr{L}} \circ [-1] + O(1).$$

(*The* $O(1)$ *bound will depend on n.*) *In particular, if* $\mathscr{L}$ *is even (i.e.* $[-1]^*\mathscr{L} \approx \mathscr{L}$), *then*

$$h_{\mathscr{L}} \circ [n] = n^2 h_{\mathscr{L}} + O(1);$$

and if $\mathscr{L}$ *is odd* ($[-1]^*\mathscr{L} \approx \mathscr{L}^{-1}$), *then*

$$h_{\mathscr{L}} \circ [n] = nh_{\mathscr{L}} + O(1).$$

(b) *If* $\mathscr{L}$ *is even, then* $h_{\mathscr{L}}$ *satisifies the parallogram law (modulo* $O(1)$),

$$h_{\mathscr{L}}(P + Q) + h_{\mathscr{L}}(P - Q) = 2h_{\mathscr{L}}(P) + 2h_{\mathscr{L}}(Q) + O(1).$$

If $\mathscr{L}$ *is odd, then* $h_{\mathscr{L}}$ *is linear (modulo* $O(1)$),

$$h_{\mathscr{L}}(P + Q) = h_{\mathscr{L}}(P) + h_{\mathscr{L}}(Q) + O(1).$$

PROOF. (a) In Theorem 4.1, put $Q = [n]P$ and $R = [-1]P$. This yields the relation

$$h_{\mathscr{L}}([n + 1]P) + h_{\mathscr{L}}([n - 1]P)$$
$$= 2h_{\mathscr{L}}([n]P) + h_{\mathscr{L}}(P) + h_{\mathscr{L}}([-1]P) + O(1).$$

Since the desired result is clearly true for $n = 0, 1$, an easy induction (both upward and downward) using the above relation gives it for all n. Finally, the two cases with $\mathscr{L}$ even or odd follow immediately from

$$h_{\mathscr{L}} \circ [-1] = h_{[-1]^*\mathscr{L}} + O(1).$$

(b) If $\mathscr{L}$ is even, then from (a), $h_{\mathscr{L}}([-1]P) = h_{\mathscr{L}}(P) + O(1)$. Putting

$R = [-1]Q$ in Theorem 4.1 and using this gives the parallogram law. Similarly, if $\mathscr{L}$ is odd, then (a) implies that $h_{\mathscr{L}}([-1]P) = -h_{\mathscr{L}}(P) + O(1)$; so putting $R = [-1](P + Q)$ in Theorem 4.1 gives the desired linearity. □

Theorem 4.1 says that the height on an abelian variety is "essentially" a quadratic function. André Néron asked whether one could find an actual quadratic function which differs from the height by a bounded amount. He constructed such a function as a sum of local "quasi-quadratic" functions. At the same time, John Tate gave a simpler global construction. We will state the main result here, and refer the reader to the literature for a proof.

Theorem 4.3. *Let $A/\bar{\mathbb{Q}}$ be an abelian variety, and let $\mathscr{L}$ be an invertible sheaf on A.*

(a) *There is a unique function*

$$\hat{h}_{\mathscr{L}}: A \to \mathbb{R}$$

with the following properties:
   (i) *$\hat{h}_{\mathscr{L}}$ is a quadratic function (i.e. the map*

$$\langle\, ,\, \rangle: A \times A \to \mathbb{R},$$
$$\langle P, Q \rangle = \hat{h}_{\mathscr{L}}(P + Q) - \hat{h}_{\mathscr{L}}(P) - \hat{h}_{\mathscr{L}}(Q)$$

   is bilinear.)
   (ii) *$\hat{h}_{\mathscr{L}} = h_{\mathscr{L}} + O(1)$ on A.*
(b) *Assume now that $\mathscr{L}$ is ample and symmetric. Then*
   (i) *$\hat{h}_{\mathscr{L}}(P) \geq 0$ for all $P \in A$.*
   (ii) *$\hat{h}_{\mathscr{L}}(P) = 0$ if and only if P is a point of finite order.*
   (iii) *More generally, $\hat{h}_{\mathscr{L}}$ is a positive definite quadratic form on $A(\bar{\mathbb{Q}}) \otimes \mathbb{R}$.*

PROOF. See [La, Chap. 5, §3, 6, 7]. □

§5. The Mordell–Weil Theorem

In this section we will use the theory of height functions to deduce the strong Mordell–Weil theorem from the weak Mordell–Weil theorem. The properties of height functions which we will use are axiomatized in the following lemma.

Lemma 5.1 (Descent Lemma). *Let A be an abelian group. Suppose that there is a height function*

$$h: A \to \mathbb{R}$$

with the following three properties:

(i) Let $Q \in A$. There is a constant C_1, depending on A and Q, so that for all $P \in A$,
$$h(P + Q) \leq 2h(P) + C_1.$$

(ii) There is an integer $m \geq 2$ and a constant C_2, depending on A, so that for all $P \in A$,
$$h(mP) \geq m^2 h(P) - C_2.$$

(iii) For every constant C_3,
$$\{P \in A : h(P) \leq C_3\}$$
is a finite set.

Suppose further that for the integer m in (ii), the quotient group A/mA is finite. Then A is finitely generated.

PROOF. Choose elements $Q_1, \ldots, Q_r \in A$ to represent the finitely many cosets in A/mA. Now let $P \in A$. The idea is to show that by subtracting an appropriate linear combination of $Q_1, \ldots, Q_r$ from P, we will be able to make the height of the resulting point less than a constant which is *independent of P*. Then $Q_1, \ldots, Q_r$ and the finitely many points with height less than this constant will generate A.

Write
$$P = mP_1 + Q_{i_1} \quad \text{for some } 1 \leq i_1 \leq r.$$

Continuing in this fashion
$$P_1 = mP_2 + Q_{i_2},$$
$$\vdots$$
$$P_{n-1} = mP_n + Q_{i_n}.$$

Now for any j, we have
$$h(P_j) \leq \frac{1}{m^2}[h(mP_j) + C_2] \quad \text{from (ii)}$$
$$= \frac{1}{m^2}[h(P_{j-1} - Q_{i_j}) + C_2]$$
$$\leq \frac{1}{m^2}[2h(P_{j-1}) + C_1' + C_2] \quad \text{from (i)},$$

where we take C_1' to be the maximum of the constants from (i) for $Q = -Q_i$, $1 \leq i \leq r$. Note that C_1' and C_2 do not depend on P.

Now use the above inequality repeatedly, starting from P_n and working back to P. This yields

$$\begin{aligned} h(P_n) &\leq \left(\frac{2}{m^2}\right)^n h(P) + \left[\frac{1}{m^2} + \frac{2}{m^4} + \frac{4}{m^6} + \cdots + \frac{2^{n-1}}{m^{2n}}\right](C_1' + C_2) \\ &< \left(\frac{2}{m^2}\right)^n h(P) + \frac{C_1' + C_2}{m^2 - 2} \\ &\leq 2^{-n} h(P) + (C_1' + C_2)/2 \quad \text{since } m \geq 2. \end{aligned}$$

It follows that by taking n sufficiently large, we will have (say)

$$h(P_n) \leq 1 + (C_1' + C_2)/2.$$

Since (from above)

$$P = m^n P_n + \sum_{j=1}^{n} m^{j-1} Q_{i_j},$$

it follows that every $P \in A$ is a linear combination of the points in the set

$$\{Q_1, \ldots, Q_r\} \cup \{Q \in A : h(Q) \leq 1 + (C_1' + C_2)/2\}.$$

From (iii), this is a finite set, which proves that A is finitely generated. □

The strong Mordell–Weil theorem is now a formal consequence of the weak Mordell–Weil theorem and the standard properties of height functions.

Theorem 5.2 (Mordell–Weil). *Let K be a number field and A/K an abelian variety. Then the group $A(K)$ is finitely generated.*

PROOF. Let $m \geq 2$ be any integer (e.g. $m = 2$), and let

$$h : A \to \mathbb{R}$$

be a height function on A corresponding to a very ample symmetric line bundle. From the weak Mordell–Weil theorem [Mil, Theorem 20.10], we know that the group $A(K)/mA(K)$ finite. Thus in order to apply the descent lemma (5.1) to the group $A(K)$, we must verify the following three properties for the height function h.

(i) Fix $Q \in A(K)$. Then

$$h(P + Q) \leq 2h(P) + O(1) \quad \text{for all } P \in A(K),$$

where the $O(1)$ constant depends only on A and Q.

(ii) $\qquad h(mP) \geq m^2 h(P) + O(1) \quad \text{for all } P \in A(K),$

where the $O(1)$ constant depends only on A and m.

(iii) For every constant C,

$$\{P \in A(K) : h(P) \leq C\}$$

is a finite set.

Now (i) is immediate from Corollary 4.2(b) (remember that h takes only non-negative values), while Corollary 4.2(a) gives something stronger than (ii). Finally, (iii) is a special case of Corollary 3.4. Therefore Lemma 5.1 is applicable, and we conclude that $A(K)$ is finitely generated. □

Heights and Metrized Line Bundles

§6. Metrized Line Bundles On Spec(R)

We fix the following notation.

K a number field.
R the ring of integers of K.
M_K the usual set of absolute values on K extending those on $\mathbb{Q}$.
$M_K^\infty(M_K^0)$ the archimedean (respectively non-archimedean) absolute values in M_K.

Recall that a line bundle (or invertible sheaf) $\mathscr{L}$ on Spec(R) is nothing other than a projective R-module of rank 1.

Definition. A *metrized line bundle on* Spec(R) is a pair $(\mathscr{L}, |\cdot|)$, where $\mathscr{L}$ is a line bundle on Spec(R), and for each archimedean absolute value $v \in M_K^\infty$, $|\cdot|_v$ is a v-adic norm (metric) on the one-dimensional K_v vector space $\mathscr{L} \otimes_R K_v$. (As usual, we let $\|\cdot\|_v = |\cdot|_v^{[K_v:\mathbb{Q}_v]}$.)

The *degree of a metrized line bundle* $(\mathscr{L}, |\cdot|)$ is defined as

$$\deg(\mathscr{L}, |\cdot|) = \log \#(\mathscr{L}/Rt) - \sum_{v \in M_K^\infty} \log \|t\|_v,$$

where we choose any $t \in \mathscr{L}$ with $t \neq 0$.

Remarks. (1) One easily checks that $\deg(\mathscr{L}, |\cdot|)$ is independent of the choice of t.

(2) If R is a P.I.D., then $\mathscr{L}$ is free, so we can choose a $t \in \mathscr{L}$ so that $\mathscr{L} = Rt$. Then $\deg(\mathscr{L}, |\cdot|) = -\sum_{v \in M_K^\infty} \log \|t\|_v$. In particular, if $R = \mathbb{Z}$, then up to ± 1 there is a unique generator t for $\mathscr{L}$, and then $\deg(\mathscr{L}, |\cdot|) = -\log |t|_\infty$.

§7. Metrized Line Bundles on Varieties

We set the following notation.

V/K a projective variety.
$\mathscr{L}$ a line bundle on V/K (often assumed to be very ample).
$\mathscr{L}_P$ the fibre (stalk) of $\mathscr{L}$ at a point P of V (it is a one-dimensional $K(P)$ vector space).

Definition. Let $v \in M_K$. A *v-adic metric on $\mathscr{L}$* consists of a (non-trivial) v-adic norm $|\cdot|_v$ on each fibre $\mathscr{L}_P \otimes K_v$ such that the norms "vary continuously with $P \in V(K_v)$." [That is, if $f \in H^0(U, \mathscr{L})$ is a section on some open set U, and if $U(K_v)$ is given the v-adic topology, then the map

$$U(K_v) \to [0, \infty), \qquad P \to |f_P|_v.$$

is continuous.]

Intuition. Let $f \in H^0(V, \mathscr{L})$ be a global section, and let $D = (f) \geq 0$ be the divisor of f. Notice that

$$|f_P|_v = 0 \iff f_P = 0 \iff P \in \mathrm{Support}(D).$$

This and the continuity condition on the metric means that we should think of $|f_P|_v$ as

$$|f_P|_v \text{ "="} \text{ the } v\text{-adic distance from } P \text{ to } D.$$

Lemma 7.1. *Let $v \in M_K^\infty$, and suppose that $|\cdot|_v$ and $|\cdot|_v'$ are two v-adic metrics on $\mathscr{L}$. Then there exist constants $c_1, c_2 > 0$ such that*

$$c_1 |\cdot|_v \leq |\cdot|_v' \leq c_2 |\cdot|_v \qquad \text{on} \quad \mathscr{L} \otimes K_v.$$

PROOF. For each $P \in V(K_v)$, choose some $f_P \in \mathscr{L}_P$ with $f_P \neq 0$. Then $|f_P|_v / |f_P|_v'$ is independent of the choice of f_P, so we obtain a well-defined map

$$F: V(K_v) \to (0, \infty), \qquad P \to |f_P|_v / |f_P|_v'.$$

But F is continuous; and since V is projective, $V(K_v)$ is compact. Therefore there are constants $c_1, c_2 > 0$ such that $c_1 \leq F(P) \leq c_2$ for all $P \in V(K_v)$, which is the desired result. □

We assume now that $\mathscr{L}$ is very ample, and fix an embedding $V \subset \mathbb{P}_K^n$ corresponding to $\mathscr{L}$ (i.e. $\mathscr{L} \approx \mathscr{O}_V(1)$.) Notice that once this is done, then any point $P \in V(K)$ extends uniquely to a point in $\mathbb{P}_{\mathbb{Z}}^n(R)$; or, what is the same thing, to a map

$$P: \mathrm{Spec}(R) \to \mathbb{P}_{\mathbb{Z}}^n.$$

Hence if we are given v-adic metrics on $\mathscr{O}(1)$ for each $v \in M_K^\infty$, then by pullback $P^*\mathscr{O}(1)$ becomes a metrized line bundle on $\mathrm{Spec}(R)$.

Proposition 7.2. *With notation as above, fix v-adic metrics on $\mathscr{O}(1)$ for each $v \in M_K^\infty$. Then*

$$\deg P^*\mathscr{O}(1) = [K:\mathbb{Q}] h_{\mathscr{L}}(P) + \mathrm{O}(1).$$

(Note that the latter $\mathrm{O}(1)$ represents a bounded function, not a line bundle.)

PROOF. Using Lemma 7.1 and the definition of the degree of a metrized line bundle, we see that if the metrics on $\mathscr{O}(1)$ are changed, then the degree of

$P^*\mathcal{O}(1)$ only changes by $O(1)$. Hence it suffices to prove the proposition for any one choice of metrics on $\mathcal{O}(1)$.

Let $x_0, \ldots, x_n$ be generators for $\mathcal{O}(1)$ (i.e. projective coordinates on $\mathbb{P}^n$.) Then for each $v \in M_K^\infty$, we define a v-adic metric on $\mathcal{O}(1)$ as follows. Let $f \in H^0(\mathbb{P}^n, \mathcal{O}(1))$ be a global section. Then for each $P \in V(K_v)$,

$$|f(P)|_v = \min_{\substack{0 \le i \le n \\ x_i(P) \ne 0}} \{|(f/x_i)(P)|_v\}.$$

Since $\mathcal{O}(1)$ is generated by the global sections $x_0, \ldots, x_n$, one easily checks that this defines v-adic metrics on $\mathcal{O}(1)$. We now prove Proposition 7.2 for this particular choice of metrics.

Let $P \in \mathbb{P}^n(K)$. By symmetry, we may assume that $x_0(P) \ne 0$. We compute the degree of the metrized line bundle $P^*\mathcal{O}(1)$ using the section $P^*x_0 = x_0(P)$. First, for $v \in M_K^\infty$,

$$|x_0(P)|_v = \min_{0 \le i \le n} \{|(x_0/x_i)(P)|_v\}.$$

On the other hand,

$$P^*\mathcal{O}(1)/Rx_0(P) \approx \left(\sum_{i=0}^n Rx_i(P)\right)/Rx_0(P) \approx \left(\sum_{i=0}^n R(x_i/x_0)(P)\right)/R;$$

so

$$\# P^*\mathcal{O}(1)/Rx_0(P) = \left|N_{K/\mathbb{Q}}\left(\sum_{i=0}^n R(x_i/x_0)(P)\right)\right|^{-1}$$

$$= \prod_{v \in M_K^0} \max_{0 \le i \le n} \|(x_i/x_0)(P)\|_v.$$

Hence for this choice of metrics on $\mathcal{O}(1)$,

$$\deg P^*\mathcal{O}(1) = \log(\# P^*\mathcal{O}(1)/Rx_0(P)) - \sum_{v \in M_K^\infty} \log \|x_0(P)\|_v$$

$$= \log\left(\prod_{v \in M_K^0} \max_{0 \le i \le n} \|(x_i/x_0)(P)\|_v\right) - \sum_{v \in M_K^\infty} \log \min_{0 \le i \le n} \|(x_0/x_i)(P)\|_v$$

$$= \sum_{v \in M_K} \log \max_{0 \le i \le n} \|(x_i/x_0)(P)\|_v$$

$$= [K:\mathbb{Q}]h([1, x_1/x_0(P), \ldots, x_n/x_0(P)])$$

$$= [K:\mathbb{Q}]h_{\mathscr{L}}(P) + O(1). \qquad \square$$

§8. Distance Functions and Logarithmic Singularities

As we have seen (Lemma 7.1), any two metrics on a line bundle over a projective variety are essentially the same. The proof of this fact used a compactness argument which depends on the projectivity of the variety.

However, we will need to work with metrics which are only defined over a Zariski-open subset of the given projective variety, and it will be necessary to describe the behavior of the metric as one approaches the boundary of the open set. For this purpose, it is convenient to have a means of specifying how far one is from the boundary. We will use the following notation in this section:

V/K a projective variety.
X/K a Zariski-closed subset of V.
U the complement of X.

Definition. Let $v \in M_K^\infty$. A *logarithmic distance function for X (with respect to v)* is a map

$$d_X: U(\bar{K}_v) \to [0, \infty)$$

with the property that if $f_1 = \cdots = f_r = 0$ are local equations for X, then the quantity

$$\left| d_X(P) - \log^+ \min_{1 \le j \le r} \{|f_j(P)|_v^{-1}\} \right|$$

extends to a bounded function on any open subset of $V(\bar{K}_v)$ on which the f_j are regular. (When dealing with several absolute values, we will use the notation $d_{X,v}$. Here $\log^+(t) = \max\{\log(t), 0\}$.)

Intuition. $d_X(P)$ "$=$" $\log$ (v-adic distance from P to X)$^{-1}$.

Remarks. (1) Using the fact that $V(\bar{K}_v)$ is compact, it is not difficult to construct a logarithmic distance function and to show that any two such functions differ by $O(1)$ (i.e. by a function which is bounded on $U(\bar{K}_v)$).

(2) If Y is a Zariski-closed subset of X, then it is clear that

$$d_X(P) \ge d_Y(P) + O(1).$$

We next give the fundamental relation between distance and height functions.

Proposition 8.1. *Let $\mathscr{L}$ be an ample line bundle on V/K. Then there exists a constant $c > 0$ such that*

$$h_\mathscr{L}(P) > c\, d_X(P) + O(1) \quad \text{for all } P \text{ in } U(K).$$

Intuition. If P is not in $X(\bar{K}_v)$, then in order for P to be close to X in the v-adic topology, it is necessary for the coordinates of P to be v-adically complicated; and this causes $h_\mathscr{L}(P)$ to increase proportionally.

PROOF. If $\mathscr{M}$ is any other ample line bundle, then there are integers m and n such that $\mathscr{L}^n \otimes \mathscr{M}^{-1}$ and $\mathscr{L}^{-1} \otimes \mathscr{M}^m$ are both very ample. It follows that

$$nh_\mathscr{L} \ge h_\mathscr{M} - O(1) \quad \text{and} \quad mh_\mathscr{M} \ge h_\mathscr{L} - O(1).$$

It thus suffices to prove the proposition for any one ample line bundle $\mathscr{L}$. Therefore by choosing a very ample effective divisor D containing X, letting $\mathscr{L} = \mathcal{O}_V(D)$, and using remark (2) above, we reduce to the case that $V = \mathbb{P}^n$, $\mathscr{L} = \mathcal{O}_\mathbb{P}(1)$, and $X = \{x_0 = 0\}$ is the hyperplane at infinity.

Now $x_0/x_i = 0$ is a local equation for X (i.e. on the open set defined by $x_i \neq 0$.) Since these open sets cover $V = \mathbb{P}^n$, a distance function d_X for a given absolute value $v \in M_K^\infty$ can be defined (globally) for P in $U(\bar{K}_v)$ by

$$d_X(P) = \max_{0 \le i \le n} \log|x_0/x_i(P)|_v^{-1} = \max_{0 \le i \le n} \log|x_i/x_0(P)|_v.$$

But by definition of the height for $\mathscr{L} = \mathcal{O}_\mathbb{P}(1)$, we have for P in $U(K)$ that

$$[K:\mathbb{Q}]h_\mathscr{L}(P) = \sum_{w \in M_K} \max_{0 \le i \le n} \log\|x_i/x_0(P)\|_w \quad \text{(note each term is } \ge 1\text{)}$$

$$\ge [K_v : \mathbb{Q}_v] \max_{0 \le i \le n} \log|x_i/x_0(P)|_v$$

$$= [K_v : \mathbb{Q}_v]\, d_X(P) - O(1).$$

Definition. Let V, X, U be as above, let $\mathscr{L}$ be a line bundle on V, and let $|\cdot|'_v$ be a v-adic metric on the restriction $\mathscr{L}|_U$. We say that $|\cdot|'_v$ has *logarithmic singularities along* X if for any v-adic metric $|\cdot|_v$ defined on all of $\mathscr{L}$, there are constants $c_1, c_2 > 0$ such that

$$\max\{|\cdot|_v/|\cdot|'_v, |\cdot|'_v/|\cdot|_v\} \le c_1(d_X + 1)^{c_2} \quad \text{on } U(\bar{K}_v).$$

If $(\mathscr{L}|_U, |\cdot|')$ is a metrized line bundle, then we say that it has *logarithmic singularities along* X if $|\cdot|'_v$ has logarithmic singularities along X for each $v \in M_K^\infty$.

Remarks. (1) Note that $|\cdot|_v/|\cdot|'_v$ gives a well-defined function $U(K_v) \to (0, \infty)$. (See the proof of Lemma 7.1.)

(2) From Lemma 7.1, we see that in order to check if a metric has logarithmic singularities, it suffices to check that it has logarithmic singularities when compared with any one metric $|\cdot|_v$ defined on all of $\mathscr{L}$.

(3) If $(\mathscr{L}|_U, |\cdot|')$ is a metrized line bundle (regardless of the singularities along X), we can define a height function on $U(K)$ by taking the degree,

$$h_{\mathscr{L}, |\cdot|'}(P) = (1/[K:\mathbb{Q}]) \deg(P^*\mathscr{L}).$$

Thus if $|\cdot|'_v$ extends to a metric on all of $\mathscr{L}$, then $h_{\mathscr{L}, |\cdot|'}$ is the usual height (Proposition 7.2).

The most important fact about metrized line bundles with logarithmic singularities is that the fundamental finiteness result (Corollary 3.4) remains true.

Proposition 8.2 (Faltings [Fa]). *With notation as above, let $(\mathscr{L}|_U, |\cdot|')$ be a very ample metrized line bundle with logarithmic singularities along X. Then for* *any constant C, the set*

$$\{P \in U(K): h_{\mathscr{L},|\cdot|'}(P) < C\}$$

is finite.

PROOF. Let $|\cdot|$ be a metric defined on all of $\mathscr{L}$, let $P \in U(K)$, and let f be a section of $\mathscr{L}$ which is defined and non-zero at P. Computing $h_{\mathscr{L},|\cdot|}(P)$ as $(1/[K:\mathbb{Q}]) \deg(P^*\mathscr{L})$ (Proposition 7.2), we see that the contributions of the non-archimedean places to $h_{\mathscr{L},|\cdot|}(P)$ and $h_{\mathscr{L},|\cdot|'}(P)$ are the same. Hence using the definition of degree and of logarithmic singularity, we compute

$$[K:\mathbb{Q}]\left|h_{\mathscr{L},|\cdot|}(P) - h_{\mathscr{L},|\cdot|'}(P)\right| = \left|\sum_{v \in M_K^\infty} \log(\|f_P\|_v / \|f_P\|_v')\right|$$

$$\leq \sum_{v \in M_K^\infty} \log c_1(d_{X,v}(P) + 1)^{c_2}$$

$$\leq c_3 \log(h_{\mathscr{L},|\cdot|}(P) + 1) + c_4$$

(Proposition 8.1).

From this it immediately follows that

$$h_{\mathscr{L},|\cdot|}(P) \leq c_5 h_{\mathscr{L},|\cdot|'}(P) + c_6$$

for $P \in U(K)$, so we are reduced to looking at sets where the usual height $h_{\mathscr{L},|\cdot|}$ is bounded. Now the desired result follows from Corollary 3.4. □

REFERENCES

[De] Deligne, P. Preuve des conjectures de Tate et Shafarevitch [d'après G. Faltings]. *Séminaire Bourbaki*, Éxposé 616, 1983/84.
[Fa] Faltings, G. Endlichkeitssätze für abelsche Varietäten über Zahlkörpern. *Invent. Math.*, **73** (1983), 349–366.
[Har] Hartshorne, R. *Algebraic Geometry*. Springer-Verlag: New York, 1977.
[La] Lang, S. *Fundamentals of Diophantine Geometry*. Springer-Verlag: New York, 1983.
[Mil] Milne, J. S. Abelian varieties, this volume, pp. 103–150.
[Mum] Mumford, D. *Abelian Varieties*. Oxford University Press: Oxford, 1974.

CHAPTER VII

Jacobian Varieties

J. S. MILNE

This chapter contains a detailed treatment of Jacobian varieties. Sections 2, 5, and 6 prove the basic properties of Jacobian varieties starting from the definition in Section 1, while the construction of the Jacobian is carried out in Sections 3 and 4. The remaining sections are largely independent of one another.

The conventions are the same as those listed at the start of Chapter V, "Abelian Varieties" (see also those at the start of Section 5 of that chapter).

§1. Definitions

Recall that for a scheme S, Pic(S) denotes the group $H^1(S, \mathcal{O}_S^\times)$ of isomorphism classes of invertible sheaves on S, and that $S \mapsto \text{Pic}(S)$ is a functor from the category of schemes over k to that of abelian groups.

Let C be a complete nonsingular curve over k. The degree of a divisor $D = \sum n_i P_i$ on C is $\sum n_i [k(P_i) : k]$. Since every invertible sheaf $\mathscr{L}$ on C is of the form $\mathscr{L}(D)$ for some divisor D, and D is uniquely determined up to linear equivalence, we can define $\deg(\mathscr{L}) = \deg(D)$. Then $\deg(\mathscr{L}^n) = \deg(nD) = n \cdot \deg(D)$, and the Riemann–Roch theorem says that

$$\chi(C, \mathscr{L}^n) = n \cdot \deg(\mathscr{L}) + 1 - g.$$

This gives a more canonical description of $\deg(\mathscr{L})$: when $\chi(C, \mathscr{L}^n)$ is written as a polynomial in n, $\deg(\mathscr{L})$ is the leading coefficient. We write $\text{Pic}^0(C)$ for the group of isomorphism classes of invertible sheaves of degree 0 on C.

Let T be a connected scheme over k, and let $\mathscr{L}$ be an invertible sheaf on $C \times T$ (by which we mean $C \times_{\text{spec}(k)} T$). Then [14, 4.2(b)] shows that

$\chi(C_t, \mathcal{L}_t^n)$, and therefore $\deg(\mathcal{L}_t)$, is independent of t; moreover, the constant degree of $\mathcal{L}_t$ is invariant under base change relative to maps $T' \to T$. Note that for a sheaf $\mathcal{M}$ on $C \times T$, $(q^*\mathcal{M})_t$ is isomorphic to $\mathcal{O}_{C_t}$ and, in particular, has degree 0. Let

$$P_C^0(T) = \{\mathcal{L} \in \mathrm{Pic}(C \times T) | \deg(\mathcal{L}_t) = 0 \text{ all } t\}/q^* \mathrm{Pic}(T).$$

We may think of $P_C^0(T)$ as being the group of families of invertible sheaves on C of degree 0 parametrized by T, modulo the trivial families. Note that P_C^0 is a functor from schemes over k to abelian groups. It is this functor that the Jacobian attempts to represent.

Theorem 1.1. *There is an abelian variety J over k and a morphism of functors $\iota: P_C^0 \to J$ such that $\iota: P_C^0(T) \to J(T)$ is an isomorphism whenever $C(T)$ is nonempty.*

Let k' be a finite Galois extension of k such that $C(k')$ is nonempty, and let G be the Galois group of k' over k. Then for every scheme T over k, $C(T_{k'})$ is nonempty, and so $\iota(T_{k'}): P_C^0(T_{k'}) \to J(T_{k'})$ is an isomorphism. As

$$J(T) \stackrel{\mathrm{df}}{=} \mathrm{Mor}_k(T, J) = \mathrm{Mor}_{k'}(T_{k'}, J_{k'})^G = J(T_{k'})^G,$$

we see that J represents the functor $T \mapsto P_C^0(T_{k'})^G$, and this implies that the pair (J, ι) is uniquely determined up to a unique isomorphism by the condition in the theorem. The variety J is called the *Jacobian variety* of C. Note that for any field $k' \supset k$ in which C has a rational point, ι defines an isomorphism $\mathrm{Pic}^0(C) \stackrel{\approx}{\to} J(k')$.

When C has a k-rational point, the definition takes on a more attractive form. A *pointed k-scheme* is a connected k-scheme S together with an element $s \in S(k)$. Abelian varieties will always be regarded as being pointed by the zero element. A *divisorial correspondence* between two pointed schemes (S, s) and (T, t) over k is an invertible sheaf $\mathcal{L}$ on $S \times T$ such that $\mathcal{L}|S \times \{t\}$ and $\mathcal{L}|\{s\} \times T$ are both trivial.

Theorem 1.2. *Let P be a k-rational point on C. Then there is a divisorial correspondence $\mathcal{M}^P$ between (C, P) and J such that, for every divisorial correspondence $\mathcal{L}$ between (C, P) and a pointed k-scheme (T, t), there exists a unique morphism $\varphi: T \to J$ such that $\varphi(t) = 0$ and $(1 \times \varphi)^* \mathcal{M}^P \approx \mathcal{L}$.*

Regard $\mathcal{M}^P$ as an element of $\mathrm{Pic}(C \times J)$; then the pair $(J, \mathcal{M}^P)$ is uniquely determined up to a unique isomorphism by the condition in (1.2). Note that each element of $\mathrm{Pic}^0(C)$ is represented by exactly one sheaf $\mathcal{M}_a$, $a \in J(k)$, and the map $\varphi: T \to J$ sends $t \in T(k)$ to the unique a such that $\mathcal{M}_a \approx \mathcal{L}_t$.

Theorem 1.1 will be proved in Section 4. Here we merely show that it implies (1.2).

Lemma 1.3. *Theorem 1.1 implies Theorem 1.2.*

Jacobian Varieties 169

Proof. Assume there is a k-rational point P on C. Then for any k-scheme T, the projection $q\colon C \times T \to T$ has a section $s = (t \mapsto (P, t))$, which induces a map $s^* = (\mathscr{L} \mapsto \mathscr{L}|\{P\} \times T)\colon \text{Pic}(C \times T) \to \text{Pic}(T)$ such that $s^* \circ q^* = \text{id}$. Consequently, $\text{Pic}(C \times T) = \text{Im}(q^*) \oplus \text{Ker}(s^*)$, and so $P_C^0(T)$ can be identified with

$$P'(T) = \{\mathscr{L} \in \text{Pic}(C \times T) | \deg(\mathscr{L}_t) = 0 \text{ all } t, \mathscr{L}|\{P\} \times T \text{ is trivial}\}.$$

Now assume (1.1). As $C(T)$ is nonempty for all k-schemes T, J represents the functor $P_C^0 = P'$. This means that there is an element $\mathscr{M}$ of $P'(J)$ (corresponding to id: $J \to J$ under ι) such that, for every k-scheme T and $\mathscr{L} \in P'(T)$, there is a unique morphism $\varphi\colon T \to J$ such that $(1 \times \varphi)^*\mathscr{M} \approx \mathscr{L}$. In particular, for each invertible sheaf $\mathscr{L}$ on C of degree 0, there is a unique $a \in J(k)$ such that $\mathscr{M}_a \approx \mathscr{L}$. After replacing $\mathscr{M}$ with $(1 \times t_a)^*\mathscr{M}$ for a suitable $a \in J(k)$, we can assume that $\mathscr{M}_0$ is trivial, and therefore that $\mathscr{M}$ is a divisorial correspondence between (C, P) and J. It is clear that $\mathscr{M}$ has the universal property required by (1.2). □

Exercise 1.4. Let $(J, \mathscr{M}^P)$ be a pair having the universal property in (1.2) relative to some point P on C. Show that J is the Jacobian of C.

We next make some remarks concerning the relation between P_C^0 and J in the case that C does not have a k-rational point.

Remark 1.5. For all k-schemes T, $\iota(T)\colon P_C^0(T) \to J(T)$ is injective. The proof of this is based on two observations. Firstly, because C is a complete variety $H^0(C, \mathcal{O}_C) = k$, and this holds universally: for any k-scheme T, the canonical map $\mathcal{O}_T \to q_*\mathcal{O}_{C \times T}$ is an isomorphism. Secondly, for any morphism $q\colon X \to T$ of schemes such that $\mathcal{O}_T \xrightarrow{\approx} q_*\mathcal{O}_X$, the functor $\mathscr{M} \mapsto q^*\mathscr{M}$ from the category of locally free $\mathcal{O}_T$-modules of finite-type to the category of locally free $\mathcal{O}_X$-modules of finite-type is fully faithful, and the essential image is formed of those modules $\mathscr{F}$ on X such that $q_*\mathscr{F}$ is locally free and the canonical map $q^*(q_*\mathscr{F}) \to \mathscr{F}$ is an isomorphism. (The proof is similar to that of [14, 5.1].)

Now let $\mathscr{L}$ be an invertible sheaf on $C \times T$ that has degree 0 on the fibres and which maps to zero in $J(T)$; we have to show that $\mathscr{L} \approx q^*\mathscr{M}$ for some invertible sheaf $\mathscr{M}$ on T. Let k' be a finite extension of k such that C has a k'-rational point, and let $\mathscr{L}'$ be the inverse image of $\mathscr{L}$ on $(C \times T)_{k'}$. Then $\mathscr{L}'$ maps to zero in $J(T_{k'})$, and so (by definition of J) we must have $\mathscr{L}' \approx q^*\mathscr{M}'$ for some invertible sheaf $\mathscr{M}'$ on $T_{k'}$. Therefore $q_*\mathscr{L}'$ is locally free of rank one on $T_{k'}$, and the canonical map $q^*(q_*\mathscr{L}') \to \mathscr{L}'$ is an isomorphism. But $q_*\mathscr{L}'$ is the inverse image of $q_*\mathscr{L}$ under $T' \to T$ (see [14, 4.2a]), and elementary descent theory (cf. (1.8) below) shows that the properties of $\mathscr{L}'$ in the last sentence descend to $\mathscr{L}$; therefore $\mathscr{L} \approx q^*\mathscr{M}$ with $\mathscr{M} = q_*\mathscr{L}$.

Remark 1.6. It is then sometimes possible to compute the cokernel to $\iota\colon P_C^0(k) \to J(k)$. There is always an exact sequence

$$0 \to P_C^0(k) \to J(k) \to \text{Br}(k),$$

where $\mathrm{Br}(k)$ is the Brauer group of k. When k is a finite extension of $\mathbb{Q}_p$, $\mathrm{Br}(k) = \mathbb{Q}/\mathbb{Z}$, and it is known (see [11, p. 130]) that the image of $J(k)$ in $\mathrm{Br}(k)$ is $P^{-1}\mathbb{Z}/\mathbb{Z}$, where P (the *period* of C) is the greatest common divisor of the degrees of the k-rational divisor classes on C.

Remark 1.7. Regard P_C^0 as a presheaf on the large étale site over C; then the precise relation between J and P_C^0 is that J represents the sheaf associated with P_C^0 (see [6, §5]).

Finally, we show that it suffices to prove (1.1) after an extension of the base field. For the sake of reference, we first state a result from descent theory. Let k' be a finite Galois extension of a field k with Galois group G, and let V be a variety over k'. A *descent datum* for V relative to k'/k is a collection of isomorphisms $\varphi_\sigma \colon \sigma V \to V$, one for each $\sigma \in G$, such that $\varphi_{\tau\sigma} = \varphi_\tau \circ \tau\varphi_\sigma$ for all σ and τ. There is an obvious notion of a morphism of varieties preserving the descent data. Note that for a variety V over k, $V_{k'}$ has a canonical descent datum. If V is a variety over k and $V' = V_{k'}$, then a descent datum on an $\mathcal{O}_{V'}$-module $\mathcal{M}$ is a family of isomorphisms $\varphi_\sigma \colon \sigma\mathcal{M} \to \mathcal{M}$ such that $\varphi_{\tau\sigma} = \varphi_\tau \circ \tau\varphi_\sigma$ for all σ and τ.

Proposition 1.8. *Let k'/k be a finite Galois extension with Galois group G.*

(a) *The map sending a variety V over k to $V_{k'}$ endowed with its canonical descent datum defines an equivalence between the category of quasi-projective varieties over k and that of quasi-projective varieties over k' endowed with a descent datum.*

(b) *Let V be a variety over k, and let $V' = V_{k'}$. The map sending an $\mathcal{O}_V$-module $\mathcal{M}$ to $\mathcal{M}' = \mathcal{O}_{V'} \otimes \mathcal{M}$ endowed with its canonical descent datum defines an equivalence between the category of coherent $\mathcal{O}_V$-modules and that of coherent $\mathcal{O}_{V'}$-modules endowed with a descent datum. Moreover, if $\mathcal{M}'$ is locally free, then so also is $\mathcal{M}$.*

PROOF. See [17, V. 20] or [19, §17]. (For the final statement, note that being locally free is equivalent to being flat, and that V' is faithfully flat over V.) □

Proposition 1.9. *Let k' be a finite separable extension of k; if (1.1) is true for $C_{k'}$, then it is true for C.*

PROOF. After possibly enlarging k', we can assume that it is Galois over k (with Galois group G, say) and that $C(k')$ is nonempty. Let J' be the Jacobian of $C_{k'}$. Then J' represents $P_{C_{k'}}^0$, and so there is a universal $\mathcal{M}$ in $P_C^0(J')$. For any $\sigma \in G$, $\sigma\mathcal{M} \in P_C^0(\sigma J')$, and so there is a unique map $\varphi_\sigma \colon \sigma J' \to J'$ such that $(1 \times \varphi_\sigma)^* \mathcal{M} = \sigma\mathcal{M}$ (in $P_C^0(\sigma J')$). One checks directly that $\varphi_{\tau\sigma} = \varphi_\tau \circ \tau\varphi_\sigma$; in particular, $\varphi_\sigma \circ \sigma\varphi_{\sigma^{-1}} = \varphi_{\mathrm{id}}$, and so the φ_σ are isomorphisms and define a descent datum on J'. We conclude from (1.8) that J' has a model J over k such that the map $P_C^0(T_{k'}) \to J(T_{k'})$ is G-equivariant for all k-schemes T. In particular, for all T, there is a map $P_C^0(T) \to P_C^0(T_{k'})^G \xrightarrow{\approx} J(k')^G = J(k)$. To see that the map is an isomorphism when $C(T)$ is nonempty, we have to show that in this case $P_C^0(T) \to P_C^0(T_{k'})^G$ is an isomorphism. Let $s \in C(T)$; then (cf. the proof of (1.3)), we can identify $P_C^0(T_{k'})$ with the set of isomorphism classes of pairs $(\mathscr{L}, \alpha)$ where $\mathscr{L}$ is an invertible sheaf on $C \times T_{k'}$ whose fibres are of degree 0 and α is an isomorphism $\mathcal{O}_{T_{k'}} \xrightarrow{\approx} (1, s)^*\mathscr{L}$. Such pairs are rigid—they have no automorphisms—and so each such pair fixed under G has a canonical descent datum, and therefore arises from an invertible sheaf on $C \times T$. □

§2. The Canonical Maps from C to its Jacobian Variety

Throughout this section, C will be a complete nonsingular curve, and J will be its Jacobian variety (assumed to exist).

Proposition 2.1. *The tangent space to J at 0 is canonically isomorphic to $H^1(C, \mathcal{O}_C)$; consequently, the dimension of J is equal to the genus of C.*

PROOF. The tangent space $T_0(J)$ is equal to the kernel of $J(k[\varepsilon]) \to J(k)$, where $k[\varepsilon]$ is the ring in which $\varepsilon^2 = 0$ (see [8, II, Ex. 2.8]). Analogously, we define the tangent space $T_0(P_C^0)$ to P_C^0 at 0 to be the kernel of $P_C^0(k[\varepsilon]) \to P_C^0(k)$. From the definition of J, we obtain a map of k-linear vector spaces $T_0(P_C^0) \to T_0(J)$ which is an isomorphism if $C(k) \neq \emptyset$. Since the vector spaces and the map commute with base change, it follows that the map is always an isomorphism.

Let $C_\varepsilon = C_{k[\varepsilon]}$; then, by definition, $P_C^0(k[\varepsilon])$ is equal to the group of invertible sheaves on C_ε whose restrictions to the closed subscheme C of C_ε have degree zero. It follows that $T_0(P_C^0)$ is equal to the kernel of $H^1(C_\varepsilon, \mathcal{O}_{C_\varepsilon}^\times) \to H^1(C, \mathcal{O}_C^\times)$. The scheme C_ε has the same underlying topological space as C, but $\mathcal{O}_{C_\varepsilon} = \mathcal{O}_C \otimes_k k[\varepsilon] = \mathcal{O}_C \oplus \mathcal{O}_C \varepsilon$. Therefore we can identify the sheaf $\mathcal{O}_{C_\varepsilon}^\times$ on C_ε with the sheaf $\mathcal{O}_C^\times \oplus \mathcal{O}_C \varepsilon$ on C, and so $H^1(C_\varepsilon, \mathcal{O}_{C_\varepsilon}^\times) = H^1(C, \mathcal{O}_C^\times) \oplus H^1(C, \mathcal{O}_C \varepsilon)$. It follows that the map $a \mapsto \exp(a\varepsilon) = 1 + a\varepsilon$, $\mathcal{O}_C \to \mathcal{O}_{C_\varepsilon}^\times$, induces an isomorphism $H^1(C, \mathcal{O}_C) \to T_0(P_C^0)$. This completes the proof. □

Let $P \in C(k)$, and let $\mathscr{L}^P$ be the invertible sheaf $\mathscr{L}(\Delta - C \times \{P\} - \{P\} \times C)$ on $C \times C$, where Δ denotes the diagonal. Note that $\mathscr{L}^P$ is symmetric and that $\mathscr{L}^P|C \times \{Q\} \approx \mathscr{L}(Q - P)$. In particular, $\mathscr{L}^P|\{P\} \times C$ and $\mathscr{L}^P|C \times \{P\}$ are both trivial, and so $\mathscr{L}^P$ is a divisorial correspondence between (C, P) and itself. Therefore, according to (1.2) there is a unique map $f^P: C \to J$ such that $f^P(P) = 0$ and $(1 \times f^P)^*\mathscr{M}^P \approx \mathscr{L}^P$. When $J(k)$ is identified with $\text{Pic}^0(C)$,

$f^P: C(k) \to J(k)$ becomes identified with the map $Q \mapsto \mathscr{L}(Q) \otimes \mathscr{L}(P)^{-1}$ (or, in terms of divisors, the map sending Q to the linear equivalence class $[Q - P]$ of $Q - P$). Note that the map $\sum n_Q Q \mapsto \sum n_Q f^P(Q) = [\sum n_Q Q]$ from the group of divisors of degree zero on C to $J(k)$ induced by f^P is simply the map defined by ι. In particular, it is independent of P, is surjective, and its kernel consists of the principal divisors.

From its definition (or from the above descriptions of its action on the points) it is clear that if P' is a second point on C, then $f^{P'}$ is the composite of f^P with the translation map $t_{[P-P']}$, and that if P is defined over a Galois extension k' of k, then $\sigma f^P = f^{\sigma P}$ for all $\sigma \in \mathrm{Gal}(k'/k)$.

If C has genus zero, then (2.1) shows that $J = 0$. From now on we assume that C has genus $g > 0$.

Proposition 2.2. *The map* $(f^P)^*: \Gamma(J, \Omega_J^1) \to \Gamma(C, \Omega_C^1)$ *is an isomorphism.*

PROOF. As for any group variety, the canonical map $h_J: \Gamma(J, \Omega_J^1) \to T_0(J)^\vee$ is an isomorphism [18, III, 5.2]. Also there is a well-known duality between $\Gamma(C, \Omega_C^1)$ and $H^1(C, \mathcal{O}_C)$. We leave it as an exercise to the reader (unfortunately rather complicated) that the following diagram commutes:

$$\begin{array}{ccc} \Gamma(J, \Omega_J^1) & \xrightarrow{f^*} & \Gamma(C, \Omega_C^1) \\ h_J \downarrow \approx & & \downarrow \approx \\ T_0(J)^\vee & \xrightarrow{\approx} & H^1(C, \mathcal{O}_C)^\vee \end{array} \qquad \text{(dual of isomorphism in (2.1)).}$$

Proposition 2.3. *The map f^P is a closed immersion (that is, its image $f^P(C)$ is closed and f^P is an isomorphism from C onto $f^P(C)$); in particular, $f^P(C)$ is nonsingular.*

PROOF. It suffices to prove this in the case that k is algebraically closed.

Lemma 2.4. *Let $f: V \to W$ be a map of varieties over an algebraically closed field k, and assume that V is complete. If the map $V(k) \to W(k)$ defined by f is injective and, for all closed points Q of V, the map on tangent spaces $T_Q(V) \to T_{fQ}(W)$ is injective, then f is a closed immersion.*

PROOF. The proof is the same as that of the "if" part of [8, II, 7.3]. (Briefly, the image of f is closed because V is complete, and the condition on the tangent spaces (together with Nakayama's lemma) shows that the maps $\mathcal{O}_{fQ} \to \mathcal{O}_Q$ on the local rings are surjective.) □

We apply the lemma to $f = f^P$. If $f(Q) = f(Q')$ for some Q and Q' in $C(k)$, then the divisors $Q - P$ and $Q' - P$ are linearly equivalent. This implies that $Q - Q'$ is linearly equivalent to zero, which is impossible if $Q \neq Q'$ because C has genus > 0. Consequently, f is injective, and it remains to show that the maps on tangent spaces $(df^P)_Q: T_Q(C) \to T_{fQ}(J)$ are injective. Because f^Q differs from f^P by a translation, it suffices to do this in the case that $Q = P$. The dual of $(df^P)_P: T_P(C) \to T_0(J)$ is clearly $\Gamma(J, \Omega^1) \xrightarrow{f^*} \Gamma(C, \Omega^1) \xrightarrow{h_C} T_P(C)^\vee$, where h_C is the canonical map, and it remains to show that h_C is surjective. The kernel of h_C is $\{\omega \in \Gamma(C, \Omega^1) | \omega(P) = 0\} = \Gamma(C, \Omega^1(-P))$, which is dual to $H^1(C, \mathscr{L}(P))$. The Riemann–Roch theorem shows that this last group has dimension $g - 1$, and so $\mathrm{Ker}(h_C) \neq \Gamma(C, \Omega^1)$: h_C is surjective, and the proof is complete. $\square$

We now assume that $k = \mathbb{C}$ and sketch the relation between the abstract and classical definitions of the Jacobian. In this case, $\Gamma(C(\mathbb{C}), \Omega^1)$ (where Ω^1 denotes the sheaf of holomorphic differentials in the sense of complex analysis) is a complex vector space of dimension g, and one shows in the theory of abelian integrals that the map $\sigma \mapsto (\omega \mapsto \int_\sigma \omega)$ embeds $H_1(C(\mathbb{C}), \mathbb{Z})$ as a lattice into the dual space $\Gamma(C(\mathbb{C}), \Omega^1)^\vee$. Therefore $J^{an} \stackrel{df}{=} \Gamma(C(\mathbb{C}), \Omega^1)^\vee / H_1(C(\mathbb{C}), \mathbb{Z})$ is a complex torus, and the pairing

$$H_1(C(\mathbb{C}), \mathbb{Z}) \times H_1(C(\mathbb{C}), \mathbb{Z}) \to \mathbb{Z}$$

defined by Poincaré duality gives a nondegenerate Riemann form on J^{an}. Therefore J^{an} is an abelian variety over $\mathbb{C}$. For each P there is a canonical map $g^P: C \to J^{an}$ sending a point Q to the element represented by $(\omega \mapsto \int_\gamma \omega)$, where γ is any path from P to Q. Define $e: \Gamma(C(\mathbb{C}), \Omega^1)^\vee \to J(\mathbb{C})$ to be the surjection in the diagram:

$$\begin{array}{ccc} \Gamma(C(\mathbb{C}), \Omega^1)^\vee & \twoheadrightarrow & J(\mathbb{C}) \\ f^{*\vee} \downarrow \approx & & \uparrow \exp \\ \Gamma(J, \Omega^1)^\vee & \xrightarrow{\approx} & T_0(J). \end{array}$$

Note that if $\Gamma(C(\mathbb{C}), \Omega^1)^\vee$ is identified with $T_P(C)$, then $(de)_0 = (df^P)_P$. It follows that if γ is a path from P to Q and $l = (\omega \mapsto \int_\gamma \omega)$, then $e(l) = f^P(Q)$.

Theorem 2.5. *The canonical surjection* $e: \Gamma(C(\mathbb{C}), \Omega^1)^\vee \twoheadrightarrow J(\mathbb{C})$ *induces an isomorphism* $J^{an} \to J$ *carrying* g^P *into* f^P.

PROOF. We have to show that the kernel of e is $H_1(C(\mathbb{C}), \mathbb{Z})$, but this follows from Abel's theorem and the Jacobi inversion theorem.

(Abel) Let $P_1, \ldots, P_r$ and $Q_1, \ldots, Q_r$ be elements of $C(\mathbb{C})$; then there is a meromorphic function on $C(\mathbb{C})$ with its poles at the P_i and its zeros at the Q_i if and only if for any paths γ_i from P to P_i and γ'_i from P to Q_i there exists a γ in $H_1(C(\mathbb{C}), \mathbb{Z})$ such that

$$\sum \int_{\gamma_i} \omega - \sum \int_{\gamma'_i} \omega = \int_\gamma \omega \quad \text{all } \omega.$$

(Jacobi) Let l be a linear mapping $\Gamma(C(\mathbb{C}), \Omega^1) \to \mathbb{C}$. Then there exist g

points $P_1, \ldots, P_g$ on $C(\mathbb{C})$ and paths $\gamma_1, \ldots, \gamma_g$ from P to P_i such that $l(\omega) = \sum \int_{\gamma_i} \omega$ for all $\omega \in \Gamma(C(\mathbb{C}), \Omega^1)$.

Let $l \in \Gamma(C(\mathbb{C}), \Omega^1)^{\vee}$; we may assume it is defined by g points $P_1, \ldots, P_g$. Then l maps to zero in $J(\mathbb{C})$ if and only if the divisor $\sum P_i - gP$ is linearly equivalent to zero, and Abel's theorem shows that this is equivalent to l lying in $H_1(C(\mathbb{C}), \mathbb{Z})$. □

§3. The Symmetric Powers of a Curve

Both in order to understand the structure of the Jacobian, and as an aid in its construction, we shall need to study the symmetric powers of C.

For any variety V, the symmetric group S_r on r letters acts on the product of r copies V^r of V by permuting the factors, and we want to define the rth symmetric power $V^{(r)}$ of V to be the quotient $S_r \backslash V^r$. The next proposition demonstrates the existence of $V^{(r)}$ and lists its main properties.

A morphism $\varphi: V^r \to T$ is said to be *symmetric* if $\varphi \circ \sigma = \varphi$ for all σ in S_r.

Proposition 3.1. *Let V be a variety over k. Then there is a variety $V^{(r)}$ and a symmetric morphism $\pi: V^r \to V^{(r)}$ having the following properties:*

(a) *as a topological space, $(V^{(r)}, \pi)$ is the quotient of V^r by S_r;*
(b) *for any open affine subset U of V, $U^{(r)}$ is an open affine subset of $V^{(r)}$ and $\Gamma(U^{(r)}, \mathcal{O}_{V^{(r)}}) = \Gamma(U^r, \mathcal{O}_{V^r})^{S_r}$ (set of elements fixed by the action of S_r).*

The pair $(V^{(r)}, \pi)$ has the following universal property: every symmetric k-morphism $\varphi: V^r \to T$ factors uniquely through π.

The map π is finite, surjective, and separable.

PROOF. If V is affine, say $V = \operatorname{spec} A$, define $V^{(r)}$ to be $\operatorname{spec}((A \otimes_k \cdots \otimes_k A)^{S_r})$. In the general case, write V as a union $\bigcup U_i$ of open affines, and construct V by patching together the $U_i^{(r)}$. See [16, II, §7, p. 66 and III, §11, p. 112] for the details. □

The pair $(V^{(r)}, \pi)$ is uniquely determined up to a unique isomorphism by the conditions of the proposition. It is called the rth *symmetric power* of V.

Proposition 3.2. *The symmetric power $C^{(r)}$ of a nonsingular curve is nonsingular.*

PROOF. We may assume that k is algebraically closed. The most likely candidate for a singular point on $C^{(r)}$ is the image of Q of a fixed point $(P, \ldots, P)$ of S_r on C^r, where P is a closed point of C. The completion $\hat{\mathcal{O}}_P$ of the local ring at P is isomorphic to $k[[X]]$, and so

$$\hat{\mathcal{O}}_{(P, \ldots, P)} \approx k[[X]] \hat{\otimes} \cdots \hat{\otimes} k[[X]] \approx k[[X_1, \ldots, X_r]].$$

It follows that $\hat{\mathcal{O}}_Q \approx k[[X_1, \ldots, X_r]]^{S_r}$ where S_r acts by permuting the variables. The fundamental theorem on symmetric functions says that, over any ring, a symmetric polynomial can be expressed as a polynomial in the elementary symmetric functions $\sigma_1, \ldots, \sigma_r$. This implies that

$$k[[X_1, \ldots, X_r]]^{S_r} = k[[\sigma_1, \ldots, \sigma_r]],$$

which is regular, and so Q is nonsingular.

For a general point $Q = \pi(P, P, \ldots, P', \ldots)$ with P occuring r' times, P' occuring r'' times, and so on,

$$\hat{\mathcal{O}}_Q \approx k[[X_1, \ldots, X_{r'}]]^{S_{r'}} \hat{\otimes} k[[X_1, \ldots, X_{r''}]]^{S_{r''}} \hat{\otimes} \ldots,$$

which the same argument shows to be regular. $\square$

Remark 3.3. The reader may find it surprising that the fixed points of the action of S_r on C^r do not force singularities on $C^{(r)}$. The following remarks may help clarify the situation. Let G be a finite group acting effectively on a nonsingular variety V, and suppose that the quotient variety $W = G\backslash V$ exists. Then $V \to W$ is ramified exactly at the fixed points of the action. A purity theorem [5, X, 3.1] says W can be nonsingular only if the ramification locus is empty or has pure codimension 1 in V. As the ramification locus of V^r over $V^{(r)}$ has pure codimension $\dim(V)$, this implies that $V^{(r)}$ can be nonsingular only if V is a curve.

Let K be field containing k. If K is algebraically closed, then (3.1a) shows that $C^{(r)}(K) = S_r\backslash C(K)^r$, and so a point of $C^{(r)}$ with coordinates in K is an unordered r-tuple of K-rational points. This is the same thing as an effective divisor of degree r on C_K. When K is perfect, the divisors on C_K can be identified with those on $C_{\bar{K}}$ fixed under the action of $\mathrm{Gal}(\bar{K}/K)$. Since the same is true of the points on $C^{(r)}$, we see again that $C^{(r)}(K)$ can be identified with the set of effective divisors of degree r on C. In the remainder of this section we shall show that $C^{(r)}(T)$ has a similar interpretation for any k-scheme. (Since this is mainly needed for the construction of J, the reader more interested in the properties of J can pass to the Section 5.)

Let X be a scheme over k. Recall [8, II, 6, p. 145] that a Cartier divisor D is *effective* if it can be represented by a family $(U_i, g_i)_i$ with the g_i in $\Gamma(U_i, \mathcal{O}_X)$. Let $\mathcal{I}(D)$ be the subsheaf of $\mathcal{O}_X$ such that $\mathcal{I}(D)|U_i$ is generated by g_i. Then $\mathcal{I}(D) = \mathcal{L}(-D)$, and there is an exact sequence

$$0 \to \mathcal{I}(D) \to \mathcal{O}_X \to \mathcal{O}_D \to 0,$$

where $\mathcal{O}_D$ is the structure sheaf of the closed subscheme of X associated with D. The closed subschemes arising from effective Cartier divisors are precisely those whose sheaf of ideals can be locally generated by a single element that is not a zero-divisor. We shall often identify D with its associated closed subscheme.

For example, let $T = \mathbb{A}^1 = \operatorname{Spec} k[Y]$, and let D be the Cartier divisor associated with the Weil divisor nP, where P is the origin. Then D is represented by $(Y^n, \mathbb{A}^1)$, and the associated subscheme is $\operatorname{Spec}(k[Y]/(Y^n))$.

Definition 3.4. Let $\pi\colon X \to T$ be a morphism of k-schemes. A *relative effective Cartier divisor* on X/T is a Cartier divisor on X that is flat over T when regarded as a subscheme of X.

Loosely speaking, the flatness condition means that the divisor has no vertical components, that is, no components contained in a fibre. When T is affine, say $T = \operatorname{spec}(R)$, then a subscheme D of X is a relative effective Cartier divisor if and only if there exists an open affine covering $X = \bigcup U_i$ and $g_i \in \Gamma(U_i, \mathcal{O}_X) = R_i$ such that:

(a) $D \cap U_i = \operatorname{spec}(R_i/g_i R_i)$;
(b) g_i is not a zero-divisor; and
(c) $R_i/g_i R_i$ is flat over R, for all i.

Henceforth all divisors will be Cartier divisors.

Lemma 3.5. *If D_1 and D_2 are relative effective divisors on X/T, then so also is their sum $D_1 + D_2$.*

PROOF. It suffices to prove this in the case that T is affine, say $T = \operatorname{spec}(R)$. We have to check that if conditions (b) and (c) above hold for g_i and g_i', then they also hold for $g_i g_i'$. Condition (b) is obvious, and the flatness of $R_i/g_i g_i' R_i$ over R follows from the exact sequence

$$0 \to R_i/g_i R_i \xrightarrow{g_i'} R_i/g_i g_i' R_i \to R_i/g_i' R_i \to 0,$$

which exhibits it as an extension of flat modules. □

Remark 3.6. Let D be a relative effective divisor on X/T. On tensoring the inclusion $\mathcal{I}(D) \hookrightarrow \mathcal{O}_X$ with $\mathcal{L}(D)$ we obtain an inclusion $\mathcal{O}_X \hookrightarrow \mathcal{L}(D)$ and hence a canonical global section s_D of $\mathcal{L}(D)$. For example, in the case that T is affine and D is represented as in the above example, $\mathcal{L}(D)|U_i$ is $g_i^{-1} R_i$ and $s_D|U_i$ is the identity element in R_i.

The map $D \mapsto (\mathcal{L}(D), s_D)$ defines a one-to-one correspondence between relative effective divisors on X/T and isomorphism classes of pairs $(\mathcal{L}, s)$ where $\mathcal{L}$ is an invertible sheaf on X and $s \in \Gamma(X, \mathcal{L})$ is such that

$$0 \to \mathcal{O}_X \xrightarrow{s} \mathcal{L} \to \mathcal{L}/s\mathcal{O}_X \to 0$$

is exact and $\mathcal{L}/s\mathcal{O}_X$ is flat over T.

Observe that, in the case that X is flat over T, $\mathcal{L}/s\mathcal{O}_X$ is flat over T if and only if, for all t in T, s does not become a zero-divisor in $\mathcal{L} \otimes \mathcal{O}_{X_t}$. (Use that an R-module M is flat if $\operatorname{Tor}_1^R(M, N) = 0$ for all finitely generated modules N, and that any such module N has a composition series whose quotients are the quotient of R by a prime ideal; therefore the criterion has only to be checked with N equal to such a module.)

Proposition 3.7. *Consider the Cartesian square*

$$\begin{array}{ccc} X & \leftarrow & X'. \\ \downarrow & & \downarrow \\ T & \leftarrow & T' \end{array}$$

If D is a relative effective divisor on X/T, then its pull-back to a closed subscheme D' of X' is a relative effective divisor on X'/T'.

PROOF. We may assume both T and T' are affine, say $T = \operatorname{spec} R$ and $T' = \operatorname{spec} R'$, and then have to check that the conditions (a), (b), and (c) above are stable under the base change $R \to R'$. Write $U_i' = U_i \times_T T'$; clearly $D' \cap U_i' = \operatorname{spec}(R_i'/g_i R_i')$. The conditions (b) and (c) state that

$$0 \to R_i \xrightarrow{g_i} R_i \to R_i/g_i R_i \to 0$$

is exact and that $R_i/g_i R_i$ is flat over R. Both assertions continue to hold after the sequence has been tensored with R'. □

Proposition 3.8. *Let D be a closed subscheme of X, and assume that D and X are both flat over T. If $D_t \stackrel{df}{=} D \times_T \{t\}$ is an effective divisor on X_t/t for all points t of T, then D is a relative effective divisor on X.*

PROOF. From the exact sequence

$$0 \to \mathscr{I}(D) \to \mathscr{O}_X \to \mathscr{O}_D \to 0$$

and the flatness of X and D over T, we see that $\mathscr{I}(D)$ is flat over T. The flatness of $\mathscr{O}_D$ implies that, for any $t \in T$, the sequence

$$0 \to \mathscr{I}(D) \otimes_{\mathscr{O}_T} k(t) \to \mathscr{O}_{X_t} \to \mathscr{O}_{D_t} \to 0$$

is exact. In particular, $\mathscr{I}(D) \otimes k(t) \xrightarrow{\approx} \mathscr{I}(D_t)$. As D_t is a Cartier divisor, $\mathscr{I}(D_t)$ (and therefore also $\mathscr{I}(D) \otimes k(t)$) is an invertible $\mathscr{O}_{X_t}$-module. We now apply the fibre-by-fibre criterion of flatness: if X is flat over T and $\mathscr{F}$ is a coherent $\mathscr{O}_X$-module that is flat over T and such that $\mathscr{F}_t$ is a flat $\mathscr{O}_{X_t}$-module for all t in T, then $\mathscr{F}$ is flat over X [2, III, 5.4]. This implies that $\mathscr{I}(D)$ is a flat $\mathscr{O}_X$-module, and since it is also coherent, it is locally free over $\mathscr{O}_X$. Now the isomorphism $\mathscr{I}(D) \otimes k(t) \xrightarrow{\approx} \mathscr{I}(D_t)$ shows that it is of rank one. It is therefore locally generated by a single element, and the element is not a zero-divisor; this shows that D is a relative effective divisor. □

Let $\pi: \mathscr{C} \to T$ be a proper smooth morphism with fibres of dimension one. If D is a relative effective divisor on $\mathscr{C}/T$, then D_t is an effective divisor on $\mathscr{C}_t$, and if T is connected, then the degree of D_t is constant; it is called the *degree* of D. Note that $\deg(D) = r$ if and only if $\mathscr{O}_D$ is a locally free $\mathscr{O}_T$-module of rank r.

Corollary 3.9. *A closed subscheme D of $\mathscr{C}$ is a relative effective divisor on $\mathscr{C}/T$ if and only if it is finite and flat over T; in particular, if $s: T \to \mathscr{C}$ is a section to π, then $s(T)$ is a relative effective divisor of degree 1 on $\mathscr{C}/T$.*

PROOF. A closed subscheme of a curve over a field is an effective divisor if and only if it is finite. Therefore (3.8) shows that a closed subscheme D of $\mathscr{C}$ is a relative effective divisor on $\mathscr{C}/T$ if and only if it is flat over T and has finite fibres, but such a subscheme D is proper over T and therefore has finite fibres if and only if it is finite over T (see [13, I, 1.10] or [8, III, Ex. 11.3]). □

If D and D' are relative effective divisors on $\mathscr{C}/T$, then we write $D \geq D'$ if $D \supset D'$ as subschemes of $\mathscr{C}$ (that is, $\mathscr{I}(D) \subset \mathscr{I}(D')$).

Proposition 3.10. *If $D_t \geq D'_t$ (as divisors on C_t) for all t in T, then $D \geq D'$.*

PROOF. Represent D as a pair $(s, \mathscr{L})$ (see 3.6). Then $D \geq D'$ if and only if s becomes zero in $\mathscr{L} \otimes \mathscr{O}_{D'} = \mathscr{L}|D'$. But $\mathscr{L} \otimes \mathscr{O}_{D'}$ is a locally free $\mathscr{O}_T$-module of finite rank, and so the support of s is a closed subscheme of T. The hypothesis implies that this subscheme is the whole of T. □

Let D be a relative effective divisor of degree r on $\mathscr{C}/T$. We shall say that D is *split* if $\mathrm{Supp}(D) = \bigcup s_i(T)$ for some sections s_i to π. For example, a divisor $D = \sum n_i P_i$ on a curve over a field is split if and only if $k(P_i) = k$ for all i.

Proposition 3.11. *Every split relative effective divisor D on $\mathscr{C}/T$ can be written uniquely in the form $D = \sum n_i s_i(T)$ for some sections s_i.*

PROOF. Let $\mathrm{Supp}(D) = \bigcup s_i(T)$, and suppose that the component of D with support on $s_i(T)$ has degree n_i. Then $D_t = (\sum n_i s_i(T))_t$ for all t, and so (3.10) shows that $D = \sum n_i s_i(T)$. □

Example 3.12. Consider a complete nonsingular curve C over a field k. For each i there is a canonical section s_i to $q: C \times C^r \to C^r$, namely, $(P_1, \ldots, P_r) \mapsto (P_i, P_1, \ldots, P_r)$. Let D_i to be $s_i(C^r)$ regarded as a relative effective divisor on $C \times C^r/C^r$, and let $D = \sum D_i$. Then D is the unique relative effective divisor $C \times C^r/C^r$ whose fibre over $(P_1, \ldots, P_r)$ is $\sum P_i$. Clearly D is stable under the action of the symmetric group S_r, and $D_{\mathrm{can}} = S_r \backslash D$ (quotient as a subscheme of $C \times C^r$) is a relative effective divisor on $C \times C^{(r)}/C^{(r)}$ whose fibre over $D \in C^{(r)}(k)$ is D.

For C a complete smooth curve over k and T a k-scheme, define $\mathrm{Div}_C^r(T)$ to be the set of relative effective Cartier divisors on $C \times T/T$ of degree r. Proposition 3.7 shows that Div_C^r is a functor on the category of k-schemes.

Theorem 3.13. *For any relative effective divisor D on $C \times T/T$ of degree r, there is a unique morphism $\varphi \colon T \to C^{(r)}$ such that $D = (1 \times \varphi)^{-1}(D_{\mathrm{can}})$. Therefore $C^{(r)}$ represents Div_C^r.*

PROOF. Let us first assume that D is split, so that $D = \sum n_i s_i(T)$ for some sections $s_i \colon T \to C \times T$. In this case, we define $T \to C^r$ to be the map $(p \circ s_1, \ldots, p \circ s_1, p \circ s_2, \ldots)$, where s_i occurs n_i times, and we take φ to be the composite $T \to C^r \to C^{(r)}$. In general, we can choose a finite flat covering $\psi \colon T' \to T$ such that the inverse image D' of D on $C \times T'$ is split, and let $\varphi' \colon T' \to C^{(r)}$ be the map defined by D'. Then the two maps $\varphi' \circ p$ and $\varphi' \circ q$ from $T' \times_T T'$ to $C^{(r)}$ are equal because they both correspond to the same relative effective divisor

$$p^{-1}(D') = (\psi \circ p)^{-1}(D) = (\psi \circ q)^{-1}(D) = q^{-1}(D')$$

on $T' \times_T T'$. Now descent theory [13, I, 2.17] shows that φ' factors through T. □

Exercise 3.14. Let E be an effective Cartier divisor of degree r on C, and define a subfunctor Div_C^E of Div_C^r by

$$\mathrm{Div}_C^E(T) = \{D \in \mathrm{Div}_C^r(T) | D_t \sim E \text{ all } t \in T\}.$$

Show that Div_C^E is representable by $\mathbb{P}(V)$ where V is the vector space $\Gamma(C, \mathscr{L}(E))$ (use [8, II, 7.12]) and that the inclusion $\mathrm{Div}_C^E \hookrightarrow \mathrm{Div}_C^r$ defines a closed immersion $\mathbb{P}(V) \hookrightarrow C^{(r)}$.

Remark 3.15. Theorem 3.13 says that $C^{(r)}$ is the Hilbert scheme $\mathrm{Hilb}_{C/k}^P$ where P is the constant polynomial r.

§4. The Construction of the Jacobian Variety

In this section, C will be a complete nonsingular curve of genus $g > 0$, and P will be a k-rational point on C. Recall (1.9), that in constructing J, we are allowed to make a finite separable extension of k.

For a k-scheme T, let

$$P_C^r(T) = \{\mathscr{L} \in \mathrm{Pic}(C \times T) | \deg(\mathscr{L}_t) = r \text{ all } t\}/\sim,$$

where $\mathscr{L} \sim \mathscr{L}'$ means $\mathscr{L} \approx \mathscr{L}' \otimes q^*\mathscr{M}$ for some invertible sheaf $\mathscr{M}$ on T. Let $\mathscr{L}_r = \mathscr{L}(rP)$; then $\mathscr{L} \mapsto \mathscr{L} \otimes p^*\mathscr{L}_r$ is an isomorphism $P_C^0(T) \to P_C^r(T)$, and so, to prove (1.1), it suffices to show that P_C^r is representable for some r. We shall do this for a fixed $r > 2g$.

Note that there is a natural transformation of functors $f \colon \mathrm{Div}_C^r \to P_C^r$ sending a relative effective divisor D on $C \times T/T$ to the class of $\mathscr{L}(D)$ (or, in other terms, $(s, \mathscr{L})$ to the class of $\mathscr{L}$).

Lemma 4.1. *Suppose there exists a section s to $f\colon \operatorname{Div}_C^r \to P_C^r$. Then P_C^r is representable by a closed subscheme of $C^{(r)}$.*

PROOF. The composite $\varphi = s \circ f$ is a natural transformation of functors $\operatorname{Div}_C^r \to \operatorname{Div}_C^r$ and Div_C^r is representable by $C^{(r)}$, and so φ is represented by a morphism of varieties. Define J' to be the fibre product,

$$\begin{array}{ccc} C^{(r)} & \leftarrow & J' \\ {\scriptstyle (1,\varphi)}\downarrow & & \downarrow \\ C^{(r)} \times C^{(r)} & \stackrel{\Delta}{\leftarrow} & C^{(r)}. \end{array}$$

Then

$$\begin{aligned} J'(T) &= \{(a,b) \in C^{(r)}(T) \times C^{(r)}(T) \mid a = b, a = \varphi b\} \\ &= \{a \in C^{(r)}(T) \mid a = \varphi(a)\} \\ &= \{a \in C^{(r)}(T) \mid a = sc, \text{ some } c \in P_C^r(T)\} \\ &\approx P_C^r(T), \end{aligned}$$

because s is injective. This shows that P_C^r is represented by J', which is a closed subscheme of $C^{(r)}$ because Δ is a closed immersion. □

The problem is therefore to define a section s or, in other words, to find a natural way of associating with a family of invertible sheaves $\mathscr{L}$ of degree r a relative effective divisor. For $\mathscr{L}$ an invertible sheaf of degree r on C, the dimension $h^0(\mathscr{L})$ of $H^0(C, \mathscr{L})$ is $r + 1 - g$, and so there is an $(r - g)$-dimensional system of effective divisors D such that $\mathscr{L}(D) \approx \mathscr{L}$. One way to cut down the size of this system is to fix a family $\gamma = (P_1, \ldots, P_{r-g})$ of k-rational points on C and consider only divisors D in the system such that $D \geq D_\gamma$, where $D_\gamma = \sum P_i$. As we shall see, this provides a partial solution to the problem.

Proposition 4.2. *Let γ be an $(r - g)$-tuple of k-rational points on C, and let $\mathscr{L}_\gamma = \mathscr{L}(\sum_{P \in \gamma} P)$.*

(a) *There is an open subvariety C^γ of $C^{(r)}$ such that, for all k-schemes T,*

$$C^\gamma(T) = \{D \in \operatorname{Div}_C^r(T) \mid h^0(D_t - D_\gamma) = 1, \text{ all } t \in T\}.$$

If k is separably closed, then $C^{(r)}$ is the union of the subvarieties C^γ.
(b) *For all k-schemes T, define*

$$P^\gamma(T) = \{\mathscr{L} \in P_C^r(T) \mid h^0(\mathscr{L}_t \otimes \mathscr{L}_\gamma^{-1}) = 1, \text{ all } t \in T\}.$$

Then P^γ is a subfunctor of P_C^r and the obvious natural transformation $f\colon C^\gamma \to P^\gamma$ has a section.

PROOF. (a) Note that for any effective divisor D of degree r on C, $h^0(D - D_\gamma) \geq 1$, and that equality holds for at least one D (for example,

$D = D_\gamma + Q_1 + \cdots + Q_g$ for a suitable choice of points $Q_1, \ldots, Q_g$; see the elementary result (5.2b) below). Let D_{can} be the canonical relative effective divisor of degree r on $C \times C^{(r)}/C^{(r)}$. Then [14, 4.2c] applied to $\mathscr{L}(D_{\text{can}} - p^{-1}D_\gamma)$ shows that there is an open subscheme C^γ of $C^{(r)}$ such that $h^0((D_{\text{can}})_t - D_\gamma) = 1$ for t in C^γ and $h^0((D_{\text{can}})_t - D_\gamma) > 1$ otherwise. Let T be a k-scheme, and let D be a relative effective divisor of degree r on $C \times T/T$ such that $h^0(D_t - D_\gamma) = 1$. Then (3.13) shows that there is a unique morphism $\varphi: T \to C^{(r)}$ such that $(1 \times \varphi)^{-1}(D_{\text{can}}) = D$, and it is clear that φ maps T into C^γ. This proves the first assertion.

Assume that k is separably closed. To show that $C = \bigcup C^\gamma$, it suffices to show that $C(k) = \bigcup C^\gamma(k)$, or that for every divisor D of degree r on C, there exists a γ such that $h^0(D - D_\gamma) = 1$. Choose a basis $e_0, \ldots, e_{r-g}$ for $H^0(C, \mathscr{L}(D))$, and consider the corresponding embedding $\iota: C \hookrightarrow \mathbb{P}^{r-g}$. Then $\iota(C)$ is not contained in any hyperplane (if it were contained in $\sum a_i X_i = 0$, then $\sum a_i e_i$ would be zero on C), and so there exist $r - g$ points $P_1, \ldots, P_{r-g}$ on C disjoint from D whose images are not contained in any linear subspace of codimension 2 (choose $P_1, P_2, \ldots$ inductively so that $P_1, \ldots, P_i$ are not contained in a linear subspace of dimension $i - 2$). The $(r - g)$-tuple $\gamma = (P_1, \ldots, P_{r-g})$ satisfies the condition because

$$H^0(C, \mathscr{L}(D - \sum P_j)) = \{\sum a_i e_i | \sum a_i e_i(P_j) = 0, j = 1, \ldots, r - g\},$$

which has dimension < 2.

(b). Let $\mathscr{L}$ be an invertible sheaf on $C \times T$ representing an element of $P^\gamma(T)$. Then $h^0(D_t - D_\gamma) = 1$ for all t, and the Riemann–Roch theorem shows that $h^1(D_t - D_\gamma) = 0$ for all t. Now [14, 4.2e] shows that $\mathscr{M} \overset{\text{df}}{=} q_*(\mathscr{L} \otimes p^*\mathscr{L}_\gamma^{-1})$ is an invertible sheaf on T and that its formation commutes with base change. This proves that P_C^γ is a subfunctor of P_C^r. On tensoring the canonical map $q^*\mathscr{M} \to \mathscr{L} \otimes p^*\mathscr{L}_\gamma^{-1}$ with $q^*\mathscr{M}^{-1}$, we obtain a canonical map $\mathscr{O}_{C \times T} \to \mathscr{L} \otimes p^*\mathscr{L}_\gamma^{-1} \otimes q^*\mathscr{M}^{-1}$. The natural map $\mathscr{L}_\gamma \to \mathscr{O}_C$ induces a map $p^*\mathscr{L}_\gamma^{-1} \to \mathscr{O}_{C \times T}$, and on combining this with the preceding map, we obtain a canonical map $s_\gamma: \mathscr{O}_{C \times T} \to \mathscr{L} \otimes q^*\mathscr{M}^{-1}$. The pair $(s_\gamma, \mathscr{L} \otimes q^*\mathscr{M}^{-1})$ is a relative effective divisor on $C \times T/T$ whose image under f in $P^\gamma(T)$ is represented by $\mathscr{L} \otimes q^*\mathscr{M}^{-1} \sim \mathscr{L}$ (see 3.6). We have defined a section to $C^\gamma(T) \to P^\gamma(T)$, and our construction is obviously functorial. □

Corollary 4.3. *The functor P^γ is representable by a closed subvariety J^γ of C^γ.*

PROOF. The proof is the same as that of (4.1). □

Now consider two $(g - r)$-tuples γ and γ', and define $P^{\gamma, \gamma'}$ to be the functor such that $P^{\gamma, \gamma'}(T) = P^\gamma(T) \cap P^{\gamma'}(T)$ for all k-schemes T. It easy to see that $P^{\gamma, \gamma'}$ is representable by a variety $J^{\gamma, \gamma'}$ such that the maps $J^{\gamma, \gamma'} \hookrightarrow J^\gamma$ and $J^{\gamma, \gamma'} \hookrightarrow J^{\gamma'}$ defined by the inclusions $P^{\gamma, \gamma'} \hookrightarrow P^\gamma$ and $P^{\gamma, \gamma'} \hookrightarrow P^{\gamma'}$ are open immersions.

We are now ready to construct the Jacobian of C. Choose tuples $\gamma_1, \ldots, \gamma_m$ of points in $C(k_s)$ such that $C^{(r)} = \bigcup C^{\gamma_i}$. After extending k, we can assume that the γ_i are tuples of k-rational points. Define J by patching together the varieties J^{γ_i} using the open immersions $J^{\gamma_i,\gamma_j} \subset J^{\gamma_i}, J^{\gamma_j}$. It is easy to see that J represents the functor P_C^r, and therefore also the functor P_C^0. Since the latter is a group functor, J is a group variety. The natural transformations $\text{Div}_C^r \to P_C^r \to P_C^0$ induce a morphism $C^{(r)} \to J$, which shows that J is complete and is therefore an abelian variety. The proof of (1.1) is complete. □

§5. The Canonical Maps from the Symmetric Powers of C to its Jacobian Variety

Throughout this section C will be a complete nonsingular curve of genus $g > 0$. Assume there is a k-rational point P on C, and write f for the map f^P defined in Section 2.

Let f^r be the map $C^r \to J$ sending $(P_1, \ldots, P_r)$ to $f(P_1) + \cdots + f(P_r)$. On points, f^r is the map $(P_1, \ldots, P_r) \mapsto [P_1 + \ldots + P_r - rP]$. Clearly it is symmetric, and so induces a map $f^{(r)}: C^{(r)} \to J$. We can regard $f^{(r)}$ as the map sending an effective divisor D of degree r on C to the linear equivalence class of $D - rP$. The fibre of the map $f^{(r)}: C^{(r)}(k) \to J(k)$ containing D can be identified with the space of effective divisors linearly equivalent to D, that is, with the linear system $|D|$. The image of $C^{(r)}$ in J is a closed subvariety W^r of J, which can also be written $W^r = f(C) + \ldots + f(C)$ (r summands).

Theorem 5.1. (a) *For all $r \leq g$, the morphism $f^{(r)}: C^{(r)} \to W^r$ is birational; in particular, $f^{(g)}$ is a birational map from $C^{(g)}$ onto J.*
(b) *Let D be an effective divisor of degree r on C, and let F be the fibre of $f^{(r)}$ containing D. Then no tangent vector to $C^{(r)}$ at D maps to zero under $(df^{(r)})_D$ unless it lies in the direction of F; in other words, the sequence*

$$0 \to T_D(F) \to T_D(C^{(r)}) \to T_a(J), \qquad a = f^{(r)}(D),$$

is exact. In particular, $(df^{(r)})_D: T_D(C^{(r)}) \to T_a(J)$ is injective if $|D|$ has dimension zero.

PROOF. For D a divisor on C, we write $h^0(D)$ for the dimension of

$$H^0(C, \mathscr{L}(D)) = \{f \in k(C) | (f) + D \geq 0\}$$

and $h^1(D)$ for the dimension of $H^1(C, \mathscr{L}(D))$. Recall that

$$h^0(D) - h^1(D) = \deg(D) + 1 - g,$$

and that $H^1(C, \mathscr{L}(D))^\vee = H^0(C, \Omega^1(-D))$, which can be identified with the set of $\omega \in \Omega^1_{k(C)/k}$ whose divisor $(\omega) \geq D$.

Lemma 5.2. (a) *Let D be a divisor on C such that $h^1(D) > 0$; then there is a nonempty open subset U of C such that $h^1(D + Q) = h^1(D) - 1$ for all closed points Q in U, and $h^1(D + Q) = h^1(D)$ for $Q \notin U$.*

(b) *For any $r \leq g$, there is an open subset U of C^r such that $h^0(\sum P_i) = 1$ for all $(P_1, \ldots, P_r)$ in U.*

PROOF. (a) If Q is not in the support of D, then $H^1(C, \mathscr{L}(D+Q))^{\vee} = \Gamma(C, \Omega^1(-D-Q))$ can be identified with the subspace of $\Gamma(C, \Omega^1(-D))$ of differentials with a zero at Q. Clearly therefore we can take U to be the complement of the zero set of a basis of $H^1(C, \mathscr{L}(D))$ together with a subset of the support of D.

(b) Let D_0 be the divisor zero on C. Then $h^1(D_0) = g$, and on applying (a) repeatedly, we find that there is an open subset U of C^r such that $h^1(\sum P_i) = g - r$ for all $(P_1, \ldots, P_r)$ in U. The Riemann–Roch theorem now shows that $h^0(\sum P_i) = r + (1-g) + (g-r) = 1$ for all $(P_1, \ldots, P_r)$ in U. □

In proving (5.1), we can assume that k is algebraically closed. If U' is the image in $C^{(r)}$ of the set U in (5.2b), then $f^{(r)}: C^{(r)}(k) \to J(k)$ is injective on $U'(k)$, and so $f^{(r)}: C^{(r)} \to W^r$ must either be birational or else purely inseparable of degree > 1. The second possibility is excluded by part (b) of the theorem, but before we can prove that we need another proposition.

Proposition 5.3. (a) *For all $r \geq 1$, there are canonical isomorphisms*

$$\Gamma(C, \Omega^1) \xrightarrow{\approx} \Gamma(C^r, \Omega^1)^{S_r} \xrightarrow{\approx} \Gamma(C^{(r)}, \Omega^1).$$

Let $\omega \in \Gamma(C, \Omega^1)$ correspond to $\omega' \in \Gamma(C^{(r)}, \Omega^1)$; then for any effective divisor D of degree r on C, $(\omega) \geq D$ if and only if ω' has a zero at D.
(b) *For all $r \geq 1$, the map $f^{(r)*}: \Gamma(J, \Omega^1) \to \Gamma(C^{(r)}, \Omega^1)$ is an isomorphism.*

PROOF. A global 1-form on a product of projective varieties is a sum of global 1-forms on the factors. Therefore $\Gamma(C^r, \Omega^1) = \oplus\, p_i^* \Gamma(C, \Omega^1)$, where the p_i are the projection maps onto the factors, and so it is clear that the map $\omega \mapsto \sum p_i^* \omega$ identifies $\Gamma(C, \Omega^1)$ with $\Gamma(C^r, \Omega^1)^{S_r}$. Because $\pi: C^r \to C^{(r)}$ is separable, $\pi^*: \Gamma(C^{(r)}, \Omega^1) \to \Gamma(C^r, \Omega^1)$ is injective, and its image is obviously fixed by the action of S_r. The composite map

$$\Gamma(J, \Omega^1) \to \Gamma(C^{(r)}, \Omega^1) \hookrightarrow \Gamma(C^r, \Omega^1)^{S_r} = \Gamma(C, \Omega^1)$$

sends ω to the element ω' of $\Gamma(C, \Omega^1)$ such that $f^{r*}\omega = \sum p_i^*\omega'$. As $f^r = \sum f \circ p_i$, clearly $\omega' = f^*\omega$, and so the composite map is f^* which we know to be an isomorphism (2.2). This proves that both maps in the above sequence are isomorphisms. It also completes the proof of the proposition except for the second part of (a), and for this we need a combinatorial lemma.

Lemma 5.4. *Let $\sigma_1, \ldots, \sigma_r$ be the elementary symmetric polynomials in $X_1, \ldots, X_r$, and let $\tau_j = \sum X_i^j\, dX_i$. Then*

$$\sigma_m \tau_0 - \sigma_{m-1}\tau_1 + \cdots + (-1)^m \tau_m = d\sigma_{m+1}, \qquad \text{all } m \leq r-1.$$

PROOF. Let $\sigma_m(i)$ be the mth elementary symmetric polynomial in the variables

$X_1, \ldots, X_{i-1}, X_{i+1}, \ldots, X_r$. Then

$$\sigma_{m-n} = \sigma_{m-n}(i) + X_i \sigma_{m-n-1}(i),$$

and on multiplying this by $(-1)^n X_i^n$ and summing over n (so that the successive terms cancel out) we obtain the identity

$$\sigma_m - \sigma_{m-1} X_i + \cdots + (-1)^m X_i^m = \sigma_m(i).$$

On multiplying this with dX_i and summing, we get the required identity. □

We now complete the proof of (5.3). First let $D = rQ$. Then $\hat{\mathcal{O}}_Q = k[[X]]$ and $\hat{\mathcal{O}}_D = k[[\sigma_1, \ldots, \sigma_r]]$ (see the proof of (3.2); by $\mathcal{O}_D$ we mean the local ring at the point D on $C^{(r)}$). If $\omega = (a_0 + a_1 X + a_2 X^2 + \cdots) dX$, $a_i \in k$, when regarded as an element of $\Omega^1_{\hat{\mathcal{O}}_Q/k}$, then $\omega' = a_0 \tau_0 + a_1 \tau_1 + \cdots$. We know that $\{d\sigma_1, \ldots, d\sigma_r\}$ is a basis for $\Omega^1_{\hat{\mathcal{O}}_D/k}$ as an $\hat{\mathcal{O}}_D$-module, but the lemma shows that $\tau_0, \ldots, \tau_{r-1}$ is also a basis. Now $(\omega) \geq D$ and $\omega'(D) = 0$ are both obviously equivalent to $a_0 = a_1 = \cdots = a_{r-1} = 0$. The proof for other divisors is similar. □

We finally prove the exactness of the sequence in (5.1). The injectivity of $(di)_D$ follows from the fact that $i: F \hookrightarrow C^{(r)}$ is a closed immersion. Moreover the sequence is a complex because $f \circ i$ is the constant map $x \mapsto a$. It remains to show that

$$\dim \text{Im}(di)_D = \dim \text{Ker}(df^{(r)})_D.$$

Identify $T_a(J)^\vee$ with $\Gamma(C, \Omega^1)$ using the isomorphisms arising from (2.1). Then (5.3) shows that ω is zero on the image of $T_D(C^{(r)})$ if and only if $(\omega) \geq D$, that is, $\omega \in \Gamma(C, \Omega^1(-D))$. Therefore the image of $(df^{(r)})_D$ has dimension $g - h^0(\Omega^1(-D)) = g - h^1(D)$, and so its kernel has dimension $r - g + h^1(D)$. On the other hand, the image of $(di)_D$ has dimension $|D|$. The Riemann–Roch theorem says precisely that these two numbers are equal, and so completes the proof. □

Corollary 5.5. *For all $r \leq g$, $f^r: C^r \to W^r$ is of degree $r!$.*

PROOF. It is the composite of $\pi: C^r \to C^{(r)}$ and $f^{(r)}$. □

Remark 5.6. (a) The theorem shows that J is the unique abelian variety birationally equivalent to $C^{(g)}$. This observation is the basis of Weil's construction of the Jacobian. (See Section 7.)

(b) The exact sequence in (5.1b) can be regarded as a geometric statement of the Riemann–Roch theorem (see especially the end of the proof). In fact it is possible to prove the Riemann–Roch theorem this way (see [12]).

(c) As we observed above, the fibre of $f^{(r)}: C^{(r)}(k) \to J(k)$ containing D can be identified with the linear system $|D|$. More precisely, the fibre of the map of functors $C^{(r)} \to J$ is the functor Div_C^D of (3.14); therefore the scheme-theoretic fibre of $f^{(r)}$ containing D is a copy of projective space of dimension $h^0(D) - 1$. Corollary 3.9 of [14] shows that conversely every copy of projective space in

$C^{(r)}$ is contained in some fibre of $f^{(r)}$. Consequently, the closed points of the Jacobian can be identified with the set of maximal subvarieties of $C^{(r)}$ isomorphic to projective space.

Note that for $r > 2g - 2$, $|D|$ has dimension $r - g$, and so $(df^{(r)})_D$ is surjective, for all D. Therefore $f^{(r)}$ is smooth (see [8, III, 10.4]), and the fibres of $f^{(r)}$ are precisely the copies of $\mathbb{P}^{r-g}$ contained in $C^{(r)}$. This last observation is the starting point of Chow's construction of the Jacobian [3].

§6. The Jacobian Variety as Albanese Variety; Autoduality

Throughout this section C will again be a complete nonsingular curve of genus $g > 0$ over a field k, and J will be its Jacobian variety.

Proposition 6.1. *Let P be a k-rational point on C. The map $f^P: C \to J$ has the following universal property: for any map $\varphi: C \to A$ from C into an abelian variety sending P to 0, there is a unique homomorphism $\psi: J \to A$ such that $\varphi = \psi \circ f^P$.*

PROOF. Consider the map $C^g \to A$, $(P_1, \ldots, P_g) \mapsto \sum \psi(P_i)$. Clearly this is symmetric, and so it factors through $C^{(g)}$. It therefore defines a rational map $\psi: J \to A$, which [14, 3.1] shows to be a morphism. It is clear from the construction that $\psi \circ f^P = \varphi$ (note that f^P is the composite of $Q \mapsto Q + (g-1)P: C \to C^{(g)}$ with $f^{(g)}: C^{(g)} \to J$). In particular, ψ maps 0 to 0, and [14, 2.2] shows that it is therefore a homomorphism. If ψ' is a second homomorphism such that $\psi' \circ f^P = \varphi$, then ψ and ψ' agree on $f^P(C) + \cdots + f^P(C)$ (g copies), which is the whole of J. □

Corollary 6.2. *Let $\mathcal{N}$ be a divisorial correspondence between (C, P) and J such that $(1 \times f^P)^* \mathcal{N} \approx \mathcal{L}^P$; then $\mathcal{N} \approx \mathcal{M}^P$ (notations as in Section 2 and (1.2)).*

PROOF. Because of [14, 6.2], we can assume k to be algebraically closed. According to (1.2) there is a unique map $\varphi: J \to J$ such that $\mathcal{N} \approx (1 \times \varphi)^* \mathcal{M}^P$. On points φ is the map sending $a \in J(k)$ to the unique b such that

$$\mathcal{M}^P | C \times \{b\} \approx \mathcal{N} | C \times \{a\}.$$

By assumption,

$$\mathcal{N} | C \times \{f^P Q\} \approx \mathcal{L}^P | C \times \{Q\} \approx \mathcal{M}^P | C \times \{f^P Q\},$$

and so $(\varphi \circ f^P)(Q) = f^P(Q)$ for all Q. Now (6.1) shows that f is the identity map. □

Corollary 6.3. *Let C_1 and C_2 be curves over k with k-rational points P_1 and P_2, and let J_1 and J_2 be their Jacobians. There is a one-to-one correspondence between $\mathrm{Hom}_k(J_1, J_2)$ and the set of isomorphism classes of divisorial correspondences between (C_1, P_1) and (C_2, P_2).*

PROOF. A divisorial correspondence between (C_2, P_2) and (C_1, P_1) gives rise to a morphism $(C_1, P_1) \to J_2$ (by 1.2), and this morphism gives rise to homomorphism $J_1 \to J_2$ (by 6.1). Conversely, a homomorphism $\psi: J_1 \to J_2$ defines a divisorial correspondence $(1 \times (f^{P_1} \circ \psi))^* \mathcal{M}^{P_2}$ between (C_2, P_2) and (C_1, P_1). □

In the case that C has a point P rational over k, define $F: C \times C \to J$ to be the map $(P_1, P_2) \mapsto f^P(P_1) - f^P(P_2)$. One checks immediately that this is independent of the choice of P. Thus, if $P \in C(k')$ for some Galois extension k' of k, and $F: C_{k'} \times C_{k'} \to J_{k'}$ is the corresponding map, then $\sigma F = F$ for all $\sigma \in \mathrm{Gal}(k'/k)$; therefore F is defined over k whether or not C has a k-rational point. Note that it is zero on the diagonal Δ of $C \times C$.

Proposition 6.4. *Let A be an abelian variety over k. For any map $\varphi: C \times C \to A$ such that $\varphi(\Delta) = 0$, there is a unique homomorphism $\psi: J \to A$ such that $\psi \circ F = \varphi$.*

PROOF. Let k' be a finite Galois extension of k, and suppose that there exists a unique homomorphism $\psi: C_{k'} \to J_{k'}$ such that $\psi \circ F_{k'} = \varphi_{k'}$. Then the uniqueness implies that $\sigma \psi = \psi$ for all σ in $\mathrm{Gal}(k'/k)$, and so ψ is defined over k. It suffices therefore to prove the proposition after extending k, and so we can assume that C has a k-rational point P. Now [14, 2.5] shows that there exist unique maps φ_1 and φ_2 from C to A such that $\varphi_1(P) = 0 = \varphi_2(P)$ and $\varphi(a, b) = \varphi_1(a) + \varphi_2(b)$ for all $(a, b) \in C \times C$. Because φ is zero on the diagonal, $\varphi_1 = -\varphi_2$. From (6.1) we know that there exists a unique homomorphism ψ from J to A such that $\varphi_1 = \psi \circ f$, and clearly ψ is also the unique homomorphism such that $\varphi = \psi \circ F$. □

Remark 6.5. The proposition says that (J, F) is the Albanese variety of C in the sense of [9, II.3, p. 45]. Clearly the pairs (J, f^P) and (J, F) are characterized by the universal properties in (6.1) and (6.4).

Assume again that C has a k-rational point P, and let $\Theta = W^{g-1}$. It is a divisor on J, and if P is replaced by a second k-rational point, Θ is replaced by a translate. For any effective divisor D on J, write

$$\mathscr{L}'(D) = m^*\mathscr{L}(D) \otimes p^*\mathscr{L}(D)^{-1} \otimes q^*\mathscr{L}(D)^{-1}$$
$$= \mathscr{L}(m^{-1}(D) - D \times J - J \times D).$$

Recall [14, 9.1 and §10], that D is ample if and only if $\varphi_{\mathscr{L}(D)}: J \to J^\vee$ is an isogeny, and then $(1 \times \varphi_{\mathscr{L}(D)})^*(\mathscr{P}) = \mathscr{L}'(D)$, where $\mathscr{P}$ is the Poincaré sheaf on $J \times J^\vee$. Write Θ^- for the image of Θ under the map $(-1)_J: J \to J$, and Θ_a for $t_a\Theta = \Theta + a$, $a \in J(k)$. Abbreviate $(\Theta^-)_a$ by Θ_a^-.

Theorem 6.6. *The map $\varphi_{\mathscr{L}(\Theta)}: J \to J^\vee$ is an isomorphism; therefore, $1 \times \varphi_{\mathscr{L}(\Theta)}$ is an isomorphism $(J \times J, \mathscr{L}'(\Theta)) \xrightarrow{\approx} (J \times J^\vee, \mathscr{P})$.*

PROOF. As usual, we can assume k to be algebraically closed. Recall [14, 12.13] that $\varphi_{\mathscr{L}(\Theta^-)} = (-1)^2 \varphi_{\mathscr{L}(\Theta)} = \varphi_{\mathscr{L}(\Theta)}$, and that $\varphi_{\mathscr{L}(\Theta_a)} = \varphi_{\mathscr{L}(\Theta)}$ for all $a \in J(k)$.

Lemma 6.7. *Let U be the largest open subset of J such that:*

(i) *the fibre of $f^{(g)}: C^{(g)} \to J$ at any point of U has dimension zero; and*
(ii) *if $a \in U(k)$ and $D(a)$ is the unique element of $C^{(g)}(k)$ mapping to a, then $D(a)$ is a sum of g distinct points of $C(k)$.*

Then $f^{-1}(\Theta_a^-) = D(a)$ (as a Cartier divisor) for all $a \in U(k)$, where $f = f^P: C \to J$.

PROOF. Note first that U can be obtained by removing the subset over which the fibres have dimension > 0, which is closed (see [18, I.6, Theorem 7]), together with the images of certain closed subsets of the form $\Delta \times C^{g-2}$. These last sets are also closed because $C^g \to J$ is proper ([18, II, 4.8]), and it follows that U is a dense open subset of J.

Let $a \in U(k)$, and let $D(a) = \sum P_i$, $P_i \neq P_j$ for $i \neq j$. A point Q_1 of C maps to a point of Θ_a^- if and only if there exists a divisor $\sum_{i=2}^g Q_i$ on C such that $f^P(Q_1) = -\sum f^P(Q_i) + a$. The equality implies $\sum_{i=1}^g Q_i \sim D$, and the fact that $|D|$ has dimension 0 implies that $\sum Q_i = D$. It follows that the support of $f^{-1}(\Theta_a^-)$ is $\{P_1, \ldots, P_g\}$, and it remains to show that $f^{-1}(\Theta_a^-)$ has degree $\leq g$ for all a.

Consider the map $\psi: C \times \Theta \to J$ sending (Q, b) to $f(Q) + b$. As the composite of ψ with $1 \times f^{g-1}: C \times C^{g-1} \to C \times \Theta$ is $f^g: C^g \to J$, and these maps have degrees $(g-1)!$ and $g!$ respectively (5.5), ψ has degree g. Also ψ is projective because $C \times \Theta$ is a projective variety (see [8, II, Ex. 4.9]). Consider $a \in U$; the fibre of ψ over a is $f^{-1}(\Theta_a^-)$ (more accurately, it is the subscheme of C associated with the Cartier divisor $f^{-1}(\Theta_a^-)$). Therefore the restriction of ψ to $\psi^{-1}(U)$ is quasi-finite and projective, and so is finite (see [8, III, Ex. 11.2]). As U is normal, this means that all the fibres of ψ over points of U are finite schemes of rank $\leq g$ (cf. [18, II.5, Theorem 6]). This completes the proof of the lemma. □

Lemma 6.8. (a) *Let $a \in J(k)$, and let $f^{(g)}(D) = a$; then $f^* \mathscr{L}(\Theta_a^-) \approx \mathscr{L}(D)$.*
(b) *The sheaves $(f \times (-1)_J)^* \mathscr{L}'(\theta^-)$ and $\mathscr{M}^P$ on $C \times J$ are isomorphic.*

PROOF. Note that (6.7) shows that the isomorphism in (a) holds for all a in a dense open subset of J. Note also that the map $C \to C \times \{a\} \to J \times J \to J$,

$$m \circ (f \times (-1)) \circ (Q \mapsto (Q, a)) = t_{-a} \circ f,$$

and so

$$(f \times (-1))^* m^* \mathscr{L}(\Theta^-) | C \times \{a\} \approx \mathscr{L}(t_{-a}^{-1} \Theta^-) | f(C) = \mathscr{L}(\Theta_a^-) | f(C)$$
$$\approx f^* \mathscr{L}(\Theta_a^-).$$

Similarly

$$(f \times (-1))^* p^* \mathscr{L}(\Theta^-)|C \times \{a\} \approx f^*\mathscr{L}(\Theta^-), \quad \text{and}$$

$$(f \times (-1))^* q^* \mathscr{L}(\Theta^-)|C \times \{a\} \text{ is trivial.}$$

On the other hand, $\mathscr{M}^P$ is an invertible sheaf on $C \times J$ such that:
(i) $\mathscr{M}^P|C \times \{a\} \approx \mathscr{L}(D - gP)$ if D is an effective divisor of degree g on C such that $f^{(g)}(D) = a$;
(ii) $\mathscr{M}^P|\{P\} \times J$ is trivial.

Therefore (a) is equivalent to $(f \times (-1))^* m^* \mathscr{L}(\Theta^-)|C \times \{a\}$ being isomorphic to $\mathscr{M}^P \otimes p^* \mathscr{L}(gP)|C \times \{a\}$ for all a. As we know this is true for all a in a dense subset of J, [14, 5.3] applied to

$$\mathscr{M}^P \otimes p^* \mathscr{L}(gP) \otimes (f \times (-1))^* m^* \mathscr{L}(\Theta^-)^{-1}$$

proves (a). In particular, on taking $a = 0$, we find that $f^* \mathscr{L}(\Theta^-) \approx \mathscr{L}(gP)$, and so $(f \times (-1))^* p^* \mathscr{L}(\Theta^-) \approx p^* \mathscr{L}(gP)$. Now [14, 5.1] shows that $(f \times (-1))^* (m^* \mathscr{L}(\Theta^-) \otimes p^* \mathscr{L}(\Theta^-)^{-1}) \approx \mathscr{M}^P \otimes q^* \mathscr{N}$ for some invertible sheaf $\mathscr{N}$ on J. On computing the restrictions of the sheaves to $\{P\} \times J$, we find that $\mathscr{N} \approx (-1)^* \mathscr{L}(\Theta^-)$, which completes the proof. □

Consider the invertible sheaf $(f \times 1)^* \mathscr{P}$ on $C \times J^\vee$. Clearly it is a divisorial correspondence, and so there is a unique homomorphism $f^\vee: J^\vee \to J$ such that $(1 \times f^\vee)^* \mathscr{M}^P \approx (f \times 1)^* \mathscr{P}$. The next lemma completes the proof of the theorem.

Lemma 6.9. *The maps* $-f^\vee: J^\vee \to J$ *and* $\varphi_{\mathscr{L}(\Theta)}: J \to J^\vee$ *are inverse.*

PROOF. Write $\psi = -\varphi_{\mathscr{L}(\Theta)} = -\varphi_{\mathscr{L}(\Theta^-)}$. We have

$$(1 \times \psi)^*(1 \times f^\vee)^* \mathscr{M}^P \approx (1 \times \psi)^*(f \times 1)^* \mathscr{P}$$

$$\approx (f \times \psi)^* \mathscr{P} \approx (f \times (-1))^*(1 \times \varphi_{\mathscr{L}(\Theta)})^* \mathscr{P}$$

$$\approx (f \times (-1))^* \mathscr{L}'(\Theta^-) \approx \mathscr{M}^P.$$

Therefore, $f^\vee \circ \psi$ is a map $\alpha: J \to J$ such that $(1 \times \alpha)^* \mathscr{M}^P \approx \mathscr{M}^P$; but the only map with this property is the identity. □

Remark 6.10. (a) Lemma 6.7 shows that $f(C)$ and Θ cross transversely at any point of U. This can be proved more directly by using the descriptions of the tangent spaces implicitly given near the end of the proof of (5.1).
(b) In (6.8) we showed that $\mathscr{M}^P \approx (f \times (-1))^* \mathscr{L}'(\Theta^-)$. This implies

$$\mathscr{M}^P \approx (f \times (-1))^*(1 \times \varphi_{\mathscr{L}(\Theta^-)})^* \mathscr{P} \approx (f \times (-1))^*(1 \times \varphi_{\mathscr{L}(\Theta)})^* \mathscr{P}$$

$$\approx (f \times (-1))^* \mathscr{L}'(\Theta).$$

Also, because $D \mapsto \varphi_{\mathscr{L}(D)}$ is a homomorphism, $\varphi_{\mathscr{L}(-\Theta)} = -\varphi_{\mathscr{L}(\Theta)}$, and so

$$\mathscr{M}^P \approx (f \times (-1))^*(1 \times \varphi_{\mathscr{L}(\Theta)})^* \mathscr{P} \approx (f \times 1)^*(1 \times \varphi_{\mathscr{L}(-\Theta)})^* \mathscr{P}$$

$$\approx (f \times 1)^* \mathscr{L}'(-\Theta).$$

(c) The map on points $J^\vee(k) \to J(k)$ defined by $f^\vee$ is induced by $f^*: \text{Pic}(J) \to \text{Pic}(C)$.

(d) Lemma 6.7 can be generalized as follows. An effective canonical divisor K defines a point on $C^{(2g-2)}$ whose image in J will be denoted κ. Let a be a point of J such that $a - \kappa$ is not in $(W^{g-2})^-$, and write $a = \sum f(P_i)$ with $P_1, \ldots, P_g$ points on C. Then W^r and $(W^{g-r})_a^-$ intersect properly, and $W^r(W^{g-r})_a^- = \sum (w_{i_1 \ldots i_r})$ where

$$w_{i_1 \ldots i_r} = f(P_{i_1}) + \cdots + f(P_{i_r})$$

and the sum runs over the $\binom{g}{r}$ combinations obtained by taking r elements from $\{1, 2, \ldots, g\}$. See [20, §39, Prop. 17].

Summary 6.11. Between (C, P) and itself, there is a divisorial correspondence $\mathscr{L}^P = \mathscr{L}(\Delta - \{P\} \times C - C \times \{P\})$.

Between (C, P) and J there is the divisorial correspondence $\mathscr{M}^P$; for any divisorial correspondence $\mathscr{L}$ between (C, P) and a pointed k-scheme (T, t), there is a unique morphism of pointed k-schemes $\varphi: T \to J$ such that $(1 \times \varphi)^* \mathscr{M}^P \approx \mathscr{L}$. In particular, there is a unique map $f^P: C \to J$ such that $(1 \times f^P)^* \mathscr{M}^P \approx \mathscr{L}^P$ and $f(P) = 0$.

Between J and $J^\vee$ there is a canonical divisorial correspondence $\mathscr{P}$ (the Poincaré sheaf); for any divisorial correspondence $\mathscr{L}$ between J and a pointed k-schemes (T, t) there is a unique morphism of pointed k-schemes $\psi: T \to J$ such that $(1 \times \psi)^* \mathscr{P} \approx \mathscr{L}$.

Between J and J there is the divisorial correspondence $\mathscr{L}'(\Theta)$. The unique morphism $J \to J^\vee$ such that $(1 \times \psi)^* \mathscr{P} \approx \mathscr{L}'(\Theta)$ is $\varphi_{\mathscr{L}(\Theta)}$, which is an isomorphism. Thus $\varphi_{\mathscr{L}(\Theta)}$ is a principal polarization of J, called the *canonical polarization*. There are the following formulas:

$$\mathscr{M}^P \approx (f \times (-1))^* \mathscr{L}'(\Theta) \approx (f \times 1)^* \mathscr{L}'(\Theta)^{-1}.$$

Consequently,

$$\mathscr{L}^P \approx (f \times f)^* \mathscr{L}'(\Theta)^{-1}.$$

If $f^\vee: J^\vee \to J$ is the morphism such that $(f \times 1)^* \mathscr{P} \approx (1 \times f^\vee)^* \mathscr{M}^P$, then $f^\vee = -\varphi_{\mathscr{L}(\Theta)}^{-1}$.

Exercise 6.12. It follows from (6.6) and the Riemann–Roch theorem [14, 13.3] that $(\Theta^g) = g!$. Prove this directly by studying the inverse image of Θ (and its translates) by the map $C^g \to J$. (Cf. [14, 8.3], but note that the map is not finite.) Hence deduce another proof of (6.6).

§7. Weil's Construction of the Jacobian Variety

As we saw in (5.6a), the Jacobian J of a curve C is the unique abelian variety that is birationally equivalent to $C^{(g)}$. To construct J, Weil used the Riemann–Roch theorem to define a rational law of composition on $C^{(g)}$ and then proved a general theorem that allowed him to construct an algebraic group out of $C^{(g)}$ and the rational law. Finally, he verified that the algebraic group so obtained had the requisite properties to be called the Jacobian of C. We give a sketch of this approach.

A *birational group* over k (or a nonsingular variety with a normal law of composition in the terminology of Weil [20, V]) is a nonsingular variety V together with a rational map $m: V \times V \dashrightarrow V$ such that

(a) m is associative (that is, $(ab)c = a(bc)$ whenever both terms are defined);
(b) the rational maps $(a, b) \mapsto (a, ab)$ and $(a, b) \mapsto (b, ab)$ from $V \times V$ to $V \times V$ are both birational.

Assume that C has a k-rational point P.

Lemma 7.1. (a) *There exists an open subvariety U of $C^{(g)} \times C^{(g)}$ such that for all fields K containing k and all (D, D') in $U(K)$, $h^0(D + D' - gP) = 1$.*
(b) *There exists an open subset V of $C^{(g)} \times C^{(g)}$ such that for all fields K containing k and all (D, D') in $V(K)$, $h^0(D' - D + gP) = 1$.*

PROOF. (a) Let D_{can} be the canonical relative effective divisor on $C \times C^{(2g)}/C^{(2g)}$ constructed in Section 3. According to the Riemann–Roch theorem, $h^0(D - gP) \geq 1$ for all divisors of degree $2g$ on C, and so [14, 4.2c] shows that the subset U of $C^{(2g)}$ of points t such that $h^0((D_{\text{can}})_t - gP) = 1$ is open. On the other hand, (5.2b) shows that there exist positive divisors D of degree g such that $h^0((D + gP) - gP) = 1$, and so U is nonempty. Its inverse image in $C^{(g)} \times C^{(g)}$ is the required set.

(b) The proof is similar to that of (a): the Riemann–Roch theorem shows that $h^0(D' - D + gP) \geq 1$ for all D and D', we know there exists a D' such that $h^0(D' - gP + gP) = h^0(D') = 1$, and [14, 4.2] applied to the appropriate invertible sheaf on $C \times C^{(g)} \times C^{(g)}$ gives the result. □

Proposition 7.2. *There exists a unique rational map $m: C^{(g)} \times C^{(g)} \dashrightarrow C^{(g)}$ whose domain of definition contains the subset U of (7.1a) and which is such that for all fields K containing k and all (D, D') in $U(K)$, $m(D, D') \sim D + D' - gP$; moreover m makes $C^{(g)}$ into a birational group.*

PROOF. Let T be an integral k-scheme. If we identify $C^{(g)}$ with the functor it represents (see (3.13)), then an element of $U(T)$ is a pair of relative effective divisors (D, D') on $C \times T/T$ such that, for all $t \in T$, $h^0(D_t + D'_t - gP) = 1$. Let $\mathscr{L} = \mathscr{L}(D + D' - g \cdot P \times T)$. Then [14, 4.2d] shows that $q_*(\mathscr{L})$ is an invertible sheaf on T. The canonical map $q^*q_*\mathscr{L} \to \mathscr{L}$ when tensored with $(q^*q_*\mathscr{L})^{-1}$ gives a canonical global section $s: \mathcal{O}_T \to \mathscr{L} \otimes (q^*q_*\mathscr{L})^{-1}$, which determines a relative effective divisor $m(D, D')$ of degree g on $C \times T/T$ (see (3.6)). The construction is clearly functorial. Therefore we have constructed a map $m: U \to C^{(g)}$ as functors of integral schemes over k, and this is represented by a map of varieties. On making the map explicit in the case that K is the spectrum of a field, one sees easily that $m(D, D') \sim D + D' - gP$ in this case.

The uniqueness of the map is obvious. Also associativity is obvious since it holds on an open subset of $U(K)$: $m((D, D'), D'') = m(D, (D', D''))$ because each is an effective divisor on C linearly equivalent to $D + D' + D'' - 2gP$, and in general $h^0(D + D' + D'' - 2gP) = 1$.

A similar argument using (7.1b) shows that there is a map $r: V \to C^{(g)}$ such that (p, r) is a birational inverse to

$$(a, b) \mapsto (a, ab): C^{(g)} \times C^{(g)} \dashrightarrow C^{(g)} \times C^{(g)}.$$

Because the law of composition is commutative, this shows that $(a, b) \mapsto (b, ab)$ is also birational. The proof is complete. □

Theorem 7.3. *For any birational group V over k, there is a group variety G over k and a birational map $f: V \dashrightarrow G$ such that $f(ab) = f(a)f(b)$ whenever ab is defined; moreover, G is unique up to a unique isomorphism.*

PROOF. In the case that $V(k)$ is dense in V (for example, k is separably closed), this is proved in [1, §2]. (Briefly, one replaces V by an open subset where m has better properties, and obtains G by patching together copies of translates of U by elements of $V(k)$.) From this it follows that, in the general case, the theorem holds over a finite Galois extension k' of k. Let $\sigma \in \text{Gal}(k'/k)$. Then $\sigma f: \sigma V_{k'} \dashrightarrow \sigma G$ is a birational map, and as $\sigma V_{k'} = V_{k'}$, the uniqueness of G shows that there is a unique isomorphism $\varphi_\sigma: \sigma G \to G$ such that $\varphi_\sigma \circ \sigma f = f$. For any $\sigma, \tau \in \text{Gal}(k'/k)$,

$$(\varphi_\tau \circ \tau \varphi_\sigma) \circ (\tau \sigma f) = \varphi_\tau \circ \tau(\varphi_\sigma \circ \sigma f) = f = \varphi_{\tau\sigma} \circ \tau\sigma f,$$

and so $\varphi_\tau \circ \tau \varphi_\sigma = \varphi_{\sigma\tau}$. Descent theory (see (1.8)) now shows that G is defined over k. □

Let J be the algebraic group associated by (7.3) to the rational group defined in (7.2).

Proposition 7.4. *The variety J is complete.*

PROOF. This can be proved using the valuative criterion of properness. (For Weil's original account, see [20, Théorème 16, et seq.].) □

Corollary 7.5. *The rational map $f: C^{(g)} \dashrightarrow J$ is a morphism. If D and D' are linearly equivalent divisors on C_K for some field K containing k, then $f(D) = f(D')$.*

PROOF. The first statement follows from [14, 3.1]. For the second, recall that if D and D' are linearly equivalent then they lie in a copy of projective space contained in $C^{(g)}$ (see (3.14)). Consequently [14, 3.9] shows that they map to the same point in J. □

We now prove that J has the correct universal property.

Theorem 7.6. *There is a canonical isomorphism of functors* $\iota\colon P_C^0 \to J$.

PROOF. As in Section 4, it suffices to show that P_C^r is representable by J for some r. In this case we take $r = g$. Let $\mathscr{L}$ be an invertible sheaf with fibres of degree g on $C \times T$. If $\dim_k \Gamma(C_t, \mathscr{L}_t) = 1$ for some t, then this holds for all points in an open neighborhood U_t of t. As in the proof of (7.2), we get a relative effective divisor $s\colon \mathcal{O}_S \to \mathscr{L} \otimes (q^* q_* \mathscr{L})^{-1}$ of degree g on U_t. This family of Cartier divisors defines a map $U_t \to C^{(g)}$ which when composed with f gives a map $\psi_{\mathscr{L}}\colon U_t \to J$. On the other hand, if $\dim_k \Gamma(C_t, \mathscr{L}_t) > 1$, then we choose an invertible sheaf $\mathscr{L}'$ of degree zero on C such that $\dim(\Gamma(C_t, \mathscr{L}_t \otimes \mathscr{L}')) = 1$, and define $\psi_{\mathscr{L}}\colon U_t \to C^{(g)}$ on a neighborhood of t to be the composite of $\psi_{\mathscr{L} \otimes p^* \mathscr{L}'}$ with t_{-a}, where $a = f(D)$ for D an effective divisor of degree g such that $\mathscr{L}(D - gP) \approx \mathscr{L}'$. One checks that this map depends only on $\mathscr{L}$, and that the maps for different t agree on the overlaps of the neighborhoods. They therefore define a map $T \to J$. □

Remark 7.7. Weil of course did not show that the Jacobian variety represented a functor on k-schemes. Rather, in the days before schemes, the Jacobian variety was characterized by the universal property in (6.1) or (6.4), and shown to have the property that $\mathrm{Pic}^0(C) \xrightarrow{\approx} J(k)$. See [20] or [9].

§8. Generalizations

It is possible to construct Jacobians for families of curves. Let $\pi\colon \mathscr{C} \to S$ be a projective flat morphism whose fibres are integral curves. For any S-scheme T of finite-type, define

$$P_{\mathscr{C}}^r(T) = \{\mathscr{L} \in \mathrm{Pic}(\mathscr{C} \times_S T) \mid \deg(\mathscr{L}_t) = r \text{ all } t\}/\sim,$$

where $\mathscr{L} \sim \mathscr{L}'$ if and only if $\mathscr{L} \approx \mathscr{L}' \otimes q^* \mathscr{M}$ for some invertible sheaf $\mathscr{M}$ on T. (The degree of an invertible sheaf on a singular curve is defined as in the nonsingular case: it is the leading coefficient of $\chi(C, \mathscr{L}^n)$ as a polynomial in n.) Note that $P_{\mathscr{C}}^r$ is a functor on the category of S-schemes of finite-type.

Theorem 8.1. *Let* $\pi\colon \mathscr{C} \to S$ *be as above; then there is a group scheme* $\mathscr{J}$ *over* S *with connected fibres and a morphism of functors* $P_{\mathscr{C}}^0 \to \mathscr{J}$ *such that* $P_{\mathscr{C}}^0(T) \to \mathscr{J}(T)$ *is always injective and is an isomorphism whenever* $\mathscr{C} \times_S T \to T$ *has a section.*

In the case that S is the spectrum of a field (but $\mathscr{C}$ may be singular), the existence of $\mathscr{J}$ can be proved by Weil's method (see [17, V]). When $\mathscr{C}$ is smooth over S, one can show as in Section 3 that $\mathscr{C}^{(r)}$ (quotient of $\mathscr{C} \times_S \ldots \times_S \mathscr{C}$ by S_r) represents the functor $\mathrm{Div}_{\mathscr{C}/S}^r$ sending an S-scheme T to the set of relative effective Cartier divisors of degree r on $\mathscr{C} \times_S T/T$. In general one can only show more abstractly that $\mathrm{Div}^r_{\mathscr{C}/S}$ is represented by a Hilbert scheme. There is a canonical map $\mathrm{Div}^r_{\mathscr{C}/S} \to P^r_{\mathscr{C}/S}$ and the second part of the proof deduces the representability of $P^r_{\mathscr{C}/S}$ from that of $\mathrm{Div}^r_{\mathscr{C}/S}$. (The only reference for the proof in the general case seems to be Grothendieck's original rather succinct account [4, Exposé 232]; we sketch some of its ideas below.)

As in the case that the base scheme is the spectrum of a field, the conditions of the theorem determine $\mathscr{J}$ uniquely; it is called the *Jacobian scheme* of $\mathscr{C}/S$. Clearly $\mathscr{J}$ commutes with base change: the Jacobian of $\mathscr{C} \times_S T$ over T is $\mathscr{J} \times_S T$. In particular, if $\mathscr{C}_t$ is a smooth curve over $k(t)$, then $\mathscr{J}_t$ is the Jacobian of $\mathscr{C}_t$ in the sense of Section 1. Therefore if $\mathscr{C}$ is smooth over S, then $\mathscr{J}$ is an abelian scheme, and we may think of it as a family of Jacobian varieties. If $\mathscr{C}$ is not smooth over S, then $\mathscr{J}$ need not be proper, even in the case that S is the spectrum of a field.

Example 8.2. Let C be complete smooth curve over an algebraically closed field k. By a *modulus* for C one means simply an effective divisor $\mathfrak{m} = \sum n_P P$ on C. Let $\mathfrak{m}$ be such a modulus, and assume that $\deg(\mathfrak{m}) \geq 2$. We shall associate with C and $\mathfrak{m}$ a new curve $C_\mathfrak{m}$ having a single singularity at a point to be denoted by Q. The underlying topological space of $C_\mathfrak{m}$ is $(C - S) \cup \{Q\}$, where S is the support of $\mathfrak{m}$. Let $\mathcal{O}_Q = k + \mathfrak{c}_Q$, where

$$\mathfrak{c}_Q = \{f \in k(C) | \mathrm{ord}(f) \geq n_P \text{ all } P \text{ in } S\},$$

and define $\mathcal{O}_{C_\mathfrak{m}}$ to be the sheaf such that $\Gamma(U, \mathcal{O}_{C_\mathfrak{m}}) = \bigcap \mathcal{O}_P$, where the intersection is over the P in U. The Jacobian scheme $J_\mathfrak{m}$ of $C_\mathfrak{m}$ is an algebraic group over k called the *generalized Jacobian of C relative to* $\mathfrak{m}$. By definition, $J_\mathfrak{m}(k)$ is the group of isomorphism classes of invertible sheaves on $C_\mathfrak{m}$ of degree 0. It can also be described as the group of divisors of degree 0 on C relatively prime to $\mathfrak{m}$, modulo the principal divisors defined by elements congruent to 1 modulo $\mathfrak{m}$ (an element of $k(C)$ is congruent to 1 modulo $\mathfrak{m}$ if $\mathrm{ord}_P(f - 1) \geq n_P$ for all P in S). For each modulus $\mathfrak{m}$ with support on S there is a canonical map $f_\mathfrak{m}: C - S \to J_\mathfrak{m}$, and these maps are universal in the following sense: for any morphism $f: C - S \to G$ from $C - S$ into an algebraic group, there is a modulus $\mathfrak{m}$ and a homomorphism $\varphi: J_\mathfrak{m} \to G$ such that f is the composite of $f_\mathfrak{m} \circ \varphi$ with a translation. (For a detailed account of this theory, see [17].)

We now give a brief sketch of part of Grothendieck's proof of (8.1). First we need the notion of the Grassmann scheme.

Let $\mathscr{E}$ be a locally free sheaf of $\mathcal{O}_S$-modules of finite rank, and, for an S-scheme T of finite-type, define $\mathrm{Grass}^{\mathscr{E}}_n(T)$ to be the set of isomorphism classes of pairs $(\mathscr{V}, h)$, where $\mathscr{V}$ is a locally free $\mathcal{O}_T$-module of rank n and h is an epimorphism $\mathcal{O}_T \otimes_k \mathscr{E} \to \mathscr{V}$. For example, if $\mathscr{E} = \mathcal{O}_S^m$, then $\mathrm{Grass}^{\mathscr{E}}_n(T)$ can be identified with the set of isomorphism classes of pairs $(\mathscr{V}, (e_1, \ldots, e_m))$ where $\mathscr{V}$ is a locally free sheaf of rank n on T and the e_i are sections of $\mathscr{V}$ over T that generate $\mathscr{V}$; two such pairs $(\mathscr{V}, (e_1, \ldots, e_m))$ and $(\mathscr{V}', (e'_1, \ldots, e'_m))$

are isomorphic if there is an isomorphism $\mathscr{V} \xrightarrow{\approx} \mathscr{V}'$ carrying each e_i to e'_i. In particular, $\mathrm{Grass}_1^{\mathcal{O}^{N+1}}(T) = \mathbb{P}_S^N(T)$ (cf. [8, II, 7.1]).

Proposition 8.3. *The functor* $T \mapsto \mathrm{Grass}_n^{\mathscr{E}}(T)$ *is representable by a projective variety* $G_n^{\mathscr{E}}$ *over* S.

PROOF. The construction of $G_n^{\mathscr{E}}$ is scarcely more difficult than that of $\mathbb{P}_S^N$ (see [7, 9.7]). □

Choose an $r > 2g - 2$ and an $m > 2g - 2 + r$. As in the case that S is the spectrum of a field, we first need to construct the Jacobian under the assumption that there is a section $s: S \to \mathscr{C}$. Let E be the relative effective divisor on $\mathscr{C}/S$ defined by s (see (3.9)), and for any invertible sheaf $\mathscr{L}$ on $\mathscr{C} \times_S T$, write $\mathscr{L}(m)$ for $\mathscr{L} \otimes \mathscr{L}(mE)$. The first step is to define an embedding of $\mathrm{Div}^r_{\mathscr{C}/S}$ into a suitable Grassmann scheme.

Let $D \in \mathrm{Div}^r_{\mathscr{C}/S}(T)$, and consider the exact sequence

$$0 \to \mathscr{L}(-D) \to \mathcal{O}_{\mathscr{C} \times T} \to \mathcal{O}_D \to 0$$

on $\mathscr{C} \times_S T$ (we often drop the S from $\mathscr{C} \times_S T$). This gives rise to an exact sequence

$$0 \to \mathscr{L}(-D)(m) \to \mathcal{O}_{\mathscr{C} \times T}(m) \to \mathcal{O}_D(m) \to 0,$$

and on applying q_* we get an exact sequence

$$0 \to q_* \mathscr{L}(-D)(m) \to q_* \mathcal{O}_{\mathscr{C} \times T}(m) \to q_* \mathcal{O}_D(m) \to R^1 q_* \mathscr{L}(-D)(m) \to \dots.$$

Note that, for all t in T, $H^1(\mathscr{C}_t, \mathscr{L}(-D)(m))$ is dual to $H^0(C_t, \mathscr{L}(K+D-mE_t))$, where E_t is the divisor $s(t)$ of degree one on $\mathscr{C}_t$. Because of our assumptions, this last group is zero, and so (see [14, 4.2e]) $R^1 q_* \mathscr{L}(-D)(m)$ is zero and we have an exact sequence

$$0 \to q_* \mathscr{L}(-D)(m) \to q_* \mathcal{O}_{\mathscr{C} \times T}(m) \to q_* \mathcal{O}_D(m) \to 0.$$

Moreover $q_* \mathcal{O}_D(m)$ is locally free of rank r, and $q_*(\mathcal{O}_{\mathscr{C} \times T}(m)) = q_* \mathcal{O}_{\mathscr{C}}(m) \otimes \mathcal{O}_T$ (loc. cit.), and so we have constructed an element $\Phi(D)$ of $\mathrm{Grass}_r^{q_* \mathcal{O}_{\mathscr{C}}(m)}(T)$.

On the other hand, suppose $a = (q_* \mathcal{O}_{\mathscr{C} \times T}(m) \twoheadrightarrow \mathscr{V})$ is an element of $\mathrm{Grass}_r^{q_* \mathcal{O}_{\mathscr{C}}(m)}(T)$. If $\mathscr{K}$ is the kernel of $q^* q_* \mathcal{O}_{\mathscr{C} \times T}(m) \to q^* \mathscr{V}$, then $\mathscr{K}(-m)$ is a subsheaf of $q^* q_* \mathcal{O}_{\mathscr{C} \times T}$, and its image under $q^* q_* \mathcal{O}_{\mathscr{C} \times T} \to \mathcal{O}_{\mathscr{C} \times T}$ is an ideal in $\mathcal{O}_{\mathscr{C} \times T}$. Let $\Psi(a)$ be the subscheme associated to this ideal. It is clear from the constructions that $\Psi\Phi(D) = D$ for any relative divisor of degree r. We have a diagram of natural transformations

$$\mathrm{Div}^r_{\mathscr{C}}(T) \xrightarrow{\Phi} \mathrm{Grass}_r^{q_* \mathcal{O}_{\mathscr{C}}(m)}(T) \xrightarrow{\Psi} \mathscr{S}(T) \supset \mathrm{Div}^r_{\mathscr{C}}(T), \quad \Psi\Phi = \mathrm{id},$$

where $\mathscr{S}(T)$ denotes the set of all closed subschemes of $\mathscr{C} \times_S T$. In particular, we see that Φ is injective.

Proposition 8.4. *The functor* Φ *identifies* $\mathrm{Div}^r_{\mathscr{C}}$ *with a closed subscheme of* $\mathrm{Grass}_r^{q_* \mathcal{O}_{\mathscr{C}}(m)}$.

PROOF. See [4, Exposé 221, p. 12] (or, under different hypotheses, [15, Lecture 15]). □

Finally, one shows that the fibres of the map $\text{Div}^r_{\mathscr{C}/S} \to P^r_{\mathscr{C}/S}$ are represented by the projective space bundles associated with certain sheaves of $\mathcal{O}_S$-modules ([4, Exposé 232, p. 11]; cf. (5.6c)) and deduces the representability of $P^r_{\mathscr{C}/S}$ (loc. cit.).

§9. Obtaining Coverings of a Curve from its Jacobian; Application to Mordell's Conjecture

Let V be a variety over field k, and let $\pi: W \to V$ be a finite étale map. If there is a finite group G acting freely on W by V-morphisms in such a way that $V = G \backslash W$, then (W, π) is said to be Galois covering of V with Galois group G. When G is abelian, then (W, π) is said to be an abelian covering of V. Fix a point P on V. Then the Galois coverings of V are classified by the (étale) fundamental group $\pi_1(V, P)$ and the abelian coverings by the maximal abelian quotient $\pi_1(V, P)^{\text{ab}}$ of $\pi_1(V, P)$. For any finite abelian group M, $\text{Hom}(\pi_1(V, P), M)$ (set of continuous homomorphisms) is equal to the set of isomorphism classes of Galois coverings of V with Galois group M. If, for example, V is nonsingular and we take P to be the generic point of V, then every finite connected étale covering of V is isomorphic to the normalization of V in some finite extension of K' of $k(P)$ contained in a fixed algebraic closure $\bar{K}$ of K; moreover, $\pi_1(V, P) = \text{Gal}(K^{\text{un}}/K)$ where K^{un} is the union of all finite extensions K' of $k(P)$ in $\bar{K}$ such that the normalization of V in K' is étale over V. The covering corresponding to a continuous homomorphism $\alpha: \text{Gal}(K^{\text{un}}/K) \to M$ is the normalization of V in $\bar{K}^{\text{Ker}(\alpha)}$. (See [13, I, 5] for a more detailed discussion of étale fundamental groups.)

Now let C be a complete nonsingular curve over a field k, and let $f = f^P$ for some P in $C(k)$. From a finite étale covering $J' \to J$ of J, we obtain an étale covering of C by pulling back relative to f:

$$\begin{array}{ccc} J' & \leftarrow & C' = C \times_J J' \\ \downarrow & & \downarrow \\ J & \stackrel{f}{\leftarrow} & C. \end{array}$$

Because all finite étale coverings of J are abelian (cf. [14, 15.3]), we only obtain abelian coverings of C in this way. The next proposition shows that we obtain all such coverings.

Henceforth, k will be separably closed.

Proposition 9.1. *If $J' \to J$ is a connected étale covering of J, then $C' = C \times_J J' \to C$ is a connected étale covering of C, and every connected abelian covering of C is obtained in this way. Equivalently, the map $\pi_1(C, P)^{\text{ab}} \to \pi_1(J, 0)$ induced by f^P is an isomorphism.*

PROOF. The equivalence of the two assertions follows from the interpretation of $\mathrm{Hom}(\pi_1(V, P), M)$ recalled above and the fact that $\pi_1(J, 0)$ is abelian. We shall prove the second assertion. For this it suffices to show that for all integers n, the map $\mathrm{Hom}(\pi_1(J, 0), \mathbb{Z}/n\mathbb{Z}) \to \mathrm{Hom}(\pi_1(C, P), \mathbb{Z}/n\mathbb{Z})$ induced by f^P is an isomorphism. The next two lemmas take care of the case that n is prime to the characteristic of k.

Lemma 9.2. *Let V be complete nonsingular variety and let P be a point of V; then for all integers n prime to the characteristic of k, $\mathrm{Hom}(\pi_1(V, P), \mathbb{Z}/n\mathbb{Z}) \approx \mathrm{Pic}(V)_n$.*

PROOF. Let D be a (Weil) divisor on V such that $nD = (g)$ for some $g \in k(V)$, and let V' be the normalization of V in the Kummer extension $k(V)(g^{1/n})$ of $k(V)$. A purity theorem [5, X.3.1] shows that $V' \to V$ is étale if, for all prime divisors Z on V, the discrete valuation ring $\mathcal{O}_Z$ (local ring at the generic point of Z) is unramified in $k(V')$. But the extension $k(V')/k(V)$ was constructed by extracting the nth root of an element g such that $\mathrm{ord}_Z(g) = 0$ if Z is not in the support of D and is divisible by n otherwise, and it follows from this that $\mathcal{O}_Z$ is unramified. Conversely, let $V' \to V$ be a Galois covering with Galois group $\mathbb{Z}/n\mathbb{Z}$. Kummer theory shows that the $k(V')/k(V)$ is obtained by extracting the nth root of an element g of $k(V)$. Let Z be a prime divisor on V. Because $\mathcal{O}_Z$ is unramified in $k(V')$, $\mathrm{ord}_Z(g)$ must be divisible by n (or is zero), and so $(g) = nD$ for some divisor D. Obviously D represents an element of $\mathrm{Pic}(V)_n$. It is easy to see now that the correspondence we have defined between coverings of V and elements of $\mathrm{Pic}(V)_n$ is one-to-one. (For a proof using étale cohomology, see [14, III, 4.11].) □

Lemma 9.3. *The map $\mathrm{Pic}(J) \to \mathrm{Pic}(C)$ defined by f induces an isomorphism $\mathrm{Pic}^0(J) \to \mathrm{Pic}^0(C)$.*

PROOF. This was noted in (6.10c). □

In the case that $n = p = \mathrm{characteristic}(k)$, (9.2) and (9.3) must be replaced by the following analogues.

Lemma 9.4. *For any complete nonsingular variety V and point P, $\mathrm{Hom}(\pi_1(V, P), \mathbb{Z}/p\mathbb{Z}) \approx \mathrm{Ker}(1 - F: H^1(V, \mathcal{O}_V) \to H^1(V, \mathcal{O}_V))$, where F is the map induced by $a \mapsto a^p: \mathcal{O}_V \to \mathcal{O}_V$.*

PROOF. See [14, p. 127] for a proof using étale cohomology as well as for hints for an elementary proof. □

Lemma 9.5. *The map $f^P: C \to J$ induces an isomorphism $H^1(J, \mathcal{O}_J) \to H^1(C, \mathcal{O}_C)$.*

PROOF. See [17, VII, Théorème 9]. (Alternatively, note that the same argument as in the proof of (2.1) gives an isomorphism $H^1(J, \mathcal{O}_J) \xrightarrow{\approx} T_0(J^\vee)$, and we know that $J \approx J^\vee$.) □

To prove the case $n = p^m$, one only has to replace $\mathcal{O}_C$ and $\mathcal{O}_J$ by the sheaves of Witt vectors of length m, $W_m \mathcal{O}_C$ and $W_m \mathcal{O}_J$. (It is also possible to use a five-lemma argument starting from the case $m = 1$.)

Corollary 9.6. *For all primes l, the map of étale cohomology groups $H^1(J, \mathbb{Z}_l) \to H^1(C, \mathbb{Z}_l)$ induced by f is an isomorphism.*

PROOF. For any variety V, $H^1(V_{et}, \mathbb{Z}/n\mathbb{Z}) = \text{Hom}(\pi_1(V, P), \mathbb{Z}/n\mathbb{Z})$ [13, III, 4]. Therefore, there are isomorphisms

$$H^1(J, \mathbb{Z}/l^m\mathbb{Z}) \xrightarrow{\approx} \text{Hom}(\pi_1(J, P), \mathbb{Z}/l^m\mathbb{Z}) \xrightarrow{\approx} \text{Hom}(\pi_1(C, P), \mathbb{Z}/l^m\mathbb{Z})$$
$$\xrightarrow{\approx} H^1(C, \mathbb{Z}/l^m\mathbb{Z}),$$

and we obtained the required isomorphism by passing to the limit. □

To obtain ramified coverings of C, one can use the generalized Jacobians.

Proposition 9.7. *Let $C' \to C$ be a finite abelian covering of C that is unramified outside a finite set Σ. Then there is a modulus $\mathfrak{m}$ with support on Σ and an étale isogeny $J' \to J_\mathfrak{m}$ whose pull-back by $f_\mathfrak{m}$ is $C' - f^{-1}(\Sigma)$.*

PROOF. See [17]. □

Example 9.8. In the case that the curve is $\mathbb{P}^1$ and $\mathfrak{m} = 0 + \infty$, we have $J_\mathfrak{m} = \mathbb{P}^1 - \{0, \infty\}$, which is just the multiplicative group GL_1, and $f_\mathfrak{m}$ is an isomorphism. For any n prime to the characteristic, there is a unique unramified covering of $\mathbb{P}^1 - \{0, \infty\}$ of degree n, namely multiplication by n on $\mathbb{P}^1 - \{0, \infty\}$. When $k = \mathbb{C}$, this covering is the usual unramified covering $z \mapsto z^n \colon \mathbb{C} - \{0\} \to \mathbb{C} - \{0\}$.

Proposition 9.9. *Let C be a curve of genus g over a number field k, and let P be a k-rational point of C. Let S be a finite set of primes of k containing all primes dividing 2 and such that C has good reduction outside S. Then there exists a field k' of degree $\leq 2^{2g}$ over k and unramified outside S, and a finite map $f_P \colon C_P \to C_{k'}$ of degree $\leq 2^{2g(g-1)+2g+1}$, ramified exactly over P, and such that C_P has good reduction outside S.*

PROOF. Sketch. Let C' be the pull-back of $2\colon J \to J$; it is an abelian étale covering of C of degree 2^{2g}, and the Hurwitz genus formula [8, IV, 2.4] shows that the genus g' of C' satisfies

$$2g' - 2 = 2^{2g}(2g - 2),$$

so that $g' = 2^{2g}(g - 1) + 1$. Let D be the inverse image of P on C'. It is a divisor of degree 2^{2g} on C', and after an extension k' of k of degree $\leq 2^{2g}$ unramified over S, some point P of D will be rational. Let $\mathfrak{m} = D - P'$, and let C'' be the pull-back of the covering $2: J_\mathfrak{m} \to J_\mathfrak{m}$ (of degree $\leq 2^{2g'}$) by $C - \Sigma \to J_\mathfrak{m}$, where $\Sigma = \mathrm{Supp}(D) - \{P\}$. Then C'' is a curve over k', and we take C_P to the associated complete nonsingular curve. □

This result has a very striking consequence. Recall that a conjecture of Shafarevich states the following:

(9.10) *For any number field k, integer g, and finite set S of primes of k, there are only finitely many isomorphism classes of curves C of genus g over k having good reduction at all primes outside S.*

Theorem 9.11. *Shafarevich's conjecture (9.10) implies Mordell's conjecture.*

PROOF. Let C be curve of genus $g \geq 2$ over k with good reduction outside a set S containing all primes of k lying over 2. There is a finite field extension K of k containing all extensions k' of k of degree $\leq 2^{2g}$ that are unramified outside S. For each k-rational point P on C, Proposition 9.9 provides a map $f_P: C_P \to C_K$ of degree $\leq$ a fixed bound $B(g)$ which is ramified exactly over P; moreover, C_P has good reduction outside S. The Hurwitz genus formula shows that

$$2g(C_P) - 2 \leq B(g)(2g - 2) + B(g) - 1.$$

Therefore Shafarevich's conjecture implies that there can be only finitely many curves C_P. A classical result of de Franchis [10, p. 223] states that for each C_P, there are only finitely many maps $C_P \to C$ (this is where it is used that $g \geq 2$). Therefore there can be only finitely many k-rational points on C, as predicted by Mordell. □

§10. Abelian Varieties Are Quotients of Jacobian Varieties

The main result in this section sometimes allows questions concerning abelian varieties to be reduced to the special case of Jacobian varieties.

Theorem 10.1. *For any abelian variety A over an infinite field k, there is a Jacobian variety J and a surjective homomorphism $J \twoheadrightarrow A$.*

Lemma 10.2. *Let $\pi: W \to V$ be a finite morphism of complete varieties, and let $\mathscr{L}$ be an invertible sheaf on V. If $\mathscr{L}$ is ample, then so also is $\pi^*\mathscr{L}$.*

PROOF. We shall use the following criterion ([8, III, 5.3]): an invertible sheaf

$\mathscr{L}$ on a complete variety is ample if and only if, for all coherent $\mathcal{O}_V$-modules $\mathscr{F}$, $H^i(V, \mathscr{F} \otimes \mathscr{L}^n) = 0$ for all $i > 0$ and sufficiently large n. Also we shall need an elementary projection formula: if $\mathscr{N}$ and $\mathscr{M}$ are coherent sheaves of modules on W and V respectively, then

$$\pi_*(\mathscr{N} \otimes \pi^*\mathscr{M}) \approx \pi_*\mathscr{N} \otimes \mathscr{M}.$$

(Locally, this says that if B is an A-algebra and N and M are modules over B and A respectively, then $N \otimes_B (B \otimes_A M) \approx N \otimes_A M$ as A-modules.)

Let $\mathscr{F}$ be a coherent $\mathcal{O}_W$-module. Because π is finite (hence affine), we have by [8, II, Ex. 4.1 or Ex. 8.2] that

$$H^i(W, \mathscr{F} \otimes \pi^*\mathscr{L}^n) \approx H^i(V, \pi_*(\mathscr{F} \otimes \pi^*\mathscr{L}^n)).$$

The projection formula shows that the second group equals $H^i(V, \pi_*\mathscr{F} \otimes \mathscr{L}^n)$, which is zero for all $i > 0$ and sufficiently large n because $\mathscr{L}$ is ample and $\pi_*\mathscr{F}$ is coherent ([8, 4.1]). The criterion now shows that $\pi^*\mathscr{L}$ is ample. □

Lemma 10.3. *Let V be a nonsingular projective variety of dimension ≥ 2 over a field k, and let Z be a hyperplane section of V relative to some fixed embedding $V \hookrightarrow \mathbb{P}^n$. Then, for any finite map π from a nonsingular variety W to V, $\pi^{-1}(Z)$ is geometrically connected (that is, $\pi^{-1}(Z)_{\bar{k}}$ is connected).*

PROOF. The hypotheses are stable under a change of the base field, and so we can assume that k is algebraically closed. It then suffices to show that $\pi^{-1}(Z)$ is connected. Because Z is an ample divisor on V, the preceding lemma shows that $\pi^{-1}(Z)$ is the support of an ample divisor on W, which implies that it is connected ([8, III, 7.9]). □

We now prove the theorem. Since all elliptic curves are their own Jacobians, we can assume that $\dim(A) > 1$. Fix an embedding $A \hookrightarrow \mathbb{P}^n$ of A into projective space. Then Bertini's theorem [8, II, 8.18] shows that there exists an open dense subset U of the dual projective space $\mathbb{P}_k^{n\vee}$ of $\mathbb{P}_k^n$ such that, for all hyperplanes H in U, $A_{\bar{k}} \cap H$ is nonsingular and connected. Because k is infinite, $U(k)$ is nonempty (consider a line L in $\mathbb{P}_k^{n\vee}$), and so there exists such an H with coordinates in k. Then $A \cap H$ is a (geometrically connected) nonsingular variety in $\mathbb{P}^n$. On repeating the argument $\dim(A) - 1$ times, we arrive at a nonsingular curve C on A that is the intersection of A with a linear subspace of $\mathbb{P}^n$. Now (10.3) applied several times shows that for any nonsingular variety W and finite map $\pi \colon W \to A$, $\pi^{-1}(C)$ is geometrically connected.

Consider the map $J \to A$ arising from the inclusion of C into A, and let A_1 be the image of the map. It is an abelian subvariety of A, and if it is not the whole of A, then there is an abelian subvariety A_2 of A such that $A_1 \times A_2 \to A$ is an isogeny (see [14, 12.1]); in particular, $A_1 \cap A_2$ is finite. As $C \subset A_1$, this implies that $C \cap A_2$ is finite. Let $W = A_1 \times A_2$ and take π to be the composite of $1 \times n_{A_2}\colon A_1 \times A_2 \to A_1 \times A_2$ with $A_1 \times A_2 \to A$, where

$n > 1$ is an integer prime to the characteristic of k. Then $\pi^{-1}(C)$ is not geometrically connected. This is a contradiction, and so A_1 must equal A.

Remark 10.4. (a) Lemma 10.2 has the following useful restatement: let V be a variety over a field k and let D be divisor on V such that the linear system $|D|$ is without base points; if the map $V \to \mathbb{P}^n$ defined by $|D|$ is finite, then D is ample.

(b) If some of the major theorems from étale cohomology are assumed, then it is possible to give a very short proof of the theorem. They show that, for any curve C on A constructed as in the above proof, the map $H^1(A, \mathbb{Z}_l) \to H^1(C, \mathbb{Z}_l)$ induced by the inclusion of C into A is injective (see [13, VI.5.6]). But $H^1(A, \mathbb{Z}_l)$ is dual to $T_l A$ and $H^1(C, \mathbb{Z}_l)$ is dual to $T_l J$, and so this says that the map $T_l J \to T_l A$ induced by $J \to A$ is surjective. Clearly this implies that J maps onto A.

Open Question 10.5. Let A be an abelian variety over an algebraically closed field k. We have shown that there is a surjection $J \twoheadrightarrow A$ with J a Jacobian variety. Let A_1 be the subvariety of J with support the identity component of the kernel of this map. Then A_1 is an abelian variety, and so there is a surjection $J_1 \twoheadrightarrow A_1$. Continuing in this way, we obtain a sequence of abelian varieties $A, A_1, A_2, \ldots$ and a complex

$$\cdots \to J_2 \to J_1 \to A \to 0.$$

Is it possible to make the constructions in such a way that the sequence terminates with 0? That is, does there exist a finite resolution (up to isogeny) of an arbitrary abelian variety by Jacobian varieties?

§11. The Zeta Function of a Curve

Let C be a complete nonsingular curve over a finite field $k = \mathbb{F}_q$. The best way to prove the Riemann hypothesis for C is to use intersection theory on $C \times C$ (see [8, V, Ex. 1.10]), but in this section we show how it can be derived from the corresponding result for the Jacobian of C. Recall [14, §19] that the characteristic polynomial of the Frobenius endomorphism π_J of J acting on $T_l J$ is a polynomial $P(X)$ of degree $2g$ with integral coefficients whose roots a_i have absolute value $q^{1/2}$.

Theorem 11.1. *The number N of points on C with coordinates in k is equal to $1 - \sum a_i + q$. Therefore, $|N - q - 1| \leq 2gq^{1/2}$.*

The proof will be based on the following analogue of the Lefschetz trace formula. A map $\alpha \colon C \to C$ induces a unique endomorphism α' of J such that $f^P \circ \alpha = \alpha' \circ f^P$ for any point P in $C(\bar{k})$ (cf. (6.1)).

Proposition 11.2. *For any endomorphism α of C,*
$$(\Gamma_\alpha \cdot \Delta) = 1 - \operatorname{Tr}(\alpha') + \deg(\alpha).$$

Recall [14, §12] that if $P_{\alpha'}(X) = \prod(X - a_i)$, then $\operatorname{Tr}(\alpha) = \sum a_i$, and that $\operatorname{Tr}(\alpha') = \operatorname{Tr}(\alpha' | T_l J)$. We now show that the proposition implies the theorem. Let $\pi_C \colon C_{\bar{k}} \to C_{\bar{k}}$ be the Frobenius endomorphism of C (see [14, §19]).

Before proving (11.2) we need a lemma.

Lemma 11.3. *Let A be an abelian variety of dimension g over a field k, and let H be the class of an ample divisor in $\operatorname{NS}(A)$. For any endomorphism α of A, write $D_H(\alpha) = (\alpha + 1)^*(H) - \alpha^*(H) - H$. Then*
$$\operatorname{Tr}(\alpha) = g \frac{(H^{g-1} \cdot D_H(\alpha))}{(H^g)}.$$

PROOF. The calculation in [14, 12.4] shows that
$$(\alpha + n)^*(H) = n(n-1)H + n(\alpha + 1)^* H - (n-1)\alpha^*(H)$$
(because $(2_A)^* H = 4H$ in $\operatorname{NS}(A)$), and so
$$(\alpha + n)^* H = n^2 H + n D_H(\alpha) + \alpha^*(H).$$
Now the required identity can be read off from the equation
$$P_\alpha(-n) = \deg(\alpha + n) = (((\alpha + n)^* H)^g)/(H^g) \quad \text{(see [14, 8.3])}$$
because $P_\alpha(-n) = n^{2g} + \operatorname{Tr}(\alpha) n^{2g-1} + \cdots$. □

We now prove (11.2). Consider the commutative diagram
$$C \times C \xrightarrow{f \times f} J \times J \xrightarrow{1 \times \alpha'} J \times J$$
$$\Delta \uparrow \qquad\qquad \Delta \uparrow$$
$$C \xrightarrow{f} J$$
where $f = f^P$ for some rational point P of C. Consider the sheaf $\mathscr{L}'(\Theta) \overset{df}{=} \mathscr{L}(m^*\Theta - \Theta \times J - J \times \Theta)$ on $J \times J$ (see Section 6). Then
$$((1 \times \alpha')(f \times f))^* \mathscr{L}'(\Theta) = ((f \times f)(1 \times \alpha))^* \mathscr{L}'(\Theta)$$
$$\approx (1 \times \alpha)^*(f \times f)^* \mathscr{L}'(\Theta) \approx (1 \times \alpha)^*(\mathscr{L}^P)^{-1}$$
by a formula in (6.11). Now
$$\Delta^*(1 \times \alpha)^* \mathscr{L}^P = \mathscr{L}(\Gamma_\alpha \cdot (\Delta - P \times C - C \times P)),$$
which has degree $(\Gamma_\alpha \cdot \Delta) - 1 - \deg(\alpha)$. We next compute the sheaf by going round the diagram the other way. As $(1 \times \alpha) \circ \Delta = (1, \alpha)$, we have
$$((1 \times \alpha) \circ \Delta)^* \mathscr{L}(m^*\Theta) \approx (1 + \alpha)^* \mathscr{L}(\Theta) \quad \text{and}$$
$$\deg f^* \mathscr{L}((1 + \alpha)^*(\Theta)) = \deg f^*(1 + \alpha)^* \Theta.$$

Similarly $\deg f^*((1 \times \alpha) \circ \Delta)^* \mathscr{L}(\Theta \times J) \approx \deg f^*\Theta$ and

$$\deg f^*((1 \times \alpha) \circ \Delta)^* \mathscr{L}(J \times \Theta) = \deg f^*(\alpha^*\theta),$$

and so we find that

$$1 - (\Gamma_\alpha \cdot \Delta) + \deg(\alpha) = \deg f^*(D_\Theta(\alpha)).$$

We know (6.12) that $(\Theta^g) = g!$, and it is possible to show that $f^*(D_\Theta(\alpha)) = (f(C) \cdot D_\Theta(\alpha))$ is equal to $(g-1)!(\Theta^{g-1} \cdot D_\Theta(\alpha))$ (see [9, IV, §3]). Therefore (11.3) completes the proof. □

Corollary 11.4. *The zeta function of C is equal to*

$$Z(C, t) = \frac{P(t)}{(1-t)(1-qt)}.$$

Remark 11.5. As we saw in (9.6), $H^1(C_{et}, \mathbb{Z}_l) = H^1(J_{et}, \mathbb{Z}_l) = (T_l J)^\vee$, and so (11.2) can be rewritten as

$$(\Gamma_\alpha \cdot \Delta) = \sum (-1)^i \operatorname{Tr}(\alpha | H^i(C_{et}, \mathbb{Z}_l)).$$

§12. Torelli's Theorem: Statement and Applications

Torelli's theorem says that a curve C is uniquely determined by its canonically polarized Jacobian (J, λ).

Theorem 12.1. *Let C and C' be complete smooth curves over an algebraically closed field k, and let $f: C \to J$ and $f': C' \to J'$ be the maps of C and C' into their Jacobians defined by points P and P' on C and C'. Let $\beta: (J, \lambda) \to (J', \lambda')$ be an isomorphism from the canonically polarized Jacobian of C to that of C'.*

(a) *There exists an isomorphism $\alpha: C \to C'$ such that $f' \circ \alpha = \pm \beta \circ f + c$, for some c in $J'(k)$.*
(b) *Assume that C has genus ≥ 2. If C is not hyperelliptic, then the map α, the sign $\pm$, and c are uniquely determined by β, P, P'. If C is hyperelliptic, the sign can be chosen arbitrarily, and then α and c are uniquely determined.*

PROOF. (a) The proof involves complicated combinatorial arguments in the W^r—we defer it to the next section.

(b) Recall [8, IV, 5] that a curve C is hyperelliptic if there is a finite map $\pi: C \to \mathbb{P}^1$ of degree 2; the fibres of such a map form a linear system on C of degree 2 and dimension 1, and this is the unique such linear system on C. Conversely if C has a linear system of degree 2 and dimension 1, then the linear system defines a finite map $\pi: C \to \mathbb{P}^1$ of degree 2, and so C is hyperelliptic; the fibres of π are the members of the linear system, and so the nontrivial automorphism ι of C such that $\pi \circ \iota = \pi$ preserves these individual members.

Now suppose that there exist α, α', c, and c' such that

$$f' \circ \alpha = +\beta \circ f + c \qquad (12.1.1)$$
$$f' \circ \alpha' = +\beta \circ f + c'$$

Then $f'(\alpha(Q)) - f'(\alpha'(Q)) = c - c'$ for all $Q \in C(k)$, which is a constant. Since the fibres of the map $\text{Div}_C^0(k) \to J(k)$ defined by f' are the linear equivalence classes (see Section 2), this implies that for all Q and Q' in $C(k)$,

$$\alpha(Q) - \alpha'(Q) \sim \alpha(Q') - \alpha'(Q'), \quad \text{or}$$
$$\alpha(Q) + \alpha'(Q') \sim \alpha'(Q) + \alpha(Q').$$

Suppose $\alpha \neq \alpha'$. Then $\alpha(Q_0) \neq \alpha'(Q_0)$ for some $Q_0 \in C(k)$ and, for a suitable Q'_0, $\alpha(Q'_0) \neq \alpha(Q'_0)$. Therefore $|\alpha(Q_0) + \alpha'(Q'_0)|$ is a linear system of dimension ≥ 1 (and degree 2) on C'. If C (hence C') is nonhyperelliptic, there is no such system, and we have a contradiction. If C is hyperelliptic, then there is a map $\pi: C \to \mathbb{P}'$ of degree 2 such that $\pi(\alpha(Q)) = \pi(\alpha'(Q'))$ for all Q, Q'. Again we have a contradiction. We conclude that $\alpha = \alpha'$, and this implies that $c = c'$.

On the other hand, suppose that the equations (12.1.1) hold with different signs, say with a plus and a minus respectively. Then the same argument shows that

$$\alpha(Q) + \alpha'(Q) \sim \alpha(Q') + \alpha'(Q'), \quad \text{all } Q, Q' \text{ in } C(k).$$

Therefore $\{\alpha(Q) + \alpha'(Q) | Q \in C(k)\}$ is a linear system on C' of dimension ≥ 1, which is impossible if C is nonhyperelliptic. (In the case C is hyperelliptic, there is an involution ι of C' such that $\iota \circ \alpha = \alpha'$.)

The case that the equations (12.1.1) hold with minus signs can be treated the same way as the first case.

Finally let C' be hyperelliptic with an involution ι such that $|Q' + \iota Q'|$ is a linear system and $f'(Q') + f'(\iota Q') = $ constant. Then if $f' \circ \alpha = \beta \circ f + c$, we have $f' \circ \iota \alpha = -\beta \circ f + c'$. $\square$

Corollary 12.2. *Let C and C' be curves of genus ≥ 2 over a perfect field k. If the canonically polarized Jacobian varieties of C and C' are isomorphic over k, then so also are C and C'.*

PROOF. Choose an isomorphism $\beta: (J, \lambda) \to (J', \lambda')$ defined over k. For each choice of a pair of points P and P' in $C(\bar{k})$ and $C'(\bar{k})$, there is a unique isomorphism $\alpha: C \to C'$ such that

$$f^{P'} \circ \alpha = \pm \beta \circ f^P + c$$

for some c in $J'(\bar{k})$ (in the case that C is hyperelliptic, we choose the sign to be $+$). Note that if (P, P') are replaced by the pair (Q, Q'), then $f^Q = f^P + d$ and $f^{Q'} = f^{P'} + e$ for some $d \in J(\bar{k})$ and $e \in J'(\bar{k})$, and so

$$f^{Q'} \circ \alpha = f^{P'} \circ \alpha + e = \pm \beta \circ f^P + c + e = \pm \beta \circ f^Q \mp \beta(d) + c + e.$$

In particular, we see that α does not depend on the choice of the pair (P, P'). On applying $\sigma \in \text{Gal}(\bar{k}/k)$ to the above equation, we obtain an equation
$$\sigma f^{P'} \circ \sigma\alpha = \pm \beta \circ \sigma f^P + \sigma c.$$
As $\sigma f^{P'} = f^{\sigma P'}$ and $\sigma f^P = f^{\sigma P}$, we see that $\sigma\alpha = \alpha$, and so α is defined over k. □

Corollary 12.3. *Let k be an algebraic number field, and let S be a finite set of primes in k. The map $C \mapsto (J_C, \lambda)$ sending a curve to its canonically polarized Jacobian variety defines an injection from the set of isomorphism classes of curves of genus ≥ 2 with good reduction outside S into the set of isomorphism classes of principally polarized abelian varieties over k with good reduction outside S.*

PROOF. Let R be the discrete valuation ring in k corresponding to a prime of k not in S. Then C extends to a smooth proper curve $\mathscr{C}$ over $\text{spec}(R)$, and (see Section 8) the Jacobian $\mathscr{J}$ of $\mathscr{C}$ has generic fibre the Jacobian of C and special fibre the Jacobian of the reduction of C. Therefore J_C has good reduction at the prime in question. The corollary is now obvious. □

Corollary 12.4. *Suppose that for any number field k, any finite set S primes of k, and any integer g, there are only finitely many principally polarized abelian varieties of dimension g over k having good reduction outside S. Then Mordell's conjecture is true.*

PROOF. Combine the last corollary with (9.11). □

Remark 12.5. Corollary 12.2 is false as stated without the condition that the genus of C is greater than 1. It would say that all curves of genus zero over k are isomorphic to $\mathbb{P}^1$ (but in general there exist conics defined over k having no rational point in k), and it would say that all curves of genus 1 are isomorphic to their Jacobians (and, in particular, have a rational point). However, it is obviously true (without restriction on the genus) that two curves over k having k-rational points are isomorphic over k if their canonically polarized Jacobians are isomorphic over k.

§13. Torelli's Theorem: The Proof

Throughout this section, C will be a complete nonsingular curve of genus $g \geq 2$ over an algebraically closed field k, and P will be a closed point of C. The maps $f^P: C \to J$ and $f^{(r)}: C^{(r)} \to J$ corresponding to P will all be denoted by f. Therefore $f(D + D') = f(D) + f(D')$, and if $f(D) = f(D')$, then
$$D \sim D' + rP \quad \text{where} \quad r = \deg(D) - \deg(D').$$
As usual, the image of $C^{(r)}$ in J is denoted by W^r. A canonical divisor K on C

defines a point on $C^{(2g-2)}$ whose image in J will be denoted by κ. For any subvariety Z of J, Z^* will denote the image of Z under the map $x \mapsto \kappa - x$

Lemma 13.1. *For all a in $J(k)$, $(W_a^{g-1})^* = W_{-a}^{g-1}$.*

PROOF. For any effective divisor D of degree $g - 1$ on C,
$$h^0(K - D) = h^1(K - D) = h^0(D) \geq 1,$$
and so there exists an effective divisor D' such that $K - D \sim D'$. Then $\kappa - f(D) - a = f(D') - a$, which shows that $(W_a^{g-1})^* \subset W_{-a}^{g-1}$. On replacing a by $-a$, we get that $(W_{-a}^{g-1})^* \subset W_a^{g-1}$, and so $W_{-a}^{g-1} = (W_{-a}^{g-1})^{**} \subset (W_a^{g-1})^*$. □

Lemma 13.2. *For any r such that $0 \leq r \leq g - 1$,*
$$W_a^r \subset W_b^{g-1} \Leftrightarrow a \in W_b^{g-1-r}.$$

PROOF. $\Leftarrow$: If $c = f(D) + a$ with D an effective divisor of degree r, and $a = f(D') + b$ with D' an effective divisor of degree $g - 1 - r$, then $c = f(D + D') + b$ with $D + D'$ an effective divisor of degree $g - 1$.

$\Rightarrow$: As $a \in W_b^{g-1}$, there is an effective divisor A of degree $g - 1$ such that $a = f(A) + b$. Let D be effective of degree r. The hypothesis states that $f(D) + a = f(\bar{D}) + b$ for some $\bar{D}$ effective of degree $g - 1$, and so $f(D) + f(A) = f(\bar{D})$ and
$$D + A \sim \bar{D} + rP.$$
Choose effective divisors A' and $\bar{D}'$ of degree $g - 1$ such that $A + A'$ and $\bar{D} + \bar{D}'$ are linearly equivalent to K (cf. the proof of (13.1)). Then
$$D + K - A' \sim K - \bar{D}' + rP, \quad \text{and so}$$
$$D + \bar{D}' \sim A' + rP.$$
As the D's form a family of dimension r, this shows that $h^0(A' + rP) \geq r + 1$. (In more detail, $|A' + rP|$ can be regarded as a closed subvariety of $C^{(r+g-1)}$, and we have shown that it projects onto the whole of $C^{(r)}$.) It follows from the Riemann–Roch theorem that $h^0(K - A' - rP) \geq 1$, and so there is an effective divisor $\bar{A}$ of degree $g - 1 + r$ such that
$$A' + \bar{A} + rP \sim K.$$
Therefore $\bar{A} + rP \sim K - A' \sim A$, and so $f(\bar{A}) = f(A')$ and $a = f(\bar{A}) + b \in W_b^{g-1-r}$. □

Lemma 13.3. *For any r such that $0 \leq r \leq g - 1$,*
$$W^{g-1-r} = \bigcap \{W_{-a}^{g-1} | a \in W^r\} \quad \text{and} \quad (W^{g-1-r})^* = \bigcap \{W_a^{g-1} | a \in W^r\}.$$

PROOF. Clearly, for a fixed a in $J(k)$,
$$W^{g-1-r} \subset W_{-a}^{g-1} \Leftrightarrow W_a^{g-1-r} \subset W^{g-1},$$

and (13.2) shows that both hold if $a \in W^r$. Therefore
$$W^{g-1-r} \subset \bigcap \{W_{-a}^{g-1} | a \in W^r\}.$$
Conversely, $c \in W_{-a}^{g-1} \Leftrightarrow a \in W_{-c}^{g-1}$, and so if $c \in W_{-a}^{g-1}$ for all $a \in W^r$, then $W^r \subset W_{-c}^{g-1}$ and $W_c^r \subset W^{g-1}$. According to (13.2), this implies that $c \in W^{g-1-r}$, which completes the proof of the first equality. The second follows from the first and the equation
$$\bigcap \{W_a^{g-1} | a \in W^r\} = \bigcap \{(W_{-a}^{g-1})^* | a \in W^r\} = (\bigcap \{W_{-a}^{g-1} | a \in W^r\})^*. \quad \square$$

Lemma 13.4. *Let r be such that $0 \leq r \leq g - 2$, and let a and b be points of $J(k)$ related by an equation $a + x = b + y$ with $x \in W^1$ and $y \in W^{g-1-r}$. If $W_a^{r+1} \not\subset W_b^{g-1}$, then $W_a^{r+1} \cap W_b^{g-1} = W_{a+x}^r \cup S$ with $S = W_a^{r+1} \cap (W_{y-a}^{g-2})^*$.*

PROOF. Write $x = f(X)$ and $y = f(Y)$ with X and Y effective divisors of degree 1 and $g - 1 - r$. If $Y \geq X$, then, because $f(X) + a = f(Y) + b$, we will have $a = f(Y - X) + b$ with $Y - X$ an effective divisor of degree $g - 2 - r$. Therefore $a \in W_b^{g-2-r}$, and so $W_a^{r+1} \subset W_b^{g-1}$ (by (13.2)). Consequently, we may assume that X is not a point of Y.

Let $c \in W_a^{r+1} \cap W_b^{g-1}$. Then $c = f(D) + a = f(D') + b$ for some effective divisors D and D' of degree $r + 1$ and $g - 1$. Note that
$$f(D) + y = f(D) + a + x - b = f(D') + x,$$
and so $D + Y \sim D' + X$.

If $D + Y = D' + X$, then $D \geq X$, and so $c = f(D) + a = f(D - X) + x + a$; in this case $c \in W_{a+x}^r$.

If $D + Y \neq D' + X$, then $h^0(D + Y) \geq 2$, and so for any point Q of $C(k)$, $h^0(D + Y - Q) \geq 1$, and there is an effective divisor $\bar{Q}$ of degree $g - 1$ such that $D + Y \sim Q + \bar{Q}$. Then
$$c = f(D) + a = f(\bar{Q}) + a - y + f(Q),$$
and so $c \in \bigcap \{W_{a-y+d}^{g-1} | d \in W^1\} = (W^{g-2})_{a-y}^*$ (by (13.3)). As $(W^{g-2})_{a-y}^* = (W_{y-a}^{g-2})^*$ and c is in W_a^{r+1} by assumption, this completes the proof that $W_a^{r+1} \cap W_b^{g-1} \subset W_{a+x}^r \cup S$.

The reverse inclusion follows from the obvious inclusions:
$$W_{a+x}^r \subset W_a^{r+1}; \quad W_{a+x}^r = W_{b+y}^r \subset W_b^{g-1}; \quad (W_{y-a}^{g-2})^* \subset (W_{y-a-x}^{g-1})^* = W_b^{g-1}.$$
$\square$

Lemma 13.5. *Let $a \in J(k)$ be such that $W^1 \not\subset W_a^{g-1}$; then there is a unique effective divisor $D(a)$ of degree g on C such that*
$$f(D(a)) = a + \kappa \tag{13.5.1}$$
and $W^1 \cdot W_a^{g-1}$, when regarded as a divisor on C, equals $D(a)$.

PROOF. We use the notations of Section 6; in particular, $\Theta = W^{g-1}$. For

$a = 0$, (13.1) says that $(\Theta^-)_\kappa = \Theta$. Therefore, on applying (6.8), we find that $W^1 \cdot W_a^{g-1} = f(C) \cdot (\Theta^-)_{a+\kappa} \stackrel{df}{=} f^{-1}((\Theta^-)_{a+\kappa}) = D$, where D is a divisor of degree g on C such that $f^{(g)}(D) = a + \kappa$. This is the required result. □

We are now ready to prove (12.1a). We use β to identify J with J', and write V^r for the image of $C'^{(r)}$ in J. As W^{g-1} and V^{g-1} define the same polarization of J, they give the same element of $\mathrm{NS}(J)$ (see [14, §12]), and therefore one is a translate of the other, say $W^{g-1} = V_c^{g-1}$, $c \in J(k)$. To prove (12.1a), we shall show that V^1 is a translate of W^1 or of $(W^1)^*$.

Let r be the smallest integer such that V^1 is contained in a translate of W^{r+1} or $(W^{r+1})^*$. The theorem will be proved if we can show that $r = 0$. (Clearly, $r < g - 1$.) Assume on the contrary that $r > 0$. We may suppose (after possibly replacing β by $-\beta$) that $V^1 \subset W_a^{r+1}$. Choose an x in W^1 and a y in W^{g-1-r}, and set $b = a + x - y$. Then, unless $W_a^{r+1} \subset W_b^{g-1}$, we have (with the notations of (13.4))

$$V^1 \cap W_b^{g-1} = V^1 \cap W_a^{r+1} \cap W_b^{g-1} = (V^1 \cap W_{a+x}^r) \cup (V^1 \cap S).$$

Note that, for a fixed a, W_{a+x}^r depends only on x and S depends only on y.

Fix an x; we shall show that for almost all y, $V^1 \not\subseteq W_b^{g-1}$, which implies that $W_a^{r+1} \not\subseteq W_b^{g-1}$ for the same y. As y runs over W^{g-1-r}, $-b$ runs over $W_{-(a+x)}^{g-1-r}$. Now, if $V^1 \subset W_b^{g-1}$ for all $-b$ in $W_{-(a+x)}^{g-1-r}$, then $V^1 \subset W_{a+x}^r$ (by (13.3)). This contradicts the definition of r, and so there exist b for which $V^1 \not\subseteq W_b^{g-1}$. Note that $V^1 \subset W_b^{g-1} (= V_{b+c}^{g-1}) \Leftrightarrow -b \in V_c^{g-2}$ (by (13.2)). Therefore $V_c^{g-2} \not\subseteq W_{-(a+x)}^{g-1-r}$, and so the intersection of these sets is a lower dimensional subset of $W_{-(a+x)}^{g-1-r}$ whose points are the $-b$ for which $V^1 \subset W_b^{g-1}$.

We now return to the consideration of the intersection $V^1 \cap W_b^{g-1}$, which equals $(V^1 \cap W_{a+x}^r) \cup (V^1 \cap S)$ for almost all y. We first show that $V^1 \cap W_{a+x}^r$ contains at most one point. If not, then as $-b$ runs over almost all points of $W_{-(a+x)}^{g-1-r}$ (for a fixed x), the element $D'(b) \stackrel{df}{=} f'^{-1}(V' \cdot W_b^{g-1})$ (cf. (13.5)) will contain at least two fixed points (because $W_{a+x}^r \subset W_{a+x-y}^{g-1} = W_b^{g-1}$), and hence $f(D'(b))$ will lie in a translate of V^{g-2}. As $f'(D'(b)) = b + \kappa'$, we would then have $(W^{g-1-r})^*$ contained in a translate of V^{g-2}, say V_d^{g-2}, and so

$$\bigcap \{V_{c-u}^{g-1} | u \in V_d^{g-2}\} \subset \bigcap \{W_{-u}^{g-1} | u \in (W^{g-1-r})^*\}.$$

On applying (13.3) to each side, we then get an inclusion of V in a translate of $(W^r)^*$, contradicting the definition of r.

Keeping y fixed and varying x, we see from (13.5.1) that $V^1 \cap W_{a+x}^r$ must contain at least one point, and hence it contains exactly one point; according to the preceding argument, the point occurs in $D'(b)$ with multiplicity one for almost all choices of y.

It is now easily seen that we can find x, x' in W^1 and y in W^{g-1-r} such that $(D'(b) =) D'(a + x - y) = Q + \bar{D}$ and $(D'(b') =) D'(a + x' - y) = Q' + \bar{D}$ where Q, Q' are in C' and $\bar{D}$ is an effective divisor of degree $g - 1$ on C' not containing Q or Q'. By equation (13.5.1), $f(Q) - f(Q') = x - x'$, and hence W^1 has two distinct points in common with some translate of V^1. Now, if x,

x' are in W^1, then $W_{-x}^{g-1} \cap W_{-x'}^{g-1} = W^{g-2} \cup (W_{x+x'}^{g-2})^*$ (by 13.4)). According to (13.3), we now get an inclusion of some translate of V^{g-2} in W^{g-2} or $(W^{g-2})^*$. Finally (13.3) shows that

$$V^1 = \bigcap \{V_{-e} | e \in V^{g-2}\}$$

which is contained in a translate of W^1 or W^{1*} according as V^{g-2} is contained in a translate of W^{g-2} or $(W^{g-2})^*$. This completes the proof. □

Bibliographic Notes for Abelian Varieties and Jacobian Varieties

The theory of abelian varieties over $\mathbb{C}$ has a long history. On the other hand, the "abstract" theory over arbitrary fields, can be said to have begun with Weil's famous announcement of the proof of the Riemann hypothesis for function fields [Sur les fonctions algébriques à corps de constantes fini, *Comp. Rendu.* **210** (1940), 592–594]. Parts of the projected proof (for example, the key "lemme important") can best be understood in terms of intersection theory on the Jacobian variety of the curve, and Weil was to spend the next six years developing the foundational material necessary for making his proof rigorous. Unable in 1941 to construct the Jacobian as a projective variety, Weil was led to introduce the notion of an abstract variety (that is, a variety that is not quasi-projective). He then had to develop the theory of such varieties, and he was forced to develop his intersection theory by local methods (rather than the projective methods used by van der Waerden [*Einführung in die algebraische Geometrie*, Springer-Verlag, 1939]). In 1944 Weil completed his book [*Foundations of Algebraic Geometry*, AMS Coll., XXIX, 1946], which laid the necessary foundations in algebraic geometry, and in 1946 he completed his two books [*Sur les Courbes algébriques et les Variétés qui s'en déduisent*, Hermann, 1948] and [20], which developed the basic theory of abelian varieties and Jacobian varieties and gave a detailed account of his proof of the Riemann hypothesis. In the last work, abelian varieties are defined much as we defined them and Jacobian varieties are constructed, but it was not shown that the Jacobian could be defined over the same field as the curve.

Chow ([Algebraic systems of positive cycles in an algebraic variety, *Amer. J. Math.*, **72** (1950), 247–283] and [3]) gave a construction of the Jacobian variety which realized it as a projective variety defined over the same ground field as the original curve. Matsusaka [On the algebraic construction of the Picard variety, *Japan J. Math.*, **21** (1951), 217–235 and **22** (1952), 51–62] gave the first algebraic construction of the Picard and Albanese varieties and demonstrated also that they were projective and had the same field of definition as the original varieties. Weil showed that his construction of a group variety starting from a birational group could also be carried out without making an extension of the ground field [On algebraic groups of transformations, *Amer. J. Math.*, **77** (1955), 355–391], and in [The field of definition of a variety, *Amer. J. Math.*, **78** (1956), 509–524] he further developed his methods of descending the field of definition of a variety. Finally Barsotti [A note on abelian varieties, *Rend. Circ. Mat. di Palermo*, **2** (1953), 236–257], Matsusaka [Some theorems on abelian varieties, *Nat. Sci. Report Ochanomizu Univ.*, **4** (1953), 22–35], and Weil [On the projective embedding of abelian varieties, in *Algebraic geometry and topology, A symposium in Honor of S. Lefschetz*, Princeton, 1957, pp. 177–181] showed that all abelian varieties are projective. In a course at the University of Chicago, 1954–55, Weil made substantial improvements to the theory of abelian varieties (the seesaw principle and the theorem of the cube, for example), and these and the results mentioned above together with Chow's theory of the "k-image" and "k-trace" [Abelian varieties over function fields, *Trans. Amer. Math. Soc.*, **78** (1955), 253–275] were incorporated by Lang in his book [9]. The main lacuna at this time (1958–59) was a satisfactory theory of isogenies of degree p and their kernels in characteristic p; for example, it was not known that the canonical map from an abelian variety to the dual of its dual was an isomorphism (its degree might have been divisible by p). Cartier [Isogenies and duality of abelian varieties, *Ann of Math.*, **71** (1960), 315–351] and Nishi [The Frobenius theorem and the duality theorem on an abelian variety, *Mem. Coll. Sc. Kyoto (A)*, **32** (1959), 333–350] settled this particular point, but the full understanding of the p-structure of abelian varieties required the development of the theories of finite group schemes and Barsotti–Tate groups. The book of Mumford [16] represents a substantial contribution to the subject of abelian varieties: it uses modern methods to give an comprehensive account of abelian varieties including the p-theory in characteristic p, and avoids the crutch of using Jacobians to prove results about general abelian varieties. (It has been a significant loss to the mathematical community that Mumford did not go on to write a second volume on the topics suggested in the introduction: Jacobians; Abelian schemes: deformation theory and moduli; the ring of modular forms and the global structure of the moduli space; the Dieudonné theory of the "fine" characteristic p structure; arithmetic theory: abelian schemes over local, global fields. We still lack satisfactory accounts of some of these topics.)

Much of the present two articles has been based on these sources. We now give some other sources and references. "Abelian Varieties" will be abbreviated by AV and "Jacobian Varieties" by JV.

The proof that abelian varieties are projective in AV, Section 7 is Weil's 1957 proof. The term "isogeny" was invented by Weil: previously, "isomorphism" had frequently been used in the same situation. The fact that the kernel of m_A has m^{2g} elements when m is prime to the characteristic was one of the main results that Weil had to check in order to give substance to his proof of the Riemann hypothesis. Proposition 11.3 of AV is mentioned briefly by Weil in [*Variétés Abéliennes. Colloque d'Algèbre et Théorie des Nombres*, 1949, pp. 125–128], and is treated in detail by Barsotti [Structure theorems for group varieties, *Annali di Mat.*, **38** (1955), 77–119]. Theorem 14.1 is folklore: it was used by Tate in [Endomorphisms of abelian varieties over finite fields, *Invent. math.*, **2** (1966), 134–144], which was one of the starting points for the work that led to Faltings's recent proof of Mordell's conjecture. The étale cohomology of an abelian variety is known to everyone who knows étale cohomology, but I was surprised not to be able to find an adequate reference for its calculation: in Kleiman [Algebraic cycles and the Weil conjectures, in *Dix Exposés sur la Cohomologie des Schémas*, North-Holland, 1968, pp. 359–386] Jacobians are used, and it was unaccountably omitted from [13]. In his 1940 announcement, Weil gives a definition of the e_m-pairing (in our terminology, $\bar{e}_m$-pairing) for divisor classes of degree zero and order m on a curve which is analogous to the explicit description at the start of Section 16 of AV. The results of that section mainly go back to Weil's 1948 monograph [20], but they were reworked and extended to the p-part in Mumford's book. The observation (see (16.12) of AV) that $(A \times A^\vee)^4$ is always principally polarized is due to Zarhin [A finiteness theorem for unpolarized Abelian varieties over number fields with prescribed places of bad reduction, *Invent. math.*, **79** (1985), 309–321]. Theorem 18.1 of AV was proved by Narasimhan and Nori [Polarizations on an abelian variety, in *Geometry and Analysis*, Springer-Verlag (1981), pp. 125–128]. Proposition 20.1 of AV is due to Grothendieck (cf. Mumford [*Geometric Invariant Theory*, Springer-Verlag, 1965, 6.1]), and (20.5) of AV (defining the K/k-trace) is due to Chow (reference above). The Mordell–Weil theorem was proved by Mordell [On the rational solutions of the indeterminate equations of the third and fourth degrees, *Proc. Cambridge Phil. Soc.*, **21** (1922), 179–192] (the same paper in which he stated his famous conjecture) for an elliptic curve over the rational numbers and by Weil [L'arithmétique sur les courbes algébriques, *Acta Math.*, **52** (1928), 281–315] for the Jacobian variety of a curve over a number field. (Weil, of course, stated the result in terms of divisors on a curve.)

The first seven sections of JV were pieced together from two disparate sources, Lang's book [9] and Grothendieck's Bourbaki talks [4], with some help from Serre [17], Mumford [15], and the first section of Katz and Mazur [*Arithmetic Moduli of Elliptic Curves*, Princeton, 1985].

Rosenlicht [Generalized Jacobian varieties, *Ann. of Math.*, **59** (1954), 505–530, and *A* universal mapping property of generalized Jacobians, *ibid.* (1957), 80–88], was the first to construct the generalized Jacobian of a curve relative to a modulus. The proof that all abelian coverings of a curve can be obtained from isogenies of its generalized Jacobians (Theorem 9.7 of JV) is due to Lang [Sur les séries L d'une variété algébrique, *Bull. SMF*, **84** (1956), 555–563]. Results close to Theorem 8.1 of JV were obtained by Igusa [Fibre systems of Jacobian varieties I, II, III, *Amer. J. Math.*, **78** (1956), 171–199, 745–760, and **81** (1959), 453–476]. Theorem 9.11 is due to Parshin [Algebraic curves over function fields, I, *Math. USSR—Izvestija*, **2** (1968), 1145–1169]. Matsusaka [On a generating curve of an abelian variety, *Nat. Sc. Rep. Ochanomizu Univ.*, **3** (1952), 1–4] showed that every abelian variety over an algebraically closed field is generated by a curve (cf. (10.1) of JV). Regarding (11.2) of JV, Hurwitz [*Math. Ann.*, **28** (1886)] was the first to show the relation between the number of fixed points of a correspondence on a Rieman surface C and the trace of a matrix describing its action on the homology of the surface (equivalently that of its Jacobian). This result of Hurwitz inspired both Lefschetz in his proof of his trace formula and Weil in his proof of the Riemann hypothesis for curves.

Proofs of Torelli's theorem can be found in Andreotti [On a theorem of Torelli, *Amer. J. Math.*, **80** (1958), 801–821], Matsusaka [On a theorem of Torelli, *Amer. J. Math.*, **80** (1958), 784–800], Weil [Zum Beweis des Torellischen Satzes, *Gott. Nachr.*, **2** (1957), 33–53], and Ciliberto [On a proof of Torelli's theorem, in *Algebraic Geometry—Open problems*, Lecture Notes in Math., 997, Springer-Verlag, 1983, pp. 113–223]. The proof in Section 13 of JV is taken from Martens [A new proof of Torelli's theorem, *Ann. Math.*, **78** (1963), 107–111]. Torelli's original paper is [Sulle varieta di Jacobi, *Rend. R. Acad. Sci. Torino*, **50** (1914–15), 439–455]. Torelli's theorem shows that the map from the moduli space of curves into that of principally polarized abelian varieties is injective on geometric points; a finer discussion of the map can be found in the paper by Oort and Steenbrink [The local Torelli problem for algebraic curves, in *Algebraic Geometry Angers 1979*, Sijthoff & Noordhoff, 1980, pp. 157–204].

Finally, we mention that Mumford [*Curves and Their Jacobians*, University of Mich] provides a useful survey of the topics in its title, and that the commentaries in Weil [*Collected Papers*, Springer-Verlag, 1979] give a fascinating insight into the origins of parts of the subject of arithmetic geometry.

REFERENCES

[1] Artin, M. Néron models, this volume, pp. 213–230.
[2] Bourbaki, N. *Algèbre Commutative*. Hermann: Paris, 1961, 1964, 1965.
[3] Chow, W.-L. The Jacobian variety of an algebraic curve. *Amer. J. Math.*, **76** (1954), 453–476.
[4] Grothendieck, A. Technique de descente et théorèmes d'existence en géométrie algébrique, I–VI. *Séminaire Bourbaki*, Éxposés 190, 195, 212, 221, 232, 236, 1959–62.
[5] Grothendieck, A. *Revêtements Étales et Groupe Fondamental* (SGA1, 1960–61). Lecture Notes in Mathematics, 224, Springer-Verlag: Heidelberg, 1971.
[6] Grothendieck, A. Le groupe de Brauer, III, in *Dix Exposés sur la Cohomologie des Schémas*. North-Holland: Amsterdam, 1968, pp. 88–188.
[7] Grothendieck, A. and Dieudonné, J. *Eléments de Géométrie Algébrique*, I. Springer-Verlag: Heidelberg, 1971.
[8] Hartshorne, R. *Algebraic Geometry*. Springer-Verlag: Heidelberg, 1977.
[9] Lang, S. *Abelian Varieties*. Interscience: New York, 1959.
[10] Lang, S. *Fundamentals of Diophantine Geometry*. Springer-Verlag: Heidelberg, 1983.
[11] Lichtenbaum, S. Duality theorems for curves over p-adic fields. *Invent. Math.*, **7** (1969), 120–136.
[12] Mattuck, A and Mayer, A. The Riemann–Roch theorem for algebraic curves. *Ann. Sci. Norm. Pisa*, **17**, (1963), 223–237.

[13] Milne, J. *Etale Cohomology*. Princeton University Press: Princeton, NJ, 1980.
[14] Milne, J. Abelian varieties, this volume, pp. 103–150.
[15] Mumford, D. *Lectures on Curves on an Algebraic Surface*. Princeton University Press: Princeton, NJ, 1966.
[16] Mumford, D. *Abelian Varieties*. Oxford University Press: Oxford, 1970.
[17] Serre, J.-P. *Groupes Algébriques et Corps de Classes*. Hermann, Paris, 1959.
[18] Shafarevich, I. *Basic Algebraic Geometry*. Springer-Verlag: Heidelberg, 1974.
[19] Waterhouse, W. *Introduction to Affine Group Schemes*. Springer-Verlag: Heidelberg, 1979.
[20] Weil, A. *Variétes Abéliennes et Courbes Algébriques*. Hermann: Paris, 1948.

CHAPTER VIII
Néron Models

M. ARTIN

This is an exposition of the main theorem of Néron's paper [2], and of Raynaud's subsequent work on the problem.

Notation and Terminology

(1) R: a Dedekind domain with field of fractions K. Often R is local (a discrete valuation ring), and then $R/p = k$.
(2) S: a normal scheme, with function field K. All schemes are assumed locally noetherian.
(3) Subscripts indicate the base scheme (or ring). Thus X_R is a scheme over Spec R, etc. Products are over the base.
   By *model* X_R (or X_S) of a variety X_K, we mean a reduced irreducible scheme whose general fibre is X_K. If X_K is projective, $X_K \subset \mathbb{P}_K^n$, then one can obtain a projective model X_S as the Zariski closure of X_K in $\mathbb{P}_S^n$.
(4) Sometimes R or S will be assumed *strictly local*, meaning henselian, with separably closed residue field k.
   A smooth scheme X_S over a strictly local base has a dense set of sections, i.e., $X_S(S)$ is dense. This allows many arguments involving points to be carried over.
(5) An extension $R \to R'$ of discrete valuation rings is called *smooth* if R' is a localization of a smooth R-scheme. This is equivalent with the three conditions (i) $pR' = p'$, (ii) k' separable over k, and (iii) tr $\deg_K K' =$ tr $\deg_k k'$.
(6) By *nonsingular* scheme X, we mean one whose local rings are regular.

§1. Properties of the Néron Model, and Examples

Let R be a Dedekind domain, with field of fractions K, and let A_K be an abelian variety over K. A *Néron model* $A_R = N(A_K)$ for A_K is a smooth group scheme over R whose general fibre is A_k, which satisfies the following universal property:

(1.1) Let X_R be smooth (over R), and let $\phi_R \colon X_R \dashrightarrow A_R$ be a rational map. Then ϕ extends uniquely to a morphism $X_R \xrightarrow{\phi} A_R$.

Clearly, this universal property characterizes the Néron model. Our object is to prove

Theorem (1.2) (Néron [2]). *The Néron model $N(A_K) = A_R$ exists, and is of finite type over R.*

The closed fibre A_K of the Néron model may have several components, so it will be an extension of a finite group by an algebraic group over k.

One should keep in mind that the Néron model does not have particularly strong functorial properties. It is true, and trivial from the definition, that $N(A \times B) = N(A) \times N(B)$. However the Néron model is not compatible with closed subgroups or with extensions of abelian varieties. It is obviously stable under smooth extensions $R \to R'$ of discrete valuation rings, but not under ramified extensions $R \to R'$. For instance, an elliptic curve may have bad reduction over R, but smooth reduction over a ramified extension. This implies that the Néron model changes.

The universal property (1.1) characterizing Néron models is made plausible by the following extension of Weil's theorem on rational maps into group varieties.

Proposition (1.3) (Weil [6]). *Let G_S be a group scheme over S, and let X_S be smooth over S. Let $f_S \colon X_S \dashrightarrow G_S$ be a rational map which is defined generically on each fibre of X_S. The set of points where f_S is not defined has pure codimension 1 in X_S.*

PROOF. We will use point notation, but the argument is scheme-theoretic.

Define a rational map $F \colon X \times X \dashrightarrow G$ by $F(x, y) = f(x) f(y)^{-1}$. Clearly, F is defined at a diagonal point (x, x) if f is defined at x, and then $F(x, x) = e$ is the identity of G in the fibre containing x. Conversely, if F is defined at (x, x) then there is a generic point η of the fibre so that F is defined at (x, η). This is because the domain of definition of F is open in $X \times X$. By assumption, f is defined at η, so the formula

$$f(x) = F(x, \eta) f(\eta)$$

defines f at x. Thus f is defined at x iff F is defined at (x, x).

Let L be the function field of the scheme $X \times X$. The rational map F

defines a map of the local ring of G at e to L:
$$\mathcal{O}_{G,e} \xrightarrow{\phi} L.$$

Since F sends (x, x) to e if defined, it follows that F is defined at (x, x) if and only if
$$\text{im } \phi \subset \mathcal{O}_{X \times X, (x,x)} \qquad (\subset L).$$

Now $X \times X$ is smooth over S, hence is a normal scheme. Therefore the local ring $\mathcal{O}_{X \times X, (x,x)}$ can be characterized as the set of functions $\alpha \in L^*$ whose polar divisor $(\alpha)_\infty = D_\alpha$ does not contain (x, x), together with the element 0. If F is not defined at (x, x), then $(x, x) \in D_\alpha$ for some $\alpha \in \text{im } \phi$. Since D_α has codimension 1 and the diagonal $\Delta \subset X \times_S X$ is a complete intersection, $D_\alpha \cap \Delta = C_\alpha$ has codimension 1 in $\Delta = X$. Clearly, F is not defined for $(y, y) \in D_\alpha$, hence f is not defined for $y \in C_\alpha$. □

Corollary (1.4). *If A_R is an abelian scheme over R, then A_R is the Néron model of its generic fibre A_K.*

PROOF. Valuative criterion and Proposition (1.3). □

Exercise (1.5). This corollary does not extend to A_S abelian over S. Explain what goes wrong with the proof, and give an example.

Corollary (1.6). *Assume R strictly local. Let A_R be a smooth group scheme of finite type extending the abelian variety A_K. Then A_R is the Néron model of A_K iff every K-valued point extends uniquely to an R-valued point:*
$$A_K(K) = A_R(R).$$

PROOF. That every K-valued point extends uniquely is Néron's property (1.1), with $X = \text{Spec } R$. Conversely, given a rational map $X_R \xrightarrow{\phi} A_R$, X smooth, we know it is defined on the general fibre ($R = K$ in Corollary (1.4)). So if ϕ is not everywhere defined, then it must be undefined on one of the irreducible closed subsets of codimension 1 which do not meet the general fibre X_K. These are the components C of the fibre X_k. Let $\bar{\Gamma}$ be the closure of the graph of ϕ in $X \times A$. The projection $\bar{\Gamma} \to X$ is birational, and its image is constructible. If the generic point η of a component C of X_k is in the image of $\bar{\Gamma}$, then ϕ is defined at η because $\mathcal{O}_{X,\eta}$ is a discrete valuation ring, and hence on all of C by Proposition (1.3). If η is not in the image of $\bar{\Gamma}$, then there is a rational point x_0 of X_k not in that image, and if x is any section of X through x_0, ϕ is not defined at x. We interpret x as a point of X with values in R, and let x_K be the corresponding K-valued point. Then $\phi(x_K) \in A_K(K)$ does not extend to $A_R(R)$. □

Example (1.7). The Néron model of the multiplicative group $\mathbb{G}_{mK}$. Let R be a discrete valuation ring. The Néron model of $\mathbb{G}_{mK}$ is obtained from the exact sequence

$$0 \to R^* \to K^* \xrightarrow{v} \mathbb{Z} \to 0,$$

where v is the valuation of R. It is not hard to find a corresponding extension of group schemes, in which R^* is replaced by $\mathbb{G}_{mR}$ and K^* by a union G of countably many translations of $\mathbb{G}_{mR}$ by rational sections:

$$G = \bigcup_{n \in \mathbb{Z}} p^n \mathbb{G}_{mR},$$

to obtain an exact sequence

$$0 \to \mathbb{G}_m \to G_R \to i_* \mathbb{Z} \to 0,$$

where $i_* \mathbb{Z}$ denotes the nonseparated scheme representing the extension over R of $\mathbb{Z}_k$ by zero.

Of course, rational maps into G need not be everywhere defined because $\mathbb{G}_{mK}$ is not proper. Also, the reasoning of Corollary (1.6) does not apply here because G is not of finite type. Nevertheless, it can be shown that the following weaker version of (1.1) holds:

(1.8) Let X_R be smooth over R, and let $X_K \xrightarrow{\phi_K} G_K$ be a morphism. Then ϕ_K extends uniquely to a morphism $\phi_R \colon X_R \to G_R$.

Theorem (1.9) (Raynaud [3]). *A Néron model $A_R = N(A_K)$ satisfying (1.8) exists whenever A_K is semi-abelian, i.e., an extension of an abelian variety by a torus.*

We are not going to prove Raynaud's extension of Néron's theorem (1.2) here.

Example (1.10). A smooth group scheme G_R which is *not* a Néron model. Such an example can be obtained from any smooth group by blowing up the origin in its closed fibre. The closed fibre gets replaced by its tangent space. We will work this out for the multiplicative group $\mathbb{G}_m = \operatorname{Spec} R[t, t^{-1}]$. Translate the identity $t = 1$ to the origin by the substitution $x + 1 = t$:

$$k[t, t^{-1}] = k[x, (x+1)^{-1}].$$

The multiplication rule in $\mathbb{G}_m$ is

$$t \rightsquigarrow t \otimes t, \quad \text{or}$$

$$x \rightsquigarrow x \otimes 1 + 1 \otimes x + x \otimes x.$$

We now blow up the ideal (p, x) in $\mathbb{G}_m$. We only need the affine piece of the blow-up generated by $z = x/p$:

$$R[x, (x+1)^{-1}][z]/(pz - x) = R[z, (pz+1)^{-1}].$$

The multiplication rule over K extends to

$$z \rightsquigarrow z \otimes 1 + 1 \otimes z + p(z \otimes z),$$

and gives $\operatorname{Spec} R[z, (pz + 1)^{-1}]$ a group structure whose closed fibre is $\operatorname{Spec} k[z]$, with law $z \rightsquigarrow z \otimes 1 + 1 \otimes z$: the additive group. Obviously, the rational map $R[z, (pz + 1)^{-1}] \dashrightarrow R[t, t^{-1}]$ contradicts (1.8) and (1.1).

Before continuing with examples of Néron models, we will review a method of Weil [6]. He showed how to reconstruct a group variety of G from a birationally equivalent variety V, if the law of composition is given as a birational map. The strategy of proof is to identify an open subset U of V which is also open in G, and then to construct G as a union of translates of U. In order to have enough translations, we assume the base scheme S strictly local. The descent to other base schemes can often be done by projective methods, using the theorem of the square (cf. Section 4 and [4]).

Let V_S be a smooth scheme of finite type with *nonempty fibres*, and let $m_S: V \times V \dashrightarrow V$ be an S-rational map. Following Weil, we call m a *normal law* if it is an associative law of composition, i.e., $(ab)c = a(bc)$ whenever both sides are defined, and if the following condition holds as well:

(1.11) The domains of definition of the rational maps $\phi, \psi: V^2 \dashrightarrow V^2$ defined by $\phi(a, b) = (a, ab)$, $\psi(a, b) = (b, ab)$ are dense in each fibre of V^2, and their restrictions to each fibre are birational.

Theorem (1.12) (Weil [6]). *Let m be a normal law on V. Assume that the Zariski-local sections of V/S are dense in each fibre, and that one of the following conditions holds:*

(i) *The geometric fibres of V are connected.*
(ii) *$S = \operatorname{Spec} R$, where R is a Dedekind domain, and the general fibre V_K is geometrically connected.*

There is a smooth group G of finite type over S and a scheme U which is open and dense in each fibre, both in V and G, such that the laws of composition on V and G agree on U. The group G is unique up to unique isomorphism.

This theorem is proved in Section 2. The connectedness assumptions and the normality of S are actually unnecessary. See [8] for a more general formulation.

Note that the hypothesis on sections is satisfied by global sections if S is strictly local.

Note: Suppose that $S = \operatorname{Spec} R$ where R is a discrete valuation ring, and that a normal law m is given on the general fibre V_K of V_R. In order for m to be a normal law on V, it is not enough that each of the maps $\phi, \phi^{-1}, \psi, \psi^{-1}$ of (1.11) have a domain of definition which is dense in the closed fibre V_k^2. In fact, if V is proper then this is automatic, by the valuative criterion. A little extra is needed to insure that the graphs of ϕ_k, ϕ_k^{-1} (and ψ_k, ψ_k^{-1}) agree. It is sufficient that ϕ_k, ψ_k be generically surjective. Zariski's Main Theorem then implies that (1.11) holds. Of course, associativity carries over automatically from the general fibre.

Exercise (1.13). Work out an example with $V_R = \mathbb{P}_R^1$ to illustrate this point.

Example (1.14). The Néron model of an elliptic curve, i.e., dim $A_K = 1$, when R is local and $k = \bar{k}$. It is known that there is a proper nonsingular model V_R of $V_K = A_K$. If V is also chosen to be a minimal proper nonsingular model, then it is unique, and so every automorphism of A_K, such as translation by a rational point, extends to an automorphism of V. The closed fibres of V have been classified by Kodaira [1] and Néron [2] into a standard list of types:

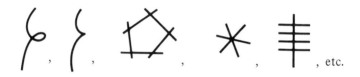

, etc.

Proposition (1.15). *Let V_R be the minimal nonsingular model of an elliptic curve A_K, and let $A \subset V$ denote the open subset of points where V is smooth. Then A is the Néron model $N(A_K)$.*

PROOF. Everything is preserved by passage to the strict localization of R, so we may assume R strictly local.

Each point of $A_K(K)$ extends to $V(R)$ by the valuative criterion. Since V is nonsingular, any section $a \in V(R)$ lies in the smooth locus of V, which is A. (For, if $(\mathcal{O}, m)$ is the local ring of V at a, then the homomorphisms $R \to \mathcal{O} \xrightarrow{a} R$ show that $P \not\subset m^2$, hence that the fibre V_k is nonsingular at P_k. Since a_k is a rational point, V_k is smooth, and also V is smooth, at a_k.) This shows that $A_K(K) = A(R)$, and by Corollary (1.6) that A is the Néron model, provided it is a group scheme.

Next, we claim that the law of composition m on A_K is a normal law on A: If $R \to R'$ is any smooth extension of discrete valuation rings, then $A_K(K') = A(R')$ too, because the construction of A is compatible with such an extension. Applying this fact to the case that R' is the localization of A^2 at a generic point of A_k^2 shows that the rational maps ϕ, ψ of (1.11) are defined at generic points of the closed fibre. Since the map ϕ is 1–1 for a fixed rational point a, the fibres of ϕ are not all positive dimensional. Therefore ϕ is generically surjective, and (1.11) holds. By Theorem (1.12), there is a group scheme B over R and a common dense open set U of A, B.

Consider the rational map $\varepsilon: B \to A$ extending the identity on U. Translation by a section $x \in B(R)$ is everywhere defined on B, and also on A, because $B_K = A_K$ and translation by any point in $A_K(K)$ extends to A. Therefore ε can be defined at any point $b \in B$ by the rule $\varepsilon = t_{-x} \circ \varepsilon \circ t_x$, where x is chosen so that $t_x(b) = b + x \in U$. Hence ε is everywhere defined, and being a dense subgroup of A, $\varepsilon(B) = A$. □

Lemma (1.16). *Let G_S be a smooth group scheme. There is an open subgroup $G_S^0 \subset G_S$, called the* connected component, *such that the geometric fibres $G_{\bar{k}}^0$ of G^0 are the connected components of the fibres $G_{\bar{k}}$.*

The proof of this lemma is not difficult, and we omit it. One way is to prove the more general "Stein factorization" below:

Lemma (1.17). *Let $X \xrightarrow{f} Y$ be a smooth map of schemes. Then f factors uniquely as*

$$X \xrightarrow{f'} Y' \xrightarrow{g} Y,$$

where f' is smooth, with geometrically connected, nonempty fibres, and g is étale.

Unfortunately, the factorization Y' exists only as algebraic space, i.e., it has an étale covering by a scheme, and it is *not* a *separated* algebraic space. Lemma (1.16) follows from (1.17) with $X = G$, $Y = S$, because $f'(e_S)$ (e_S the identity section) is locally closed in Y', and $G^0 = f'^{-1}f(e_S)$.

Exercise (1.18). Describe the factorization (1.17) when $X = G$.

Example (1.19) (The Néron Model of a Jacobian). Let V_K be a smooth curve of genus $g \geq 2$, with rational point, and let A_K be the Jacobian of V_K. Assume that K is the field of fractions of a discrete valuation ring R. Let $V = V_R$ be a regular proper model for V_K. It is known that the relative Picard functor $\mathscr{P}ic\ V/R$ is representable as algebraic space.

Proposition (1.20). *The connected components $(\mathscr{P}ic\ V/R)^0$ and $\mathrm{N}(A_K)^0$ are canonically isomorphic.*

This is not difficult to show, assuming the existence of $\mathscr{P}ic\ V/R$, but we will not do it. The key point is that $\mathscr{P}ic\ V/R$ has a property analogous to (1.6), i.e., every class α in $\mathscr{P}ic\ V_K(K)$ extends to $\mathscr{P}ic\ V(R)$.

To get a canonical extension, one simply chooses a divisor D representing α, and takes its Zariski closure $\bar{D}$ in V. Then $\bar{D}$ is a Cartier divisor because V is nonsingular.

The exact relation of $\mathscr{P}ic\ V/R$ and $A = \mathrm{N}(A_K)$ is described as follows. We assume k algebraically closed. The closed fibre V_k of V is a Cartier divisor which can be written as a linear combination $V_k = \sum r_i C_i$, where r_i are positive integers, and $\{C_1, \ldots, C_n\}$ are the irreducible components of V_k. Let $D = \mathbb{Z}^n/\mathbb{Z}V_k$ denote the group of divisors supported on V_k, modulo V_k. Its dual $D^* = \mathrm{Hom}_{\mathbb{Z}}(D, \mathbb{Z})$ is the subgroup $(V_k)^{\perp}$ of $\mathbb{Z}^n$. Let $P \subset \mathscr{P}ic\, V/R$ denote the subgroup of bundles of total degree zero, so that $\mathscr{P}ic\, V/R \approx P \oplus \mathbb{Z}$. There is an exact diagram

(1.21)
$$\begin{array}{ccccccccc}
& & A^0 & = & A^0 & & & & \\
& & \downarrow & & \downarrow & & & & \\
0 & \longrightarrow & i_* D & \xrightarrow{\mathrm{cl}} & P & \xrightarrow{\mathrm{Ner}} & A & \longrightarrow & 0 \\
& & \parallel & & \downarrow \mathrm{deg} & & \downarrow & & \\
0 & \longrightarrow & i_* D & \xrightarrow{(C_i \cdot C_j)} & i_* D^* & \longrightarrow & i_* F & \longrightarrow & 0 \\
& & & & \downarrow & & \downarrow & & \\
& & & & 0 & & 0 & &
\end{array}$$

where cl sends a divisor to its class, Ner is the canonical map (1.1), deg sends a bundle L to the vector $(\deg L|_{C_1}, \ldots, \deg L|_{C_n})$, $(C_i \cdot C_j)$ is the intersection matrix, and i_* is the extension by zero outside of the closed point $\mathrm{Spec}\, k$. Thus the finite group F of components of A_K is identified as the cokernel of $(C_i \cdot C_j)$.

Definition (1.22). A group scheme A_S is *semi-abelian* if it is smooth, and if every geometric fibre is semi-abelian, i.e., is an extension of an abelian variety by a torus:

$$0 \to \mathbb{G}^r_{m, \bar{k}} \to A_{\bar{k}} \to (\text{abelian})_{\bar{k}} \to 0.$$

Proposition (1.23). *Assume $A = A_R$ semi-abelian, and A_K abelian. Then A_R is the connected component of the Néron model $\mathrm{N}(A_K)$.*

PROOF (Assuming Existence of $\mathrm{N}(A_K)$). The proof reduces immediately to the case that R is strictly local. The universal property (1.1) defines a map $A \xrightarrow{\phi} \mathrm{N}(A_K)$ which is a homomorphism, and an isomorphism on the general fibre. It suffices to show that its kernel on the special fibre is finite, for then ϕ is quasi-finite and birational, hence an open immersion by Zariski's Main Theorem, etc. ....

To show $\ker \phi_k$ finite, we use points of order l^ν, l prime to char k. The scheme of points of order l^ν in A is closed and étale, and similarly the corresponding scheme in $\mathrm{N}(A_K)$ is étale. Hence this scheme does not meet $\ker \phi_k$. On the other hand, points of order $l^\nu (\nu = 1, \ldots)$ are dense in any semi-abelian scheme B_k, hence in the connected component of $\ker \phi_k$. Thus the connected component has dimension 0. Since $\ker \phi_k$ is closed in A, it is finite. □

§2. Weil's Construction: Proof

We will now prove Theorem (1.12). Suppose first that V has connected (nonempty) fibres. We may as well assume S local, and then the assumption on local sections implies that the global sections are dense in each fibre.

By assumption, there are open sets $X_\phi, Y_\phi \subset V^2$, dense in each fibre, such that the rational map ϕ defines an isomorphism $X_\phi \xrightarrow{\sim} Y_\phi$. Let X be the intersection of the four sets $X_\phi, Y_\phi, X_\psi, Y_\psi$, where X_ψ, Y_ψ are defined analogously. This set X is also dense in each fibre.

Lemma (2.1). *Replacing V by an open subset dense in each fibre if necessary, we may assume that the two projections $\mathrm{pr}_i: X \to V (i = 1, 2)$ are surjective.*

PROOF. Clearly, $\mathrm{pr}_i X$ is open and dense in each fibre of V. So we will try to replace V by $V' = \mathrm{pr}_1 X \cap \mathrm{pr}_2 X$. Let $C = V - V'$ and $Z = (C \times V) \cup (V \times C)$. The subset X'_ϕ of V'^2 corresponding to X_ϕ is the complement in X_ϕ of $S = (X_\phi \cap Z) \cup \phi^{-1}(Y_\phi \cap Z)$. It is easily seen that the only fibres of $\mathrm{pr}_1: X_\phi \to V$ which are contained in S are in $C \times V$. Thus $\mathrm{pr}_1: X'_\phi \to V'$ is surjective. Similarly, $\mathrm{pr}_i: W' \to V'$ is surjective for each of the sets $W' = X'_\phi, Y'_\phi, X'_\psi, Y'_\psi$ and for $i = 1, 2$. Since the fibres of $\mathrm{pr}_i: V'^2 \to V'$ are connected, any open subset which is nonempty in each fibre is dense in each fibre. Hence, denoting by X' the intersection of the four sets W', the maps $\mathrm{pr}_i: X' \to V'$ are also surjective. □

Having replaced V as in the lemma, the theorem will now be proved with $V = U$. We can state the conclusion of the lemma this way:

(2.2) For every $a \in V$, the rational maps $\phi^{\pm 1}, \psi^{\pm 1}$ are defined at (x, a) and at (a, x), provided x is generic (i.e., lies in a dense open set) in the fibre of V containing a.

Lemma (2.3). *Assume that (2.2) holds. Let Γ denote the closure in V^3 of the graph of the law of composition. Then the three maps $\mathrm{pr}_{ij}: \Gamma \to V^2 (1 \le i < j \le 3)$ are open immersions, dense in each fibre of V^2.*

This lemma asserts that each of the rational maps $V^2 \dashrightarrow V$ obtained from Γ is defined at a point, or else the correspondence is empty there.

We can write this symbolically as follows: If $(a_0, b_0, c_0) \in \Gamma$, then the three maps

$$(a, b) \rightsquigarrow ab = c,$$

$$(a, c) \rightsquigarrow a^{-1}c = b,$$

$$(b, c) \rightsquigarrow cb^{-1} = a,$$

are defined at (a_0, b_0, c_0). It follows that ϕ is also defined at (a_0, b_0), etc. . . . .

PROOF. By Zariski's Main Theorem, it suffices to show that the maps pr_{ij}: $\Gamma \to V^2$ are set-theoretically injective. Let x be a section of V. The two rational maps $\Gamma \dashrightarrow V$ defined by

$$(a, b, c) \rightsquigarrow (xa)b$$
$$\rightsquigarrow xc$$

are equal, and so they agree whenever both terms are defined. This follows from the associative law for m. Let $(a, b, c), (a, b, c') \in \Gamma$ and choose x generic. Then $xc = (xa)b = xc'$, and xc is generic. Hence $\phi(x, c) = \phi(x, c')$, and so $c = c'$. This shows that pr_{12} is injective. The other projections are treated in the same way. □

We will now expand V by gluing translations, to get the group we want. Let s be a section of V, and let V_s be another copy of V, thought of as the "translate" $V_s = \{as | a \in V\}$. The subset

$$W_s = V \times s \times V \cap \Gamma$$

is closed in $V \times s \times V \approx V^2$, and the two projections $W_s \to V$ are open immersions. This follows from the previous lemma. Therefore W_s defines gluing data and yields a scheme

$$V' = V \cup_{W_s} V_s.$$

Lemma (2.4). *V is dense in each fibre of V', and V' has property* (2.2).

PROOF. We omit the verification that V is dense in each fibre of V'. To show (2.2), we have to show that for $a' \in V'$ and x generic in the fibre, ϕ is defined at (a', x), etc. .... This is clear if $a' \in V$. If $a' \in V_s$, then choose y generic so that $s^{-1}y$ and ay are defined. Let $x = s^{-1}y$. Then the rule

$$(a', x) \rightsquigarrow (a', ay)$$

defines ϕ at (a', x), as required. □

The above lemma allows us to replace V by V', hence to expand V whenever there is a section s such that s is not defined for all $a \in V$. Let V' be the result of finitely many such expansions, and let $W \subset V^2 \times V'$ be the closure of Γ. By Lemma (2.3) applied to V', $\mathrm{pr}_{12}: W \to V^2$ is an open immersion. Its image is the set of points $(a, b) \in V^2$ such that $m: V^2 \to V'$ is defined at (a, b). If $V \times s \not\subset W$ for some sections s of V, then replacing V' by $V' \cup V'_s$ increases V' and W. By noetherian induction, we may assume $V \times s \subset W$ for all sections s of V. Then $W = V^2$. Namely, given $(a, b) \in V^2$, $c = a(bx^{-1})$ is defined for generic x, $(c, x) \in V \times x \subset W$, and $(a, b) \rightsquigarrow cx$ defines $m: V^2 \dashrightarrow V'$ at (a, b).

Lemma (2.5). *Let V, V', W be as above. If $W = V^2$, then the law of composition is defined everywhere on V', and V' is a group scheme.*

PROOF. Let $(a', b') \in V'^2$, and let x be a generic section. Then $a'x$ and $x^{-1}b'$ are both defined and in the dense open V, so

$$(a', b') \rightsquigarrow (a'x)(x^{-1}b')$$

defines m at (a', b'). Similarly,

$$(b', c') \rightsquigarrow (b'^{-1}x)(x^{-1}c'),$$
$$(a', c') \rightsquigarrow (c'x^{-1})(xa^{-1}),$$

define the laws $b^{-1}c$, ca^{-1} at any point of V'^2. This shows that ϕ, ψ extend to automorphisms of V'^2.

We leave the verification of the group axioms and the uniqueness of the group G as an exercise. For example, to find the identity element, let a be any section of V'. Then ϕ defines an isomorphism $a \times V' \to a \times V'$. Hence $ae = a$ for some section e, and for every b, $be = (ba^{-1})(ae) = (ba^{-1})a = b$, etc. .....

It remains to consider case (ii) of (1.12). We use the Stein factorization (1.17):

$$V \xrightarrow{f'} S' \xrightarrow{g} S,$$

so that f' has connected, nonempty fibres, and g is étale. Since V is assumed to have a dense set of local sections, S' is actually a prescheme, and the normal law on V induces a structure of étale group on S'/S. The identity section e of S' is open, and $f'^{-1}(e) = V^0$ is an open subset of V with connected fibres, and with a normal law induced by m. By what has been shown, V^0 is birational to a group G^0, and the required group G is easily constructed as a union of translates of G^0. □

§3. Existence of the Néron Model: R Strictly Local

We will first construct $N(A_K)$ in case R is strictly local. Then we can use Weil's construction (1.12) of a group from birational data and the characterization (1.6) of $N(A_K)$ in terms of sections.

If we had $N(A_K) = A$, then the components C_i of the closed fibre A_k would provide us with a finite set of prime divisors (see §5) of the function field $F = \text{fract}(A_K)$, namely

(3.1) $\qquad R'_i = \mathcal{O}_{A,\eta_i}, \qquad \eta_i = $ general point of C_i.

These prime divisors are smooth extensions of R; let us call such prime divisors *smooth*.

Conversely, if we knew the prime divisors R'_i, then we would have the Néron model birationally, and could use Weil's construction (1.12).

Lemma (3.2). *Assume R strictly local. Let $R'_1, \ldots, R'_n$ be smooth prime divisors of $F = \text{fract}(A_K)$. Assume that the finite set $\{R'_i\}$ is permuted by every rational*

translation $A_K \overset{t_a}{\to} A_K$, $a \in A_K(K)$. Then the Néron model $N(A_K) = A$ exists, and the R'_i are the local rings (3.1) of the components of A_k.

PROOF. Choose a model V_R for A_K on which R'_i are of the first kind (cf. (5.2)). Let C_i be the corresponding components of V_k. Since R'_i is smooth, V is generically smooth along C_i. So, replacing V by an open subset if necessary, we may assume V smooth and that $V_k = \bigcup C_i$.

Let a be a section of V. Translation by a_K permutes $\{R'_i\}$, and therefore defines a birational map $V \to V$ whose domain of definition is dense in V_k. Since V is smooth, its sections are dense in V_k, and from this one concludes that the rational map $\phi: V^2 \to V^2$ (1.11) has a domain of definition which is dense in V_k^2. It is generically surjective because translation by a permutes the generic points of V_k. Thus (1.11) holds, and the law of composition on A_K is a normal law on V. By (1.12), there is a smooth group A realizing this law, and the components of A_k are the centers of the prime divisors R'_i.

Since translation by $a \in A_K(K)$ permutes $\{R'_i\}$, it is defined generically on each fibre of A, hence everywhere defined by (1.3). So $a = t_a(e)$ is in $A(R)$, and A is the Néron model by (1.6). $\square$

We now have to find the finite set of prime divisors $\{R'_i\}$ as in (3.2). Following Néron, we choose a nonzero holomorphic g-form α on A_k, where $g = \dim A_K$. This form α is unique up to constant factor, is translation-invariant, and is nowhere zero. Given any smooth prime divisor R' of F, choose a model $V = V_R$ on which R' is of the first kind and such that V is smooth (restrict to an open neighborhood of Spec R'). Then α can be viewed as a rational section of $\Omega^g_{V/R}$, and we set

(3.3) $\qquad v_{R'}(\alpha) = $ order of zero of α at R'.

Lemma (3.4). *Assume R local. Among all smooth prime divisors R' of A_K, the integer $v_{R'}(\alpha)$ takes on its minimum value finitely often. The corresponding prime divisors are called Ω-minimal. If $R \to R_1$ is an étale local extension, then the R_1-minimal prime divisors of $R_1 \otimes_R F$ are the extensions of the Ω-minimal prime divisors of F.*

The last assertion will be clear from the proof, and so we will not mention it further.

Lemma (3.4) will complete the proof, because since α is translation-invariant, the set of Ω-minimal prime divisors is translation-invariant, and (3.2) applies.

A FALSE START AT THE PROOF. Assume for the moment that there is a *proper nonsingular* model $V = V_R$ for $A_K = V_K$. Then the Ω-minimal prime divisors are of the first kind on V, hence form a finite set. This is because differentials acquire zeros on the exceptional divisor of a blowing-up, and we can make any prime divisor of the first kind by blowing up (5.2). This is seen as follows:

The center y of a smooth prime divisor R' on V exists because V is proper, and it has these properties:

(3.5)   (i) The residue field $k(y)$ is separable over k; and
        (ii) V is smooth at y.

The first is because R' is smooth, hence $k' = R'/p'$ is separable over k, and $k(y) \subset k'$. The second follows from (i) because p is a local parameter at the (nonsingular) point y, since $pR' = p'$ and $m_y^2 R' = (m_y R')^2$.

We now study the effect of blowing up the closure Y of y. Locally for the étale topology, we put the pair V, Y into standard position:

$$V \xrightarrow{\text{étale}} \mathbb{A}_R^n, \quad \text{coordinates } (x_1, \ldots, x_r; y_1, \ldots, y_s)$$
$$\cup \qquad\qquad \cup$$
$$Y \xrightarrow{\text{étale}} \mathbb{A}_k^s, \quad \text{the locus } \{p = x = 0\}.$$

Blowing up $\{p = x = 0\}$ in $\mathbb{A}_R^n$ induces the blowing-up of Y in V by pullback, and is defined (on an affine piece) by the substitution $pz = x$. Hence

$$dx \wedge dy \rightsquigarrow p^r \, dz \wedge dy \qquad (dx = dx_1 \wedge \cdots \wedge dx_r, \text{ etc.}).$$

Our differential α, being holomorphic and nonzero on V_K, has the form

$$\alpha = (\text{unit}) p^v \, dx \wedge dy,$$

where $v = $ order of zero of α along the component of V_k which contains Y. Therefore by induction,

$$v_{R'}(\alpha) \geq v + r,$$

which shows that R' is not Ω-minimal unless $r = 0$, i.e., R' is of the first kind on V.

This argument provides the proof of Lemma (3.4), assuming resolution of singularities. Unfortunately, resolution is not known in general, and so Néron proved what he needed:

Lemma (3.6). *Let X_K be proper and smooth over K. There is a finite set $V_R^{(i)}$ of models of X_K such that every point x' of $X_K(K')$, where K' is the fraction field of a smooth extension of discrete valuation rings $R \to R'$, extends to a smooth point of $V^{(i)}(R')$ on at least one model $V^{(i)}$.*

The proof is similar to Zariski's proof of (5.2), but must be done carefully so as to give a uniform bound for all points. We first introduce Néron's measure l of singularity along a point x' of a model with values in R'. Choose an affine presentation of a neighborhood of x' in V:

(3.7) $R[x]/(f), \qquad x = x_1, \ldots, x_n; \quad f = f_1, \ldots, f_m.$

Let M run over $(n - g)$-rowed minors of the Jacobian matrix $\partial f_i / \partial x_j$, where $g = \dim X_K$. The elements let M generate the unit ideal in $K[x]/(f)$ because

X_K is smooth. Let us denote the coordinates of the point x' by $x'_1, \ldots, x'_n$. Then

(3.8) $$l(x') \underset{\text{defn}}{=} \min_M \{v_{R'}(\det M(x'))\}.$$

It is easily seen that $l(x')$ is independent of the presentation (3.7), and is bounded above for all points x' (with R' smooth). Moreover, $l(x') = 0$ if and only if V is smooth at x'.

Lemma (3.6) follows by induction on l together with noetherian induction on V_k, from the following lemma of Raynaud. This lemma is a significant simplification of Néron's original proof.

Lemma (3.9) (Raynaud [3]). *Let $V = V_R$ be a model of the smooth scheme $V_K = X_K$. Let $Y \subset V_k$ be an irreducible closed subset such that the points x' of V with values in smooth extensions R' of R, which pass through Y, are dense in Y. Let V_1 be the blowing-up of Y in V. There is an open dense set $U \subset Y$, so that if x' passes through U and x'_1 is the lifting of x' to V_1, then*

$$l(x'_1) < l(x').$$

PROOF. (a) The existence of a dense set of points x passing through Y implies that Y is generically smooth over k. Therefore if we work locally at the generic point of Y, we can put the pair V, Y into standard position:

(3.10) $V: \{f(x, y) = 0\}, \quad x = x_1, \ldots, x_r, \quad y = y_1, \ldots, y_s, \quad f = f_1, \ldots, f_m,$

$Y: \{p = x = 0\}.$

Actually, this is a local description in the *étale topology*, which means that the equations $f(x, y) = 0$ are defined implicitly: $f(x, y)$ solves a polynomial equation $F(x, y, u) = 0$ with nonzero partial derivative $\partial F/\partial u$, i.e., V is the locus $\{F = u = 0\}$, and Y is the pull-back of $\{p = x = 0\}$ to V. This modification will not change our computation, so we suppress it. Think analytically.

(b) If some partial derivative $\partial f_i/\partial x_j$ or $\partial f_i/\partial y_j$ does not vanish identically on Y, then we can eliminate a variable and an equation. Hence we may assume that

(3.11) $$\left(\frac{\partial f}{\partial x}, \frac{\partial f}{\partial y}\right) \equiv 0 \pmod{(p, x)}.$$

(c) Next, we may work with a particular minor M whose determinant has minimal valuation on a dense set. We may assume that M involves the partial derivatives $\partial f_i/\partial x_j$ only. This can be achieved locally by a change of variable $x = \tilde{x}, y = \tilde{y} + c\tilde{x}$, which results in

$$\frac{\partial f}{\partial \tilde{x}} = \frac{\partial f}{\partial x} + c\frac{\partial f}{\partial y}; \quad \frac{\partial f}{\partial \tilde{y}} = \frac{\partial f}{\partial y}.$$

So, re-indexing, we assume M is the minor $\partial f_i/\partial x_j$, $1 \le i, j \le n - g$.

(d) The equations $f_{n-g+1}, \ldots, f_m$ could help lower l after blowing-up, but we do not need them. So, assume $V = \{f_1 = \ldots f_{n-g} = 0\}$.

(e) By Taylor expansion,
$$f(x, y) = f(0, y) + \frac{\partial f}{\partial x}(0, y)x + \mathcal{O}(x^2).$$

We have a dense set of points of V passing through Y, i.e., whose coordinates (x', y') have the property $x' \equiv 0 \pmod{p}$. For these points, $f(x', y') = 0$, and $(\partial f/\partial x)(0, y') \equiv 0 \pmod{p}$ by (3.11). Hence
$$0 = f(0, y') + \mathcal{O}(p^2).$$

Using (3.11) again, we can conclude that

(3.12) $\qquad f(0, y) \equiv 0 \pmod{p^2}$

for all y, i.e., identically. This is because a dense set of values y'' of $y \pmod{p^2}$ can be written in the form $y'' = y' + ph$, where (x', y') is as above. Then
$$f(0, y'') = f(0, y') + \frac{\partial f}{\partial x}(0, y')ph + p^2\mathcal{O}(h^2) \equiv 0 \pmod{p^2},$$

by (3.11).

(f) Blow up Y by the substitution $pz = x$. The points (x', y') with $x' \equiv 0 \pmod{p}$ lift, and
$$f(pz, y) = f(0, y) + \frac{\partial f}{\partial x}(0, y)pz + p^2\mathcal{O}(z^2) \equiv 0 \pmod{p^2}.$$

Write $f(pz, y) = p^2 g(z, y)$. Then
$$p\frac{\partial g}{\partial z} = \frac{\partial f}{\partial x},$$

which shows that l decreases by blowing-up. This completes the proof of Lemma (3.9), and of the existence of the Néron model when R is strictly local. □

§4. Projective Embedding

Proposition (4.1). (a) *Let $A = A_R$ be a smooth group scheme of finite type, with abelian general fibre A_K. Let D be a divisor on A such that:*

(i) *D_K is ample on A_K, and D is the closure of D_K.*
(ii) *There are only finitely many points $a_0 \in A_k(\bar{k})$ such that translation by a_0 in $A_{\bar{k}}$ carries $D_{\bar{k}}$ to itself.*
(iii) *$D = D^-$.*

Then D is ample on A.

(b) *For any bundle L on A, $H^0(A, L)$ is a finite R-module.*

PROOF. We may assume R strictly local (flat base change) and also that D_K is very ample. Let $a \in A(R)$. By the theorem of the square, $E = D_a + D_{-a} - 2D$ is linearly equivalent to zero on A_K, hence E_K is the divisor of a function ϕ on A_K. Therefore the divisor of ϕ on A_R has the form

(4.2) $$(\phi) = D_a + D_{-a} - 2D + \sum r_i C_i,$$

where $r_i \in \mathbb{Z}$ and C_i are the components of A_k. If $a \in A^0(R)$ (is in the connected component), then it follows by continuity that the r_i are all equal, and that

(4.3) $$D_a + D_{-a} - 2D \sim 0 \quad \text{on } A, \quad \text{if} \quad a \in A^0(R).$$

Since sections a can be found through any rational point of A_k, it follows that the linear system $|2D|$ has no base points.

To show that $|2D|$ separates points of A_k, assume $b_0 \neq c_0 \in A(k)$. Then using (ii), (iii), one can find a section $a \in A^0(R)$ such that $b_0 \in D_a$ but $c_0 \notin D_a + D_{-a}$. This shows that $|2D|$ separates rational points. That it separates algebraic points follows by pull-back to a (ramified) finite extension $R \to R_1$. Thus the map $A \xrightarrow{\phi} \mathbb{P}_R$ defined by $|2D|$ is quasi-finite, and an embedding on the general fibre.

Let V be the closure of the image of ϕ, and let $\bar{V} \xrightarrow{\pi} V$ be the normalization of V. Then ϕ lifts to $A \xrightarrow{\bar{\phi}} \bar{V}$ because A is normal, and $\bar{\phi}$ is an open immersion by Zariski's Main Theorem. The line bundle $\bar{L} = \pi^* \mathcal{O}_V(1)$ is ample on V, and if $\bar{L}^{\otimes n}$ is very ample on $\bar{V}$, it follows easily that $|2nD|$ defines a (locally closed) embedding $A \to \mathbb{P}_R$.

We omit the proof that $H^0(A, L)$ is a finite R-module. It is done by reduction to dimension 1, but we will not really need it, since in any case only finitely many sections are needed to define an embedding. □

PROOF OF THEOREM (1.2). Using Proposition (4.1), we can derive the existence of the Néron model over an arbitrary Dedekind domain R, as follows. Clearly, we may assume R local. Since the Néron model exists over the strict localization of R, it also exists over some étale local extension $\tilde{R}$. Also, the Ω-minimal prime divisors exist over R (3.4). So, there is a smooth model V_R for A_K such that the birational correspondences $V_{\tilde{R}} \leftrightarrow A_{\tilde{R}}$ are generically defined on each fibre.

Lemma (4.4). *There is a divisor D_K on A_K, such that the closure $\tilde{D}$ of $D_{\tilde{K}}$ in $A_{\tilde{R}}$ satisfies (i)–(iii) of Proposition (4.1).*

PROOF. We first note that if D_0 is a positive divisor on $A_{\bar{k}}$ which is stabilized by only finitely many translations, then the same thing is true for any $D_0' \geq D_0$. For, let G be the stabilizer of D_0'. The subgroup of G which stabilizes every component of D_0' has finite index in G, and it stabilizes D_0. Using this fact, it is easy to see that there exist divisors on $A_{\bar{k}}$ whose stabilizer is finite.

Next, let D_0 be any divisor on $A_{\bar{k}}$. There exists a rational function f on $A_{\tilde{R}}$ which vanishes on D_0 but does not vanish on any component of $A_{\bar{k}}$. Replacing f by a norm, we may assume that it is a rational function on A_K. Thus there exists a divisor D'_K such that $\tilde{D}'$ satisfies condition (ii).

Since very abelian variety is projective, it is clear that a divisor E_K may be found so that $\tilde{E}$ satisfies (i). Then for large n, $\tilde{D} = n\tilde{E} + \tilde{D}'$ satisfies (i), (ii), and $D + D^-$ satisfies (iii) as well. □

Now choose a divisor D_K as in Lemma (4.4). Let n be an integer such that $|n\tilde{D}|$ is very ample on $A_{\tilde{R}}$, and let D_1 be the closure of D_K in V_R.

Lemma (4.5). (i) $|nD_1|$ has no base points.
(ii) Let $W_R \subset \mathbb{P}_R$ denote the closure of the image of the map $V_R \to \mathbb{P}_R$ defined by $|nD_1|$. Then $|nD|$ defines an open immersion $A_{\tilde{R}} \to W_{\tilde{R}}$, and its image is the set of smooth points of $W_{\tilde{R}}$.
(iii) Let A_R be the set of smooth points of W_R. Then A_R is the Néron model of A_K.

PROOF. We can describe $H^0(V_R, \mathcal{O}(nD_1))$ as the submodule of sections of $H^0(A_K, \mathcal{O}(nD_K))$ which are regular at every smooth prime divisor of the first kind on V_R. A similar description works for $H^0(V_{\tilde{R}}, \mathcal{O}(n\tilde{D}_1))$ and $H^0(A_{\tilde{R}}, \mathcal{O}(n\tilde{D}))$. Hence

(4.6) $$H^0(V_{\tilde{R}}, \mathcal{O}(n\tilde{D}_1)) = H^0(A_{\tilde{R}}, \mathcal{O}(n\tilde{D})).$$

By assumption, the left side defines an embedding of $A_{\tilde{R}}$. Since $V_{\tilde{R}}$ is smooth, the resulting map $V_{\tilde{R}} \dashrightarrow A_{\tilde{R}}$ is everywhere defined, which shows that $|n\tilde{D}_1|$ has no base points. By flat base change, $|nD_1|$ has no base points. This proves (i), and most of (ii). That the image of $A_{\tilde{R}}$ in $W_{\tilde{R}}$ is the set of smooth points follows again from the universal property (1.1). Finally, (iii) follows easily from (ii). □

§5. Appendix: Prime Divisors

Let $(\mathcal{O}, m)$ be a local domain with field of fractions F. A *prime divisor* of $\mathcal{O}$ is a discrete valuation ring (R', p') of F such that $\mathcal{O} \subset R'$, $p' \cap \mathcal{O} = m$ (i.e., Spec $R' \to$ Spec $\mathcal{O}$ is a local map), and one of the following equivalent conditions holds:

(5.1) (i) tr deg$_{\mathcal{O}/m} R'/p' = $ dim $\mathcal{O} - 1$.
(ii) R' is a localization of an $\mathcal{O}$-algebra of finite type.

If X is an integral scheme with function field F, then a *prime divisor* R' of X is by definition a prime divisor of the local ring $\mathcal{O}_{X,x}$ at a point $x \in X$, which is called the *center* of R'. This means that Spec R' maps to X, the map sends Spec R'/p' to x, and in addition (5.1) holds. A prime divisor is said to be *of the first kind* on X if $\mathcal{O}_{X,x} = R'$, i.e., R' is the local ring of X at the general point of a closed subset of codimension 1, and X is normal at x.

Suppose that X_R is proper over R, with function field F. Then a prime divisor of X is also called a *prime divisor of the function field F*. By the valuative criterion, this notion depends only on F, and not on X.

Assume that the prime divisor R' is not of the first kind on X, and let Z be the closure of its center x. Let $X_1 \xrightarrow{\pi_1} X$ be the blowing-up of Z in X. Then R' has a center x_1 in X_1 by the valuative criterion for properness. Also, dimension theory shows that (5.1) (i) holds at $\mathcal{O}_{X_1, x_1}$ if and only if it holds at $\mathcal{O}_{X,x}$ hence that R' is a prime divisor of X_1 if and only if it is a prime divisor of X.

The following is a classical result, which as a corollary proves the equivalence of (5.1) (i) and (ii):

Theorem (5.2) (Zariski [7]). *Let R' be a prime divisor of X in the sense of* (5.1) (i). *If R' is not of the first kind on X, replace X by the blowing-up X_1, as above. After finitely many repetitions of this procedure, R' will be of the first kind.*

PROOF. Let x, x_1 be the centers of R' on X, X_1, and let $\mathcal{O} = \mathcal{O}_{X,x}$, $\mathcal{O}_1 = \mathcal{O}_{X_1, x_1}$. Then $\mathcal{O}_1$ is a localization of the ring $\mathcal{O}[x_1/x_0, \ldots, x_n/x_0]$, where $\{x_0, \ldots, x_n\}$ is a minimal set of generators of m_x and x_0 is chosen so that $v_{R'}(x_0)$ is minimal. Since $x_i/x_0 \notin \mathcal{O}$, the blowing-up is not an isomorphism at x unless $n = 0$, in which case $\mathcal{O}$ is a discrete valuation ring, necessarily equal to R'.

If dim $\mathcal{O} = 1$, then R' is the normalization of $\mathcal{O}$ and $R'/\mathcal{O}$ has finite length. The theorem follows by induction from $\mathcal{O} \subsetneq \mathcal{O}_1 \subset R'$.

Assume dim $\mathcal{O} > 1$, and choose $\alpha \in R'$ whose residue in R'/p' is transcendental over $\mathcal{O}/m$. Write $\alpha = a/b$, where $a, b \in \mathcal{O}$. Since α has a transcendental residue, $\alpha \notin \mathcal{O}$, hence $b \in m \subset p$. We use induction on $v_R(b)$. In the ring $\mathcal{O}_1$, we can factor x_0 out of a and b: $a = x_0 a_1$, $b = x_0 b_1$. Then $\alpha = a_1/b_1$, and $v_{R'}(b_1) < v_{R'}(b)$. So eventually α will become algebraic. Induction on dim $\mathcal{O}$ completes the proof. □

REFERENCES

[1] Kodaira, K. On compact analytic surfaces, II. *Ann. Math.*, **77** (1963), 563–626.
[2] Néron, A. Modèles minimaux des variétés abéliennes sur les corps locaux et globaux. *Publ. Math. I.H.E.S.*, **21** (1964).
[3] Raynaud, M. Modèles de Néron. *C.R. Acad. Sci. Paris*, **262** (1966), 413–416.
[4] Raynaud, M. *Faisceaux Amples sur les Schémas en groupes et les Espaces Homogènes*. Springer Lecture Notes, 119. Springer-Verlag: Berlin, 1970.
[5] Raynaud, M. Spécialization du foncteur de Picard. *Publ. Math. I.H.E.S.*, **38**, (1971).
[6] Weil, A. *Variétés Abéliennes et Courbes Algébriques*. Hermann, Paris, 1948.
[7] Zariski, O. The reduction of singularities of an algebraic surface. *Ann. Math.*, **40** (1939), 639–689.
[8] SGA3. *Séminaire de Geométrie Algébrique 1964: Schémas en Groupes*, exposé XVIII. Lecture Notes in Mathematics, 152. Springer-Verlag: Berlin, 1970.

CHAPTER IX

Siegel Moduli Schemes and Their Compactifications over $\mathbb{C}$

CHING-LI CHAI*

The purpose of this chapter is to give a brief introduction to the moduli spaces of abelian varieties and their compactification. Only the geometric aspects of the theory are discussed. The arithmetic side is left untouched. The Satake and toroidal compactification are described within the realm of matrices. Although the theory looks more elementary and explicit, this approach also tends to obscure its group-theoretic nature (see [B–B], [SC] for the general case). The readers interested in a deeper pursuit of this subject may find more references in [GIT] and [Fr].

§0. Notations and Conventions

0.1. $\mathbb{Z}, \mathbb{Q}, \mathbb{R}, \mathbb{C}$ denote integers, rational numbers, real numbers, and complex numbers respectively. $\mathbb{N} = \mathbb{Z}_{\geq 0}$, $\mathbb{N}_+ = \mathbb{Z}_{>0}$.

0.2. All rings are commutative with identity, all schemes are locally noetherian, unless otherwise stated.

0.3. For any ring R, $m, n \in \mathbb{N}_+$, $M_{m,n}(R) = m \times n$ matrices with entries in R, $M_n(R) = M_{n,n}(R)$, $GL_n(R)$ = invertible elements in $M_n(R)$. Thus $M_{m,n}$, M_n, GL_n are all schemes over $\mathbb{Z}$. For any $A \in M_{m,n}(R)$, ${}^tA \in M_{n,m}(R)$ denotes the transpose of A.

0.4. For all $g \in \mathbb{N}_+$, Sp_{2g} is a subgroup scheme of GL_{2g} over $\mathbb{Z}$, such that for

---

* Supported in part by NSF grant MCS-8108814 (A03).

any ring R,

$$\mathrm{Sp}_{2g}(R) = \left\{\gamma \in M_{2g}(R) \middle| {}^t\gamma \begin{pmatrix} 0 & 1_g \\ -1_g & 0 \end{pmatrix} \gamma = \begin{pmatrix} 0 & 1_g \\ -1_g & 0 \end{pmatrix}\right\}$$

$$= \left\{\gamma = \begin{pmatrix} A & B \\ C & D \end{pmatrix} \in M_{2g}(R) \middle| A^t B = B^t A, C^t D = D^t C, A^t D - B^t C = 1_g\right\},$$

where

$$1_g = \begin{pmatrix} 1 & & 0 \\ & \ddots & \\ 0 & & 1 \end{pmatrix} \in \mathrm{GL}_{2g}(R).$$

For any $\gamma \in M_{2g}(R)$, $\gamma \in \mathrm{Sp}_{2g}(R)$ if and only if ${}^t\gamma \in \mathrm{Sp}_{2g}(R)$. Thus we get an equivalent set of identities in A, B, C, D characterizing Sp_{2g}.

0.5. For all $g, n \in \mathbb{N}_+$, $\Gamma_g(n) := \{\gamma \in \mathrm{Sp}_{2g}(\mathbb{Z}) | \gamma \equiv 1_{2g} \pmod{n}\}$ is the principal congruence subgroup of level n. $\Gamma_g(1) = \mathrm{Sp}_{2g}(\mathbb{Z})$ is also denoted by Γ_g, called the full modular group.

0.6. For all $r \in \mathbb{N}_+$, $C_r := \{A \in M_r(\mathbb{R}) | A = {}^tA, A \text{ positive-definite}\}$, $\bar{C}_r :=$ "rational closure" of C_r in $M_r(\mathbb{R}) = \{A \in M_r(\mathbb{R}) | A = {}^tA, A \text{ positive, semi-definite radical of } A \text{ is defined over } \mathbb{Q}\}$.

0.7. For all $g \geq 1$, $g \in \mathbb{Z}$, $\mathfrak{H}_g = \{\Omega \in M_g(\mathbb{C}) | \Omega = {}^t\Omega, \mathrm{Im}\,\Omega \gg 0$ (i.e. imaginary part of Ω is positive-definite)$\}$. $\mathrm{Sp}_{2g}(\mathbb{R})$ acts on $\mathfrak{H}_g$ via

$$\gamma = \begin{pmatrix} A & B \\ C & D \end{pmatrix}: \Omega \mapsto (A\Omega + B) \cdot (C\Omega + D)^{-1}.$$

0.8. $\mathbb{G}_m = \mathrm{GL}_1$. For any $n \in \mathbb{Z}, n \neq 0$, $\boldsymbol{\mu}_n = $ the finite flat subgroup scheme of $\mathbb{G}_m$ consisting of nth root of unity = Cartier dual of the constant group scheme $\mathbb{Z}/n\mathbb{Z}$.

0.9. Schemes = the category of (locally noetherian) schemes. **Sets** = the category of sets. Given a scheme X, let h_X be the contravariant functor from **Schemes** to **Sets** defined by $h_X(S) = $ the set of all morphisms from S to X for any scheme S.

§1. The Moduli Functors and Their Coarse Moduli Schemes

1.1. We first recall some basic definitions. All schemes here are locally noetherian.

1.1.1. An *abelian scheme* is a smooth, proper group scheme $X \underset{\varepsilon}{\overset{\pi}{\rightrightarrows}} S$ of finite type with connected geometric fibres. (We can delete the adjective "geometric".) For any $n \in \mathbb{Z}$, denoted by $[n]_X$ the morphism "multiplication by n", and let $X_{[n]} = \text{Ker}[n]_X$. $X_{[n]}$ is a finite flat group scheme over S, and is étale over $S[1/n]$.

1.1.2. A *polarization* of an abelian scheme $X \underset{\varepsilon}{\overset{\pi}{\rightleftarrows}} S$ is an S-homomorphism $\lambda: X/S \to X^\vee/S = \text{Pic}^0(X/S)$ such that for any geometric point s of S, the induced homomorphism between fibres $\lambda_s: X_s \to X_s^\vee$ is of the form $\lambda_s = \phi_{\mathscr{L}_s}$ for some ample invertible sheaf $\mathscr{L}_s$ on X_s (notation as in [Mi]).

1.1.3. Let $\lambda: X/S \to X^\vee/S$ be a polarization as above. Then $\lambda_* \mathcal{O}_X$ is a locally free $\mathcal{O}_{X^\vee}$-module. Its rank is constant over each connected component of S, and is always a perfect square. Call it the *degree* of λ. λ is a principal polarization if $\deg \lambda = 1$, i.e. if λ is an isomorphism.

1.1.4. Given a polarization λ as above, we can produce a relatively ample invertible sheaf $\mathscr{L}^\Delta(\lambda)$ on X by setting $\mathscr{L}^\Delta(\lambda) = (\text{id}_X, \lambda)^* P$, where P is the Poincaré sheaf on $X \times_S X^\vee$. It follows from the theorem of cube that $\phi_{\mathscr{L}^\Delta(\lambda)} = 2\lambda$.

1.2. The basic problem we address here is to classify polarized abelian varieties, especially principally polarized ones. The right way to formulate this problem is to use categorical language. Let us first introduce the most important moduli functor $\mathbf{A}_g$ classifying principally polarized abelian varieties.

1.2.1. Define $\mathbf{A}_g$: **Schemes** $\to$ **Sets** to be the contravariant functor which assigns to any scheme S the set $\mathbf{A}_g(S) :=$ the set of isomorphism classes of principally polarized abelian schemes over S of relative dimension g.

1.2.2. Remark. In more sophisticated terms, the category $\mathbf{PPAV}_g$ of principally polarized abelian schemes of relative dimension g has a natural structure as a fibred category over **Schemes**. It turns out that $\mathbf{PPAV}_g$ is an algebraic stack, and one can use M. Artin's method (see [Ar 1], [Ar 2]) to study the moduli problem. Here we use only moduli functors to keep things simple. But there is no doubt that algebraic stacks are the right thing to consider when there are nontrivial isotropy subgroups.

1.2.3. We can define similarly contravariant functors $\mathbf{A}_{g,d,n}$: **Schemes** $\to$ **Sets** to take degree of polarization and level structure into account. For any (locally noetherian) scheme S, $\mathbf{A}_{g,d,n}(S)$ is by definition the set of isomorphism classes of triples $(X/S, \lambda, \sigma)$, where X/S is an abelian scheme, λ is a polarization of degree d^2, and $\sigma = (\sigma_1, \ldots, \sigma_{2g})$ consists of $2g$ elements of $X_{[n]}(S)$ which induce an isomorphism $X_{[n]} \simeq (\mathbb{Z}/n\mathbb{Z})^{2g}/S$. Note that $\mathbf{A}_{g,1,1} = \mathbf{A}_g$. Since every abelian variety over an algebraically closed field is isogenous to a principally polarized variety, $\mathbf{A}_{g,1,n}$ is of special importance. Observe that if some point of S has residue characteristic p dividing n, then $\mathbf{A}_{g,d,n}(S) = \phi$.

There is a variant $\mathbf{A}_{g,n}^*$ of $\mathbf{A}_{g,1,n}$. By definition, for any scheme S, $\mathbf{A}_{g,n}^*(S) = $ isomorphism classes of tiples $(X/S, \lambda, \alpha)$, where X/S is an abelian scheme, λ is a principle polarization, and α is an isomorphism $X_{[n]} \tilde{\to}_{/S} (\mathbb{Z}/n\mathbb{Z})^g \times \mu_n^g$ such that α^* (standard symplectic pairing) = Weil pairing. If k is a field, $\text{char}(k) = p | n$, $\mathbf{A}_{g,n}^*(k)$ consists of ordinary abelian varieties. Note that $\mathbf{A}_{g,1}^* = \mathbf{A}_g$.

1.3. Given a contravariant functor $\mathbf{A} := \mathbf{Schemes} \to \mathbf{Sets}$, a natural question is whether it is *representable* or not, i.e. whether it is isomorphic to h_A for some scheme A.

1.3.1. $\mathbf{A}$ is representable by A means that there exists $a \in \mathbf{A}(A)$ such that for any scheme S, $x \in \mathbf{A}(S)$, there is a unique morphism $f: S \to A$ with $f^*(x) = a$. A is unique up to isomorphism in this case, and is called the *fine moduli scheme* of $\mathbf{A}$.

1.3.2. It is a fact of life that many of the moduli functors we encounter are not representable, e.g. $\mathbf{A}_g$. But they are "almost representable" in the following sense: there is a scheme A and a morphism $F: \mathbf{A} \to h_A$ of functors such that

(a) any morphism $G: \mathbf{A} \to h_X$ for some scheme X factors through F via a unique morphism $g: A \to X$.
(b) for any algebraically closed field k, $F(\text{spec } k): \mathbf{A}(k) \to A(k)$ is a bijection.

In this case, A is unique up to isomorphism, and is called the *coarse moduli scheme of* $\mathbf{A}$.

1.4. Theorem. (a) *For any g, d, $n \in \mathbb{N}_+$, the coarse moduli scheme $\mathscr{A}_{g,d,n}$ of $\mathbf{A}_{g,d,n}$ exists. It is faithfully flat over $\text{Spec } \mathbb{Z}[1/n]$ and quasi-projective over $\text{Spec } \mathbb{Z}[1/np]$ for any prime number p. Furthermore, if $n \geq 3$, then $\mathscr{A}_{g,d,n}$ actually represents $\mathbf{A}_{g,d,n}$, and is smooth over $\text{Spec } \mathbb{Z}[1/nd]$.*
(b) *Similarly, the coarse moduli scheme $\mathscr{A}_{g,n}^*$ of $\mathbf{A}_{g,n}^*$ always exists. It is faithfully flat over $\text{Spec } \mathbb{Z}$ and quasi-projective over $\text{Spec } \mathbb{Z}[1/p]$ for any prime number p. If $n \geq 3$, $\mathscr{A}_{g,n}^*$ is a fine moduli scheme, and is smooth over $\text{Spec } \mathbb{Z}[1/n]$.*

We denote $\mathscr{A}_{g,1,1} = \mathscr{A}_{g,1}^*$ by $\mathscr{A}_g$.

1.4.1. Exercise. (i) Use the existence of the fine moduli scheme $\mathscr{A}_{g,n}^*$ to prove the nonexistence of a fine moduli scheme for $\mathbf{A}_g$. (Hint: Use a principally polarized abelian variety with nontrivial group of automorphisms.)

(ii) For any automorphism τ of $\mathbb{C}$, let X^τ be $X \times_{(\text{Spec } \mathbb{C}, \tau)} \text{Spec } \mathbb{C}$. If there are only finitely many isomorphism classes among all X^τ's, then X is defined over a number field.

1.5. One way to prove the existence theorem is to use geometric invariant theory (another way is to use Artin's method in [Ar 1], [Ar 2]). Roughly, the problem reduces via the theory of Hilbert schemes to the problem of taking the quotient of a quasi-projective scheme (actually, a locally closed subscheme of some Hilbert scheme) by PGL_N for some N. As Hilbert and Mumford taught us, in general, one has to throw away some "bad points", and take only the so-called "stable points", in order to get good quotients. In practice, this amounts to checking some stability conditions using explicit geometric information. The quasi-projectivity is a by-product of this method. See ([GIT, Chap. 7]) for details. [G] contains a nice short exposition of geometric invariant theory.

1.6. There is a very concrete approach to moduli of abelian varieties via theta constants. Anticipating the transcendental uniformization in the next section, the right level structure to use is given by the subgroups

$$\Gamma_g(n, 2n) := \left\{ \begin{pmatrix} A & B \\ C & D \end{pmatrix} \in \mathrm{Sp}_{2g}(\mathbb{Z}) \middle| A \equiv D \equiv 1_g \pmod{n}, \right.$$

$$\left. B \equiv C \equiv 0 \pmod{n}, \mathrm{diag}\, B \equiv \mathrm{diag}\, C \pmod{2n} \right\},$$

$$n \equiv 0 \pmod{2}.$$

The classical theta constants were extensively studied by Igusa, see [I 1], [I 2]. In [Mu 1], algebraic theta functions were introduced and used to *construct* moduli schemes. See also [Mu 3], [Mu 4], [Ch]. The algebraic theta constants are also useful for compactification (cf. 4.3.1.).

§2. Transcendental Uniformization of the Moduli Spaces

2.1. As we saw in [R], the isomorphism classes of $\{X, (\alpha_i)_{0 \leq i \leq 2g}\}$ of abelian varieties X over $\mathbb{C}$ of dimension g together with a symplectic basis $\alpha_i \in H_1(X(\mathbb{C}), \mathbb{Z})$ are parametrized by points of $\mathfrak{H}_g$. Here $\alpha_j \cdot \alpha_k = \alpha_{g+j} \cdot \alpha_{g+k} = 0$, $\alpha_j \cdot \alpha_{g+k} = -\alpha_{g+j} \cdot \alpha_k = \delta_{j,k}$ for all $0 \leq j, k \leq g$. Now we reformulate it: Consider the space $\mathfrak{H}_g \times \mathbb{C}^g$. $\mathbb{Z}^{2g}$ acts on $\mathfrak{H}_g \times \mathbb{C}^g$ via

$$\begin{pmatrix} n_1 \\ n_2 \end{pmatrix} : (\Omega, z) \to (\Omega, z + \Omega n_1 + n_2), \quad \forall \begin{pmatrix} n_1 \\ n_2 \end{pmatrix} \in \mathbb{Z}^{2g}, \quad \forall (\Omega, z) \in \mathfrak{H}_g \times \mathbb{C}^g.$$

Let $\mathcal{X}_g = \mathbb{Z}^{2g} \backslash \mathfrak{H}_g \times \mathbb{C}^g$. The natural projection $\mathcal{X}_g \to \mathfrak{H}_g$ defines a "universal" holomorphic family of principally polarized varieties with symplectic basis of H_1.

2.2. $\mathrm{Sp}_{2g}(\mathbb{Z})$ acts on $\mathfrak{H}_g \times \mathbb{C}^g$ via

$$\begin{pmatrix} A & B \\ C & D \end{pmatrix} : (\Omega, z) \to ((A\Omega + B) \cdot (C\Omega + D)^{-1}, {}^t(C\Omega + D)^{-1} z),$$

$$\forall \begin{pmatrix} A & B \\ C & D \end{pmatrix} \in \mathrm{Sp}_{2g}(\mathbb{Z}), \quad \forall (\Omega, z) \in \mathfrak{H}_g \times \mathbb{C}^g.$$

The action of $\mathrm{Sp}_{2g}(\mathbb{Z})$ and $\mathbb{Z}^{2g}$ on $\mathfrak{H}_g \times \mathbb{C}^g$ do not commute, but we put them together and get an action of a semi-direct product $\mathrm{Sp}_{2g}(\mathbb{Z}) \ltimes \mathbb{Z}^{2g}$ on $\mathfrak{H}_g \times \mathbb{C}^g$, in which $\mathbb{Z}^{2g}$ is a normal subgroup and $\mathrm{Sp}_{2g}(\mathbb{Z})$ acts on $\mathbb{Z}^{2g}$ by matrix multiplication. The $\mathrm{Sp}_{2g}(\mathbb{Z})$ action on $\mathfrak{H}_g \times \mathbb{C}^g$ descends to $\mathscr{X}_g$, and the map $\mathscr{X}_g \to \mathfrak{H}_g$ is clearly $\mathrm{Sp}_{2g}(\mathbb{Z})$-equivariant. From the formula

$$((A\Omega + B) \cdot (C\Omega + D)^{-1}, 1_g) = {}^t(C\Omega + D)^{-1} \cdot (\Omega, 1_g) \cdot \begin{pmatrix} {}^tA & {}^tC \\ {}^tB & {}^tD \end{pmatrix},$$

we see that the action of $\gamma \in \mathrm{Sp}_{2g}(\mathbb{Z})$ on $\mathfrak{H}_g$ corresponds, up to the conjugation by $\begin{pmatrix} 0 & 1_g \\ -1_g & 0 \end{pmatrix}$, to changing symplectic basis of H_1 by γ via the description of $\mathfrak{H}_g$ in 2.1 above.

2.3. Let $\Gamma \subseteq \mathrm{Sp}_{2g}(\mathbb{Z})$ be a subgroup of finite index (which necessarily contains some congruence subgroup if $g \geq 2$) acting freely on $\mathfrak{H}_g$. For instance, this is the case if $\Gamma \subseteq \Gamma_g(n)$ for some $n \geq 3$. Then the holomorphic family $\mathscr{X}_g \to \mathfrak{H}_g$ descends to $\mathscr{X}_{g,\Gamma} = \Gamma \backslash \mathscr{X}_g \to \Gamma \backslash \mathfrak{H}_g$.

2.4. From the discussion above, we see easily that for all $n \in \mathbb{N}_+$, $\mathscr{A}_{g,n}^*(\mathbb{C}) \simeq \Gamma_g(n) \backslash \mathfrak{H}_g$ as analytic spaces. In particular, $\mathscr{A}_g(\mathbb{C}) \simeq \Gamma_g \backslash \mathfrak{H}_g$. Note that $\mathscr{A}_{g,1,n}(\mathbb{C})$ is a disjoint union of copies of $\Gamma_g(n) \backslash \mathfrak{H}_g$.

Actually, we can do even better. In fact, the theory of Satake compactification, to be discussed later, gives $\Gamma_g(n) \backslash \mathfrak{H}_g$ the structure of a (unique) quasi-projective variety, $A_{g,\Gamma(n)}$. Furthermore, results in [Bo] shows that for any scheme S over $\mathbb{C}$, any holomorphic family of polarized abelian variety over $S(\mathbb{C})$ is algebraic. It follows that $\mathscr{A}_{g,n}^* \times_{\mathrm{Spec}\, \mathbb{Z}} \mathrm{Spec}\, \mathbb{C} \simeq \mathscr{A}_{g,\Gamma_g(n)}$ for all $n \in \mathbb{N}_+$. In particular, $\mathscr{A}_g \times_{\mathrm{Spec}\, \mathbb{Z}} \mathrm{Spec}\, \mathbb{C} \simeq \mathscr{A}_{g,\Gamma_g}$.

2.5. We now define Siegel modular forms.

2.5.1. The isotropy subgroup K at $\sqrt{-1}\, 1_g \in \mathfrak{H}_g$ is isomorphic to the unitary group $U_g(\mathbb{R})$. Actually,

$$K = \left\{ \begin{pmatrix} A & B \\ -B & A \end{pmatrix} \in M_{2g}(\mathbb{R}) \,\middle|\, A^t B = B^t A,\ A^t A + B^t B = 1_g \right\},$$

and the isomorphism is given by $\begin{pmatrix} A & B \\ -B & A \end{pmatrix} \to A + \sqrt{-1}\, B$. K is a maximal compact subgroup of $\mathrm{Sp}_{2g}(\mathbb{R})$, and its complexification $K_{\mathbb{C}}$ is isomorphic to $\mathrm{GL}_g(\mathbb{C})$. Clearly, $\mathrm{Sp}_{2g}(\mathbb{R})/K \simeq \mathfrak{H}_g$.

2.5.2. Let $\rho: \mathrm{GL}_g(\mathbb{C}) = K_\mathbb{C} \to \mathrm{GL}(V_\rho)$ be a finite-dimensional representation of K. There is a natural one-to-one correspondence between $\{C^\infty\text{-functions } f: \mathfrak{H}_g \to V_\rho\}$ and $\{C^\infty\text{-functions } \phi: \mathrm{Sp}_{2g}(\mathbb{R}) \to V_\rho | \phi(gk) = \rho(k)^{-1}\phi(g), \forall k \in K, \forall g \in G\}$. Note that the latter is the space of C^∞-sections of the vector bundle $\mathbb{E}_\rho = \mathrm{Sp}_{2g}(\mathbb{R}) \times_{(K,\rho)} V_\rho$. The correspondence is given by

$$f \mapsto \phi_f, \quad \phi_f(g) = \rho(C\sqrt{-1} + D)^{-1} f(g\sqrt{-1}\, 1_g), \quad \forall g = \begin{pmatrix} A & B \\ C & D \end{pmatrix} \in \mathrm{Sp}_{2g}(\mathbb{R}),$$

$$\phi \mapsto f_\phi, f_\phi(\Omega) = \rho(C\sqrt{-1} + D)\phi(g_\Omega)$$

for any $\Omega \in \mathfrak{H}_g$ and any

$$g_\Omega = \begin{pmatrix} A & B \\ C & D \end{pmatrix} \in \mathrm{Sp}_{2g}(\mathbb{R}) \quad \text{such that} \quad g_\Omega(\sqrt{-1}\, 1_g) = \Omega.$$

2.5.3. Let $\rho: \mathrm{GL}_g(\mathbb{C}) = K_\mathbb{C} \to \mathrm{GL}(V_\rho)$ be a finite-dimensional represenation of $K_\mathbb{C}$, and $\Gamma \subset \mathrm{Sp}_{2g}(\mathbb{Z})$ be a subgroup of finite index. We define a (vector valued) *Siegel modular form of type ρ* with respect to Γ to be a holomorphic function $f: \mathfrak{H}_g \to V_\rho$ such that:

(i) $f(\gamma\Omega) = \rho(C\Omega + D)f(\Omega), \forall \gamma = \begin{pmatrix} A & B \\ C & D \end{pmatrix} \in \Gamma_g, \forall \Omega \in \mathfrak{H}_g$;

(ii) f is holomorphic at all cusps if $g = 1$.

The condition (i) means that f defines a Γ-invariant holomorphic section of the homogeneous vector bundle $\mathbb{E}_\rho = \mathrm{Sp}_{2g}(\mathbb{R}) \times_{(K,\rho)} V_\rho$ on $\mathfrak{H}_g$. When $g \geq 2$, Koecher's principle says that (i) already implies that f is holomorphic at all cusps, so we do not need the second condition.

2.5.4. In the special case $\rho = (\det)^k$ for some $k \in \mathbb{N}$, the transformation law becomes

$$f(\gamma\Omega) = \det(C\Omega + D)^k f(\Omega), \quad \forall \gamma = \begin{pmatrix} A & B \\ C & D \end{pmatrix} \in \Gamma_g, \quad \forall \Omega \in \mathfrak{H}_g.$$

Any such f is called a *Siegel modular form of weight k* with respect to Γ. The $\mathbb{C}$-vector space of such f's will be denoted $R_k(\Gamma)$.

2.6. Theorem. *For any subgroup Γ of finite index in $\mathrm{Sp}_{2g}(\mathbb{Z})$:*

(i) *the graded $\mathbb{C}$-algebra $R(\Gamma) = \bigoplus_{k \in \mathbb{N}} R_k(\Gamma)$ is finitely generated over $\mathbb{C}$;*

(ii) *trans. $\deg_\mathbb{C} R(\Gamma) = \binom{g+1}{2} + 1 = \frac{1}{2}(g+1)g + 1$;*

(iii) *$\dim R_k(\Gamma) < +\infty, \forall k$, and $\dim_\mathbb{C} R_k(\Gamma) = O(k^{\binom{g+1}{2}})$;*

(iv) *$R(\Gamma)$ embeds $\Gamma \backslash \mathfrak{H}_g$ in $\mathrm{Proj}(R(\Gamma))(\mathbb{C})$ as an open dense subvariety in Zariski topology.*

Note that (iii) *is a consequence of* (i) *and* (ii).

2.6.1. Siegel proved (iii) by estimating Fourier coefficients of nonzero cusp forms. The theory of compactification and Eisenstein series gives (i)–(iv) simultaneously.

2.7. The Siegel modular forms are defined function-theoretically. Now we discuss their geometric meaning.

2.7.1. Let $\mathcal{X}_g \overset{\pi}{\underset{\varepsilon}{\rightleftarrows}} \mathfrak{H}_g$ be the universal family of principally polarized abelian varieties defined in 2.1. The rank g vector bundle $\mathbb{H}_g = \pi_*(\Omega^1_{\mathcal{X}_g/\mathfrak{H}_g}) = \varepsilon^*(\Omega^1_{\mathcal{X}_g/\mathfrak{H}_g})$ can be canonically identified with $\mathfrak{H}_g \times (\mathbb{C}\, dz_1 \oplus \cdots \oplus \mathbb{C}\, dz_g)$. The action of $\mathrm{Sp}_{2g}(\mathbb{Z})$ on the holomorphic sections $dz_1, \ldots, dz_g$ is given by

$$\gamma = \begin{pmatrix} A & B \\ C & D \end{pmatrix}: (\Omega, (dz_1, \ldots, dz_g)) = (\gamma\Omega, (dz_1, \ldots, dz_g) \cdot (C\Omega + D)).$$

If std is the standard representation of $\mathrm{GL}_g(\mathbb{C})$ on $\mathbb{C}^g$, we see that the vector bundle $\mathbb{H}_g$ is just $\mathbb{E}_{\mathrm{std}}$.

2.7.2. Let $\omega_g = \Lambda^g \mathbb{H}_g$. This is a holomorphic line bundle on $\mathfrak{H}_g$ isomorphic to $\mathbb{E}_{\mathrm{det}}$. For any subgroup $\Gamma \subset \mathrm{Sp}_{2g}(\mathbb{Z})$ of finite index, we see that $R_k(\Gamma)$ is canonically isomorphic to Γ-invariant holomorphic sections of $\omega_g^{\otimes k}$. When Γ has no fixed point on $\mathfrak{H}_g$, we have a holomorphic family of principally polarized abelian varieties $\mathcal{X}_{g,\Gamma} \overset{\pi_\Gamma}{\underset{\varepsilon_\Gamma}{\rightleftarrows}} \Gamma\backslash\mathfrak{H}_g$. Define $\mathbb{H}_{g,\Gamma} = \pi_{\Gamma*}(\Omega^1_{\mathcal{X}_{g,\Gamma}/(\Gamma\backslash\mathfrak{H}_g)}) = \varepsilon_\Gamma^*(\Omega^1_{\mathcal{X}_{g,\Gamma}/(\Gamma\backslash\mathfrak{H}_g)})$; then we see that $\omega_{g,\Gamma} := \Lambda^g \mathbb{H}_{g,\Gamma} \simeq \Gamma\backslash\omega_g$. Thus $R_k(\Gamma)$ is canonically isomorphic to the space of holomorphic sections of $\omega_{g,\Gamma}^{\otimes k}$ over $\Gamma\backslash\mathfrak{H}_g$.

2.7.3. Let $\gamma = \begin{pmatrix} A & B \\ C & D \end{pmatrix} \in \mathrm{Sp}_{2g}(\mathbb{R})$, $\Omega \in \mathfrak{H}_g$, and identify the holomorphic tangent space at Ω with $\{W \in M_g(\mathbb{C}) |\, {}^tW = W\}$. Then the Jacobian of γ at Ω is given by $W \mapsto {}^t(C\Omega + D)^{-1} \cdot W \cdot (C\Omega + D)^{-1}$. From this we deduce that the cotangent bundle of $\mathfrak{H}_g$ is canonically isomorphic to $\mathbb{E}_{S^2(\mathrm{std})}$, where $S^2(\mathrm{std})$ is the second symmetric product of the standard representation. Easy linear algebra then shows that the canonical bundle $K_{\mathfrak{H}_g}$ is isomorphic to $\mathbb{E}_{\mathrm{det}^{\otimes(g+1)}}$. Thus if Γ is a torsion-free discrete subgroup of $\mathrm{Sp}_{2g}(\mathbb{R})$, the space of holomorphic sections of $K_{(\Gamma\backslash\mathfrak{H}_g)}^{\otimes k}$ is identified with $R_{(g+1)k}(\Gamma)$.

§3. The Satake Compactification

3.1. Given a subgroup $\Gamma \subseteq \mathrm{Sp}_{2g}(\mathbb{Z})$ of finite index, the quotient $\Gamma\backslash\mathfrak{H}_g$ is always noncompact. As stated in 2.6, $\Gamma\backslash\mathfrak{H}_g$ is canonically embedded in $\mathrm{Proj}\, R(\Gamma)(\mathbb{C})$. Thus $\mathrm{Proj}\, R(\Gamma)$ is a natural compactification of $\Gamma\backslash\mathfrak{H}_g$, and will turn out to be the Satake compactification to be discussed in this section. But the structure of $R(\Gamma)$ is hard to analyze. Finding generators of $R(\Gamma)$ and their relations is still largely an open question. Thus it is not easy to understand

Proj $R(\Gamma)$ algebraically. The theory of Satake compactification provides a description of Proj $R(\Gamma)(\mathbb{C})$. This is good enough for most purposes by GAGA.

Some History. Satake ([Sa]) first introduced the topological compactification and defined analytic structures. Then Baily ([Ba 2]) proved projectivity. Finally, Baily and Borel ([B–B]) extended the theory to any arithmetic quotient of bounded symmetric domains.

3.2. The Siegel Subsets

3.2.1. Any $\Omega \in \mathfrak{H}_g$ can be written uniquely as $\Omega = X + \sqrt{-1}Y$, with $X, Y \in M_g(\mathbb{R})$, $X = {}^t X$, $Y = {}^t Y$, $Y \gg 0$. Any such Y can be uniquely expressed as $Y = {}^t BDB$, where $D = \begin{pmatrix} d_1 & 0 \\ & \ddots & \\ 0 & & d_g \end{pmatrix}$ is a diagonal matrix, each $d_i > 0$, and $B = \begin{pmatrix} 1 & & b_{ij} \\ & \ddots & \\ 0 & & 1 \end{pmatrix} \in M_g(\mathbb{R})$ is unipotent upper triangular. $Y = {}^t BDB$ is called the Jacobi decomposition of Y.

3.2.2. For each $u > 0$, define the *Siegel subset* $\mathscr{F}_g(u) \subset \mathfrak{H}_g$ by
$$\mathscr{F}_g(u) = \{X + \sqrt{-1}Y \in \mathfrak{H}_g \mid |x_{ij}| < u, \forall i, j, |b_{ij}| < u, \forall 1 \leq i < j \leq g,$$
$$1 < ud_1, d_i < ud_{i+1}, \forall i, 1 \leq i \leq g-1\}.$$

These Siegel subsets have the following properties:
(a) $\bigcup_{u > 0} \mathscr{F}_g(u) = \mathfrak{H}_g$.
(b) $\mathrm{Sp}_{2g}(\mathbb{Z}) \mathscr{F}_g(u) = \mathfrak{H}_g$ if u is sufficiently large.
(c) For all $u > 0$, the set $\{\gamma \in \mathrm{Sp}_{2g}(\mathbb{Z}) \mid \gamma \mathscr{F}_g(u) \cap \mathscr{F}_g(u) = \phi\}$ is finite.

3.3. How we choose and fix a sufficiently large u_0 such that 3.2.2(b) is satisfied. The Satake topology will not depend on the choice of u_0. Let $\mathscr{F}_g^* := \overline{\mathscr{F}_g(u_0)} \amalg \overline{\mathscr{F}_{g-1}(u_0)} \amalg \cdots \amalg \overline{\mathscr{F}_0(u_0)}$, where $\overline{\mathscr{F}_r(u_0)}$ is the closure of $\mathscr{F}_r(u_0)$ in $\mathfrak{H}_r$ for each $r \geq 1$, $\overline{\mathscr{F}_0(u_0)}$ denotes a point, and $\amalg$ means set-theoretic disjoint union. We will define a topology for $\mathscr{F}_g^*$.

3.3.1. First we introduce some auxiliary subsets. For any $r \in \mathbb{N}$, any open subset $U \subseteq \overline{\mathscr{F}_r(u_0)}$, any $s \in \mathbb{N}$ with $0 \leq r \leq s \leq g$, and any $c \in \mathbb{R}_{>0}$
$$W_{r,s}(U, c) = \left\{ \Omega = \begin{pmatrix} \Omega_1 & * \\ * & * \end{pmatrix} \begin{matrix} r \\ s-r \end{matrix} = X + \sqrt{-1}Y \in \mathscr{F}_s(u_0) \middle| \Omega_1 \in U, \right.$$
$$\text{and } d_{r+1} > c \text{ in the Jacobi decomposition}$$
$$\left. Y = {}^t BDB \text{ of } Y, D = \begin{pmatrix} d_1 & 0 \\ & \ddots & \\ 0 & & d_s \end{pmatrix} \right\}.$$

3.3.2. Now we can define a topology on $\mathscr{F}_g^*$. For each $\Omega \in \overline{\mathscr{F}_r(u_0)} \subset \mathscr{F}_g^*$, a fundamental system of neighborhoods of Ω is given by $\{\bigcup_{s, r \leq s \leq g} W_{r,s}(U, c)\}$, where U ranges through neighborhoods of Ω, and c runs through $\mathbb{R}_{>0}$.

In other words, a sequence

$$\left\{ \Omega_n = \begin{pmatrix} \Omega_{1,n} & * & * \\ * & * & \\ & & \end{pmatrix} \begin{matrix} r \\ g-r \end{matrix}, \quad n \in \mathbb{N} \right\}$$
$$\quad\quad\quad\quad r \quad g-r$$

in $\overline{\mathscr{F}_g(u_0)}$ converges to a point $\Omega_0 \in \mathscr{F}_r(u_0)$ if and only if $\lim_{n \to \infty} \Omega_{1,n} = \Omega_0$ and $\lim_{n \to \infty} d_{r+1,n} = \infty$, where d_{r+1} is the $(r+1)$st diagonal entry of D in the Jacobi decomposition $\operatorname{Im} \Omega_n = {}^tBDB$.

3.4. We will define a topological space $\mathfrak{H}_g^*$ extending $\mathfrak{H}_g$, such that $\operatorname{Sp}_{2g}(\mathbb{Q})$ acts on it extending the action on $\mathfrak{H}_g$. Then the Satake compactification of $\Gamma \backslash \mathfrak{H}_g$ will be $\Gamma \backslash \mathfrak{H}_g^*$ as a topological space.

3.4.1. For any $r \in \mathbb{N}$, $0 \leq r \leq g$, let $N_{r,g} \subset \operatorname{Sp}_{2g}$ be the parabolic subgroup scheme such that for any ring R

$$N_{r,g}(R) = \left\{ \begin{pmatrix} A_1 & 0 & B_1 & B_{12} \\ A_{21} & A_2 & B_{21} & B_2 \\ C_1 & 0 & D_1 & D_{12} \\ 0 & 0 & 0 & D_2 \end{pmatrix} \in \operatorname{Sp}_{2g}(R) \right\}.$$

There is a natural homomorphism $p_{r,g} \colon N_{r,g} \to \operatorname{Sp}_{2r}$, which sends a typical element as above to

$$\begin{pmatrix} A_1 & B_1 \\ C_1 & D_1 \end{pmatrix} \in \operatorname{Sp}_{2r}(R).$$

3.4.2. We define $\mathfrak{H}_g^*$ set-theoretically as follows: $\mathfrak{H}_g^* = \{(\gamma, \Omega) \mid \gamma \in \operatorname{Sp}_{2g}(\mathbb{Q}), \Omega \in \mathfrak{H}_r \text{ for some } r \in \mathbb{N}, 0 \leq r \leq g\}$ modulo an equivalence relation R, where $(\gamma_1, \Omega_1) \sim_R (\gamma_2, \Omega_2)$ for $\Omega_1 \in \mathfrak{H}_r$, $\Omega_2 \in \mathfrak{H}_{r'}$ if and only if $r = r'$ and $\Omega_2 = p_{r,g}(\gamma_2^{-1}\gamma_1)\Omega_1$. $\operatorname{Sp}_{2g}(\mathbb{Q})$ acts on $\mathfrak{H}_g^*$ by multiplying the first component of any representative. For any $r \in \mathbb{N}$, $0 \leq r \leq g$, $\mathfrak{H}_r$ is embedded in $\mathfrak{H}_g^*$ by $\Omega \mapsto (1, \Omega)$ for $\Omega \in \mathfrak{H}_r$. This gives a set-theoretic embedding $\mathscr{F}_g^* \subset \mathfrak{H}_g^*$. The image of an embedded $\mathfrak{H}_r$ under some $\gamma \in \operatorname{Sp}_{2g}(\mathbb{Q})$ is called a (rational) boundary component of $\mathfrak{H}_g^*$.

3.4.3. We can now define the *Satake topology* on $\mathfrak{H}_g^*$. For any point $x \in \mathfrak{H}_g^*$, a fundamental system of neighborhoods of x is given by the family consisting of subsets $U \subset \mathfrak{H}_g^*$ such that:

(a) For all $\gamma \in \operatorname{Sp}_{2g}(\mathbb{Q})$, $\gamma U \cap \mathscr{F}_g^*$ is an open neighborhood of x in $\mathscr{F}_g^*$.
(b) For all $\gamma \in \operatorname{Sp}_{2g}(\mathbb{Z})$ such that $\gamma x = x$, we have $\gamma U = U$.

3.5. Proposition. *Given any subgroup Γ of finite index in $\text{Sp}_{2g}(\mathbb{Z})$, the Satake topology can be characterized as the unique topology $\mathcal{T}$ extending the usual metric topology of $\mathfrak{H}_g$ which satisfies the following properties:*

(i) *$\mathcal{T}$ induces on $\mathscr{F}_g^*$ the topology defined in 3.3.2.*
(ii) *Any $\gamma \in \text{Sp}_{2g}(\mathbb{Q})$ acts on $\mathfrak{H}_g^*$ as a homeomorphism with respect to $\mathcal{T}$.*
(iii) *If x, $x' \in \mathfrak{H}_g^*$ are not equivalent with respect to Γ, then there exists a neighborhood U, U' of x, x' such that $\Gamma \cdot U \cap U' = \phi$.*
(iv) *For all $x \in \mathfrak{H}_g^*$, there exists a fundamental system of neighborhoods $\{U_i\}_{i \in I}$ of x such that $\gamma U_i = U_i$ if $\gamma x = x$, and $\gamma U_i \cap U_i = \phi$ if $\gamma x \neq x$.*

3.6. Theorem. *For any subgroup Γ of finite index in $\text{Sp}_{2g}(\mathbb{Z})$:*

(i) *$\Gamma \backslash \mathfrak{H}_g^*$ is a compact Hausdorff space.*
(ii) *$(\Gamma \backslash \mathfrak{H}_g^*) \smallsetminus (\Gamma \backslash \mathfrak{H}_g)$ has a natural finite stratification. Each stratum is a quotient of some $\mathfrak{H}_r$ ($0 \leq r \leq g$) by some subgroup of finite index in $\mathfrak{H}_r$.*
(iii) *Each stratum of $\Gamma \backslash \mathfrak{H}_g^*$ has a natural structure of normal (complex) analytic space.*

3.7. In view of 3.6(iii), in order to give $\Gamma \backslash \mathfrak{H}_g^*$ the structure of a normal analytic space, we only have to "piece the strata together". There is a theorem of H. Cartan on prolongation of analytic spaces which is exactly what we need. So we digress to discuss it in some detail.

3.7.1. Theorem. *Let V be a locally compact, second countable Hausdorff space, which is a set-theoretical disjoint, locally finite union of countably many subspaces $V_0, V_1, \ldots$, such that each V_i is an irreducible normal analytic space. Let $\mathcal{O}_V$ be the sheaf of germs of analytic functions on V, where restriction to each stratum (as continuous function on that stratum) is again continuous. Assume*

(1) *$\dim V_i < \dim V_0$ for all $i > 0$, and V_0 is dense in V.*
(2) *For all $d \in \mathbb{N}$, $\bigcup_{\dim V_i \leq d} V_i$ is closed in V.*
(3) *$\mathcal{O}_V|_{V_i} = \mathcal{O}_{V_i}$, the structure sheaf of V_i. Here $\mathcal{O}_V|_{V_i}$ is the restriction of continuous functions on V to continuous functions on V_i.*
(4) *For all $x \in V$ there exists a fundamental system of open neighborhoods $\{U_\alpha\}$ of x such that each $U_\alpha \cap V_0$ is connected.*
(5) *For all $x \in V$ there exists an open neighborhood U_x of x such that $\Gamma(U_x, \mathcal{O}_V)$ separates points of U_x.*

The $(V, \mathcal{O}_V)$ is an irreducible normal analytic space, and for every $d \leq \dim V_0$, $\bigcup_{\dim V_i \leq d} V_i$ is a closed analytic subspace of V with dimension equal to $\max_{\dim V_i \leq d} \dim V_i$.

3.7.2. Let us try to explain why the theorem is true for the case where there are only two strata: $V = V_0 \amalg V_1$. Obviously, if V has the structure of normal analytic space, the sheaf $\mathcal{O}_V$ defined above is the structure sheaf of V.

The main tool to prove the theorem is the following theorem of Remmert and Stein:

(*) *Let Y be an analytic subset of an open subset D in $\mathbb{C}^N$ with dimension $\leq p$. Suppose W is a closed analytic subset of $D\setminus Y$ purely of dimension n. If $n > p$, then the closure $\overline{W}$ of W in D is an analytic subset of D, purely of dimension n.*

For any point $x \in V_1$, the assumptions allow one to find a small neighborhood U of z_0 in V, and a proper continuous map $f: U \to U'$, where U' is open in $\mathbb{C}^N$, such that:

(a) the coordinate functions $f_1, \ldots, f_N$ of f are in $\Gamma(U, \mathcal{O}_V)$.
(b) $f^{-1}(0) \cap U = \{x\}$.
(c) f induces an isomorphism of the reduced analytic subspace $V_1 \cap U$ to its image Z in U'.
(d) $K := f^{-1}(Z') \cap U \cap V_0$ is an analytic subset of $U \cap V$, and $\dim_x K < n$ for all $x \in K$.

Now, $W := f(U \cap V_0 \setminus K)$ is an analytic subset of $U' \setminus Z$ because f is proper. $f: (U \cap V_0) \setminus K \to W$ is a finite morphism because f is proper and $\Gamma(U, \mathcal{O}_V)$ separates points of $U \cap V_0$. Thus $\dim W = \dim V_0$ and one can apply the theorem of Remmert–Stein to conclude that $f(U)$ is an analytic subset of U'. Then it is easy to see that f induces an isomorphism from U to the normalization of $f(U)$.

3.8. We would like to apply Theorem 3.7.1. to give $\Gamma \backslash \mathfrak{H}_g^*$ the structure of a normal analytic space. Conditions (1), (2) are obviously satisfied. Condition (4) is topological and not hard to verify. In order to check the analytic conditions (3) and (5), we need the Poincaré–Eisenstein series and Φ-operator. The theory of Poincaré–Eisenstein series produces enough modular form, but we will not give any details here.

3.8.1. The $\mathrm{Sp}_{2g}(\mathbb{Q})$-equivariant line bundle ω on $\mathfrak{H}_g$ extends to an $\mathrm{Sp}_{2g}(\mathbb{Q})$-equivariant line bundle ω_g^* on $\mathfrak{H}_g^*$. A modular form f of weight k with respect to Γ extends to a Γ-invariant section f^* of $\omega^{*\otimes k}$. Let F be a boundary component of $\mathfrak{H}_g^*$, i.e. the image of $\{(\gamma, \Omega) | \Omega \in \mathfrak{H}_r\}$ for some $\gamma \in \mathrm{Sp}_{2g}(\mathbb{Q})$ and $0 \leq r \leq g$. The Φ-operator Φ_F sends f to the restriction of f^* on F. These operators help one to verify condition (3) of 3.7.1, and projectivity.

3.8.2. Theorem. (i) *$\Gamma \backslash \mathfrak{H}_g^*$ has the structure of compact normal analytic space. The strata of $\Gamma \backslash \mathfrak{H}_g^*$ are all (locally closed) analytic subspaces, where are all of the form $\Gamma' \backslash \mathfrak{H}_r, 0 \leq r \leq g$.*
(ii) *$R(\Gamma)$ defines an isomorphism $\Gamma \backslash \mathfrak{H}_g^* \simeq \mathrm{Proj}\, R(\Gamma)(\mathbb{C})$ as analytic spaces. Thus $\Gamma \backslash \mathfrak{H}_g^*$ is projective.*

3.8.3. Remarks. (a) It is not always possible to descend ω^* to a line bundle $\omega_{g,\Gamma}^*$ on $\Gamma \backslash \mathfrak{H}_g^*$. In fact, if $\gamma \in \Gamma$ fixes $x \in \mathfrak{H}_g^*$, then γ acts on the fibre of $\omega_{g,\Gamma}^*$ by a root of 1. If we raise ω_g^* to some power n which kills all such roots of 1, then $\omega_g^{*\otimes n}$ descends to $\Gamma \backslash \mathfrak{H}_g^*$.

(b) If $g \geq 2$, the "boundary" $(\Gamma\backslash\mathfrak{H}_g^*)\backslash(\Gamma\backslash\mathfrak{H}_g)$ has codimension $g \geq 2$. This explains, *a posteriori*, why Koecher's principle is true.

(c) One can show that $R(\Gamma)$ is spanned by the subspace $R(\Gamma)_\mathbb{Q}$ of modular forms with all Fourier coefficients in $\mathbb{Q}$. Thus $R(\Gamma) = R(\Gamma)_\mathbb{Q} \otimes \mathbb{C}$, and $\text{Proj}(R(\Gamma)_\mathbb{Q})$ gives a $\mathbb{Q}$-structure on $\Gamma\backslash\mathfrak{H}_g$. When $\Gamma = \Gamma_g(n)$, the q-expansion principle says that this $\mathbb{Q}$-structure coincides with the one given by $\mathscr{A}_{g,n}^* \times_{\text{Spec }\mathbb{Z}} \text{Spec }\mathbb{Q}$.

§4. Toroidal Compactification

The Satake compactification $\Gamma\backslash\mathfrak{H}_g^*$ is in some sense the minimal compactification of $\Gamma\backslash\mathfrak{H}_g$. Although it is a natural object to study, its complicated singularities at the boundary present severe obstacles. In [SC], Mumford and his coworkers constructed an explicit desingularization of the Baily–Borel compactification of an arithmetic quotient of bounded symmetric domains. We will describe this construction in the Siegel case.

4.1. The local coordinates of toroidal compactification are very easy to describe. It suffices to do so for the standard rational boundary components $\mathscr{F}_r$ = image of $1 \times \mathfrak{H}_r$ in $\mathfrak{H}_g^*$, $0 \leq r < g$.

4.1.1. Given r, $0 \leq r < g$, write an element $\Omega \in \mathfrak{H}_g$ in $(r, g-r)$-block form:

$$\Omega = \begin{pmatrix} t & w \\ {}^tw & \tau \end{pmatrix} \begin{matrix} r \\ g-r \end{matrix}$$
$$\ \ r \ \ g-r$$

Let

$$D_r = \left\{ \begin{pmatrix} t & w \\ {}^tw & \tau \end{pmatrix} \in M_g(\mathbb{C}) \,\bigg|\, t \in \mathfrak{H}_r, \tau \in M_{g-r}(\mathbb{C}), {}^t\tau = \tau \right\},$$

$$\mathfrak{H}_g = \left\{ \begin{pmatrix} t & w \\ {}^tw & \tau \end{pmatrix} \in D_r \,\bigg|\, \text{Im } \tau - {}^t(\text{Im } w)(\text{Im } t)^{-1}(\text{Im } w) \gg 0 \right\}.$$

4.1.2. Let $U_r \subseteq \text{Sp}_{2g}$ be the $\mathbb{Q}$-subgroup scheme such that

$$U_r(\mathbb{Q}) = \left\{ \begin{pmatrix} 1_r & 0 & 0 & 0 \\ 0 & 1_{g-r} & 0 & b \\ 0 & 0 & 1_r & 0 \\ 0 & 0 & 0 & 1_{g-r} \end{pmatrix} \,\bigg|\, b \in M_{g-r}(\mathbb{Q}), b = {}^tb \right\}.$$

Clearly U_r is a vector group, and can be identified with its Lie algebra. We give U_r the $\mathbb{Z}$-structure such that $U_r(\mathbb{Z}) = \Gamma \cap U(\mathscr{F}_r)(\mathbb{Q})$. Let $U_r(\mathbb{Z})^* := \text{Hom}_\mathbb{Z}(U_r(\mathbb{Z}), \mathbb{Z})$.

4.1.3. Inside $U_r(\mathbb{R})$ lies the positive cone C_{g-r} and its closure $\overline{C_{g-r}}$. Let $\sigma \in \overline{C_{g-r}}$ be a top-dimensional cone generated by a $\mathbb{Z}$-basis $\{\xi_1, \ldots, \xi_n\}$ of $U_r(\mathbb{Z})$, $n = \binom{g-r+1}{2}$, i.e. $\sigma = \sum_{i=1}^n \mathbb{R}_{\geq 0} \xi_i$. Let $\xi_1, \ldots, \xi_j \in C_{g-r}$, $\xi_{j+1}, \ldots, \xi_n \in \overline{C_{g-r}} \setminus C_{g-r}$. Let $\{l_1, \ldots, l_n\} \in U_r(\mathbb{Z})^*$ be the dual basis of $\{\xi_i\}$.

4.1.4. $U_r(\mathbb{Z}) \otimes_{\mathbb{Z}} \mathbb{C}^* = \mathrm{Hom}_{\mathbb{Z}}(U_r(\mathbb{Z})^*, \mathbb{C}^*)$ is an algebraic form isomorphic to $(\mathbb{C}^*)^n$. A coordinate system of $U_r(\mathbb{Z}) \otimes_{\mathbb{Z}} \mathbb{C}^*$ is given by $z_i = \exp(2\pi\sqrt{-1}\, l_i)$. Let $\exp: U_r(\mathbb{C}) \to U_r(\mathbb{Z}) \otimes_{\mathbb{Z}} \mathbb{C}^*$ be the exponential map.

4.1.5. Identify $M_{r, g-1}(\mathbb{C})$ with $\mathbb{C}^k$, $k = r(g-r)$, and D_r with $U_r(\mathbb{C}) \times \mathbb{C}^k \times \mathscr{F}_r$. Let (τ, w, t) be the coordinates on D_r, (z, w, t) be the coordinates on $(U_r(\mathbb{Z}) \otimes \mathbb{C})^* \times \mathbb{C}^k \times \mathscr{F}_r$. We have

$$\begin{array}{ccc} \mathfrak{H}_g & \subset U_r(\mathbb{C}) \times \mathbb{C}^k \times \mathscr{F}_r & (\tau, w, t) \\ \downarrow & \downarrow \exp \times \mathrm{id} \times \mathrm{id} & \downarrow \\ U_r(\mathbb{Z})\backslash \mathfrak{H}_g & \subset (U_r(\mathbb{Z}) \otimes \mathbb{C}^*) \times \mathbb{C}^k \times \mathscr{F}_r & (\exp(2\pi\sqrt{-1}\, l_i(\tau)), w, t) \\ \downarrow & & \\ \Gamma\backslash \mathfrak{H}_g & & \end{array}$$

4.1.6. Define $(U_r(\mathbb{Z})\backslash \mathfrak{H}_g)^{\sim} = \{x \in (U_r(\mathbb{Z}) \otimes \mathbb{C}^*) \times \mathbb{C}^k \times \mathscr{F}_r) \mid$ there exists a neighborhood U of x such that $U \cap (U_r(\mathbb{Z}) \otimes \mathbb{C}^*) \times \mathbb{C}^k \times \mathscr{F}_r \subseteq U_r(\mathbb{Z})\backslash \mathfrak{H}_g\} = $ interior of closure of $U_r(\mathbb{Z})\backslash \mathfrak{H}_g$ in $(U_r(\mathbb{Z}) \otimes \mathbb{C}^*) \times \mathbb{C}^k \times \mathscr{F}_r$. Note that $\bigcup_{i=1}^j \{(z, w, t) \mid z_i = 0\}$ is contained in $(U_r(\mathbb{Z})\backslash \mathfrak{H}_g)^{\sim}$. $(U_r(\mathbb{Z})\backslash \mathfrak{H}_g)^{\sim}$ determines a local coordinate system of a typical toroidal embedding $\overline{(\Gamma\backslash \mathfrak{H}_g)}$. (If $\Gamma \subseteq \Gamma_g(n)$ for some $n \geq 3$, there is a natural étale map $(U_r(\mathbb{Z})\backslash \mathfrak{H}_g)^{\sim} \to \overline{(\Gamma\backslash \mathfrak{H}_g)}$.)

4.2. We give a very rough description of the toroidal compactification. Assume that $\Gamma \subset \Gamma_g(n)$ for some $n \geq 3$. As we saw in the above construction, given a boundary component $\mathscr{F}_r$ and a rational polyhedral cone σ in some cone $\overline{C(\mathscr{F}_r)}$ naturally attached to $\mathscr{F}_r$, there is a space $X_{\mathscr{F}_{r,\sigma}} = (U_r(\mathbb{Z})\backslash \mathfrak{H}_g)^{\sim}$ giving a "partial completion". The theory of toroidal embeddings [TE] tells us that if we fix $\mathscr{F}_r$, a systematic way to produce compatible $X_{\mathscr{F}_{r,\sigma}}$'s for gluing is to take a decomposition of $\overline{C_r} = \overline{C(\mathscr{F}_r)}$ into polyheral cones $\{\sigma_\alpha^{\mathscr{F}_r}\}_{\alpha \in I_\alpha}$.

4.2.1. Let $G_l(\mathscr{F}_r)$ be the $\mathbb{Q}$-subgroup scheme of Sp_{2g} whose $\mathbb{Q}$-points are

$$\left\{ \begin{pmatrix} 1_r & 0 & 0 & 0 \\ 0 & u & 0 & 0 \\ 0 & 0 & 1_r & 0 \\ 0 & 0 & 0 & {}^t u^{-1} \end{pmatrix} = u \in \mathrm{GL}_{g-r}(\mathbb{Q}) \right\},$$

which acts trivially on $\mathscr{F}_r$, and acts on U_r and $\overline{C_r}$ via conjugation. Explicitly,

$u \in G_l(\mathcal{F}_r)(\mathbb{Q})$ sends $b \in U_r(\mathbb{Q})$ to ub^tu. Since $\Gamma \cap G_l(\mathcal{F}_r)(\mathbb{Q})$ acts on C_r, we had better choose $\{\sigma_\alpha^{\mathcal{F}_r}\}$ to be invariant under $\Gamma \cap G_l(\mathcal{F}_r)(\mathbb{Q})$ (requiring $\{\sigma_\alpha^{\mathcal{F}_r}\}$ to be invariant under $\mathrm{GL}_{g-r}(\mathbb{Z})$ will guarantee this). The classical reduction theory for positive definite quadratic forms says that such Γ-invariant polyhedral cone decompositions $\{\sigma_\alpha^{\mathcal{F}_r}\}$ always exist, and modulo $\Gamma \cap G_l(\mathcal{F}_r)(\mathbb{Q})$ there are only finitely many cones. Using such a "Γ-admissible" rational polyhedral cone decomposition $\Sigma_{\mathcal{F}} = \{\sigma_\alpha\}$ for $\overline{C(\mathcal{F}_r)}$, we can glue the $X_{\mathcal{F}_r, \sigma_\alpha}$'s together to form a space $X'_{\Sigma_{\mathcal{F}_r}}$ and a map $X'_{\Sigma_{\mathcal{F}_r}} \to U_{\mathcal{F}_r}(\mathbb{Z}) \backslash \mathfrak{H}_g^*$. (The assumption in 4.1.3 that σ is top-dimensional and is generated by a $\mathbb{Z}$-basis is not essential. The theory of toroidal embeddings tells us what to do for a general rational polyhedral cone.)

4.2.2. Let N_r be the $\mathbb{Q}$-subgroup scheme of Sp_{2g} whose $\mathbb{Q}$-points are

$$\left\{ \begin{pmatrix} A_{11} & 0 & B_{11} & * & r \\ * & A_{22} & * & * & g-r \\ C_{11} & 0 & D_{11} & * & r \\ 0 & 0 & 0 & {}^tA_{22}^{-1} & g-r \\ r & g-r & r & g-r & \end{pmatrix} \in \mathrm{Sp}_{2g}(\mathbb{Q}) \right\}.$$

Let $\Gamma_{\mathcal{F}_r} = \Gamma \cap N_r$. A basic fact is that $\Gamma_{\mathcal{F}_r}/U_r(\mathbb{Z})$ acts properly discontinuously on $X'_{\Sigma_{\mathcal{F}_r}}$. Let $X_{\Sigma_{\mathcal{F}_r}}$ be the quotient. Thus we have a diagram

$$\begin{array}{ccc} \Gamma_{\mathcal{F}_r} \backslash \mathfrak{H}_g & \to & \Gamma \backslash \mathfrak{H}_g \\ \cap & & \cap \\ X_{\Sigma_{\mathcal{F}_r}} & \to & \Gamma \backslash \mathfrak{H}_g^*. \end{array}$$

4.2.3. We can do the same thing for any rational boundary component. So we choose a Γ-admissible rational polyhedral cone decomposition $\Sigma_{\mathcal{F}} = \{\sigma_\alpha^{\mathcal{F}}\}$ of $\overline{C(\mathcal{F})} \subseteq U_{\mathcal{F}}(\mathbb{R})$ for each rational boundary component $\mathcal{F}$ of $\mathfrak{H}_g^*$. We require this collection of $\Sigma_{\mathcal{F}}$'s to be compatible with the natural Γ-action, and call such a collection again a Γ-admissible polyhedral cone decomposition.

Let $W = \coprod_{\mathcal{F}} X_{\Sigma_{\mathcal{F}}}$ (disjoint union). There is a natural Γ action on W. Also, for any two boundary components $\mathcal{F}, \mathcal{F}'$ with $\mathcal{F} \subset \mathcal{F}'$, there is an étale map $X_{\Sigma_{\mathcal{F}'}} \to X_{\Sigma_{\mathcal{F}}}$. Together with the Γ-action, we get an equivalence relation R on W. R is in fact represented by a closed graph $\Delta \subset W \times W$.

Passing to the quotient, we get a Hausdorff analytic variety $\tilde{X}_\Sigma$. This is the toroidal compactification.

4.3. Theorem (Notation as above).

(i) $\tilde{X}_\Sigma$ is the unique Hausdorff analytic space containing $\Gamma \backslash \mathfrak{H}_g$ as an open dense subset such that:

(a) *for every rational boundary component $\mathscr{F}$ of $\mathfrak{H}_g^*$, there is an open analytic morphism $\pi_\mathscr{F}$ making the following diagram commutative*

$$\begin{array}{ccc} U_\mathscr{F}(\mathbb{Z})\backslash\mathfrak{H}_g & \hookrightarrow & X'_{\Sigma_\mathscr{F}} \\ \downarrow & & \downarrow \pi_\mathscr{F} \\ \Gamma\backslash\mathfrak{H}_g & \hookrightarrow & \tilde{X}_\Sigma \end{array}$$

(b) *every point of $\tilde{X}_\Sigma$ is the image of at least one of the maps $\pi_\mathscr{F}$.*
(ii) *$\tilde{X}_\Sigma$ is a compact normal algebraic space.*
(iii) *there exists a natural morphism from $\tilde{X}_\Sigma$ to $\Gamma\backslash\mathfrak{H}_g^*$ inducing the identity morphism on $\Gamma\backslash\mathfrak{H}_g$.*

4.3.1. Remark. $\tilde{X}_\Sigma$ is not a projective variety in general. However, a theorem of Y. S. Tai shows that if Σ has certain convexity properties, then $\tilde{X}_\Sigma$ is projective. In fact, $\tilde{X}_\Sigma$ is the blowing-up of $\Gamma\backslash\mathfrak{H}_g^*$ along the coherent sheaf of ideals defined by cusp forms vanishing along the boundary to sufficiently divisible order. Of course, the order of vanishing is defined in terms of Σ. In [Ch] this blowing-up is carried out over $\mathbb{Z}[\frac{1}{2}]$, via algebraic theta constants, to *construct* an arithmetic version of the toroidal compactification over $\mathbb{Z}[\frac{1}{2}]$.

4.4. Now we have two compactifications: Satake and toroidal. A natural question is: What is the moduli-theoretic meaning of the boundary points? We will indicate a (partial) answer via degeneration.

4.4.1. Let $\Delta = \{t \in \mathbb{C} \,|\, |t| < 1\}$, $\Delta^* = \Delta\backslash\{0\}$. Let $f: \Delta \to \Gamma_g\backslash\mathfrak{H}_g^*$ be a holomorphic map such that $f(\Delta) \subset \Gamma_g\backslash\mathfrak{H}_g$. After a base change $t \mapsto t^n$, we may (and do) assume that:

(a) there is a family of principally polarized abelian variety $\mathscr{X}^* \xrightarrow{\pi^*} \Delta^*$ giving rise to $f|_{\Delta^*}$.
(b) $\mathscr{X}^* \to \Delta^*$ has semi-stable reduction, i.e. we can extend $\mathscr{X}^*$ (uniquely) to a holomorphic family of algebraic groups $\mathscr{X} \xrightarrow{\pi} \Delta$, such that $X_0 = \pi^{-1}(0)$ is a semi-abelian variety, i.e. an extension of an abelian variety B_0 of dimension r by an algebraic torus T of dimension $g - r$.

It turns out that B_0 inherits a principal polarization, and the corresponding point in $\Gamma_r\backslash\mathfrak{H}_r$ is just $f(0)$. This gives an interpretation of boundary points of $\Gamma_g\backslash\mathfrak{H}_g^*$. In other words, principally polarized semi-abelian varieties with isomorphic abelian parts are identified under the Satake compactification.

The $\Gamma_g(n)\backslash\mathfrak{H}_g^*$'s can be treated similarly. The precise formulation is left to the reader.

4.4.2. It is not easy to give a really satisfactory modular interpretation for the toroidal compactifications. After all, they depend on a choice of the combinatorial data "Γ-admissible rational polyhedral cone decomposition", hence are far from been canonical. But one can construct a family of semi-abelian varieties over the toroidal compactification, and this can actually be done over $\mathbb{Z}$ (cf. [Fa 2]). We will try to give some idea about the relation between the toroidal coordinates and degeneration of abelian varieties. For a Hodge theoretic point of view, see [C–C–K].

4.4.3. We use the notation in 4.4.1. Let $\mathscr{X}^* \xrightarrow{\pi^*} \Delta^*$ and $\mathscr{X} \xrightarrow{\pi} \Delta$ satisfy 4.4.1(a), (b). There exists a family of semi-abelian varieties $\tilde{\mathscr{X}} \xrightarrow{\tilde{\pi}} \Delta$ of constant tori rank $g - r$ such that:

(i) $\tilde{\pi}^{-1}(0)$ is (canonically identified with) $\pi^{-1}(0)$;
(ii) there is a family of discrete subgroups $\Lambda \subset \tilde{\mathscr{X}}|_{\Delta^*}$, $\Lambda(t) \cong \mathbb{Z}^{g-r}$ for all $t \in \Delta$, such that $\mathscr{X}^* \to \Delta^*$ is the quotient of $\tilde{\mathscr{X}}|_{\Delta^*}$ by Λ.

The semi-abelian family $\tilde{\mathscr{X}} \to \Delta$ fits uniquely into an exact sequence $0 \to T \to \tilde{\mathscr{X}} \to B \to 0$, where B is an abelian family and T is a family of algebraic torus. Clearly, the fibre of B over $0 \in \Delta$ is the B_0 in 4.4.1. Since an algebraic torus has no moduli, T is isomorphic to $\mathbb{G}_m^{g-r}$ over Δ. Thus $\tilde{\mathscr{X}}$ is determined by the abelian family B and an extension of B by $\mathbb{G}_m^{g-r}$.

4.4.4. It is now easy to relate the above data and toroidal coordinates. We use the expression $\Omega = \begin{pmatrix} t & w \\ {}^t w & \tau \end{pmatrix} \begin{matrix} r \\ g-r \end{matrix}$ as in 4.1. The variable $t \in \mathfrak{H}_r$ corresponds to the moduli of B. The variable w gives the extension class. Finally, the variable τ gives the "periods" Λ. How about the cone σ? Well, it determines the "direction" of degeneration.

4.4.5. Remark. A naive illustration: take a family of g periods $\lambda_1(t), \ldots, \lambda_g(t)$ in $(\mathbb{C}^*)^g$, $t \in \Delta^*$, such that $(\mathbb{C}^*)^g / \langle \lambda_i(t) \rangle$ is a (principally polarized) abelian family $\mathscr{X}^*$. Assume that exactly $g - r$ periods $\lambda_{r+1}(t), \ldots, \lambda_g(t)$ degenerate. Then $(\mathbb{C}^*)^g / \langle \lambda_1(t), \ldots, \lambda_r(t) \rangle$ extends to a semi-abelian family $\tilde{X}$ of constant tori rank $g - r$ over Δ. $\lambda_{r+1}(t), \ldots, \lambda_g(t)$ determine $g - r$ periods of $\mathscr{X}^*|_{\Delta^*}$, and the quotient $(\mathscr{X}^*|_{\Delta^*}) / \langle \lambda_{r+1}, \ldots, \lambda_g \rangle$ is $\mathscr{X}^*$.

§5. Modular Heights

5.1. Let us first recall Faltings' definition of height of an abelian variety defined over a number field. Let X_K be an abelian variety of dimension g over a number field K, and let $N(X_K) \xrightarrow{\varepsilon} \text{Spec } \mathcal{O}_K$ be its Néron model. Let $\omega_{X_K} = \varepsilon^*(\Omega^g_{N(X_K)/\mathcal{O}_K})$. ω_{X_K} is an invertible sheaf on $\text{Spec } \mathcal{O}_K$, i.e. a projective rank 1 $\mathcal{O}_K$-module. $\omega_{X_K} \otimes_{\mathcal{O}_K} K$ is canonically isomorphic to $\Gamma(X_K, \Omega^g_{X_K/K})$. For each infinite prime v, put a metric on $\omega_{X_K} \otimes K_v$ by $\|\alpha\|_v^2 = 2^{-g} \int_{X_v(\mathbb{C})} |\alpha \wedge \bar{\alpha}|$ for all $\alpha \in \omega_{X_K} \otimes K$. The general theory of heights [Si] then allows us to define $h(\omega_{X_K})$. The height of X_K is by definition equal to $h(\omega_{X_K})$. Explicitly, pick $\alpha \in \omega_{X_K}, \alpha \neq 0$, then

$$h(X_K) = 1/[K:\mathbb{Q}]\{\log \#(\omega_{X_K}/\alpha \mathcal{O}_K) - \sum_{\infty | v} \varepsilon_v \log \|\alpha\|_v\},$$

where $\varepsilon_v = 1$ or 2 according to whether ε_v is real or complex.

5.1.1. Remark. Since the operation of forming Néron models does not commute with base change of number fields in general, we cannot expect $h(X_K)$ to be invariant under base change. However, for any finite extension L of K, we have $h(X_L) \leq h(X_K)$, where $X_L = X_K \times_{\text{Spec } K} \text{Spec } L$. The reason: by the universal property of Néron model, there is a canonical map $N(X_K) \times_{\text{Spec } \mathcal{O}_K} \text{Spec } \mathcal{O}_L \xrightarrow{f} N(X_L)$. Thus we obtain an injection $\omega_{X_K} \xrightarrow{f^*} \omega_{X_K} \otimes_{\mathcal{O}_K} \mathcal{O}_L$ of projective rank-1 $\mathcal{O}_L$-modules. The inequality $h(X_L) \leq h(X_K)$ follows.

5.1.2. Remark. If X_K has semi-stable reduction over $\mathcal{O}_K$, then for any finite extension L of K, $N(X_K)^0 \times_{\text{Spec } \mathcal{O}_K} \text{Spec } \mathcal{O}_L$ is canonically isomorphic to $N(X_L)^0$, see [Ar 3]. Hence $h(X_L) = h(X_K)$.

5.1.3. Let X_K be an abelian variety over a number field. By the semi-stable reduction theorem, there is a finite extension L of K such that $X_L := X_K \times_{\text{Spec } K} \text{Spec } L$ has semi-stable reduction. Then we can define the geometric height (or modular height) of X_K to be $h_{\text{geom}}(X_K) := h(X_L)$. This definition is independent of the choice of L by 5.1.2.

5.2. Theorem 1 of [Fa 1] asserts that given any number field K, any number $C \in \mathbb{R}$, there are only finitely many isomorphism classes of principally polarized abelian varieties X over K with $h_{\text{geom}}(X) \leq C$. By 5.1.1, the above statement remains true if we change $h_{\text{geom}}(X)$ to $h(X)$.

Faltings used moduli of stable curves and abelian varieties to prove this theorem. A more natural proof would not involve the moduli of curves. What is needed is a nice compactification theorem over $\mathbb{Z}$. Since such results were not available at the time, Faltings resorted to the compact moduli of stable curves. More recently Faltings has described a way to compactify the moduli stack of principally polarized abelian varieties, together with a semi-abelian family over it (cf. [Fa 2]).

We will show later in this section that the metric of the invertible sheaf ω on $\mathscr{A}_g$ (defined by integration as in 5.1) has only a logarithmic singularity along the boundary. (Actually, ω is not defined on $\mathscr{A}_g(\mathbb{C})$, only some power $\omega^{\otimes n}$ of ω is defined on $\mathscr{A}_g(\mathbb{C})$. So we really should say that the metric of any such $\omega^{\otimes n}$ has a logarithmic singularity.)

We would like to explain the nice compactification theory and logarithmic singularity which imply the finiteness theorem.

SIEGEL MODULI SCHEMES AND THEIR COMPACTIFICATIONS OVER $\mathbb{C}$ 249

5.3. We first summarize what we need from the theory of arithmetic compactification. Since we do not want to scare the general readers away with algebraic stacks, we will pretend that we have:

(a) A proper scheme $\bar{\mathscr{A}}_g$ over Spec $\mathbb{Z}$, containing $\mathscr{A}_g$ as a dense open subscheme. (Actually, $\bar{\mathscr{A}}_g$ is an arithmetic version of toroidal compactification. In particular, $\bar{\mathscr{A}}_g(\mathbb{C}) = \tilde{X}_\Sigma$ for some Γ_g-admissible polyhedral cone decomposition. But we do not need this here.)

(b) A semi-abelian scheme $\bar{\mathscr{X}}_g \overset{\pi}{\underset{\varepsilon}{\rightleftarrows}} \bar{\mathscr{A}}_g$ (i.e. each fibre is an extension of an abelian variety of a torus) extending the universal family over $\mathscr{A}_g$.

Let $\omega = \varepsilon^*(\Lambda^g \Omega^1_{\bar{\mathscr{X}}_g/\bar{\mathscr{A}}_g})$. The global sections of powers of ω define a morphism $\bar{\mathscr{A}}_g \overset{f}{\to} \mathscr{A}_g^* = \mathrm{Proj}(\bigoplus_{k \in \mathbb{N}} \Gamma(\bar{\mathscr{A}}_g, \omega^{\otimes k}))$. Note that $\mathscr{A}_g^*(\mathbb{C}) = \Gamma \backslash \mathfrak{H}_g^*$. Thus $\mathscr{A}_g^*$ is a $\mathbb{Z}$-model of the Satake compactification. As before, although ω does not descend to $\mathscr{A}_g^*$, some power $\omega^{\otimes n}$ does descend to an invertible sheaf $\mathscr{L}_n$ on $\mathscr{A}_g^*$ (i.e. $f^*\mathscr{L}_n = \omega^{\otimes n}$). $\mathscr{L}_n$ is ample on $\mathscr{A}_g^*$.

Let X_K be a principally polarized abelian variety over a number field K with semi-stable reduction over $\mathcal{O}_K$. X_K defines a K-point $[X_K]$ of $\mathscr{A}_g$. Because $\bar{\mathscr{A}}_g$ is proper, $[X_K]$ extends to an $\mathcal{O}_K$-valued point $\langle X_K \rangle$: Spec $\mathcal{O}_K \to \bar{\mathscr{A}}_g$. The pull-back of $\bar{\mathscr{X}}_g$ by $\langle X_K \rangle$ is a semi-abelian scheme extending X_K. Hence it is just $N(X_K)^0$ (see [Ar 3]). Thus $\omega_{X_K} = \langle X_K \rangle^* \omega$. Since f is defined by ω, we deduce that $h_{\mathrm{geom}}(X_K) = (1/n)(h_{\mathscr{L}_n}([X_K]))$. Since $\mathscr{L}_n$ has only logarithmic singularity, Proposition 8.2 of [Si] concludes the finiteness theorem.

5.4. We come back to the logarithmic singularity of $\mathscr{L}_n$.

5.4.1. First, we make a general remark. Let Z be a closed subvariety of an analytic space V such that $U = V \backslash Z$ is smooth. Let $\mathscr{L}$ be a line bundle on V with a hermitian metric h on U. Let $\tilde{V}$ be another analytic space and $f: \tilde{V} \to$ be a morphism inducing an isomorphism over U. Then $(\mathscr{L}, h)$ has logarithmic singularity along Z if and only if $(f^*\mathscr{L}, h)$ has logarithmic singularity along $f^{-1}(Z)$.

5.4.2. The remark above says that we only have to check logarithmic singularities on the toroidal compactifications. (This is a big advantage. Otherwise, how can we find defining equations for the boundary of $(\Gamma_g \backslash \mathfrak{H}_g^*)$?) Also, it is clear that we only have to check for standard cusps (i.e. images of standard boundary components). Let $r \in \mathbb{N}$, $0 \le r < g$, $x \in$ image of $\mathscr{F}_r$ in $\Gamma_g \backslash \mathfrak{H}_g^*$. We use the notation in 2.1. The basic observation is that $(dz_1 \wedge \cdots \wedge dz_g)^{\otimes n}$ extends to an invertible analytic section near x. This is a fact we mentioned in Section 3, and is essentially a consequence of the theory of Poincaré–Eisenstein series. A direct way to prove this is to use a degenerating family of abelian variety over a neighborhood of a, see [Mu 3], [Ch], [Fa 2].

5.4.3. Now we compute

$$2^{-g} \int_{\mathbb{C}^g/\mathbb{Z}^g \oplus \Omega \mathbb{Z}^g} |dz_1 \wedge \cdots \wedge dz_g \wedge \overline{dz}_1 \wedge \cdots \wedge \overline{dz}_g|$$

$$= \int_{\mathbb{C}^g/\mathbb{Z}^g \oplus \Omega \mathbb{Z}^g} dx_1 \wedge dy_1 \wedge \cdots \wedge dx_g \wedge dy_g$$

$$= \text{area of fundamental domain of } \mathbb{Z}^g \oplus \Omega \mathbb{Z}^g$$

$$= \det(\text{Im } \Omega).$$

Looking at the toroidal coordinates described in 4.1, one sees visibly that $\mathscr{L}_n$ has logarithmic singularities along the boundary.

REFERENCES

[Ar 1] Artin, M, Algebraization of formal moduli, I, in *Global Analysis* (in honor of K. Kodaira), Princeton Univ. Press: Princeton, NJ, 1970, pp. 21–71; II. *Ann. Math.*, **91** (1970), 88–135.

[Ar 2] Artin, M. Versal deformations and algebraic stacks. *Invent. Math.*, **27** (1974), 165–189.

[Ar 3] Artin, M. Néron Models, this volume, pp. 213–230.

[Ba 1] Baily, W. L. On Satake's compactification of V_n. *Amer. J. Math.*, **80** (1958) 348–364.

[Ba 2] Baily, W. L. On the theory of theta functions, the moduli of abelian varieties and the moduli of curves. *Ann. Math.*, **75** (1962), 342–381.

[B–B] Baily, W. L. and Borel, A. Compactification of arithmetic quotients of bounded symmetric domains. *Ann. Math.*, **84** (1966), 442–528.

[Bo] Borel, A. Some metric properties of arithmetic quotients of symmetric spaces and an extension theorem. *J. Diff. Geom.*, **6** (1972), 543–560.

[Ca] Cartan, H. et al. Fonctions automorphes. *Séminaire H. Cartan*, 1957/58.

[C–C–K] Carlson, J., Cattani, E. H. and Kaplan, A. G. Mixed Hodge structures and compactification of Siegel's space, in *Algebraic Geometry*, Angèrrs, 1979, pp. 77–105.

[Ch] Chai, C.-L. *Compactification of Siegel Moduli Schemes*. London Mathematical Society Lecture Notes Series 107, Cambridge University Press: Cambridge, 1985.

[Fa 1] Faltings, G. Endlichkeitssätze für abelsche Varietäten über Zahlkörpern. *Invent. Math.*, **73** (1983), 349–366.

[Fa 2] Faltings, G. *Arithmetische Kompaktifizierung des Modulraums der Abelschen Varietäten*. Lecture Notes in Mathematics, 1111, Springer-Verlag: New York, 1985, pp. 321–383.

[Fr] Freitag, E. *Siegelsche Modulfunktionen*. Springer-Verlag: New York.

[G] Gieseker, D. *Geometric Invariant Theory and Application to Moduli Problems*. Lecture Notes in Mathematics, 996. Springer-Verlag: New York, 1983, pp. 45–73.

[GIT] Mumford, D. and Fogarty, J. *Geometric Invarient Theory*, 2nd ed. Ergebnisse der Mathematik und ihrer Grenzgebiete, 34 (1982).

[I 1] Igusa, J. I. On the graded ring of theta constants, I, II. *Amer. J. Math.*, **86** (1974), 219–246; **88** (1966), 221–236.

[I 2] Igusa, J. I. *Theta Functions*, Die Grundlehren der mathematischen Wissenshaften, 194. Springer-Verlag: New York, 1972.

[Mi] Milne, J. S. Abelian varieties and Jacobian varieties, this volume, pp. 103–148 and 167–212.

[Mu 1] Mumford, D. On the equations defining abelian varieties, I, II, III. *Invent. Math.*, **1** (1966) 287–354; **3** (1967), 71–135, 215–244.

[Mu 2] Mumford, D. *Abelian Varieties*. Tata Institute of Fundamental Research, Studies in Mathematics, 5, Oxford University Press: Oxford, 1970.

[Mu 3] Mumford, D. An analytic construction of degenerate abelian varieties over complete rings. *Compos. Math.*, **24**, Fasc. 3 (1972), 239–272.

[Mu 4] Mumford, D. *Tata Lectures on Theta*, I, II. Birkhäuser-Verlag: Boston, 1983, 1984. (III to appear.)

[N 1] Namikawa, Y. A new compactification of the Siegel space and degeneration of abelian varieties, I, II. *Math. Ann.*, **221** (1976), 97–141, 201–241.

[N 2] Namikawa, Y. *Toroidal Compactification of Siegel Spaces*. Lecture Notes in Mathematics, 812. Springer-Verlag: New York, 1980.

[R] Rosen, M. Abelian varieties over $\mathbb{C}$, this volume, pp. 79–101.

[Sa] Satake, I. On the compactification of the Siegel space. *J. Indian Math. Soc.*, **20** (1956), 259–281.

[SC] Ash, R., Mumford, D., Rapoport, M. and Tai, Y.-S. *Smooth Compactification of Locally Symmetric Varieties*. Mathematical Science Press, 1975.

[Si] Silverman, J. The theory of heights, this volume, pp. 151–166.

[TE] Kempf, G., Knudson, F., Mumford, D. and Saint-Donald, B. *Toroidal Embeddings*, I. Lectures Notes in Mathematics, 338. Springer-Verlag: New York, 1973.

CHAPTER X

Heights and Elliptic Curves

JOSEPH H. SILVERMAN

Many of the deep results involving heights of abelian varieties become quite transparent in the case of elliptic curves. In this chapter we propose to prove some of these theorems for elliptic curves by using explicit Weierstrass equations. We will also point out how the height of an elliptic curve appears in various other contexts in arithmetical geometry.

We start in Section 1 by giving a formula for the height $h(E/K)$ of an elliptic curve E/K in terms of the minimal discriminant of E/K and the various periods of E. Next, in Section 2, we give an estimate for $h(E/K)$ in terms of the j-invariant of E. This allows us to show that there are only finitely many elliptic curves with bounded height. Section 3, as a change of pace, deals with strong Weil curves $E/\mathbb{Q}$. In particular, we relate the height of $E/\mathbb{Q}$ to the degree of its Weil parametrization $X_0(N) \to E$; and then using the results from Section 2, we show that this degree grows fairly rapidly as E is varied. Finally, in Section 4, we show how a conjecture of Serge Lang can be formulated using the notion of the height of an elliptic curve; and we reprove a special case of this conjecture using arithmetic intersection theory.

We have not felt it necessary to give extensive references for "standard" facts about elliptic curves. The reader might profitably consult any (or all) of [3], [4], [11] and [13]. A nice introduction to Weil curves is given in [12].

§1. The Height of an Elliptic Curve

We set the following notation for this section and the next:

$K/\mathbb{Q}$ a number field with ring of integers R.
M_K a complete set of inequivalent absolute values on K, normalized as usual (cf. [10, §1]).

M_K^∞, M_K^0 the archimedean (respectively non-archimedean) absolute values in M_K.
E/K an elliptic curve.
j_E the j-invariant of E.
$\Delta_{E/K}$ the minimal discriminant of E/K (this is an integral ideal of K).
$\Delta(\tau)$ $= \dfrac{1}{(2\pi)^{12}} q_\tau \prod_{n=1}^{\infty} (1 - q_\tau^n)^{24}$, where $q_\tau = e^{2\pi i \tau}$.
$j(\tau)$ $= 1728(4g_2(\tau))^3/\Delta(\tau) = 1728\{q_\tau^{-1} + 744 + \cdots\}$ the modular j-function.

For each $v \in M_K^\infty$, choose a $\tau_v \in \mathbb{H}$ ($\mathbb{H}$ = the upper half-plane) so that

$$E(\bar{K}_v) \cong \mathbb{C}/(\mathbb{Z} + \mathbb{Z}\tau_v).$$

Proposition 1.1. *With notation as above, the height of E/K is given by the formula*

$$[K : \mathbb{Q}]h(E/K) = \tfrac{1}{12}\{\log|N_{K/\mathbb{Q}}\Delta_{E/K}| - \sum_{v \in M_K^\infty} n_v \log(|\Delta(\tau_v)|(\operatorname{Im}\tau_v)^6)\},$$

where

$$n_v = [K_v : \mathbb{Q}_v] = 1 \ (resp.\ 2) \quad \text{if } v \text{ is real (resp. complex).}$$

Before giving the proof, we make several remarks.

Remark 1.2. (1) The height $h(E/K)$ in Proposition 1.1 is *not* the stabilized height obtained by going to an extension over which E has everywhere semistable (i.e. good or multiplicative) reduction.

(2) Since $\Delta(\tau)$ is a modular form of weight 12 for $SL_2(\mathbb{Z})$, one easily checks that the quantity $|\Delta(\tau)|(\operatorname{Im}\tau)^6$ is $SL_2(\mathbb{Z})$ invariant. Hence the expression in Proposition 1.1 is independent of the choice of the τ_v's.

(3) Given a Weierstrass equation for E/K, one can use Tate's algorithm [14] (or [6] if K has class number 1) to calculate $\Delta_{E/K}$; and numerical integration (or, more rapidly, Gauss' arithmetic-geometric mean) to compute the periods τ_v. Thus it is quite feasible to compute $h(E/K)$ numerically to any desired accuracy.

(4) The intuition is that $h(E/K)$ measures the "arithmetic complexity of E/K." Thus $N_{K/\mathbb{Q}}\Delta_{E/K}$ gives information about the primes of bad reduction and how "bad" the reduction is. On the other hand, $\operatorname{Im}\tau_v$ is the area of a fundamental parallelogram for $E/\bar{K}_v$.

PROOF OF PROPOSITION. 1.1. Let

$$E: y^2 + a_1 xy + a_3 y = x^3 + a_2 x^2 + a_4 x + a_6 \tag{1}$$

be any Weierstrass equation for E/K, and let $\Delta = \Delta(a_1, \ldots, a_6)$ be the discriminant of the equation. For each $v \in M_K^0$, choose a minimal Weierstrass

equation
$$E: y_v^2 + a_{1,v} x_v y_v + a_{3,v} y_v = x_v^3 + a_{2,v} x_v^2 + a_{4,v} x_v + a_{6,v} \tag{2}$$
for E at v, and let Δ_v be its discriminant. (So $v(\Delta_v) = v(\Delta_{E/K})$.) Finally, for each $v \in M_K^\infty$, we take for E the Weierstrass equation
$$E: Y_v^2 = 4X_v^3 - g_2(\tau_v) X_v - g_3(\tau_v) \tag{3}$$
parametrized by the Weierstrass $\wp$-function,
$$\mathbb{C}/(\mathbb{Z} + \mathbb{Z}\tau_v) \xrightarrow{\sim} E(\bar{K}_v),$$
$$z \to (\wp(z, \tau_v), \wp'(z, \tau_v)).$$
We note that the discriminant for this equation is
$$\Delta_v = g_2(\tau_v)^3 - 27 g_3(\tau_v)^2 = \Delta(\tau_v).$$

The height $h(E/K)$ is defined as $1/[K:\mathbb{Q}]$ times the degree of the metrized line bundle $\omega_{\mathscr{E}/R}$, where $\mathscr{E}/R$ is a Néron model for E/K. It is more convenient to work instead with the line bundle $(\omega_{\mathscr{E}/R})^{\otimes 12}$, so our final answer will be $12 h(E/K)$. To compute its degree, we choose the (meromorphic) section
$$\beta = \Delta \cdot \left(\frac{dx}{2y + a_1 x + a_3} \right)^{\otimes 12},$$
where x, y, a_1, a_3, Δ are the quantities associated with equation (1). The advantage of working with β is that it is invariant under the usual change of coordinates for a Weierstrass equation. (Cf. the formulas in [14].)

Now let $v \in M_K^0$. Then equation (2) gives a model for the connected component of the Néron model of E of v. (More precisely, (2) gives a scheme over $\mathrm{Spec}(R_v)$ whose smooth part is a model for $\mathscr{E}^0/R_v$.) Hence the bundle $\omega_{\mathscr{E}/R_v}$ has the holomorphic, non-vanishing section
$$\alpha_v = dx_v/(2y_v + a_{1,v} x_v + a_{3,v});$$
so we can compute
$$\frac{\omega_{\mathscr{E}/R_v}^{\otimes 12}}{R_v \beta} \approx \frac{R_v \alpha_v^{\otimes 12}}{R_v \beta} \approx \frac{R_v}{R_v \Delta_v}.$$
(Note that since β is invariant under change of coordinates, we have $\beta = \Delta_v \alpha_v^{\otimes 12}$.) Hence the contribution from v to $\deg(\omega_{\mathscr{E}/R}^{\otimes 12})$ is
$$\log \#(R_v/R_v \Delta_v) = \log \|N_{K_v/\mathbb{Q}_v} \Delta_v\|_v = \log \|N_{K/\mathbb{Q}} \Delta_{E/K}\|_v.$$
Adding these up over all $v \in M_K^0$ gives $\log |N_{K/\mathbb{Q}} \Delta_{E/K}|$, so the contribution to $h(E/K)$ is
$$\frac{1}{12[K:\mathbb{Q}]} \log |N_{K/\mathbb{Q}} \Delta_{E/K}|.$$

It remains to compute the archimedean contribution. Let $v \in M_K^\infty$ and, to ease notation, let $\tau = \tau_v$. Using the change of variable formula, we have as

above
$$\beta = \Delta(\tau)(dX_v/Y_v)^{\otimes 12};$$
and now the uniformization $X_v = \wp(z, \tau)$, $Y_v = \wp'(z, \tau)$ lets us write
$$\beta = \Delta(\tau)(dz)^{\otimes 12}$$
on $\mathbb{C}/(\mathbb{Z} + \mathbb{Z}\tau)$. Letting
$$\alpha = \Delta(\tau)^{1/12}\, dz,$$
the contribution to $h(E/K)$ is then
$$\frac{n_v}{[K:\mathbb{Q}]} \log\left(\frac{i}{2}\int_{\mathbb{C}/(\mathbb{Z}+\mathbb{Z}\tau)} \alpha \wedge \bar{\alpha}\right)^{1/2}.$$

We compute the integral:
$$\frac{i}{2}\int \alpha \wedge \bar{\alpha} = \frac{i}{2}\int |\Delta(\tau)|^{1/6}\, dz \wedge d\bar{z}$$
$$= |\Delta(\tau)|^{1/6} \int dx \wedge dy$$
$$= |\Delta(\tau)|^{1/6}\, \mathrm{Im}(\tau),$$

since the last integral is just the area of a fundamental parallelogram for $\mathbb{C}/(\mathbb{Z} + \mathbb{Z}\tau)$. Hence the contribution to $h(E/K)$ from $v \in M_K^\infty$ is
$$-(n_v/12[K:\mathbb{Q}])\log(|\Delta(\tau_v)|\mathrm{Im}(\tau_v)^6);$$
and adding over $v \in M_K^\infty$ completes the proof of Proposition 1.1. $\square$

§2. An Estimate for the Height

We now turn to estimating $h(E/K)$ in terms of more readily calculable quantities. For this purpose, we choose our τ_v's in the usual fundamental domain $\mathscr{F}$ for $\mathbb{H}/SL_2(\mathbb{Z})$, so in particular $\mathrm{Im}(\tau_v) \geq \sqrt{3}/2$. Then for any such τ, $|q_\tau| \leq e^{-\pi\sqrt{3}}$, so
$$\left|\prod_{n=1}^{\infty}(1 - q_\tau^n)\right|$$
is bounded above and below (away from 0). Hence if $\tau \in \mathscr{F}$, then
$$\log|\Delta(\tau)| = \log|q_\tau| + O(1), \tag{4}$$
where here and in what follows the $O(1)$ constants are absolute.

EXERCISE
Show that for $\tau \in \mathscr{F}$, $\log|\Delta(\tau)| = \log|q_\tau/(2\pi)^{12}| + A_\tau$ with $|A_\tau| \leq 1/9$.

Next, from the q-expansion for $j(\tau)$, we would like to say that for $\tau \in \mathcal{F}$,
$$\log|j(\tau)| = \log|1/q_\tau| + O(1).$$
This is alright as $\tau \to i\infty$, but of course $j(e^{2\pi i/3}) = 0$. So we split $\mathcal{F}$ by a horizontal line (such as $y = 1$), and then
$$\log|j(\tau)| = \begin{cases} \log|1/q_\tau| + O(1) & \text{for } \tau \in \mathcal{F}, \operatorname{Im}(\tau) \geq 1, \\ O(1) & \text{for } \tau \in \mathcal{F}, \operatorname{Im}(\tau) \leq 1. \end{cases}$$
Hence for all $\tau \in \mathcal{F}$, we have the relation
$$\log(\max\{|j(\tau)|, 1\}) = \log|1/q_\tau| + O(1). \tag{5}$$
Combining equations (4) and (5) yields
$$-\log|\Delta(\tau)| = \log(\max\{|j(\tau)|, 1\}) + O(1). \tag{6}$$
Further, since $\log|q_\tau| = -2\pi \operatorname{Im}(\tau)$, from equation (5) we obtain
$$\log(\operatorname{Im} \tau) = \log\log\max\{|j(\tau)|, e\} + O(1). \tag{7}$$

We are now ready to calculate $h(E/K)$. We use equations (6) and (7) to evaluate the archimedean terms in Proposition 1.1. Remembering that $j(\tau_v) = j_E$, we obtain
$$12[K:\mathbb{Q}]h(E/K) = \log|N_{K/\mathbb{Q}}\Delta_{E/K}| + \sum_{v \in M_K^\infty} n_v(\log\max\{|j_E|_v, 1\} \tag{8}$$
$$- 6\log\log\max\{|j_E|_v, e\} + O(1)).$$
Next write
$$(j_E) = \mathfrak{A}\mathfrak{D}^{-1}$$
as a quotient of relatively prime integral ideals $\mathfrak{A}$ and $\mathfrak{D}$. We note that if E/K has everywhere semistable reduction, then $\mathfrak{D} = \Delta_{E/K}$; and in all cases $\mathfrak{D}$ divides $\Delta_{E/K}$. Let us define the *unstable minimal discriminant of* E/K by
$$\Upsilon_{E/K} = \Delta_{E/K}\mathfrak{D}^{-1},$$
where $(j_E) = \mathfrak{A}\mathfrak{D}^{-1}$ as above. Then we can rewrite equation (8) as
$$12[K:\mathbb{Q}]h(E/K) = \log|N_{K/\mathbb{Q}}\mathfrak{D}| + \sum_{v \in M_K^\infty} n_v \log\max\{|j_E|_v, 1\}$$
$$+ \log|N_{K/\mathbb{Q}}\Upsilon_{E/K}|$$
$$+ \sum_{v \in M_K^\infty} n_v(-6\log\log\max\{|j_E|_v, e\} + O(1)). \tag{9}$$
Now if $a \in K^*$ and $(a) = \mathfrak{A}\mathfrak{B}^{-1}$ with $\mathfrak{A}$ and $\mathfrak{B}$ relatively prime integral ideals, then one easily checks ([5, p. 53]) that
$$H_K(a) = (N_{K/\mathbb{Q}}\mathfrak{B}) \prod_{v \in M_K^\infty} \max\{|a|_v^{n_v}, 1\}.$$
Thus the first two terms in equation (9) give precisely
$$\log|N_{K/\mathbb{Q}}\mathfrak{D}| + \sum_{v \in M_K^\infty} n_v \log\max\{|j_E|_v, 1\} = \log H_K(j_E) = [K:\mathbb{Q}]h(j_E). \tag{10}$$

Next, we must estimate the log log terms. Letting $d = [K:\mathbb{Q}]$ and using the arithmetic-geometric inequality, we compute (all sums and products are over $v \in M_K^\infty$; note $\sum n_v = d$)

$$0 \le \sum n_v \log \log \max\{|j_E|_v, e\}$$
$$= \log \prod (\log \max\{|j_E|_v, e\})^{n_v}$$
$$\le \log(\sum \log \max\{|j_E|_v, e\}/d)^d \qquad (11)$$
$$\le d \log(1 + (1/d) \sum \log \max\{|j_E|_v, 1\})$$
$$\le d \log(1 + h(j_E)).$$

Finally, we note that

$$\sum_{v \in M_K^\infty} n_v O(1) = [K:\mathbb{Q}] O(1). \qquad (12)$$

Now combining equations (9), (10), (11), and (12), and dividing by $[K:\mathbb{Q}]$, we have proven the following result.

Proposition 2.1. *Let K be a number field and E/K an elliptic curve. Then*

$$O(1) \le \{h(j_E) + \frac{1}{[K:\mathbb{Q}]} \log N_{K/\mathbb{Q}} \Upsilon_{E/K}\} - 12h(E/K)$$
$$\le 6 \log(1 + h(j_E)) + O(1),$$

where $\Upsilon_{E/K}$ is the unstable minimal discriminant of E/K as defined above and the $O(1)$'s are absolute constants. In particular, if E/K is semistable (i.e. no additive reduction), then

$$|h(j_E) - 12h(E/K)| \le 6 \log(1 + h(j_E)) + O(1).$$

EXERCISE

Normalize $v \in M_K^0$ so that $v(K^*) = \mathbb{Z}$. Prove that

$$v(\Upsilon_{E/K}) < 12 + 12v(2) + 6v(3).$$

Remark 2.2. Note that Proposition 2.1 is an explicit version (for elliptic curves) of Faltings' theorem [2], that the height of a semistable abelian variety is a multiple of the height of the corresponding point in moduli space, up to a logarithmic error term.

For elliptic curves over $\mathbb{Q}$, the estimate in Proposition 2.1 can be rewritten (slightly weakened) as follows:

Corollary 2.3. *Let $E/\mathbb{Q}$ be an elliptic curve with j-invariant j. Take a minimal Weierstrass equation for $E/\mathbb{Q}$, and let $\Delta, c_4, c_6 \in \mathbb{Z}$ be the associated quantities. Then for any $\varepsilon > 0$,*

$$h(E/\mathbb{Q}) + O(1) \le \tfrac{1}{12} \log \max\{|j|, |\Delta j|\} \le (1 + \varepsilon)h(E/\mathbb{Q}) + O_\varepsilon(1).$$

Similarly,
$$h(E/\mathbb{Q}) + O(1) \le \tfrac{1}{12} \log \max\{|c_4|^3, |c_6|^2\} \le (1+\varepsilon) h(E/\mathbb{Q}) + O_\varepsilon(1).$$

(*Here the $O(1)$'s are absolute constants, and the $O_\varepsilon(1)$'s depend only on ε.*)

PROOF. Write $j = a/d \in \mathbb{Q}$ in lowest terms. Then by definition of unstable minimal discriminant, $\Upsilon = |\Delta/d|$. Hence
$$h(j) + \log \Upsilon = \log\max\{|a|, |d|\} + \log|\Delta/d| = \log\max\{|j\Delta|, |\Delta|\}.$$

Now Proposition 2.1 gives the estimate
$$O(1) \le \log\max\{|j\Delta|, |\Delta|\} - 12 h(E/\mathbb{Q}) \le 6 \log(1 + h(j)) + O(1).$$

Since
$$h(j) \le \log\max\{|j\Delta|, |\Delta|\},$$
this is stronger than the first inequality in Corollary 2.3. (That is, we are using the easy estimate
$$6 \log(1 + \log t) \le \varepsilon \log t + O_\varepsilon(1) \text{ for } t \ge 1.)$$

Finally, since
$$\Delta = (c_4^3 - c_6^2)/1728 \quad \text{and} \quad j = c_4^3/\Delta,$$
the second inequality in Corollary 2.3 follows immediately from the first. □

Remark 2.4. From Corollary 2.3 it is immediate that there are only finitely many elliptic curves $E/\mathbb{Q}$ with bounded height. Indeed, if $h(E/\mathbb{Q})$ is bounded, then there are only finitely many possible values for c_4 and c_6. The analogous statement for number fields follows from Proposition 2.1 in a similar fashion, as we now prove. (However, it is not true that there are finitely many pairs (K, E) consisting of a number field K of bounded degree and an elliptic curve E/K of bounded height. See [18] for details.)

Corollary 2.5. *Fix a number field $K/\mathbb{Q}$ and a constant C. Then there are only finitely many K-isomorphism classes of elliptic curves E/K satisfying*
$$h(E/K) \le C.$$

PROOF. From Proposition 2.1, we see that bounding $h(E/K)$ has the effect of bounding $h(j_E)$. Now [10, Theorem 2.1] implies that a number field K has only finitely many elements of bounded height. Hence the elliptic curves E/K with $h(E/K) \le C$ give only finitely many j-invariants, and so only finitely many $\bar{K}$-isomorphism classes of elliptic curves.

We are now reduced to the following problem. Fix a number field $K/\mathbb{Q}$ and an elliptic curve E/K. Then up to K-isomorphism, there are only finitely many elliptic curves E'/K satisfying
$$j_{E'} = j_E \quad \text{and} \quad h(E'/K) \le C.$$

But again using Proposition 2.1, this time for a fixed field and j-invariant, we see that bounding $h(E'/K)$ is the same as bounding $N_{K/\mathbb{Q}}\Upsilon_{E'/K}$.

We now choose a finite set of places $S \subset M_K$ containing the following:

(i) M_K^∞;
(ii) all $v \in M_K^0$ at which E has bad reduction;
(iii) all $v \in M_K^0$ with $v(6) \neq 0$.

We further enlarge S so that

(iv) the ring of S-integers R_S is a P.I.D.

Then we can find a Weierstrass equation

$$E: y^2 = x^3 + Ax + B$$

for E/K with $A, B \in R_S$ and $\Delta = -16(4A^3 + 27B^2) \in R_S^*$. (We will now assume that $j_E \neq 0, 1728$, so $AB \neq 0$ and $\text{Aut}(E) = \{\pm 1\}$. The other cases are done similarly.) The elliptic curves over K isomorphic over $\bar{K}$ to E are the *twists* given by equations

$$E_D: Dy^2 = x^3 + Ax + B, \qquad D \in K^*;$$

and $E_D \cong E_{D'}$ over K if and only if $(D/D') \in (K^*)^2$. Further, for $v \notin S$, E_D has additive reduction at v if and only if $v(D) \equiv 1 \pmod 2$. Thus

$$N_{K/\mathbb{Q}}\Upsilon_{E_D/K} \geq \prod_{\substack{v \in M_K - S \\ v(D) \equiv 1(2)}} N_{K/\mathbb{Q}} v,$$

so bounding $N_{K/\mathbb{Q}}\Upsilon_{E_D/K}$ has the effect of bounding the set of $v \in M_K - S$ for which $v(D) \equiv 1(2)$. Hence enlarging S yet again, we may restrict attention to those twists E_D/K for which $v(D) \equiv 0(2)$ for all $v \in M_K - S$. It thus remains to show that

$$\{D \in K^*: v(D) \equiv 0(2) \text{ for all } v \in M_K - S\}/(K^*)^2$$

is a finite set. But this follows immediately from the finiteness of the class number and Dirichlet's unit theorem. □

§3. Weil Curves

We now turn our attention to elliptic curves defined over $\mathbb{Q}$. More precisely, we let $E/\mathbb{Q}$ be a strong Weil curve of geometric conductor N, and let

$$\phi_E: X_0(N) \to E$$

be the corresponding Weil parametrization. Take a minimal Weierstrass equation for $E/\mathbb{Q}$, and let

$$\alpha_E = dx/(2y + a_1 x + a_3)$$

be the usual Néron differential. Then

$$\phi_E^* \alpha_E = c_E f_E(z)\, dz, \tag{13}$$

where $f_E(z)$ is a normalized weight 2 newform for $\Gamma_0(N)$, and $c_E \in \mathbb{Q}^*$.

Note that since α_E is defined using a minimal Weierstrass equation, it generates the sheaf of 1-forms on every fibre of the Néron model of $E/\mathbb{Q}$. Hence the height of $E/\mathbb{Q}$ is given entirely by the archimedean term:

$$h(E/\mathbb{Q}) = -\frac{1}{2}\log \frac{i}{2}\int_{E(\mathbb{C})} \alpha_E \wedge \bar{\alpha}_E. \tag{14}$$

We also recall that the *Petersson norm* of $f_E(z)$ is defined by

$$\|f_E\|^2 = \frac{i}{2}\int_{X_0(N)} f_E(z)\, dz \wedge \overline{f_E(z)\, dz}. \tag{15}$$

We now integrate both sides of equation (13), using equations (14), (15), and the change of variable formula:

$$|c_E|^2 \|f_E\|^2 = \frac{i}{2}\int_{X_0(N)} \phi_E^* \alpha_E \wedge \overline{\phi_E^* \alpha_E}$$

$$= (\deg \phi_E)\frac{i}{2}\int_{E(\mathbb{C})} \alpha_E \wedge \bar{\alpha}_E$$

$$= (\deg \phi_E) e^{-2h(E/\mathbb{Q})}.$$

Taking logarithms gives the following result.

Proposition 3.1. *With notation as above,*

$$\tfrac{1}{2}\log \deg \phi_E = h(E/\mathbb{Q}) + \log\|f_E\| + \log|c_E|.$$

Of the four terms in Proposition 3.1, the least interesting is the one involving c_E; in fact, one guesses that it is negligible.

Conjecture 3.2 (Manin). $c_E = \pm 1$.

In any case, one has the following.

Theorem 3.3. (a) (Mazur and Swinnerton-Dyer [7]). *If N is square-free, then $|c_E|$ is a power of 2.*
(b) (Raynaud). *If N is square-free, then $\log|c_E|$ is bounded.*

For any particular curve $E/\mathbb{Q}$, the quantities $h(E/\mathbb{Q})$ and $\|f_E\|$ can be computed numerically to any desired degree of accuracy, each being essentially the integral of a certain 2-form over a Riemann surface. In this way one can use Proposition 3.1 (and the assumption that $|c_E| = 1$) to compute the more intractable quantity $\deg \phi_E$. This has been done by Zagier [15], who also proves a result similar to the following proposition.

Proposition 3.4. *Let $\phi_E: X_0(N) \to E$ be a strong Weil parametrization, and let $c_4, c_6 \in \mathbb{Z}$ be the usual quantities associated to a minimal Weierstrass equation for $E/\mathbb{Q}$. Assume the quantity c_E in equation (13) satisfies $|c_E| \geq 1$. Then for every $\varepsilon > 0$,*

$$\deg \phi_E \geq C_\varepsilon \max\{|c_4|^{1/4}, |c_6|^{1/6}\}^{2-\varepsilon}$$

for a constant $C_\varepsilon > 0$ depending only on ε. [That is, $\deg \phi_E$ grows "polynomially with weight 2" in the coefficients of the minimal Weierstrass equation for E.]

PROOF. From Corollary 2.3 we have the estimate

$$h(E/\mathbb{Q}) \geq (1 - \varepsilon)\log \max\{|c_4|^{1/4}, |c_6|^{1/6}\} + O_\varepsilon(1); \tag{16}$$

and by assumption

$$\log|c_E| \geq 0. \tag{17}$$

It remains to estimate the Petersson norm $\|f_E\|$. To do this, we note that for any N, one can choose a fundamental domain for $X_0(N)$ containing the region

$$\mathcal{R} = \{z \in \mathbb{C}: |x| < 1/2 \text{ and } y > 1\}.$$

Hence writing $f_E(z) = \sum_{n \geq 1} a_n e^{2\pi i n z}$, we have

$$\|f_E\|^2 \geq \int_\mathcal{R} |f_E(z)|^2 \, dx \, dy$$

$$= \sum_{m,n \geq 1} a_m \bar{a}_n \int_{-1/2}^{1/2} e^{2\pi i(m-n)x} \, dx \int_1^\infty e^{-2\pi(m+n)y} \, dy$$

$$= \sum_{n \geq 1} |a_n|^2 e^{-4\pi n}/(4\pi n).$$

But $f_E(z)$ is a normalized cusp-form (i.e. $a_1 = 1$), so just using the $n = 1$ term of this sum gives

$$\|f_E\|^2 \geq e^{-4\pi}/4\pi. \tag{18}$$

Now using the lower bounds provided by equations (16), (17), and (18) in the equation for $\deg \phi_E$ given in (3.1) yields

$$2 \log \deg \phi_E \geq (1 - \varepsilon)\log \max\{|c_4|^{1/4}, |c_6|^{1/6}\} + O_\varepsilon(1),$$

which is the desired result. □

§4. A Relation with the Canonical Height

We return now to the case of a number field K and an elliptic curve E/K, and let

$$\hat{h}\colon E(K) \to (0, \infty)$$

be the canonical height on E relative to the divisor (O). (See e.g. [4, Chap. IV] or [11, VIII, §9].) Then for $P \in E(\bar{K})$, $\hat{h}(P) \geq 0$, with equality if and only if P is a torsion point. Serge Lang has conjectured a uniform lower bound for the height of non-torsion points; the following is a slight generalization.

Conjecture 4.1 (Lang, [4, p. 92]). *There is a constant $c > 0$, depending only on K, such that for any elliptic curve E/K and any non-torsion point $P \in E(K)$,*

$$\hat{h}(P) \geq c\,h(E/K).$$

The following special case, originally proven in [8] using other methods, can now be given a short demonstration by adapting an analogous argument for function fields due to Tate. (But it is worth mentioning that the argument in [8] shows that the constant c in Conjecture 4.1 may be chosen depending only on the degree $[K:\mathbb{Q}]$, and not otherwise on K. This added uniformity is important in some applications.)

Theorem 4.2. *Lang's Conjecture 4.1 is true if one restricts attention to elliptic curves with integral j-invariant.*

PROOF. Let $\mathscr{E}/R$ be a Néron model for E/K, and let $P \in E(K)$ be a non-torsion point. Then P induces a map $P\colon \operatorname{Spec} R \to \mathscr{E}$; and we identify P with its image $P(\operatorname{Spec} R)$, which is a divisor on $\mathscr{E}$. Since j_E is integral, the group of components of every fibre of $\mathscr{E}$ has order dividing 12 (cf. [14]). Thus replacing P by $12P$, we may assume that $P \in \mathscr{E}^0(R)$. (That is, P hits the identity component of each fibre.) Since $\hat{h}(12P) = 144\hat{h}(P)$, we need merely divide our final answer by 144.

Now for $P \in \mathscr{E}^0(R)$, the canonical height can be computed using Arakelov (arithmetic) intersection theory as follows. (See [1, Theorem 5.1(ii)]. We take $D = (P) - (O)$, and note that $P \cdot P = O \cdot O$.)

$$[K:\mathbb{Q}]\hat{h}(P) = P \cdot O - O \cdot O. \tag{19}$$

(Here $O \in \mathscr{E}^0(R)$ is the identity section.) Since P is non-torsion, we can use the pigeon-hole principle (cf. [8]) or a Fourier averaging process (cf. [16, Ch. VI, Thm. 5.1]) to show that

$$P \cdot O \geq -c_1[K:\mathbb{Q}] \tag{20}$$

for an *absolute* constant c_1; and the adjunction formula [1, Prop. 4.1] reads

$$O \cdot O + (\omega_{\mathscr{E}/R}) \cdot O = \log N_{K/\mathbb{Q}} D_{K/K} = 0, \tag{21}$$

where $(\omega_{\mathscr{E}/R})$ is the canonical (metrized) divisor class on $\mathscr{E}/R$ and $D_{K/\mathbb{Q}}$ is the absolute discriminant of $K/\mathbb{Q}$. But by definition,

$$(\omega_{\mathscr{E}/R}) \cdot O = \deg O^* \omega_{\mathscr{E}/R} = [K:\mathbb{Q}]h(E/K). \tag{22}$$

Combining equations (19), (20), (21), and (22) yields

$$\hat{h}(P) \geq h(E/K) - c_1.$$

Since there are only finitely many curves E/K with bounded $h(E/K)$, this gives the desired result. □

A lower bound such as (4.2) has many applications in Diophantine geometry. For example, it is one of the tools used in the proof of the following result, which gives a quantitative relationship between the Mordell–Weil theorem and Siegel's theorem.

Theorem 4.3 ([9]). *Let K be a number field, E/K an elliptic curve with integral j-invariant, and*

$$E: y^2 = x^3 + Ax + B$$

a "quasi-minimal" equation for E. (That is, $|N_{K/\mathbb{Q}}(4A^3 + 27B^2)|$ is minimal subject to A and B being integral.) Let R_S be the ring of S-integers of K for some finite set of places S. Then

$$\#\{P \in E(K): x(P) \in R_S\} \leq C^{\operatorname{rank} E(K) + \#S + 1}$$

for a constant C depending only on K.

REFERENCES

[1] Chinberg, T. An introduction to Arakelov intersection theory, this volume, pp. 289–307.
[2] Faltings, G. Finiteness theorems for abelian varieties, this volume, pp. 9–27.
[3] Lang, S. *Elliptic Functions*. Addison-Wesley: Reading, MA, 1973.
[4] Lang, S. *Elliptic Curves: Diophantine Analysis*. Springer-Verlag: New York, 1978.
[5] Lang, S. *Fundamentals of Diophantine Geometry*. Springer-Verlag: New York, 1983.
[6] Laska, M. An algorithm for finding a minimal Weierstrass equation for an elliptic curve. *Math. Comput.*, **38** (1982), 257–260.
[7] Mazur, B. and Swinnerton-Dyer, H. P. F. Arithmetic of Weil curves. *Invent. Math.*, **25** (1974), 1–61.
[8] Silverman, J. Lower bound for the canonical height on elliptic curves. *Duke Math. J.*, **48** (1981), 633–648.
[9] Silverman, J. A quantitative version of Siegel's theorem. *J. Reine Angew. Math.*, **378** (1987), 60–100.
[10] Silverman, J. The theory of height functions, this volume, pp. 151–166.
[11] Silverman, J. *The Arithmetic of Elliptic Curves*. Graduate Texts in Mathematics, 106. Springer-Verlag: New York, 1986.

[12] Swinnerton-Dyer, H. P. F. and Birch, B. J. Elliptic curves and modular functions, in *Modular Functions of One Variable*, IV. Lecture Note in Mathematics, 476. Springer-Verlag: New York, pp. 2–32.
[13] Tate, J. The arithmetic of elliptic curves. *Invent. Math.*, **23** (1974), 179–206.
[14] Tate, J. Algorithm for determining the type of a singular fiber in an elliptic pencil, in *Modular Functions of One Variable*, IV. Lecture Note in Mathematics, 476. Springer-Verlag: New York, pp. 33–52.
[15] Zagier, D. Modular parametrizations of elliptic curves. *Can. Math. Bull.*, **28** (1985), 372–384.
[16] Lang, S. *Introduction to Arakelov Theory*. Springer-Verlag: New York, 1988.
[17] Hindry, M. and Silverman, J. The canonical height and integral points on elliptic curves. *Invent. Math.*, **93** (1988), 419–450.
[18] Silverman, J. Elliptic curves of bounded degree and height, *Proc. AMS*, **105** (1989), 540–545.

Addendum to the Second Printing. For further material on Lang's Conjecture 4.1, including its relation with Szpiro's Conjecture and an unconditional proof over function fields, see [17].

CHAPTER XI

Lipman's Proof of Resolution of Singularities for Surfaces

M. Artin

§1. Introduction

This is an exposition of Lipman's beautiful proof [9] of resolution of singularities for two-dimensional schemes. His proof is very conceptual, and therefore works for arbitrary excellent schemes, for instance arithmetic surfaces, with relatively little extra work. (See [4, Chap. IV] for the definition of excellent scheme.)

We want to thank Lipman for a number of helpful comments on our manuscript.

We will call a noetherian, normal, connected and excellent scheme X of dimensional 2 a *surface*. By *point* of X we mean a point of codimension 2, necessarily a closed point. A surface X has finitely many *singular* points, at which the local ring is not regular [4, IV, 7.8.6 (iii)].

Define a sequence

$$(*) \qquad X = X_0 \xleftarrow{f_1} X_1 \xleftarrow{f_2} X_2 \leftarrow \cdots$$

of surfaces X_i inductively as follows: Let $S_i \subset X_i$ be the (reduced) singular locus. Then X_{i+1} is the normalization of the blowing-up of S_i in X_i, or equivalently, the normalization of the scheme obtained by blowing up the maximal ideals of the points of S_i in succession. Each X_i is a surface, and the maps f_i are proper.

The main result of [9] is

Theorem (1.1). *The scheme X_n is nonsingular, if n is sufficiently large.*

We will prove it under the following additional hypothesis, which can, however, be removed [9].

(1.2) If X is a scheme of characteristic $p > 0$, then the residue fields $k = k(p)$ at points of X have the property $[k : k^p] < \infty$.

Let us reserve the notation

(1.3) $$X' \xrightarrow{f} X$$

for a birational map of surfaces, such that X' is projective over X. The structure sheaves of X and X' will be denoted by $\mathcal{O} = \mathcal{O}_X$ and $\mathcal{O}' = \mathcal{O}_{X'}$. We will call such a map a *modification* of X. For example, the maps f_i appearing in (*) are modifications. A modification will be an isomorphism outside of a finite set of points of X, often except at a single point p. Its fibre over p will be a connected scheme of dimension 1, for which we use the notation

$$E = \bigcup C_i, \quad i = 1, \ldots, n,$$

where the C_i are the irreducible components. Each of them is a projective curve over the residue field $k = k(p)$.

A point $p \in X$ is called a *rational singularity* if for every modification (1.3), the stalk of $R^1 f_* \mathcal{O}'$ at p is zero.

Theorem (1.1) will be proved in the following three steps:

Step 1. Reduction to rational singularities.
Step 2. Assuming rational singularities, reduction to rational double points.
Step 3. Resolution of rational double points.

The first two steps are done using the dualizing sheaf $\omega = \omega_X$. Since a normal surface is Cohen–Macaulay, the dualizing complex consists of a single sheaf ω, which is a rank 1, reflexive $\mathcal{O}$-module. If a point p is not a rational singularity of X, then there is a local section α of ω, and a modification (1.3), such that α has poles along some component C_i of E (cf. (3.5), [7]). Roughly speaking, the first step consists in blowing-up so that all such polar divisors appear, and what is required is to show that there are only finitely many of them. Once this has been done, step 2 consists in blowing up the module ω, so that it becomes locally free. Among rational singularities, only the points of multiplicity $\mu \leq 2$ have locally free dualizing sheaves (5.3), i.e., are Gorenstein. The main difficulty here is to prove that if $X' \xrightarrow{f} X$ denotes the blowing-up of ω, then the $\mathcal{O}'$-module generated by ω is the dualizing sheaf $\omega' = \omega_{X'}$ (cf. (5.1)). The proof of the final step 3 is the least conceptual one, but rational double points are very special, and the explicit analysis is not too unpleasant.

For reference, we recall the basic duality theorem [5] for a proper map $Y \xrightarrow{f} X$. Let F be a bounded complex of sheaves on Y. Then in the derived category

(1.4) $$Rf_*(R\,\mathcal{H}om(F, \omega_Y)) \approx R\,\mathcal{H}om(Rf_* F, \omega_X).$$

We need to expand this duality in the case that f is the modification (1.3) at a point $p \in X$ and that F consists of a single reflexive ($=$ Cohen–Macaulay) module M'. For this purpose, we recall the following facts:

(1.5) (a) For any coherent $\mathcal{O}'$-module M', $R^q f_* M' = 0$ if $q > 1$ and $R^1 f_* M'$ is a finite length $\mathcal{O}$-module concentrated at p.
(b) If M' is reflexive, then $\mathcal{E}xt^q_{\mathcal{O}'}(M', \omega') = 0$ for $q \neq 0$, and $M'^D = \mathcal{H}om_{\mathcal{O}'}(M', \omega')$ is reflexive.
(c) For any finite length $\mathcal{O}$-module ε, $\mathcal{E}xt^q_{\mathcal{O}}(\varepsilon, \omega) = 0$ if $q \neq 2$, and $\mathcal{E}xt^2_{\mathcal{O}}(\varepsilon, \omega)$ has the same length as ε.

Entering this information into (1.4), we obtain an exact sequence

(1.6) $\quad 0 \to f_*(M'^D) \to (f_* M')^D \to \mathcal{E}xt^2_{\mathcal{O}}(R^1 f_* M', \omega) \to R^1 f_*(M'^D) \to 0$,

which expresses the essential content of (1.4). In this sequence, the second term is reflexive and the last two have finite length.

Our assumption that the surface X is an excellent scheme is used at several places. Mainly, it enables us to replace X by its completion at a singular point p, which is permissible because of the following proposition. Of course, substitutes for this unaesthetic operation can be found in various special cases.

Proposition (1.7). *Let X be a surface and let $\bar{X} = \text{Spec } \hat{\mathcal{O}}_p$ be its completion at a point p.*

(i) *The sequence $(*)$ for X induces the analogous sequence for $\bar{X}$ by base change: $\bar{X}_n \simeq \bar{X} \times_X X_n$.*
(ii) *Let $X' \xrightarrow{f} X$ be a modification (1.2). Then $\bar{X}' = \bar{X} \times_X X'$ is a modification of $\bar{X}$, and X' is nonsingular above p if and only if $\bar{X}'$ is.*
(iii) *Every modification $\bar{X}' \to \bar{X}$ arises by base change from a modification $X' \to X$.*

The proof is an exercise, starting with [4, IV, 7.8.3].

As a consequence, we may assume that $\mathcal{O}$ is a quotient of a regular local ring R. Then the dualizing module is $\omega \approx \mathcal{E}xt^{n-2}_R(\mathcal{O}, R)$, which shows that ω exists.

Another technical point is that at some places general position arguments require a sufficiently large residue field at a point $p \in X$. This can be accomplished if we replace X by a suitable étale extension. Let $\tilde{k}$ be a finite separable extension of $k = k(p)$. There exists an étale map $\tilde{X} \to X$ and a point $\tilde{p}$ of $\tilde{X}$ lying over p, such that $k(\tilde{p}) = \tilde{k}$ [4, IV, 18.1.1]. The following proposition allows us to replace X by $\tilde{X}$, and we will do this when necessary, without further comment:

Proposition (1.8). *With the above notation, the sequence $(*)$ induces the analogous sequence for $\tilde{X}$ by base change: $\tilde{X}_n \simeq \tilde{X} \times_X X_n$.*

For it shows that $\tilde{X}_n$ is étale over X_n. Therefore $\tilde{X}_n$ is nonsingular (or rational, etc. ...) at a point $\tilde{q}$ if and only if X_n is nonsingular (rational, ...) at its image q [4, IV, 18.4.10].

We omit the proof of (1.8).

Proposition (1.9). (i) *Any two modifications f_i (1.3) are dominated by a third one.*
(ii) *Any modification $X' \xrightarrow{f} X$ is dominated by a sequence of normalized blowings-up.*

Here domination of X' by $X'' \xrightarrow{g} X$ means, of course, that the birational correspondence $X'' \xrightarrow{\pi} X'$ determined by f and g is regular. Note that the sequence (∗) does not dominate every modification, because one can blow up nonsingular points.

To prove (i), it suffices to take for X'' the normalization of the join of X_1', X_2', i.e., of the unique component of $X_1' \times_X X_2'$ which maps birationally to X.

For (ii), let C_i be the one-dimensional components of the fibres of f. The local ring R_i of X' at the general point of C_i is a discrete valuation ring. We will show that there is a sequence of normalized blowings-up so that, on the modification $X'' \xrightarrow{g} X$ obtained from the sequence, every such valuation ring R_i is a local ring. Then the birational correspondence $X'' \xrightarrow{\pi} X'$ is an isomorphism at every point of X' of codimension 1, and Zariski's Main Theorem, together with the fact that X'' is a surface, shows that π is regular.

Since there are finitely many fibre components, it suffices to treat a single one, say, with $f(C) = p$. The following proof is due to Zariski [13]. Let M be the maximal ideal of R. Note that $R/M = k(C)$ is of transcendence degree 1 over $\mathcal{O}_p/m_p = k$. Choose an element $r \in R$ which has a transcendental residue over k, and write it as $r = a/b$, where $a, b \in \mathcal{O}_p$. Since its residue is transcendental and $M \cap \mathcal{O}_p = m_p$, the element r is not in $\mathcal{O}_p$. Therefore b is not a unit, i.e., $b \in m_p \subset M$. Similarly, $r^{-1} \notin \mathcal{O}_p$, and so $a \in M$. We use induction on the orders of zero $v(a), v(b)$, where v denotes the valuation of R. They are positive integers.

Let $X_1 \xrightarrow{f_1} X$ denote the normalized blowing-up of p in X. Since f_1 is proper, the valuative criterion shows that Spec R maps to X_1. Its closed point will have an image in X_1 which is either the general point η of a fibre component, or else is a closed point p_1 of the fibre. In the first case $\mathcal{O}_{X_1, \eta}$ is a discrete valuation ring contained in R. Therefore it is equal to R and we are done. In the second case, choose a standard affine of X_1 containing p_1. If $x_1, \ldots, x_n$ generate m_p, then the affine coordinate ring A_1 will be the normalization of a ring $A[x_1', \ldots, x_n']$, where A is some affine ring of X, and say $x_i' = x_i/x_n$. In A_1, we can write $a = a_1 x_n$, $b = b_1 x_n$. Then $r = a_1/b_1$. Since $x_n \in M$, $v(a_1) < v(a)$ and $v(b_1) < v(b)$. This completes the proof. □

§2. Proper Intersection Numbers and the Vanishing Theorem

By *divisor* on a surface X', we mean Weil divisor, i.e., a linear combination $Z = \sum r_i Y_i$ of irreducible closed subsets Y_i of X' of codimension 1. The notions of divisor of a rational function and of linear equivalence are the usual ones.

Also, the sheaf $\mathcal{O}'(Z)$ of functions with pole $\leq Z$ is a rank 1 reflexive $\mathcal{O}'$-module. It is locally free if and only if Z is a locally principal (Cartier) divisor. Similarly, if M' is a reflexive $\mathcal{O}'$-module, then $M'(Z)$ denotes the sheaf of sections of M' with pole $\leq Z$, which is the reflexive hull of $M' \otimes_{\mathcal{O}'} \mathcal{O}'(Z)$.

Let $X' \xrightarrow{f} X$ be a modification, and let $C = C_i$ be a one-dimensional component of some fibre $f^{-1}(p)$. Denote by $\bar{C}$ the normalization of C. Given a divisor Z on X', we consider the sheaf

(2.1) $$\mathcal{O}_{\bar{C}}(Z) \underset{\text{defn}}{=} [\mathcal{O}'(Z) \otimes_{\mathcal{O}'} \mathcal{O}_{\bar{C}}]/(\text{torsion}).$$

Since $\bar{C}$ is a nonsingular curve, this sheaf is locally free, of rank 1. We define the *proper intersection number* of Z and C to be

(2.2) $$[Z \cdot C] = \text{degree } \mathcal{O}_{\bar{C}}(Z),$$

the degree (or Chern class) being computed with respect to the ground field $k = k(p)$. *This symbol is not symmetric.*

As justification for the terminology, we will describe a method of calculating this number, which is however not needed elsewhere. If $q \in X'$ is a point at which a positive cycle Z is locally principal, for example, a nonsingular point of X', and if C is not a component of Z, then $(Z \cdot C)_q$ denotes the usual local intersection number: Let $\{\phi = 0\}$ be a local equation for Z at q. Then

(2.3) $$(Z \cdot C)_q = \dim_k (\mathcal{O}_{C,q}/(\phi)).$$

Proposition (2.4). *Suppose that $\mathcal{O}'(Z)$ is generated by global sections. Then for a sufficiently general divisor Z_1 linearly equivalent to Z and positive in a neighborhood of C,*

$$[Z \cdot C] = \sum_{\substack{q \text{ nonsing.} \\ \text{on } X'}} (Z_1 \cdot C)_q.$$

In other words, intersections of Z and C which are forced by the fact that Z is not locally principal at some singular point do not count in $[Z \cdot C]$. If $\mathcal{O}'(Z)$ is not generated by global sections, this description can be modified by allowing divisors Z' which are not positive.

PROOF OF (2.4). A generating set of global sections of $\mathcal{O}'(Z)$ will induce a generating set for $\mathcal{O}_{\bar{C}}(Z)$. Localizing X and X' in a neighborhood of p and applying (1.8) if necessary, we may choose a generic global section s_1 such that the induced section $\bar{s}$ of $\mathcal{O}_{\bar{C}}(Z)$ does not vanish at any point whose image in X' is singular. Then the divisor Z_1 of s_1 is the required one. For, on the one hand the degree of $\mathcal{O}_{\bar{C}}(Z)$ is the degree of the zero-cycle $\{\bar{s} = 0\}$ on $\bar{C}$, and on the other hand, this degree can be calculated locally as the intersection number $(Z \cdot C)_q$, provided q is nonsingular:

(2.5) $$(Z_1 \cdot C)_q = \sum_{\bar{q} \text{ over } q} \text{length } (\mathcal{O}_{\bar{C}}(Z)/\bar{s}\, \mathcal{O}_{\bar{C}}(Z))_{\bar{q}}.$$

The proposition follows. □

Several formal properties of the symbol $[Z \cdot C]$ are summed up in the proposition below:

Proposition (2.6). (i) $[Z \cdot C] = [Z' \cdot C]$ *if Z and Z' are linearly equivalent. If Z is locally principal, then $[Z \cdot C]$ is equal to the usual intersection number $(Z \cdot C)$.*
(ii) *Positivity. Assume that $Z \geq 0$ and that C is not a component of Z. Then $[Z \cdot C] \geq 0$. If in addition $Z \cap C$ contains a nonsingular point of X', then $[Z \cdot C] > 0$.*
(iii) *Sublinearity. $[A \cdot C] + [B \cdot C] \leq [A + B \cdot C]$, with equality if B is locally principal.*

These properties are very elementary. We will only indicate the proof of (iii), which results from consideration of the diagram

$$\begin{array}{ccc} \mathcal{O}'(A) \otimes_{X'} \mathcal{O}'(B) & \xrightarrow{\alpha} & \mathcal{O}'(A+B) \\ \downarrow & & \downarrow \\ \mathcal{O}_{\bar{C}}(A) \otimes_{\bar{C}} \mathcal{O}_{\bar{C}}(B) & \xrightarrow{\beta} & \mathcal{O}_{\bar{C}}(A+B). \end{array}$$

The top arrow α is multiplication; and it induces β functorially. Since α is an isomorphism except at singular points of X', the map β is not zero. It follows from the existence of this map that

$$[A \cdot C] + [B \cdot C] = \deg(\mathcal{O}_{\bar{C}}(A) \otimes \mathcal{O}_{\bar{C}}(B)) \leq \deg \mathcal{O}_{\bar{C}}(A+B) = [A+B \cdot C]. \quad \square$$

Let $C_1 \cup \cdots \cup C_n = E$ be the fibre of our map $X' \xrightarrow{f} X$ over p. If X' is nonsingular, the intersection matrix $\|(C_i \cdot C_j)\|$ is negative definite. The following proposition extends that fact to the singular case.

Proposition (2.7). *Let $A = \sum a_i C_i$ be a positive divisor supported on E. Then for some index j, $[A \cdot C_j] < 0$.*

PROOF. Choose a rational function g on X' whose divisor has the form $(g) = Z + D$, where

(2.8) (i) $Z = \sum r_i C_i$, and $r_i > 0$ for all i.
(ii) $D > 0$, and for all i, $D \cap C_i$ contains a nonsingular point of X'.

It is clear that such a function exists. Namely, choose any positive divisor D_0 meeting each C_i at a regular point, and take for g any nonzero element of $\mathcal{O}_{X,p}$ which vanishes on $f(D_0)$.

The only property of the configuration E we need is the existence of the function g. Note that the hypotheses (2.8) carry over to subsets of $\{C_1, \ldots, C_n\}$. So we can proceed by induction on the number of components C_i. Choose positive integers l, m so that $lA - mZ = B \geq 0$, but that

Supp $B < C_1 \cup \cdots \cup C_n$. Sublinearity implies that for every $C = C_i$,
$$l[A \cdot C] \leq [lA \cdot C] = [B + mZ \cdot C] < [B + mZ \cdot C] + [mD \cdot C]$$
$$\leq [B + m(g) \cdot C] = [B \cdot C].$$
By induction, $[B \cdot C_j] \leq 0$ for some j. Therefore $[A \cdot C_j] < 0$. □

Theorem (2.9) (Vanishing Theorem [3]). *Let $X' \xrightarrow{f} X$ be a modification and let $E = f^{-1}(p)$ be the fibre of f at p.*

(i) *Let ω' be the dualizing sheaf on X'. Then*
$$R^q f_* \omega' = 0 \quad \text{if } q \neq 0.$$

(ii) *For every positive divisor Y with support on E, the map $R^1 f_* \mathcal{O}' \to R^1 f_* \mathcal{O}'(Y)$ is injective.*

PROOF. (ii). By Proposition (2.7), $[Y \cdot C] < 0$ for some $C = C_j$. It follows, in the notation of (2.2), that $H^0(\bar{C}, \mathcal{O}_{\bar{C}}(Y)) = 0$, hence $H^0(C, \mathcal{O}_C(Y)) = 0$ too, where $\mathcal{O}_C(Y)$ is the restriction of $\mathcal{O}'(Y)$ to C, which is defined by the exact sequence

(2.10) $$0 \to \mathcal{O}'(Y - C) \to \mathcal{O}'(Y) \to \mathcal{O}_C(Y) \to 0.$$

Hence the map $H^1(X', \mathcal{O}'(Y - C)) \to H^1(X', \mathcal{O}'(Y))$ is injective. By induction on Y, $H^1(X', \mathcal{O}') \to H^1(X', \mathcal{O}'(Y))$ is injective. This is equivalent with (2.9) (ii).

(i) We may assume that f is a modification at p. Consider the sequence
$$0 \to \omega'(-Y) \to \omega' \to \omega' \otimes \mathcal{O}_Y \to 0,$$
and the associated sequence

(2.11) $$R^1 f_* \omega'(-Y) \xrightarrow{\phi_Y} R^1 f_* \omega' \to R^1 f_* (\omega' \otimes \mathcal{O}_Y) \to 0.$$

By the Holomorphic Functions Theorem and the fact that $R^1 f_* \omega'$ has finite length,

(2.12) $$R^1 f_* \omega' \approx \widehat{R^1 f_* \omega'} \approx \mathop{\text{proj lim}}_Y R^1 f_* (\omega' \otimes \mathcal{O}_Y).$$

The maps in the inverse system on the right are surjective (vanishing of $R^2 f_*$). It follows that the map ϕ_Y of (2.11) is zero for large Y.

Now substitute $\mathcal{O}'$ and $\mathcal{O}'(Y)$ for M' into the sequence (1.6), to obtain a diagram

(2.13)
$$\begin{array}{ccccccc} 0 \to & f_* \omega' & \to \omega \to & \text{Ext}^2(R^1 f_* \mathcal{O}', \omega) & \longrightarrow & R^1 f_* \omega' & \to 0 \\ & \uparrow & \| & \uparrow \psi_Y & & \uparrow \phi_Y & \\ 0 \to & f_* \omega'(-Y) & \to \omega \to & \text{Ext}^2(R^1 f_* \mathcal{O}'(Y), \omega) & \to & R^1 f_* \omega'(-Y) & \to 0. \end{array}$$

Being dual to the injective map (2.9) (ii), ψ_Y is surjective (cf. (1.5) (c)). Since ϕ_Y is zero for large Y, it follows that $R^1 f_* \omega' = 0$, as required. □

For future reference, we note the adjunction formula:

Proposition (2.14). *Let Y be a positive divisor on X. The dualizing module ω_Y fits into an exact sequence*

$$0 \to \omega \to \omega(Y) \to \omega_Y \to 0.$$

This follows by duality from the exact sequence

$$0 \to \mathcal{O}(-Y) \to \mathcal{O} \to \mathcal{O}_Y \to 0,$$

which also shows that $\omega_Y = \mathscr{E}xt^1(\mathcal{O}_Y, \omega)$. □

§3. Step 1. Reduction to Rational Singularities

Let X be a surface, and consider a diagram

(3.1)
$$X'' \xrightarrow{\pi} X'$$
$$g \searrow \swarrow f$$
$$X$$

of modifications of X.

Proposition (3.2). (i) *There is a natural exact sequence*

$$0 \to R^1 f_* \mathcal{O}_{X'} \to R^1 g_* \mathcal{O}_{X''} \to f_* R^1 \pi_* \mathcal{O}_{X''} \to 0.$$

(ii) *If X has rational singularities, so does X'.*
(iii) *Nonsingular points $p \in X$ are rational singularities.*
(iv) *To prove step 1, it suffices to show that for every singular point $p \in X$, the length of $R^1 f_* \mathcal{O}_{X'}$ at p is bounded independently of the modification f.*

PROOF. (i) This follows from an analysis of the spectral sequence $R^p f_* R^q \pi_* \mathcal{O}_{X''} \Rightarrow R^{p+q} g_* \mathcal{O}_{X''}$ and (1.5) a.
 (ii) This follows from (i).
 (iii) By (1.9) and (ii), it is enough to show that $R^1 f_* \mathcal{O}_{X'} = 0$ when $X' \xrightarrow{f} X$ is the blowing-up of p. Since p is nonsingular, the maximal ideal $m_p \subset \mathcal{O}_p$ is generated by two elements u_0, u_1, and the construction of the blowing-up represents it locally as a subscheme $X' \xhookrightarrow{i} \mathbb{P}^1_X = \mathbb{P}$. (In fact, X' can be defined by the equation $u_0 Z_1 - u_1 Z_0 = 0$.) Thus $i_* \mathcal{O}_{X'}$ is a quotient of $\mathcal{O}_\mathbb{P}$. Let $\bar{f}$ denote the projection $\mathbb{P} \to X$. The relative dimension of $\mathbb{P}$ over X is 1, and so $R^1 \bar{f}_* R$ is a right exact functor of $\mathcal{O}_\mathbb{P}$-modules F. Also, $R^1 \bar{f}_* \mathcal{O}_\mathbb{P} = 0$. Therefore $R^1 \bar{f}_* (i_* \mathcal{O}_{X'}) = R^1 f_* \mathcal{O}_{X'} = 0$ too.
 (iv) Assume the length bounded, and let $X' \xrightarrow{f} X$ be a modification such that the maximum is achieved. Then the sequence (i) shows that for any diagram (3.1), $f_* R^1 \pi_* \mathcal{O}_{X''} = 0$. Since $R^1 \pi_* \mathcal{O}_{X''}$ has zero-dimensional support, it follows that $R^1 \pi_* \mathcal{O}_{X''} = 0$. This being true for every π, X' has rational singularities. Now we apply (1.9) to dominate $X' \to X$ by a sequence of

normalized blowings-up, say by $X'' \xrightarrow{\pi} X'$. Then (ii) implies that X'' has rational singularities. Nonsingular points may have been blown up in this process, but by (iii) their blowing-up was not necessary to reduce to rational singularities. Therefore the sequence (∗), which dominates any sequence of normalized blowings-up of singular points, leads to rational singularities too. □

We begin the proof of step 1 by applying the vanishing theorem (2.9) (i). Substitution of $\mathcal{O}'$ for M' in (1.6) yields the exact sequence

(3.3) $\qquad 0 \to f_*\omega' \to \omega \to \mathscr{E}xt^2(R^1f_*\mathcal{O}', \omega) \to 0.$

Since $\varepsilon \leadsto \mathscr{E}xt^2(\varepsilon, \omega)$ is a duality for finite length $\mathcal{O}$-modules (1.5) (c), we obtain

Corollary (3.4). (i) *The finite length $\mathcal{O}$-modules $R^1f_*\mathcal{O}'$ and $\omega/f_*\omega'$ are dual, via $\mathscr{E}xt^2(\cdot, \omega)$.*
(ii) *The point p is a rational singularity if and only if for every modification (1.2) at p, $f_*\omega' = \omega$.*
(iii) *To complete step 1, it suffices to show that the length of $\omega/f_*\omega'$ is bounded independently of f, or equivalently, that some fixed power of the maximal ideal $\mathfrak{m}$ annihilates $\omega/f_*\omega'$.*

The last assertion follows from (3.2)(iv) and (3.4)(i).

Suppose that X is of finite type over a perfect field k. Then ω is the reflexive hull of $\Omega^2 = \Lambda^2\Omega^1_{X/k}$. This is because Ω^2 and ω are isomorphic at smooth points of X, and ω is reflexive.

Thus Ω^2 maps to ω, and the map is an isomorphism except at singular points of X.

Proposition (3.5). *If X is of finite type over a perfect field k, then for any modification (1.3) the canonical map $\Omega^2 \to \omega$ factors through $f_*\omega'$. Therefore $\omega/f_*\omega'$ is bounded by $\omega/\operatorname{im} \Omega^2$.*

This follows from the fact that Ω^2 is contravariant, so that $\Omega^2_X \to f_*\Omega^2_{X'} \to f_*\omega'$, and completes the proof of step 1 in that case. A similar argument can be given if $X = \operatorname{Spec} R$, where R is a complete local k-algebra with residue field k, and k is perfect.

In general, Lipman presents X as finite over a two-dimensional regular scheme Y. This can always be done if $X = \operatorname{Spec} R$ and R is a complete local ring, which we may assume (1.7). Let $q \in Y$ be the image of the singular point $p \in X$. If $Y' \xrightarrow{g} Y$ is a modification of Y, we will use the notation

(3.6)
$$\begin{array}{ccc} X' & \xrightarrow{f} & X \\ \pi' \downarrow & & \downarrow \pi \\ Y' & \xrightarrow{g} & Y \end{array}$$

where X' denotes the normalization of $Y' \times_Y X$.

We omit the proof of the following lemma, which is similar to that of (1.9) (ii).

Lemma (3.7). *The modifications f which arise in the above way are cofinal among modifications of X at p.*

To prove it, let R be as in the proof of (1.9) (ii), and apply the method used there to $R \cap K(Y)$.

Let us drop the symbol π_* when regarding $\mathcal{O}_X$-modules as sheaves on Y. Note that $\mathcal{O}_X$ has depth 2, hence is locally free over $\mathcal{O}_Y$. By duality (1.4), $\omega_X \approx \mathcal{H}om_Y(\mathcal{O}_X, \omega_Y)$. Since Y is nonsingular, $\omega_Y \approx \mathcal{O}_Y$ locally, and we choose such a local isomorphism. Then

$$(3.8) \qquad \omega_X \approx \mathcal{H}om_Y(\mathcal{O}_X, \mathcal{O}_Y).$$

Also, consider the trace from $\mathcal{O}_X$ to $\mathcal{O}_Y$. It defines a pairing

$$(3.9) \qquad \mathcal{O}_X \otimes \mathcal{O}_X \to \mathcal{O}_Y$$

$$a \otimes b \rightsquigarrow \operatorname{tr} ab.$$

This pairing and (3.8) define a map

$$(3.10) \qquad \mathcal{O}_X \xrightarrow{\phi} \omega_X.$$

which vanishes at points of Y where the discriminant d of the trace pairing in (3.9) is zero.

Lemma (3.11). *For any modification (1.3), d annihilates $\omega/f_*\omega'$.*

PROOF. The point is that the trace pairing is contravariant. Since a nonsingular point is a rational singularity, there is an injection $f^*\omega_Y \subset \omega_{Y'}$ adjoint to (3.4)(ii). Combining it with the chosen isomorphism $\mathcal{O}_Y \cong \omega_Y$ gives us an injection $\mathcal{O}_{Y'} \subset \omega_{Y'}$. We now replace Y by Y' in (3.0), to obtain a pairing

$$\mathcal{O}_{X'} \otimes \mathcal{O}_{X'} \to \mathcal{O}_{Y'} \subset \omega_{Y'},$$

hence a map $\mathcal{O}_{X'} \to \omega_{X'}$. Applying g_*, we find $\mathcal{O}_X = f_*\mathcal{O}_{X'} \subset$. Thus $\phi(\mathcal{O}_X) \subset f_*\omega_{X'}$. On the other hand, $d\omega_X \subset \phi(\mathcal{O}_X)$. $\square$

If X is generically separable over Y, so that $d \neq 0$, we can complete the proof now. This includes the cases that X has unequal characteristic, or (by suitable choice of π) that the residue field k is perfect [11]. Choose an element $u \in \mathcal{O}_X$ which vanishes on $\{d = 0\}$, but which generates a radical ideal. Lemma (3.14) below shows that this is always possible locally, and then d divides some power u^ν of u.

Given a modification (1.3), define D to be the locus of zeros of u on X', and D' to be the locus of zeros of u on X'. Then the normalization $\bar{D}$ of D lies over D', i.e., there are natural maps $\bar{D} \to D' \to D$. So, viewing $\omega_{\bar{D}}$ as an $\mathcal{O}_D$-module,

$$(3.12) \qquad \omega_{\bar{D}} \subset f_*\omega_{D'} \to \omega_D.$$

Therefore $\omega_D/f_*\omega_{D'}$ has length bounded independently of f, and is annihilated by a fixed power m^N of the maximal ideal of p in X.

We construct a diagram

(3.13)
$$\begin{array}{ccccccccc} 0 & \to & f_*\omega' & \xrightarrow{u} & f_*\omega' & \to & f_*\omega_{D'} & \to & 0 \\ & & \downarrow & & \downarrow & & \downarrow & & \\ 0 & \to & \omega & \xrightarrow{u} & \omega & \to & \omega_D & \to & 0 \\ & & \downarrow & & \downarrow & & \downarrow & & \\ & & \omega/f_*\omega' & \xrightarrow{u} & \omega/f_*\omega' & \to & \omega_D/f_*\omega_{D'} & \to & 0 \end{array}$$

with exact rows and columns. The middle row comes from (2.14) and the fact that $u = 0$ defines D, and the top row from (2.14) and (2.9)(i). Since u^v is divisible by d, Lemma (3.14) shows that u^v annihilates $\omega/f_*\omega'$. The bottom row implies that m^{vN} annihilates $\omega/f_*\omega'$, which therefore has bounded length.

Lemma (3.14). *Let $X = \text{Spec } \mathcal{O}$, where $\mathcal{O}$ is a normal local ring, and let $Y < X$ be a closed subscheme. There exists a nonzero element $u \in \mathcal{O}$ which vanishes on Y and generates a radical ideal.*

PROOF. First of all, if $u \neq 0$ is an element of m, there is an integer N so that the multiplicity $\mu(\mathcal{O}/(u'))$ [14, p. 294] is bounded for every $u' \equiv u$ (modulo m^N). This is because $\mu = \mu(\mathcal{O}/(u'))$ is bounded by $l = \text{length } (\mathcal{O}/(u, x))$, where $x = \{x_1, \ldots, x_{d-1}\}$ are elements chosen so that the ideal (u, x) is m-primary. This length will not change if u is replaced by u' and $N \gg 0$.

Now choose any nonzero u vanishing on Y. Write its divisor in the form $(u) = V + U$, where V is the sum of those prime divisors having multiplicity 1. If $U = 0$ we are done. In any case, $\mu = \mu(V) + \mu(U)$, and we proceed by induction on $\mu(V)$. By the Chinese Remainder Theorem we may choose $w \in m^N$ (N as above), so that its divisor is $2V + U_0 + W$, where $U_0 = U_{\text{red}}$ and where W has no component in common with $V + U_0$. Then $u' = u + w$ vanishes to order 1 on each component of V and U_0. Since $\mu(V) < \mu(V + U_0) \leq l$, induction completes the proof. $\square$

To complete step 1, it remains to consider the case that no generically separable map π (3.6) exists. We have to show that $R^1 f_* \mathcal{O}'$ is bounded in this case too, and we may assume (1.7) that $X = \text{Spec } R$, where R is a complete normal local ring of characteristic $p > 0$. Then R is a quotient of a power series ring:

(3.15) $$R \approx k[\![x_1, \ldots, x_n]\!]/(g_1, \ldots, g_m) = k[\![x]\!]/(g).$$

Let us call such a quotient *separable* if it is finite and generically étale over a power series ring in d variables, $d = \dim R$. It is known [11] that this is true if and only if Fract R is a separable field extension of k, but we will not use that fact.

We now use the hypothesis (1.2) that $[k:k^p]$ is finite. Choose any representation of R as a finite algebra over $k[\![y]\!] = k[\![y_1, y_2]\!]$. Let $k_v = k^{1/p^v}$, and $y_i^v = y_i^{1/p^v}$. Then $k[\![y]\!]^{1/p_v} = k_v[\![y^v]\!]$ is a finite $k[\![y]\!]$-algebra. It follows from field theory that the normalization R_v of $(k[\![y]\!]^{1/p^v} \otimes_{k[\![y]\!]} R)_{\text{red}}$ is separable over $k[\![y]\!]^{1/p^v}$ for large v, and so (3.2) (iv) has already been shown for those rings. Also, R_v is a finite, purely inseparable R-algebra.

Denote by $\mathcal{O}_v$ the coherent sheaf of $\mathcal{O}$-module defined by R_v, and by U the complement of the closed point of X. By Lemma (3.17) below, the kernel $\mathcal{J}$ of the map

$$H^1(U, \mathcal{O}) \to H^1(U, \mathcal{O}_v)$$

is a finite-dimensional k-vector space.

Now, given a modification $f: X' \to X$, we construct a modification $f_v: X'_v \to X_v = \operatorname{Spec} R_v$ by letting X'_v be the normalization of $(X' \times_X X_v)_{\text{red}}$. Since $H^1(X'_v, \mathcal{O}'_v)$ is bounded, we may assume that f is "large enough" so that this group is maximized, at least among modifications f_v arising in the above way. Consider the sequence

(3.16) $$0 \to \mathcal{K} \to H^1(X', \mathcal{O}') \to H^1(X'_v, \mathcal{O}'_v),$$

where $\mathcal{K}$ is the kernel. We may assume that X' dominates the blowing-up of p in X, so that $\mathcal{O}'m = I$ is a locally principal ideal. Then the inclusion $j: U \to X$ is an affine map, and so

$$H^1(U, \mathcal{O}_U) \approx H^1(X', j_*\mathcal{O}_U) \approx \operatorname{inj\,lim}_Y H^1(X', \mathcal{O}'(Y)).$$

By (2.9) (ii), $H^1(X', \mathcal{O}') \subset H^1(U, \mathcal{O})$. So $\mathcal{K}$ is contained in $\mathcal{J}$, hence is bounded, and we may assume f large enough so that $\mathcal{K}$ is maximized. Then the sequence (3.16) shows that $H^1(X', \mathcal{O}')$ is maximized too. Therefore $H^1(X', \mathcal{O}')$ is bounded, as required. □

Lemma (3.17). *Let $M \to N$ be an injective map of finite R-modules. Then the kernel of the map*

$$H^1(U, \tilde{M}) \to H^1(U, \tilde{N})$$

is a finite-dimensional k-vector space.

PROOF. Let $P = N/M$. The kernel is a quotient of P, and also it is m-torsion. Therefore it has finite length. □

§4. Basic Properties of Rational Singularities

This section contains background material and should be skipped by people familiar with rational singularities. Most of it is taken from [8].

We assume throughout this section that p is a *rational singularity* and $X' \xrightarrow{f} X$ is a modification of X at p.

Proposition (4.1). *Let M' be an $\mathcal{O}'$-module generated by global sections. Then $R^q f_* M' = 0$ for all $q > 0$.*

PROOF. The only question is $q = 1$. To say that M' is generated by global sections means that M' is a quotient of a free module F. Since p is a rational singularity, $R^1 f_* F = 0$. Also, $R^1 f_*$ is right exact. Thus $R^1 f_* M' = 0$. □

Proposition (4.2). *Let L be a reflexive $\mathcal{O}$-module, and denote by $\mathcal{O}'L$ the torsion-free $\mathcal{O}'$-module it induces, i.e., $\mathcal{O}'L = \mathcal{O}' \otimes_{\mathcal{O}} L/(\text{torsion})$. Then $\mathcal{O}'L$ is reflexive.*

PROOF. The point to note is that $\mathcal{O}'L$ is generated by sections above any affine open of X. Therefore $R^1 f_* \mathcal{O}'L = 0$, by (4.1). Let L' denote the reflexive hull of $\mathcal{O}'L$, so that there is an exact sequence

$$0 \to \mathcal{O}'L \to L' \to \varepsilon \to 0,$$

where ε has support of dimension zero. Applying f_* gives an exact sequence, since $R^1 f_* \mathcal{O}'L = 0$. Also, $f_* \mathcal{O}'L = f_* L' = L$ because L is reflexive. Therefore $f_* \varepsilon = 0$, and since ε has zero-dimensional support, $\varepsilon = 0$. □

Now suppose that X' dominates the blowing-up of the maximal ideal m at p. This is equivalent with saying that $I = \mathcal{O}'m$ is locally principal. Let $Z = V(I)$ be the divisor defined by I, so that $I = \mathcal{O}'(-Z)$. The ideal I is generated by global sections, namely by elements of m, locally above p. Also the relative dimension of X' over X is 1, and I is locally principal. Therefore one or the other of two generic local sections $x, y \in m$ will generate I at any point of Z. We may localize X so that x, y generate everywhere: $I = (x, y)\mathcal{O}'$.

Proposition (4.3). *With the above notation, let M be a reflexive $\mathcal{O}'$-module. There is an exact sequence*

$$0 \to M(Z) \xrightarrow{\alpha} M \oplus M \xrightarrow{\beta} M(-Z) \to 0,$$

the maps are defined by

$$\alpha = (y, -x): a \rightsquigarrow (ya, -xa),$$

$$\beta = \begin{pmatrix} x \\ y \end{pmatrix}: (u, v) \rightsquigarrow xu + yv.$$

PROOF. β is surjective because $M(-Z) = IM = (x, y)M$. The only part which is not obvious is exactness in the middle. Let q be a point of X', and say that x generates I there, so that $y = zx$ locally. Let $(m', m) \in \ker \beta$, i.e., $xm' = -ym$. Cancel x to obtain $m' = -zm$. Let $a = x^{-1}m$. This is a local section of $M(Z)$, and $\alpha(a) = (m', m)$ as required. □

Proposition (4.4). *With the equation (4.3), $f_* I^r = m^r$ and $R^q f_* I^r = 0$, for all $r \geq 0$, $q > 0$.*

PROOF. The vanishing of $R^q f_* I^r$ for $q > 0$ follows from (4.1). Also, the equality $f_* I^r = m^r$ is clear for $r = 0, 1$. If $r \geq 1$, set $M = I^r = \mathcal{O}'(-rZ)$ in (4.3), to obtain

(4.5) $$0 \to I^{r-1} \to I^r \oplus I^r \xrightarrow{(x, y)} I^{r+1} \to 0.$$

Apply f_* and vanishing of $R^1 f_* I^{r-1}$ to conclude that

$$f_* I^{r+1} = (x, y) f_* I^r \subset m f_* I^r.$$

Since also $m^{r+1} \subset f_* I^{r+1}$, it follows by induction that $f_* I^r = m^r$ for all r, and that $m^{r+1} = (x, y) m^r$ if $r \geq 1$. The assertions for I^r/I^{r+1} follow immediately. □

Proposition (4.6). *Let p be a rational singular point of X of multiplicity μ. Then* $\dim_k m^r/m^{r+1} = r\mu + 1$.

PROOF. Let $d_r = \dim_k m^r/m^{r+1}$. Proposition (4.4) shows that

(4.7) $$f_*(I^r/I^{r+1}) \approx m^r/m^{r+1} \quad \text{and} \quad R^1 f_* I^r/I^{r+1} = 0.$$

The sequences (4.5) form a nested family. Taking successive quotients and applying f_*, we obtain the relation

$$d_{r+1} + d_{r-1} - 2d_r = 0.$$

This relation, with $d_0 = 1, d_1 = d$, has the unique solution $d_r = r(d - 1) + 1$. By definition [14, p. 294], $d - 1 = \mu$ is the multiplicity of X at p. □

Corollary (4.8). *With the notation of (4.3), let $Z = \text{Spec } \mathcal{O}'/I$ be the scheme-theoretic fibre of f. Then $\chi(Z, \mathcal{O}_Z) = 1$, and $(Z \cdot Z) = -\mu$.*

PROOF. By (4.7), $f_* \mathcal{O}'/I = \mathcal{O}/m$, and $R^1 f_* \mathcal{O}'/I = 0$. In other words, $H^0(Z, \mathcal{O}_Z) = k$ and $H^1(Z, \mathcal{O}_Z) = 0$, which shows that $\chi(Z, \mathcal{O}_Z) = 1$. Next, $-(Z \cdot Z) = $ degree $\mathcal{O}_Z(-Z)$, and by Riemann–Roch on Z, $\chi(Z, \mathcal{O}_Z(-Z)) = 1 - (Z \cdot Z)$. On the other hand, $\mathcal{O}_Z(-Z) \approx I/I^2$. By (4.7), $H^0(Z, \mathcal{O}_Z(-Z)) = m/m^2$ has dimension $\mu + 1$, and $H^1(Z, \mathcal{O}_Z(-Z)) = 0$. Thus $-(Z \cdot Z) = \mu$. □

Theorem (4.9) (Lipman [8], Mattuck [10]). *Let $X_1 \xrightarrow{\pi} X$ denote the blowing-up of the maximal ideal m at a rational singular point p of X. Then X_1 is a normal surface.*

PROOF. We may assume that $X = \text{Spec } R$ is affine. Let $A = R[mt] = R \oplus mt \oplus m^2 t^2 \oplus \cdots$ be the subalgebra of $R[t]$ generated by mt. This is a graded ring, and Proj $A = X_1$.

The integral closure $\bar{A}$ of A is a graded subring of $R[t]$, and an element $st^v \in Rt^v$ is in $\bar{A}$ if and only if it satisfies a homogeneous integral equation. Canceling powers of t, we obtain an equation

(4.10) $$s^n + c_1 s^{n-1} + \cdots + c_n = 0,$$

where $c_j \in m^{vj}$. Now by definition, an ideal I is called integrally closed if every element $s \in R$ satisfying an equation (4.10) in which $c_j \in I^i$, lies in I. Thus in our case, s lies in the integral closure M_ν of the ideal m^ν, and $\bar{A} = R \oplus M_1 \oplus \cdots$. To show X normal, it suffices to show that each m^ν is integrally closed.

Let X' be the normalization of X_1, and let $I^\nu = \mathcal{O}'m^\nu$. Since I^ν is locally principal and X' is normal, I^ν is an integrally closed ideal sheaf. Thus $\mathcal{O}'M_\nu = I^\nu$. By Proposition (4.4), $H^0(X', I^\nu) = m^\nu$. Since $m^\nu \subset M_\nu \subset H^0(X', I^\nu)$, we have $m^\nu = M_\nu$, as required. $\square$

§5. Step 2: Blowing Up the Dualizing Sheaf

In this section we consider a surface X with *rational singularities*. The reduction to rational double points is done by the following three propositions:

Proposition (5.1). *Let $p \in X$ be a rational singularity of multiplicity $\mu > 1$, and let $X' \xrightarrow{f} X$ be the blowing-up of p. Then $\omega' = \mathcal{O}'\omega$.*

Proposition (5.2). *Consider the sequence (*) of Section 1. If n is large, then $\mathcal{O}_{X_n}\omega$ is locally free.*

Proposition (5.3). *Let p be a rational singularity of X of multiplicity μ. If ω is locally free at p, then $\mu \le 2$.*

We will prove (5.3) in a more general form due to Wahl [12].

Proposition (5.4). *Let p be a rational singularity of multiplicity $\mu > 1$. Then with the usual notation, $\dim_k(\omega/m\omega) = \mu - 1$.*

Since $\dim(\omega/m\omega)$ is the number of local generators, (5.3) follows from (5.4).

PROOF OF PROPOSITION (5.4). Let x, y be generic elements of m. Then the multiplicity μ of X at p is the multiplicity of the zero-dimensional ring $\bar{\mathcal{O}} = \mathcal{O}_p/(x, y)$, which is just its length (cf. [14, p. 296]). Let $\bar{m}$ be the maximal ideal of $\bar{\mathcal{O}}$, i.e., $\bar{m} = m/(x, y)$. Note that since $\dim_k m/m^2 = \mu + 1$ (4.6), we have $\dim_k \bar{m}/\bar{m}^2 \ge \mu - 1$. Therefore there is no room in $\bar{\mathcal{O}}$ for $\bar{m}^2$, so $\bar{m}^2 = 0$ and $\dim \bar{m} = \mu - 1 > 0$.

Next, the dualizing module $\bar{\omega}$ of $\bar{\mathcal{O}}$ is isomorphic to $\omega/(x, y)\omega$ because $\{x, y\}$ is a regular sequence in $\mathcal{O}_p$. Since $\bar{\mathcal{O}}$ is zero-dimensional, the functor $N \rightsquigarrow N^D = \mathrm{Hom}_{\bar{\mathcal{O}}}(N, \bar{\omega})$ is a perfect duality on finite length $\bar{\mathcal{O}}$-modules. Therefore $k^D = \mathrm{Socle}(\bar{\omega})$ has length 1, and so the dual of the sequence

$$0 \to \bar{m} \to \bar{\mathcal{O}} \to k \to 0$$

has the form

$$0 \to k \to \bar{\omega} \to k^{\mu-1} \to 0.$$

It follows that $k^{\mu-1} \approx \bar{\omega}/\bar{m}\bar{\omega} \approx \omega/m\omega$, as required. $\square$

PROOF OF PROPOSITION (5.1). We know that X' is normal, by (4.9). Also $\omega \approx f_*\omega'$ (3.5). This isomorphism is adjoint to a map $f^*\omega \to \omega'$, hence

(5.5) $$\mathcal{O}'\omega \subset \omega'.$$

Assume that $\mathcal{O}'\omega < \omega'$. Then since both modules are reflexive, $\omega' = \mathcal{O}'\omega(Y)$ for some positive divisor $Y = \sum a_i C_i$ supported on E. We revert to the notation of the previous section, in which $Z = \mathscr{S}pec\, \mathcal{O}'/I$ is the scheme-theoretic fibre. The proof will be completed by contradiction, using the inequality

(5.6) $$\chi(Z, \mathcal{O}'\omega \otimes \mathcal{O}_Z) \leq \chi(Z, \omega' \otimes \mathcal{O}_Z),$$

together with induction and

Lemma (5.7). *Let L be any reflexive module on X', and let C be a component of Z. Then*

$$\chi(Z, L \otimes \mathcal{O}_Z) > \chi(Z, L(C) \otimes \mathcal{O}_Z).$$

To prove (5.6), we compute $\chi(Z, \omega' \otimes \mathcal{O}_Z)$ using the adjunction formula $\omega_Z = \omega'(Z) \otimes \mathcal{O}_Z$ (2.14), and (4.8). Riemann–Roch on Z gives

$$\chi(Z, \omega' \otimes \mathcal{O}_Z) = \chi(Z, \omega_Z) - (Z^2) = \mu - 1.$$

Next, consider the exact sequence

$$0 \to I\mathcal{O}'\omega \to \mathcal{O}'\omega \to \mathcal{O}'\omega \otimes \mathcal{O}_Z \to 0.$$

The middle term is reflexive by (4.2), and so is the one on the left, because I is locally principal. By (4.1), R^1f_* vanishes for the first and second terms, hence for the third as well. Also $f_*(\mathcal{O}'m\omega) \supseteq m\omega$. Applying (5.4), we find

$$\chi(Z, \mathcal{O}'\omega \otimes \mathcal{O}_Z) = h^0(Z, \mathcal{O}'\omega \otimes \mathcal{O}_Z) \leq \dim_k \omega/m\omega = \mu - 1,$$

as required. □

PROOF OF LEMMA (5.7). Consider the diagram

$$\begin{array}{ccccccc}
& & 0 & & 0 & & \\
& & \downarrow & & \downarrow & & \\
0 \to & & L(-Z) & \to & L & \to L \otimes \mathcal{O}_Z \to 0 \\
& & \downarrow & & \downarrow & & \\
0 \to & & L(C-Z) & \to & L(C) & \to L(C) \otimes \mathcal{O}_Z \to 0 \\
& & \downarrow & & \downarrow & & \\
& & L(C-Z) \otimes \mathcal{O}_C & \to & L(C) \otimes \mathcal{O}_C & & \\
& & \downarrow & & \downarrow & & \\
& & 0 & & 0. & &
\end{array}$$

It shows that

$$\chi(Z, L(C) \otimes \mathcal{O}_Z) - \chi(Z, L \otimes \mathcal{O}_Z) = \chi(C, L(C) \otimes \mathcal{O}_C)$$
$$- \chi(C, L(C - Z) \otimes \mathcal{O}_C).$$

Also, since Z is locally principal, the right-hand side of this equation is easily identified as $(Z \cdot C)$, and $(Z \cdot C) < 0$ because $\mathcal{O}'(-Z)$ is ample on the blowing-up X'. This proves the lemma. $\square$

PROOF OF PROPOSITION (5.2). There exists a modification $X' \xrightarrow{f} X$ such that $\mathcal{O}'\omega$ is locally free. Namely, one can blow up the module ω, for instance by realizing ω locally as an ideal, and then normalize. (It can be shown *a posteriori that the blowing up of ω is already normal, and that its singularities are rational double points.*) By (1.9), this modification X' can be dominated by a sequence of blowings-up. Now since the dualizing sheaf (or any reflexive module) is locally free at a nonsingular point, it is not necessary to blow such a point up. Therefore the sequence (∗) leads to a locally free sheaf $\mathcal{O}_{X_n}\omega = \omega_{X_n}$. $\square$

§6. Step 3: Resolution of Rational Double Points

Now suppose that the only singularities of X are *rational double points*. At this point the dualizing sheaf has done most of its work, and a closer analysis is required. We will consider only those modifications $X' \xrightarrow{f} X$ which are obtained by a sequence of blowings-up at singular points. Then (5.1) $\omega' = \mathcal{O}'\omega$, hence ω' is locally free, and is in fact free above a neighborhood of any point p of X. By (3.2)(ii) and (5.3), the singularities of X' are rational double points.

Lemma (6.1). *Let $Y \subset X$ be a nonsingular closed subscheme of codimension 1, and let p be a point of Y. Then p is a nonsingular point of X if and only if the ideal $\mathcal{O}(-Y)$ of Y in X is locally principal at p.*

PROOF. The maximal ideal of $\mathcal{O}_{Y,p}$ is locally principal. Therefore, if $\mathcal{O}(-Y)$ is locally principal, then the maximal ideal of $\mathcal{O}_{X,p}$ is generated by two elements, and so X is nonsingular at p. The converse follows from the unique factorization property of regular local rings. $\square$

The following lemma is proved at the end of the section.

Lemma (6.2). *Let $Y \subset X$ be a nonsingular closed subscheme of codimension 1. Let Y_n denote the proper transform of Y on the surface X_n of the sequence (∗). Then X_n is nonsingular at every point of Y_n, if $n \gg 0$.*

Now consider the case of a single blowing-up $X_1 \xrightarrow{f_1} X$ of a double point p. The scheme-theoretic fibre Z_1 is the projectivized tangent cone to X at p. Since the multiplicity is 2 and $\dim_k m/m^2 = 3$ (4.6), Z_1 is represented as a divisor of degree 2 in $\mathbb{P}_k^2$. We will use the following lemma without proof. (Caution: the case of characteristic 2 needs to be considered carefully.)

Lemma (6.3). *Let k be a field and $Z \subset \mathbb{P}_k^2$ a divisor of degree 2. Then Z is of one of the following types:*

(i) *A reduced nonsingular scheme.*
(ii) *A cone of k-degree 2 over a reduced subscheme of $\mathbb{P}_k^1$; the vertex of the cone is its unique singular point.*
(iii) *A double line: $Z = 2C$.*

Lemma (6.4). *Let $X_1 \xrightarrow{f_1} X$ be the blowing-up of X at p, and let Z_1 be the fibre over p.*

(i) *If a point p_1 is nonsingular on Z_1, it is also nonsingular on X_1.*
(ii) *Assume that there is some nonsingular subscheme $Y \subset X$ of codimension 1, passing through p. Let Y' be its proper transform on X_1, and let $p_1 = Y' \cap Z_1$. If p_1 is singular on Z_1, then it is singular on X_1.*

PROOF. (i) The scheme Z_1 is defined by the locally principal ideal $I = \mathcal{O}_1 m$ in X_1. So we may apply Lemma (6.1).

(ii) It follows from the fact that Y is nonsingular, that Y' meets Z_1 transversally, i.e., that the scheme-theoretic intersection $Y' \cap Z_1$ is the reduced point p_1. If p_1 is singular on Z_1, then the tangent space to $Y' \cup Z_1$ at p_1 has dimension ≥ 3. Therefore so does the tangent space to X_1 at p_1, and so X_1 is singular at P_1. □

Lemma (6.5). *Let X_1 be the blowing-up of p in X, and suppose that the fibre has the form $Z_1 = 2C$, where C is a line. Let $p_1, \ldots, p_n$ be the points and infinitely near points which must be blown up on X_1 and its blowings-up to remove the singularities from C, as in (6.2). Then*

$$\sum_i [k(p_i) : k] = 3.$$

PROOF. By *conormal bundle* to C in X_1 we mean the locally free sheaf $\mathcal{O}_C(-C) = \mathcal{O}_{X_1}(-C)/\mathcal{O}_{X_1}(-2C)$. Its degree is, by definition (2.2), $[-C \cdot C]$. In our case, since $2C$ is isomorphic to a double line in $\mathbb{P}^2$, the degree is the same as for such a line, i.e., $[-C \cdot C] = -1$.

Next, blowing up a point q on C has the obvious effect on the conormal bundle. That is, if C' denotes the proper transform of C, then

$$[-C' \cdot C'] = [-C \cdot C] + [k(q) : k].$$

This is because blowing up q leads to an exact sequence
$$0 \to \mathcal{O}_C(-C) \to \mathcal{O}_{C'}(-C') \to k(q) \to 0$$
for the conormal bundles.

Finally, let $X'' \xrightarrow{f} X_1$ denote a modification so that X'' is nonsingular along the proper transform C'', and obtained by a sequence of blowings-up at singular points. Since C'' is a fibre component of X'' over X, ω'' is free along C''. Therefore the genus formula $\frac{1}{2}(C'' \cdot C'') + 1 = 0$ (which follows from (2.14)) shows that $[-C'' \cdot C''] = (-C'' \cdot C'') = 2$. The lemma follows. □

We now proceed with the proof that the sequence (∗) leads to a resolution of rational double points. We may assume that X has one singular point p. Let $X_1 \xrightarrow{f_1} X$ denote its blowing-up.

Case 1. The fibre Z_1 is a nonsingular curve (6.3) (i). In this case, X_1 is nonsingular by (6.4) (i), and we are done.

Case 2. Z_1 is a reduced cone (6.3)(ii). Then the vertex p_1 of the cone is the only point which may be singular on X_1, by (6.4)(i). Assume that it is singular, and let $X_2 \xrightarrow{f_2} X_1$ be the blowing-up of p_1. It induces a blowing-up $Z_1' \to Z_1$ of the cone. The fibre P' of $Z_1' \to Z_1$ is a reduced, zero-dimensional subscheme of $\mathbb{P}^2$ of degree 2, which lies on the line L determined by the two-dimensional tangent space to Z_1 at p_1. Let Z_2 denote the fibre of X_2 over X_1. Then $Z_2 \cap L = P'$ scheme-theoretically. This shows that Z_2 is nonsingular at P'. Therefore Z_2 is reduced. Thus, by (6.4), X_2 has at most one singular point p_2, and it is not on P'. If p_2 is singular we blow it up and repeat the argument.

Suppose that this process were to continue indefinitely. Then we would obtain a sequence $p_1, p_2, \ldots$ of points infinitely near to p. Since p_1 is the vertex of a cone defined over $k(p)$, $k(p_1) = k(p)$, and similarly $k(p_n) = k(p)$ for all n. Also, p_2 does not lie on the proper transform Z_1' of Z_1, i.e., is not a satellite point [15, p. 8], and the same is true of each p_n. Such a sequence of infinitely near points, defines a nonsingular *branch* Y, i.e., a closed codimension 1 subscheme of the completion of X at p. (We leave the proof of this fact to the reader.) Replace X by its completion, and apply Lemma (6.2). A finite sequence of blowings-up separates Y from the singular locus. This contradicts the construction of the sequence $\{p_n\}$, and completes the proof in case 2.

It remains to consider the possibility that Z is a double line, and we split this up into cases according to the degree of the singular locus of X_1. Define $\delta = \sum_q [k(q):k]$, where q runs over the singular points of X_1. So, $\delta \le 3$ by Lemma (6.5).

Case 3. $Z_1 = 2C$ and $\delta = 3$. Let X_2 be the result of blowing up the singular locus. Lemma (6.5) shows that X_2 is nonsingular at every point of the proper transform C' of C. Therefore by (6.4) (ii) the exceptional fibres of $X_2 \to X_1$ are reduced, and we are back to case 1 or 2.

Case 4. $Z_1 = 2C$ and $\delta = 2$. Let $X_2 \to X_1$ denote the blowing-up of the singular locus. By (6.5), one more blowing-up at a rational point is required to separate the proper transform C' from the singular locus. This implies that the singular locus on C consists of two rational points, and that the exceptional fibre of X_2 over one of them is reduced (cases 1, 2). Let p_1 denote the other point. We need only consider the case that the fibre Z_2 of X_2 over X_1 at p_1 is not reduced. Also, we know that blowing up X_2 at $Z'_1 \cap Z_2$ leads to a reduced exceptional fibre, by (6.4)(ii), hence back to cases 1, 2 again. The only situation which does not lead back to previous cases is that Z_2 contains precisely one rational singular point p_2 besides $Z'_1 \cap Z_2$. Then we may repeat the above considerations to construct a sequence $p_1, p_2 \ldots$ of infinitely near points, which is settled as in case 2.

Case 5. $Z_1 = 2C$ and $\delta = 1$. Let $X_2 \to X_1$ be the blowing-up at the unique rational singular point p. This reduces to previous cases unless $Z_2 = 2C_2$, and X_2 has a unique rational singular point p_2 at $Z'_1 \cap Z_2$. If so, let $X_3 \to X_2$ denote the blowing-up at p_2. Then (6.5) predicts that X_3 has at least two singular points, namely at $Z''_1 \cap Z_3$ and $Z''_2 \cap Z_3$, where Z''_i are the proper transforms of Z_i on X_3. Thus we are back to a previous case, and the proof is complete. □

PROOF OF LEMMA (6.2). This argument is due to Giraud. Note that by (4.9) the blowings-up are automatically normal, and so normalization is not needed in the construction of the sequence (∗). The proof given here makes no further use of rationality. It applies to any local scheme X and nonsingular subscheme Y of dimension 1 such that X is nonsingular at the generic point of Y.

Let $\mathcal{O} \to \mathcal{O}'$ denote the map of local rings given by the blowing up of X at some point $p \in Y$. Denote by P the ideal of Y in $\mathcal{O}$ and by P' the ideal of its proper transform Y' in $\mathcal{O}'$. Let $t \in \mathcal{O}$ be an element whose residue in $\mathcal{O}/P$ generates the maximal ideal and let $\mathrm{gr}_p \mathcal{O}$ denote the graded ring $\mathcal{O}/P \oplus P/P^2 \oplus \cdots$. Giraud constructs a surjective homomorphism

(6.6) $$\phi: \mathrm{gr}_p \mathcal{O} \to \mathrm{gr}_{p'} \mathcal{O}'$$

whose kernel contains all elements killed by t. If ϕ is also injective, then $\mathrm{gr}_p \mathcal{O}$ is torsion free, and P/P^2 is a free module over the discrete valuation ring $\mathcal{O}/P$. It has rank $d - 1$ ($d = \dim \mathcal{O}$) because X is nonsingular at the generic point of Y. Therefore $m = (P, t)$ is generated by d elements, and X is nonsingular at p. In any case, $\mathrm{gr}_p \mathcal{O}$ is noetherian, and so the kernel will be zero after a finite sequence of blowings-up. Thus X_n is nonsingular at P_n if $n \gg 0$, as required.

It remains to construct the map (6.6). Choose generators so that $m = (y_1, \ldots, y_{n-1}; t)$, where $P = (y_1, \ldots, y_{n-1})$. Let $y'_i = t^{-1} y_i$. Then $\mathcal{O}'$ is the localization of $\mathcal{O}[y'_1, \ldots, y'_{n-1}]$ at $(y', t) = m'$. Note that $P' = (y'_1, \ldots, y'_{n-1}) = t^{-1} P\mathcal{O}'$. Therefore we can define, for all $r \geq 0$, $\mathcal{O}$-linear maps $P^r \to P'^r$, by

$x \rightsquigarrow t^{-r}x$. These maps send $P^{r+1} \to P'^{r+1}$. Therefore they define the required homomorphism ϕ (6.6). Since Y is normal and of dimension 1, $\mathcal{O}/P \approx \mathcal{O}'/P'$. This, together with the fact that $P' = t^{-1}P\mathcal{O}'$, implies that $P/P^2 \xrightarrow{\phi} P'/P'^2$ is surjective. Therefore ϕ is surjective. Its kernel contains all elements killed by t. For, if (say) $x \in P^r$ and $tx \in P^{r+1}$, so that the residue $\bar{x}$ of x in $\operatorname{gr}_P \mathcal{O}$ is annihilated by t, then $t^{-r}x = t^{-r-1}(tx) \in P'^{r+1}$, hence $\phi(x) = 0$. This completes the proof of Lemma (6.2). □

REFERENCES

[1] Abhyankar, S. S. Resolution of singularities of algebraic surfaces, in *Algebraic Geometry*. Oxford University Press: London, 1969, pp. 1–11.

[2] Abhyankar, S. S. Quasirational singularities. *Amer. J. Math.*, **101** (1979), 267–300.

[3] Grauert, H. and Riemenschneider, O. Verschwindugssätze fur analytische Kohomologiegruppen. *Invent. Math.*, **11** (1970), 263–292.

[4] Grothendieck, A. and Dieudonné J., Éléments de géométrie algébrique, *Publ. Math. I.H.E.S.*, **8, 11, 17, 20, 24, 28, 32** (1961–67).

[5] Hartshorne, R. *Residues and Duality*. Lecture Notes in Mathematics, 20. Springer-Verlag: Heidelberg, 1966.

[6] Hironaka, H. Desingularization of excellent surfaces. *Seminar in Algebraic Geometry*, Bowdoin, 1967 (mimeographed notes).

[7] Laufer, H. On rational singularities. *Amer. J. Math.*, **94** (1972), 597–608.

[8] Lipman, J. Rational singularities, *Publ. Math. I.H.E.S.*, **36** (1969), 195–279.

[9] Lipman, J. Desingularization of two-dimensional schemes. *Ann. Math.*, **107** (1978), 151–207.

[10] Mattuck, A. Complete ideals and monoidal transforms. *Proc. Amer. Math. Soc.*, **26** (1970), 555–560.

[11] Nagata, M. Note on complete local integrity domains. *Mem. Coll. Sci. Univ. Kyoto, Ser. A Math.*, **28**, no. 3 (1954), 271–278.

[12] Wahl, J. Equations defining rational singularities. *Ann. Sci. École Norm. Sup.*, **10** (1977), 231–264.

[13] Zariski, O. Introduction to the problem of minimal models, in *Collected Papers*, Vol. II. M.I.T. Press: Cambridge, MA, 1973, p. 325.

[14] Zariski, O. and Samuel, P. *Commutative Algebra*, Vol. II. Van Nostrand: Princeton, NJ, 1960.

[15] Zariski, O. *Algebraic Surfaces*, 2nd edn. Springer-Verlag: Berlin, 1971.

CHAPTER XII

An Introduction to Arakelov Intersection Theory

T. CHINBURG

In this chapter we review the basic definitions of Arakelov intersection theory, and then sketch the proofs of some fundamental results of Arakelov, Faltings and Hriljac. Many interesting topics are beyond the scope of this introduction, and may be found in the references [2], [3], [8], [12], [20] and their bibliographies.

In section 4 we give a proof of the Arakelov adjunction formula based on the Faltings Riemann-Roch formula. We thank P. Hriljac for showing us the statement of the adjunction formula. We also thank S. Lang for pointing out in [13, p. 100, Remark] an error in the sketch of the proof Theorem 4.1 in the 1986 version of this paper. For a different approach to Theorem 4.1, see [13].

§1. Definition of the Arakelov Intersection Pairing

Throughout this chapter, K will be a number field with ring of integers $\mathfrak{O}_K$ and $\mathscr{X}$ will be a regular curve over $S = \mathrm{Spec}(\mathfrak{O}_K)$. We will assume that K is algebraically closed in $K(\mathscr{X})$. By [14], [7, Lemma 7.1(c)], the fibre of $\mathscr{X}$ over the generic point of S is geometrically irreducible.

Let S_{inf} denote the set of infinite places of K. An Arakelov divisor (cf. [2]) of $\mathscr{X}$ is a finite formal linear combination

$$D = D_{\mathrm{fin}} + D_{\mathrm{inf}} = \sum_i k_i C_i + \sum_{\infty \in S_{\mathrm{inf}}} \lambda_\infty F_\infty,$$

in which the k_i are integers, each C_i is an irreducible closed subscheme of $\mathscr{X}$ of codimension one in $\mathscr{X}$, the λ_∞ are real numbers, and F_∞ is a symbol standing for the fibre of $\mathscr{X}$ above ∞. If $\pi(C_i) = S$, then C_i is called horizontal.

Otherwise, $\pi(C_i)$ is a closed point of S and C_i is called vertical. The fibres F_∞ for $\infty \in S_{\inf}$ will be called vertical and irreducible.

For $\infty \in S_{\inf}$, let ds_∞^2 be a hermitian metric on the Riemann surface $\mathscr{X}_\infty = X(\bar{K}_\infty)$, where $\bar{K}_\infty \cong \mathbb{C}$. Let du_∞ be the volume element associated to ds_∞^2. We assume that ds_∞^2 has been normalized so that

$$\int_{\mathscr{X}_\infty} du_\infty = 1.$$

The Arakelov divisor of a function $f \in K(\mathscr{X})$ is

$$(f) = (f)_{\text{fin}} + \sum_{\infty \in S_{\inf}} v_\infty(f) F_\infty,$$

where (f) is the usual Weil divisor of f, and where

$$v_\infty(f) = \int_{\mathscr{X}_\infty} -\log|f|_\infty \, du_\infty$$

is a real number. Here $|\ |_\infty$ is the Euclidean absolute value on $\bar{K}_\infty \cong \mathbb{C}$. (In [2] Arakelov uses the embeddings of K into $\mathbb{C}$ rather than $S_{\inf}$. To translate between places and embeddings, one should view F_∞ as the sum of the fibres of $\mathscr{X}$ over the embeddings of K into $\mathbb{C}$ which correspond to the place ∞.)

The Arakelov intersection $[D_1, D_2]$ of two distinct irreducible Arakelov divisors D_1 and D_2 is defined in the following way.

Suppose first that $D_2 = F_\infty$ for some $\infty \in S_{\inf}$. Let $e_\infty = 1$ (resp. 2) if ∞ is real (resp. complex). Define $[D_1, F_\infty] = [F_\infty, D_1]$ to be 0 (resp. $e_\infty m$) if D_1 is vertical (resp. if D_1 is horizontal and of degree m on the fibre of $\mathscr{X}$ over the generic point of S).

Suppose now that D_1 and D_2 are distinct, finite irreducible divisors. We let

$$[D_1, D_2] = [D_1, D_2]_{\text{fin}} + [D_1, D_2]_{\inf},$$

where $[D_1, D_2]_{\text{fin}}$ is the usual intersection multiplicity of Weil divisors on $\mathscr{X}$, and where $[D_1, D_2]_{\inf}$ will be defined below. (Recall that

$$[D_1, D_2]_{\text{fin}} = \sum \log \#(\mathscr{O}_{\mathscr{X}, P}/(f_{1,P}, f_{2,P})),$$

where the summation is over the points P of $\mathscr{X}$ such that $\dim \mathscr{O}_{\mathscr{X}, P} = 2$, and $f_{i,P}$ is a local equation for D_i at P. Compare [14, §I.1] and [7, Prop. 4.1].)

Define $[D_1, D_2]_{\inf} = 0$ if either D_1 or D_2 is a component of a vertical fibre.

Suppose now that D_1 and D_2 are irreducible and horizontal. The function field $K(D_i)$ of D_i is then a finite extension of K. The generic point of D_i determines a point P_i on $\mathscr{X}(K(D_i))$. Fix $\infty \in S_{\inf}$ and an embedding τ_∞ of K into $\mathbb{C}$ inducing ∞. Each P_i determines a collection $\{P_{i,j}^\infty : j = 1, \ldots, [K(D_i):K]\}$ of points on $\mathscr{X}_\infty$ corresponding to the various embeddings of $K(D_i)$ into $\mathbb{C}$ which extend τ_∞. Define

$$[D_1, D_2]_{\inf} = \sum_{\infty \in S_{\inf}} e_\infty \sum_{j,k} -\log G_\infty(P_{1,j}^\infty, P_{2,k}^\infty),$$

where $G_\infty(P, z)$ is the unique function specified by the following proposition for the Riemann surface $\mathscr{X}_\infty$ and the hermitian metric ds_∞^2.

Proposition 1.1 (Arakelov [2]). *Let X be a Riemann surface and let P be a point on X. Let ds^2 be a hermitian metric on X with associated volume form du. There is a unique function $G(P, z)$ of $z \in X$ having the following properties:*

(1.1) $G(P, z)$ *is a smooth nonnegative real-valued function of z which has a unique zero at $z = P$, this zero being of first order.*

(1.2) $$\frac{1}{\pi i} \frac{\partial}{\partial z} \frac{\partial}{\partial \bar{z}} \log G(P, z) \, dz \wedge d\bar{z} = du \qquad \text{for } z \neq P.$$

(1.3) $$\int_X \log G(P, z) \, du = 0.$$

One has $G(P, z) = G(z, P)$ for all z and P on X.

The existence and uniquencess of $G(P, z)$ is shown in [2, §2]; the function $g(P, Q) = \log G(P, Q)$ will be called Green's function of X with respect to ds^2. The first main result of [2] is

Theorem 1.1 (Arakelov). *The pairing $[D_1, D_2]$ extends in a unique way to a symmetric, bilinear pairing on all Arakelov divisors which depends only on the linear equivalence classes of D_1 and D_2.*

EXERCISE

Let $K = \mathbb{Q}$ and let $\mathscr{X}$ be $P_\mathbb{Z}^1$. Let ds_∞^2 have volume form

$$du_\infty = \frac{i \, dz \wedge d\bar{z}}{2\pi(1 + |z|^2)^2}.$$

Verify that

$$G(P, z) = \frac{e^{1/2}|P - z|}{(1 + |P|^2)^{1/2}(1 + |z|^2)^{1/2}},$$

where $|\ |$ is the Euclidean absolute value. Let $D(\infty)$ (resp. $D(a)$ for $a \in A_\mathbb{Q}^1$) be the irreducible horizontal divisor on $\mathscr{X}$ which is the Zariski closure of the point at infinity (resp. at a) on the general fibre $P_\mathbb{Q}^1$ of $\mathscr{X}$. Show

$$[D(a), D(\infty)] = \sum_p \max(0, \log|a|_p) + \log|1 + |a|^2| - 1/2,$$

where $|\ |_p$ is the usual normalized absolute value at the prime p.

Remark 1.1 (cf. [6]). The expression $[D(a), D(\infty)] + 1/2$ above is the capacity, in the sense of Cantor [5] and Rumely [18], of the adelic points of $\mathscr{X}$ which are as close to $D(a)$ as to $D(\infty)$ in every adelic fibre of $\mathscr{X}$, where the distance function on each adelic fibre is the spherical one defined in [5], [18] and [6].

A useful technique for studying $[D_1, D_2]$ is to pass to a finite extension L of the base field K. Let $S(L) = \mathrm{Spec}(\mathfrak{O}_L)$ and let $\mathscr{X}_L = \mathscr{X} \times_{S(K)} S(L)$. In [11, p. 7], Hriljac shows that $\mathscr{X}_L$ is a curve over $S(L)$, which amounts to checking that $\mathscr{X}_L$ is integral. If L/K is ramified at some places of K over which the fibres of $\mathscr{X}$ are reducible, it is possible that $\mathscr{X}_L$ is not regular. By results of Abhyankar [1] and Lipman [15], [4], one may perform a finite sequence of blow-ups and normalizations to arrive at a regular curve $\mathscr{X}_L^{\mathrm{reg}}$ over $S(L)$ together with a proper birational $S(L)$-morphism $\mathscr{X}_L^{\mathrm{reg}} \to \mathscr{X}_L$. Let $\sigma: \mathscr{X}_L^{\mathrm{reg}} \to \mathscr{X}$ be the induced $S(K)$-morphism. Suppose ∞' is an infinite place of L, and that ∞ is the place of K under ∞'. We may identify $\bar{L}_{\infty'}$ in a natural way with $\bar{K}_\infty$, and in this way $\mathscr{X}_{L_{\infty'}}^{\mathrm{reg}} = \mathscr{X}_L^{\mathrm{reg}}(\bar{L}_{\infty'})$ may be identified with $\mathscr{X}_\infty$. Let $\{du_{\infty'}\}_{\infty'}$ be the set of hermitian metrics on the $\mathscr{X}_{L_{\infty'}}^{\mathrm{reg}}$ which result from the metrics $\{du_\infty\}$ on the $\mathscr{X}_\infty$. Let σ^* be the linear map on Arakelov divisors which is induced by the pull-back of Weil divisors from $\mathscr{X}$ to $\mathscr{X}_L^{\mathrm{reg}}$ and by $\sigma^*(F_\infty) = \sum_{\infty'|\infty} F_{\infty'}$ for ∞ an infinite place of K.

The following result is proved by Hriljac in [11, Prop. 10].

Proposition 1.2. *Let L/K, $\sigma: \mathscr{X}_L^{\mathrm{reg}} \to \mathscr{X}$ and $\{du_{\infty'}\}_{\infty'}$ be as above. Then*

$$[\sigma^*(D_1), \sigma^*(D_2)]_L = [L:K][D_1, D_2]$$

for all Arakelov divisors D_1 and D_2 on $\mathscr{X}$, where $[\ ,\ ]_L$ is the Arakelov intersection pairing on $\mathscr{X}_L^{\mathrm{reg}}$.

Remark 1.2. A finite irreducible horizontal divisor D on $\mathscr{X}$ has the form $\varepsilon(\mathrm{Spec}(\mathfrak{O}_{K(D)}))$ for some morphism $\varepsilon: \mathrm{Spec}(\mathfrak{O}_{K(D)}) \to \mathscr{X}$ over $S(K)$. If $K(D) = K$ then ε is a section of $\pi: \mathscr{X} \to S(K)$ and D is nonsingular; otherwise D may be singular. Let L be the Galois closure of $K(D)$ over K. The divisor $\pi_L^*(D)$ on $\mathscr{X}_L^{\mathrm{reg}}$ will be a sum of sections of the projection $\pi_L: \mathscr{X}_L^{\mathrm{reg}} \to S(L)$ and of irreducible components of the fibres of π_L.

§2. Metrized Line Bundles

A hermitian line bundle L on a Riemann surface X is a line bundle together with a hermitian inner product on the fibre of L over each point x of X, these inner products varying smoothly with x. Let $|\ |$ be the induced norm on the fibres of L. Let s be a meromorphic section of L, and let z be a local coordinate on X. The $(1,1)$-form

$$\mathrm{curv}(L) = \frac{\partial^2}{\partial z\, \partial \bar{z}} \log(|s|^2)\, dz \wedge d\bar{z}$$

is independent of the choice of s and z, and is called the curvature form of L. This form satisfies

$$\int_X \mathrm{curv}(L) = 2\pi i \deg(L).$$

Definition 2.1. The hermitian line bundle L on X is admissible with respect to the hermitian metric ds^2 on X if $\operatorname{curv}(L) = 2\pi i \deg(L)\, du$, where du is the $(1, 1)$-form which is the volume form of ds^2.

Proposition 2.1 (Arakelov [2]). *Each line bundle on X has an admissible hermitian metric with respect to ds^2, which is unique up to multiplication by a scalar.*

The admissible metric of this proposition is connected in the following way to the Green's function $g(P, Q)$ of X with respect to ds^2. Let Q be a point of X. The constant function 1 is a distinguished section of the line bundle $\mathcal{O}_X(Q)$ on X. We can put a unique hermitian metric $|\ |$ on $\mathcal{O}_X(Q)$ by letting

$$|1|(P) = G(P, Q) = \exp(g(P, Q)),$$

where $|1|(P)$ is the norm of the section 1 at the point P of X, and where $G(P, Q)$ is the function of Proposition 1.1. This $|\ |$ is admissible because of (1.1) and (1.2); property (1.3) distinguishes $|\ |$ from its positive scalar multiples. By tensoring we arrive at an admissible metric on $\mathcal{O}_X(C)$ for each divisor C of X, which we will call the Green's metric on $\mathcal{O}_X(C)$.

Suppose now that $\mathscr{X}$ is a curve over the ring of integers $\mathfrak{O}_K$ of a number field K, and that for each $\infty \in S_{\inf}$, we have a hermitian metric ds_∞^2 on the Riemann surface $\mathscr{X}_\infty$. A metrized line bundle on $\mathscr{X}$ is a line bundle $\mathscr{L}$, together with a hermitian metric $|\ |_\infty$ on the line bundle $\mathscr{L}_\infty$ induced by $\mathscr{L}$ on $\mathscr{X}_\infty$ for each $\infty \in S_{\inf}$. We will call $\mathscr{L}$ admissible if each $\mathscr{L}_\infty$ is admissible with respect to ds_∞^2.

An admissible line bundle $\mathcal{O}_{\mathscr{X}}(D)$ may be associated to each Arakelov divisor

$$D = D_{\text{fin}} + \sum_{\infty \in S_{\inf}} \lambda_\infty F_\infty$$

of $\mathscr{X}$ in the following way. Define $\mathcal{O}_{\mathscr{X}}(D)$ to have $\mathcal{O}_{\mathscr{X}}(D_{\text{fin}})$ as its underlying bundle, and to have the metric at $\infty \in S_{\inf}$ which is the Green's metric on the induced line bundle on $\mathscr{X}_\infty$ multiplied by the factor $\exp(-\lambda_\infty)$. Note that $\mathcal{O}_{\mathscr{X}}(D)$ has the distinguished meromorphic section 1.

Suppose now that $\mathscr{L}$ is any admissible metrized line bundle on $\mathscr{X}$, and that s is a meromophic section of $\mathscr{L}$. The divisor of s is defined to be

$$\operatorname{div}(s) = (s)_{\text{fin}} + \sum v_\infty(s) F_\infty,$$

where $(s)_{\text{fin}}$ is the usual divisor of s on $\mathscr{X}$, and where

$$v_\infty(s) = \int_{\mathscr{X}_\infty} -\log |s|_\infty \, du_\infty.$$

From the uniqueness of admissible metrics on $\mathscr{L}$ up to scalar multiples, we conclude that $\mathscr{L}$ is isometric to the line bundle $\mathcal{O}_{\mathscr{X}}(\operatorname{div}(s))$ constructed above, with the meromorphic section s of $\mathscr{L}$ corresponding to the meromorphic section 1 of $\mathcal{O}_{\mathcal{X}}(\text{div}(s))$. The finite divisor $(s)_{\text{fin}}$ determines s up to multiplication by a unit of $\mathfrak{O}_K$, while $\text{div}(s)$ determines s up to multiplication by a root of unity. If $\mathcal{L}'$ is another admissible metrized line bundle on $\mathcal{X}$, and s' is a section of $\mathcal{L}'$, then $\mathcal{L}$ and $\mathcal{L}'$ are isometric iff $\text{div}(s) - \text{div}(s')$ is a principal Arakelov divisor. Thus we may unambiguously define

$$[\mathcal{L}, \mathcal{L}'] = [\text{div}(s), \text{div}(s')]$$

if s and s' are any sections of $\mathcal{L}$ and $\mathcal{L}'$, respectively. One also concludes that the group of isometry classes of admissible metrized line bundles on $\mathcal{X}$ is naturally isomorphic to $\text{Div}^{\wedge}(\mathcal{X})/P^{\wedge}(\mathcal{X})$, where $\text{Div}^{\wedge}(\mathcal{X})$ (resp. $P^{\wedge}(\mathcal{X})$) is the group of Arakelov divisors (resp. principal Arakelov divisors) on $\mathcal{X}$. (Since we have used infinite places rather than embeddings of K into $\mathbb{C}$, these groups are different than the ones appearing in [2].)

Suppose that $s: S = \text{Spec}(\mathfrak{O}_K) \to \mathcal{X}$ is a section of the projection $\pi: \mathcal{X} \to S$. Let $D_1 = s(S)$, and let D_2 be any Arakelov divisor on $\mathcal{X}$. The hermitian metrics of $L_2 = \mathcal{O}_{\mathcal{X}}(D_2)$ make $s^*(L_2)$ a metrized line bundle on S. We recall from [19] that $s^*(L_2)$ therefore corresponds to a projective rank one $\mathfrak{O}_K$ module P, together with hermitian metrics at the infinite places of K. The degree of $s^*(L_2)$ is defined to be

$$\deg(P) = \log(\#(P/\mathfrak{O}_K p)) - \sum_{\infty \in S_{\text{inf}}} e_{\infty} \log|p|_{\infty}$$

if p is any nonzero element of P, where $|\ |_{\infty}^{e_{\infty}}$ is the norm on $K_{\infty} \otimes_{\mathfrak{O}_K} P$ which corresponds to the normalized absolute value of K at ∞. If the finite part of D_2 is effective, the following formula for $[D_1, D_2]$ results on letting $p = s^*(1)$. The formula holds for arbitrary D_2 because $[D_1, D_2]$ and $\deg(s^*(L_2))$ are linear in D_2.

Proposition 2.2. *Let* $s: S \to \mathcal{X}$ *be a section of* $\pi: \mathcal{X} \to S$, *and let* $D_1 = s(S)$. *Then*

$$[D_1, D_2] = \deg(s^*(\mathcal{O}_{\mathcal{X}}(D_2)))$$

for all Arakelov divisors D_2.

§3. Volume Forms

In this section X will be a Riemann surface of genus $g > 0$. Let Ω_X^1 be the sheaf of holomorphic 1-forms on X. There is a natural hermitian inner product on $\Gamma(X, \Omega_X^1)$ given by

$$\langle w_1, w_2 \rangle = (i/2) \int_X w_1 \wedge \bar{w}_2.$$

Let $w_1, \ldots, w_g$ be an orthonormal basis for $\Gamma(X, \Omega_X^1)$ with respect to this inner product. We will fix a hermitian metric ds^2 on X by requiring it to correspond to the (1, 1)-form

(3.1) $$du = (i/2g) \sum_{j=1}^{g} w_j \wedge \bar{w}_j$$

on X. Viewing du as a volume form, one has $\int_X du = 1$. Choosing a basepoint P_0 on X, we have the usual embedding

$$P \to \left(\int_{P_0}^{P} w_1, \ldots, \int_{P_0}^{P} w_g \right)$$

of X into its Jacobian $J(X) = \mathbb{C}^g/\Lambda$, where Λ is the period lattice of X. The metric on $J(X)$ induced by the usual Euclidean metric on $\mathbb{C}^g$ will be called the flat metric on $J(X)$. This metric induces the (1, 1)-form $g\, du$ on X.

The metric ds^2 has the following useful property. Let P be a point of X. Give $\mathbb{C}$ the canonical hermitian metric, and give $\mathcal{O}_X(P)$ Green's metric defined in the paragraph following Proposition 2.1. The residue of a differential at P gives an isomorphism from the fibre at P of the line bundle

$$\Omega_X^1(P) = \Omega_X^1 \otimes \mathcal{O}_X(P)$$

to $\mathbb{C}$. We give Ω_X^1 the hermitian metric such that for all P, this residue map is an isometry.

Proposition 3.1 (Arakelov [2]). *The above metric on Ω_X^1 is admissible if ds^2 is determined by equation (3.1).*

Let L be a line bundle on X. The classical Riemann–Roch formula gives an expression for the Euler characteristic

$$\chi(L) = \dim_\mathbb{C} H^0(X, L) - \dim_\mathbb{C} H^1(X, L)$$

of L. Suppose now that $\mathscr{L}$ is a metrized line bundle on a curve $\mathscr{X}$ over a number field. To develop an analogous metrized Euler characteristic $\chi(\mathscr{L})$, one would like to replace $\dim_\mathbb{C} H^j(X, L)$ for $j = 0$ and 1 by the negative of the logarithm of the covolume of the lattice $H^j(\mathscr{X}, \mathscr{L})$ in the vector space $H^j(\mathscr{X}, \mathscr{L}) \otimes_\mathbb{Z} \mathbb{R}$ with respect to some natural volume form on $H^j(\mathscr{X}, \mathscr{L}) \otimes_\mathbb{Z} \mathbb{R}$.

The object of this section will be to discuss volume forms due to Faltings [8] on the formal difference $H^0(X, L) - H^1(X, L)$ when L is an admissible metrized line bundle on a Riemann surface X. In the next section we discuss how these volume forms give rise to an Euler characteristic $\chi(\mathscr{L})$ with desirable properties, e.g. for which one has a Riemann–Roch formula.

If V is a complex vector space of dimension d, define $\Lambda(V) = \Lambda^d(V)$. To give a volume form on V is the same as giving a hermitian metric on V. For L a line bundle on a Riemann surface X, let

$$\lambda(R\Gamma(X, L)) = \mathrm{Hom}_\mathbb{C}(\Lambda(H^1(X, L)), \Lambda(H^0(X, L))).$$

We will also denote $\lambda(R\Gamma(X, L))$ by $\Lambda(H^0(X, L)) \otimes \Lambda(H^1(X, L))^{\otimes(-1)}$. A hermitian metric on $\lambda(R\Gamma(X, L))$ will be called a volume form on the formal difference $H^0(X, L) - H^1(X, L)$.

Let D and D_1 be divisors on X such that $D = D_1 + P$ for some point P on X. One then has an exact sequence

(3.2) $\quad\quad\quad\quad 0 \to \mathcal{O}_X(D_1) \to \mathcal{O}_X(D) \to \mathcal{O}(D)[P] \to 0$

in which $\mathcal{O}(D)[P]$ is a skyscraper sheaf supported on P. The previously defined Green's metrics on $\mathcal{O}_X(D_1)$ and $\mathcal{O}_X(D)$ give rise to a Green's metric on $\mathcal{O}(D)[P]$. On identifying $\mathcal{O}(D)[P]$ with the fibre of $\mathcal{O}_X(D)$ over P, this metric is simply the restriction of the metric on $\mathcal{O}_X(D)$ to the fibre over P.

One has $H^1(X, \mathcal{O}(D)[P]) = 0$, and $H^0(X, \mathcal{O}(D)[P])$ may be identified with $\mathcal{O}(D)[P]$. The long exact sequence in cohomology of (3.2) now gives an isomorphism

$$\lambda(R\Gamma(X, \mathcal{O}_X(D))) = \lambda(R\Gamma(X, \mathcal{O}_X(D_1))) \otimes \mathcal{O}(D)[P].$$

The following result is Theorem 1 of [8].

Theorem 3.1 (Faltings). *There is a unique way to assign to each admissible hermitian line bundle L on X a hermitian metric on $\lambda(R\Gamma(X, L))$ such that the following properties hold.*

(i) *An isometry of hermitian line bundles induces an isometry of the corresponding $\lambda(R\Gamma(X, L))$.*

(ii) *If the metric on L is changed by a factor $\alpha > 0$ then the metric on $\lambda(R\Gamma(X, L))$ is changed by $\alpha^{\chi(L)}$, where $\chi(L) = \deg(L) + 1 - g$.*

(iii) *The metrics on the $\lambda(R\Gamma(X, L))$ are compatible with the Green's metrics on the $\mathcal{O}(D)[P]$, in the following sense. Suppose D_1 and D are divisors on X such that $D = D_1 + P$ for some point P of X. Then the isomorphism*

$$\lambda(R\Gamma(X, \mathcal{O}_X(D))) = \lambda(R\Gamma(X, \mathcal{O}_X(D_1))) \otimes \mathcal{O}(D)[P]$$

is an isometry.

(iv) *The metric on $\lambda(R\Gamma(X, \Omega_X^1)) = \Lambda^g(\Gamma(X, \Omega_X^1))$ is the one determined by the canonical scalar product on $\Gamma(X, \Omega_X^1)$.*

SKETCH OF THE PROOF. We first claim that metrics may be put on the $\lambda(R\Gamma(X, L))$ in a unique way so that properties (ii), (iii) and (iv) hold. Let D be a divisor on X such Ω_X^1 and $\mathcal{O}_X(D)$ are isomorphic as line bundles. The Green's metric on $\mathcal{O}_X(D)$ is a scalar multiple of the metric we have fixed on Ω_X^1. Conditions (iv) and (ii) now determine the metric on $\lambda(R\Gamma(X, \mathcal{O}_X(D)))$. Condition (iii) shows how one may determine the metric on $\lambda(R\Gamma(X, \mathcal{O}_X(D')))$ for any divisor D' by adding or subtracting points from D. This may be done explicitly by writing the norm of the image of the meromorphic section 1 of $\mathcal{O}_X(D'')$ in $\mathcal{O}_X(D'')[P]$ for the appropriate divisors D'' and points P in terms of the Green's function of X. One sees in this way that the equality $G(P, Q) =$

An Introduction to Arakelov Intersection Theory

$G(Q, P)$ for P and Q on X implies that the order in which one adds or subtracts points does not alter the resulting metric on $\lambda(R\Gamma(X, \mathcal{O}_X(D')))$. This implies our first claim, since every admissible L is isometric to the line bundle obtained by multiplying the Green's metric on $\mathcal{O}_X(D')$ for some D' by a positive scalar.

One must now show that an isometry $\mathcal{O}_X(D) \cong \mathcal{O}_X(D')$ induces an isometry $\lambda(R\Gamma(X, \mathcal{O}_X(D))) \cong \lambda(R\Gamma(X, \mathcal{O}_X(D')))$. By adding or subtracting points, one may reduce to the case in which D and D' both have degree $g - 1$ and are of the form

$$E - (P_1 + \cdots + P_r)$$

for some fixed divisor E and some points $P_1, \ldots, P_r$ on X.

For $\mathscr{P} = (P_1, \ldots, P_r) \in X^r$ define $L(\mathscr{P})$ to be the line bundle $\mathcal{O}_X(E - (P_1 + \cdots + P_r))$. We wish to construct a line bundle N on X^r whose fibre at $\mathscr{P}$ is naturally identified with $\lambda(R\Gamma(X, L(\mathscr{P})))$.

Define $Y = X^r$ and $Z = Y \times_{\mathbb{C}} X$. Let D_j be the divisor on Z which is the graph of the morphism $Y \to X$ sending $\mathscr{P} = (P_1, \ldots, P_r)$ to P_j. The sum of the D_j is a divisor $\mathbb{D}$ on Z whose intersection with the fibre $Z_\mathscr{P}$ of Z over $\mathscr{P} \in Y$ is just $(P_1 + \cdots + P_r)$. The line bundle $F = \mathcal{O}_Z(Y \times_{\mathbb{C}} E - \mathbb{D})$ on Z is flat over Y ([9, p. 261]) and the fibre $F_\mathscr{P}$ over $Z_\mathscr{P}$ is isomorphic to $L(\mathscr{P})$ when we identify $Z_\mathscr{P}$ with X.

We are interested in the cohomology groups $H^j(Z_\mathscr{P}, F_\mathscr{P})$ as $\mathscr{P}$ varies over Y. Let $U = \text{Spec}(A)$ be an affine open neighborhood of a fixed point $\mathscr{P}$ of Y. Hartshorne shows in [9, Chap. III, Prop. 12.2 and Lemma 12.3] how to construct a bounded complex of A-modules

$$L^*: L^0 \xrightarrow{\delta_0} L^1 \xrightarrow{\delta_1} \cdots \xrightarrow{\delta_{n-1}} L^n$$

with the following properties. Each L^i is a finitely generated free A-module. For all A-modules M, let $h^i(L^* \otimes_A M)$ be the ith cohomology group of the complex $L^* \otimes_A M$ obtained by tensoring L^* with M. Then

$$h^i(L^* \otimes_A M) = H^i(Z, F \otimes_A M).$$

The relevant M for us is the residue field $k(\mathscr{P}) = \mathbb{C}$ of the point $\mathscr{P}$ of Y. Then

(3.3) $\qquad h^i(L^* \otimes_A k(\mathscr{P})) = H^i(Z_\mathscr{P}, F_\mathscr{P}) = H^i(X, L(\mathscr{P})).$

Let $r(i) = \text{rank}_A(L^i)$. Define B to be the free rank 1 A-module

$$B = \bigotimes_{i=0}^{n} (\Lambda^{r(i)}(L^i))^{\otimes (-1)^i}$$

Over $U = \text{Spec}(A)$, define N to be the line bundle associated to B, and then patch to define N over all of Y. From (3.3) and the fact that $H^i(X, L(\mathscr{P})) = 0$

if $i > 1$ we find that the fibre of N over $\mathscr{P}$ is

$$B \otimes_A k(\mathscr{P}) = \bigotimes_{i=0}^{n} (\Lambda^{r(i)}(L^i \otimes_A k(\mathscr{P})))^{\otimes(-1)^i}$$
$$= \Lambda^{d(0)} H^0(X, L(\mathscr{P})) \otimes (\Lambda^{d(1)} H^1(X, L(\mathscr{P})))^{\otimes -1}$$
$$= \lambda(R\Gamma(X, L(\mathscr{P}))),$$

where $d(i) = \dim_{\mathbb{C}} H^i(X, L(\mathscr{P}))$.

We now determine the isomorphism class of N, using the following distinguished meromorphic section of N. Let $K(A)$ be the fraction field of the ring A above. Define rank $\delta_i = \dim_{K(A)} \delta_i(L^i) \otimes_A K(A)$, where δ_i is the ith boundary map in the complex L^*. Let $\phi: Y \to \text{Pic}_{g-1}(X)$ send $\mathscr{P}$ to $L(\mathscr{P})$. The divisor Θ in $\text{Pic}_{g-1}(X)$ consists of those line bundles having global sections. Suppose $\phi(\mathscr{P}) \notin \Theta$. By Riemann–Roch, $H^0(X, L(\mathscr{P})) = H^1(X, L(\mathscr{P})) = 0$. The complex

$$L^0 \otimes_A k(\mathscr{P}) \xrightarrow{\delta_0(\mathscr{P})} L^1 \otimes_A k(\mathscr{P}) \xrightarrow{\delta_1(\mathscr{P})} \cdots \to L^n \otimes_A k(\mathscr{P})$$

induced by tensoring L^* with $k(\mathscr{P})$ is then exact, and this is clearly necessary and sufficient for $\phi(\mathscr{P})$ to not be in Θ. By Riemann–Roch, we may assume r has been taken large enough so that ϕ is surjective. Then $\phi(\mathscr{P}) \notin \Theta$ for a Zariski-dense set of $\mathscr{P}$, and we conclude that these $\mathscr{P}$ are precisely those for which rank $\delta_i = \text{rank}_{\mathbb{C}} \delta_i(\mathscr{P})$ for all i, where $\text{rank}_{\mathbb{C}} \delta_i(\mathscr{P}) = \dim_{\mathbb{C}} \text{image}(\delta_i(\mathscr{P}))$. Thus we have a meromorphic section

$$s = \bigotimes_{i=0}^{n-1} (\Lambda^{\text{rank } \delta_i} \delta_i)^{\otimes(-1)^{i+1}}$$

of the bundle N over $U = \text{Spec}(A)$, where s is a finite nonzero section over all $\mathscr{P} \in U$ for which $\phi(\mathscr{P}') \notin \Theta$. For such $\mathscr{P}'$, one can view the image $s_{\mathscr{P}'}$ of s in the fibre $N_{\mathscr{P}'}$ as the constant section 1 of $\lambda(R\Gamma(X, L(\mathscr{P}')))$.

With respect to the Zariski topology on $\phi^{-1}(\Theta)$, there is a dense open set of points $\mathscr{P}'' \in \phi^{-1}(\Theta)$ such that

$$\dim_{\mathbb{C}} H^0(X, L(\mathscr{P}'')) = \dim_{\mathbb{C}} H^1(X, L(\mathscr{P}'')) = 1.$$

Suppose that $\mathscr{P}' \in U - \phi^{-1}(\Theta)$ tends toward such a $\mathscr{P}''$. Since $\text{rank}_{\mathbb{C}} \delta_0(\mathscr{P}'') = \text{rank}_{\mathbb{C}} \delta_0(\mathscr{P}') - 1$ and $\text{rank}_{\mathbb{C}} \delta_1(\mathscr{P}'') = \text{rank}_{\mathbb{C}} \delta_1(\mathscr{P}')$ if $i > 0$, we conclude that $s_{\mathscr{P}'}$ has a first-order pole as $\mathscr{P}'$ tends toward $\mathscr{P}''$. We may now conclude that there is an isomorphism of line bundles

$$N = \phi^* \mathcal{O}(-\Theta)$$

in which the constant meromorphic section 1 of $\mathcal{O}(-\Theta)$ goes to the meromorphic section s of N.

Since the fibre $N_{\mathscr{P}}$ has been identified with $\lambda(R\Gamma(X, L(\mathscr{P})))$, the metrics we have put on the $\lambda(R\Gamma(X, L(\mathscr{P})))$ give a metric $|\ |_N$ on N. Let $|\ |_{\Theta}$ be the metric on N which is the pull-back via ϕ of the canonical metric on $\mathcal{O}(-\Theta)$. Suppose we show there is a positive scalar α such that

(3.4) $$\|\ \|_N = \alpha |\ |_\Theta.$$

It will then follow that the metric on $\lambda(R\Gamma(X, L(\mathscr{P})))$ depends only on the image of $\mathscr{P}$ in $\mathrm{Pic}_{g-1}(X)$, i.e. only on the isometry class of $L(\mathscr{P})$. This will complete the proof. To show (3.4), it is sufficient to show that the curvature forms of $(N, |\ |_N)$ and $(N, |\ |_\Theta)$ agree.

This calculation is carried out in [8, p. 397] and will not be repeated here. As in [8, pp. 400–401], one can give an explicit normalization of $|\ |_\Theta$ via classical theta functions. The constant $\alpha = \alpha(X)$ in (3.4) is a new invariant of the Riemann surface X; some properties of it are discussed and conjectured by Faltings in [8, pp. 401–403].

§4. The Riemann–Roch Theorem and the Adjunction Formula

As in Section 1, let K be a number field and let $\mathscr{X}$ be a regular curve over $S = \mathrm{Spec}(\mathfrak{O}_K)$ such that K is algebraically closed in $K(\mathscr{X})$. We assume, unless stated otherwise, that the general fibre of $\mathscr{X}$ has genus $g > 0$. For each infinite place ∞ of K, fix the metric ds_∞^2 on the Riemann surface $\mathscr{X}_\infty$ to be the one determined by equation (3.1). Define $w_{\mathscr{X}/S}$ to be the relative dualizing sheaf of $\mathscr{X}$ over S. On $\mathscr{X}_\infty$, $w_{\mathscr{X}/S}$ induces the sheaf $\Omega^1_{\mathscr{X}_\infty}$. Hence we may give $w_{\mathscr{X}/S}$ the admissible metrics at the infinite places S_{inf} of K which are specified in Proposition 3.1.

Suppose M is a finitely generated $\mathbb{Z}$-module for which we have a Haar measure on $M \otimes_\mathbb{Z} \mathbb{R}$. Define

$$\chi(M, \mathbb{Z}) = -\log(\mathrm{vol}(M \otimes_\mathbb{Z} \mathbb{R}/M)/\# M_{\mathrm{tor}}).$$

There is a natural isomorphism

$$\mathfrak{O}_K \otimes_\mathbb{Z} \mathbb{C} = \prod_{\sigma: K \to \mathbb{C}} K \otimes_\sigma \mathbb{C},$$

where the product is over the distinct embeddings of K into $\mathbb{C}$. The normalized Haar measure on $\mathfrak{O}_K \otimes_\mathbb{Z} \mathbb{C}$ is defined to be the one induced by this isomorphism and by Euclidean Haar measure on each factor $K \otimes_\sigma \mathbb{C} = \mathbb{C}$. The normalized Haar measure μ_K on $\mathfrak{O}_K \otimes_\mathbb{Z} \mathbb{R}$ is defined to be one which induces the normalized Haar measure on $\mathfrak{O}_K \otimes_\mathbb{Z} \mathbb{C} = \mathfrak{O}_K \otimes_\mathbb{Z} \mathbb{R} + \mathfrak{O}_K \otimes_\mathbb{Z} \mathbb{R}i$.

EXERCISE
Show that $\mu_K(\mathfrak{O}_K \otimes_\mathbb{Z} \mathbb{R}/\mathfrak{O}_K) = |d_{K/\mathbb{Q}}|^{1/2}$. Conclude that μ_K is the Haar measure on

$$\mathfrak{O}_K \otimes_\mathbb{Z} \mathbb{R} = K \otimes_\mathbb{Q} \mathbb{R} = \prod_{v \in S_{\mathrm{inf}}} K_v,$$

which results from ε_v times Euclidean Haar measure on K_v for each $v \in S_{\mathrm{inf}}$.

Suppose now that M is a finitely generated $\mathfrak{O}_K$ module. Define

$$\chi(M, \mathfrak{O}_K) = \chi(M, \mathbb{Z}) - \mathrm{rank}_{\mathfrak{O}_K}(M)\chi(\mathfrak{O}_K, \mathbb{Z}).$$

The Euler characteristics $\chi(M, \mathbb{Z})$ and $\chi(M, \mathfrak{O}_K)$ are additive on exact sequences of modules which are also volume exact, in the sense of having compatible Haar measures when tensored with $\mathbb{R}$.

Let $\mathscr{L}$ be a metrized line bundle on $\mathscr{X}$ such that the metric on the induced bundle $\mathscr{L}_\infty$ on $\mathscr{X}_\infty$ for $\infty \in S_{\inf}$ is admissible with respect to the $(1, 1)$-forms of equation (3.1). We will call such $\mathscr{L}$ admissible with respect to the Jacobian metrics of $\mathscr{X}$ at infinity. The metric on $\lambda(R\Gamma(\mathscr{X}_\infty, \mathscr{L}_\infty))$ specified in Theorem 3.1 gives Haar measures on $H^0(\mathscr{X}_\infty, \mathscr{L}_\infty)$ and $H^1(\mathscr{X}_\infty, \mathscr{L}_\infty)$ which are defined up to a common factor. As in Section 3, we view these measures as a measure on the formal difference

$$H^0(\mathscr{X}_\infty, \mathscr{L}_\infty) - H^1(\mathscr{X}_\infty, \mathscr{L}_\infty).$$

One has $H^j(\mathscr{X}_\infty, \mathscr{L}_\infty) = H^j(\mathscr{X}, \mathscr{L}) \otimes_{\mathfrak{O}_K} \bar{K}_\infty$, and the above Haar measures are compatible with complex conjugation if $K_\infty = \mathbb{R}$. We thus obtain a Haar measure on

$$H^0(\mathscr{X}, \mathscr{L}) \otimes_\mathbb{Z} \mathbb{R} - H^1(\mathscr{X}, \mathscr{L}) \otimes_\mathbb{Z} \mathbb{R}$$

$$= \prod_{\infty \in S_{\inf}} H^0(\mathscr{X}, \mathscr{L}) \otimes_{\mathfrak{O}_K} K_\infty - \prod_{\infty \in S_{\inf}} H^1(\mathscr{X}, \mathscr{L}) \otimes_{\mathfrak{O}_K} K_\infty.$$

Definition 4.1 (Faltings [8]). Let $\mathscr{L}$ be an admissible metrized line bundle on $\mathscr{X}$ with respect to the Jacobian metrics on $\mathscr{X}$ at infinity. Define

$$\chi(\mathscr{L}) = \chi(H^0(\mathscr{X}, \mathscr{L}), \mathfrak{O}_K) - \chi(H^1(\mathscr{X}, \mathscr{L}), \mathfrak{O}_K).$$

Theorem 4.1 (Faltings [8]). *For $\mathscr{L}$ an admissible metrized line bundle on $\mathscr{X}$ one has the following Riemann–Roch formula:*

$$\chi(\mathscr{L}) = (1/2)[\mathscr{L}, \mathscr{L} - w_{\mathscr{X}/S}] + \chi(\mathscr{O}_\mathscr{X}).$$

SKETCH OF THE PROOF. We will show $\tau(D) = \mathscr{X}(\mathscr{O}_\mathscr{X}(D)) - \mathscr{X}(\mathscr{O}_\mathscr{X}) - [D, D - w_{\mathscr{X}/S}]/2$ is 0 for all Arakelov divisors D. We claim (i) the map $D \to \tau(D)$ factors through a homomorphism from $\mathrm{Pic}(\mathscr{X})$ to $\mathbb{R}$, and $\tau(D) = 0$ if D is fibral; (ii) if D_1 and D_2 are horizontal Weil divisors on $\mathscr{X}$ which induce the same Cartier divisor on the reductions of all singular fibers of $\mathscr{X}$ over S, then $\tau(D_1) = \tau(D_2)$. To prove (i) and (ii), let K' be a finite extension of K and $S' = \mathrm{Spec}(\mathfrak{O}_{K'})$. Let $g : \mathscr{C} = \mathscr{X} \times_S S' \to \mathscr{X}$ be the base-change morphism. Let $h : \mathscr{C}^\sharp \to \mathscr{C}$ be the normalization morphism, and let $f : \mathscr{X}' \to \mathscr{C}^\sharp$ be the minimal desingularization [4, 14]. There is a fibral Weil divisor T on $\mathscr{X}'$ such that $w_{\mathscr{X}'/S'} \otimes (\pi^* w_{\mathscr{X}/S})^{-1}$ is isometric to $\mathscr{O}_{\mathscr{X}'}(T)$ when $\pi = g \circ h \circ f$. Using the Leray spectral sequence, the projection formula and [4, pp. 267–269], one shows

(4.1) $\tau(\pi^* D) - [K' : K]\tau(D) = \eta(D) - \eta(0) + [\pi^* D, T]/2,$

where $\eta(D) = \mathcal{X}(h_*h^*g^*\mathcal{O}_{\mathcal{X}}(D)) - \mathcal{X}(g^*\mathcal{O}_{\mathcal{X}}(D))$ relative to the Faltings volume forms resulting from identifying the general fibers of $\mathscr{C}$ and $\mathcal{X}'$. As in [8, pp. 405–406], one shows that $\tau(D)$ does not change if we add to D (iii) a section of $\mathcal{X} \to S$, (iv) a real multiple of an infinite fiber of $\mathcal{X}$, or (v) a geometrically irreducible component of a finite fiber of $\mathcal{X} \to S$. (For this one uses Proposition 3.1, the adjunction formula of [10, pp. 142–144] for geometrically irreducible curves over a finite field, and Theorem 3.1 (ii).) Choosing K' to be an extension of K unramified over a large finite subset of S with large residue fields over this subset, we can use (4.1) to change (v) to (v') any fibral divisor. Now choose K' large enough so that π^*D becomes a linear combination of Arakelov divisors of types (iii), (iv) and (v'). We conclude $\tau(\pi^*D) = \tau(0) = 0$ in (4.1). By using the finiteness of h, the projection formula and the fact that η is real valued, one can show that $\eta(D + D') - \eta(D)$ is independent of the divisor D when D' contains no irreducible component of a fiber, and that this function depends only on the Cartier divisor D' induces on the reduced singular fibers of $\mathcal{X}$. This, (4.1), and (iii), (iv) and (v') are sufficient to prove (i) and (ii). To finish the proof of Theorem 4.1, one now uses a moving Lemma to show that (i) and (ii) imply $\tau(D) = \mu \cdot \deg(D_K)$ for some $\mu \in \mathbb{R}$, where $\deg(D_K)$ is the degree of $D \times_S \mathrm{Spec}(K)$. (For this step, one uses that (vi) if $\deg(D_K) = 0$, then a non-zero multiple of D plus a fibral divisor has degree 0 on each irreducible component of a fiber of $\mathcal{X}$, and (vii) the group of Cartier divisor classes on the union of the reduced singular fibers of $\mathcal{X}$ which have degree 0 on each irreducible fiber component is finite.) Now $\mu = 0$ follows from metrized Serre duality (viii) $\mathcal{X}(\mathcal{O}_{\mathcal{X}}(D)) = \mathcal{X}(w_{\mathcal{X}/S} \otimes \mathcal{O}_{\mathcal{X}}(D)^{-1})$ for all Arakelov divisors D. One proves (viii) by showing the volume forms in Theorem 3.1 respect duality and by using [10, Chap. III, Thm. 11.1, Remark 3].

Proposition 4.1 (Adjunction Formula). *Let D be an irreducible horizontal Arakelov divisor on $\mathcal{X}$. The field L of functions on D is a finite extension of K, and $D = f(S')$ for some morphism $f: S' = \mathrm{Spec}(\mathfrak{O}_L) \to \mathcal{X}$. For $\infty \in S_{\mathrm{inf}}$, fix an embedding τ_∞ of K into $\bar{K}_\infty$, and let $\{P_i^\infty : i = 1, \ldots, [L:K]\}$ be the set of points on $\mathcal{X}_\infty = \mathcal{X}(\bar{K}_\infty)$ which are determined by D and the extensions of τ_∞ to L. Then*

$$[D + w_{\mathcal{X}/S}, D] = \log(\#w_{D/S}) - \sum_{\infty \in S_{\mathrm{inf}}} \varepsilon_\infty \sum_{i \neq j} \log G_\infty(P_i^\infty, P_j^\infty).$$

Here $\#w_{D/S} = \mathrm{Norm}_{K/\mathbb{Q}}(\mathrm{disc}(f^*(\mathcal{O}_D)/\mathfrak{O}_K))$, where $f^*(\mathcal{O}_D)$ is an $\mathfrak{O}_K$ order in $\mathfrak{O}_L$, and disc denotes the relative discriminant over $\mathfrak{O}_K$.

PROOF. Let $S(L) = \mathrm{Spec}(L)$, and let $\sigma: S(L) \to D$ be the inclusion of the generic point of D into D. One has an exact sequence

$$0 \to \mathcal{O}_D \to \sigma_* \mathcal{O}_{S(L)} \to F \to 0$$

of sheaves on D, where $H^0(D, F) = L/f^*(\mathcal{O}_D)$ when one views $f^*(\mathcal{O}_D)$ as an

$\mathfrak{O}_K$ order in $\mathfrak{O}_L$. On taking cohomology, one has

(4.2) $$H^1(D, \mathcal{O}_D) = 0.$$

Consider now the exact sequence of sheaves on $\mathscr{X}$

(4.3) $$0 \to \mathcal{O}_\mathscr{X}(-D) \to \mathcal{O}_\mathscr{X} \to \mathcal{O}_D \to 0.$$

By Faltings' theorem, one has volume forms on the formal difference

$$H^0(\mathscr{X}, \mathscr{L}) \otimes_\mathbb{Z} \mathbb{R} - H^1(\mathscr{X}, \mathscr{L}) \otimes_\mathbb{Z} \mathbb{R}$$

for $\mathscr{L} = \mathcal{O}_\mathscr{X}(-D)$ and $\mathscr{L} = \mathcal{O}_\mathscr{X}$. In view of (4.2) and (4.3), these volume forms induce a Haar measure μ' on

$$H^0(D, \mathcal{O}_D) \otimes_\mathbb{Z} \mathbb{R} = L \otimes_\mathbb{Q} \mathbb{R}$$

with respect to which the cohomology of (4.3) is volume exact. We will show below that

(4.4) $$\mu' = G\mu_L,$$

where μ_L is the normalized Haar measure on $L \otimes_\mathbb{Z} \mathbb{R}$, and where G is the constant

$$G = \prod_{\infty \in S_{\inf}} \prod_{i<j} (G_\infty(P_i^\infty, P_j^\infty))^{-\varepsilon_\infty}.$$

Let us first complete the proof from this.

Since $H^1(D, \mathcal{O}_D) = 0$, the cohomology of (4.3) gives

$$\chi(\mathcal{O}_\mathscr{X}) - \chi(\mathcal{O}_\mathscr{X}(-D)) = \chi(f^*(\mathcal{O}_D), \mathfrak{O}_K),$$

where $f^*(\mathcal{O}_D)$ is a projective $\mathfrak{O}_K$ module of rank $[L:K]$ with Haar measure μ' on $f^*(\mathcal{O}_D) \otimes_\mathbb{Z} \mathbb{R} = L \otimes_\mathbb{Q} \mathbb{R}$. By Riemann–Roch, the left-hand side is

$$-[D, D + w_{\mathscr{X}/S}]/2.$$

The right-hand side is

$$-\log \mu'(L \otimes_\mathbb{Q} \mathbb{R}/f^*(\mathcal{O}_D)) + [L:K](\log d_{K/\mathbb{Q}})/2$$
$$= -\log G - \log \mu_L(L \otimes_\mathbb{Q} \mathbb{R}/f^*(\mathcal{O}_D)) + [L:K](\log d_{K/\mathbb{Q}})/2$$
$$= -\log G - (\log \mathrm{Norm}_{K/\mathbb{Q}} \mathrm{disc}(f^*(\mathcal{O}_D)/\mathfrak{O}_K))/2.$$

Since $2\log(G)$ is the Green's function term in the right-hand side of the adjunction formula, the proof will be complete once we demonstrate (4.4).

Fix $\infty \in S_{\inf}$. For simplicity, we will write P_j for the point P_j^∞ of $\mathscr{X}_\infty$. Define D_0 to be the trivial divisor on $\mathscr{X}_\infty$. For $1 \leq j \leq n$, let D_j be the divisor

$$P_1 + \cdots + P_j$$

on $\mathscr{X}_\infty$. For $0 \leq j < n$ we have an exact sequence

$$0 \to \mathcal{O}_{\mathscr{X}_\infty}(-D_{j+1}) \to \mathcal{O}_{\mathscr{X}_\infty}(-D_j) \to \mathcal{O}(-D_j)[P_{j+1}] \to 0$$

of sheaves on $\mathscr{X}_\infty$, where $\mathcal{O}(-D_j)[P_{j+1}]$ is a skyscraper sheaf supported on P_{j+1}. Theorem 3.1 (iii) now gives an isometry

$$(4.5) \quad \lambda(R\Gamma(\mathscr{X}_\infty, \mathcal{O}_{\mathscr{X}_\infty})) = \lambda(R\Gamma(\mathscr{X}_\infty, \mathcal{O}_{\mathscr{X}_\infty}(-D_n))) \otimes \left(\bigotimes_{0 \le j < n} \mathcal{O}(-D_j)[P_{j+1}] \right).$$

The metric on $\mathcal{O}(-D_j)[P_{j+1}]$ results from identifying $\mathcal{O}(-D_j)[P_{j+1}]$ as the fibre of $\mathcal{O}_{\mathscr{X}_\infty}(-D_j)$ at P_{j+1}. In particular, the norm of the image in $\mathcal{O}(-D_j)[P_{j+1}]$ of the meromorphic section 1 of $\mathcal{O}_{\mathscr{X}_\infty}(-D_j)$ is

$$G(\infty, j) = \prod_{1 \le i \le j} G_\infty(P_i, P_{j+1})^{-1}.$$

Another way of saying this is that by identifying the images of the meromorphic section 1, we have an isomorphism

$$\mathcal{O}(-D_j)[P_{j+1}] \cong \mathcal{O}(D_0)[P_{j+1}],$$

which is an isometry once we multiply the metric on the right-hand side by $G(\infty, j)$. Hence from (4.5) we have the following conclusion:

(4.6) The isomorphism

$$\lambda(R\Gamma(\mathscr{X}_\infty, \mathcal{O}_{\mathscr{X}_\infty})) = \lambda(R\Gamma(\mathscr{X}_\infty, \mathcal{O}_{\mathscr{X}_\infty}(-D_n))) \otimes \left(\bigotimes_{0 \le j < n} \mathcal{O}(D_0)[P_{j+1}] \right)$$

is an isometry once the metric on the right is multiplied by $G(\infty) = \prod_{1 \le j < n} G(\infty, j)$.

We now return to metrized line bundles on $\mathscr{X}$. Recall that the normalized Haar measure μ_L on $H^0(D, \mathcal{O}_D) \otimes_\mathbb{Z} \mathbb{R} = L \otimes_\mathbb{Q} \mathbb{R}$ is the one which induces on

$$H^0(D, \mathcal{O}_D) \otimes_\mathbb{Z} \mathbb{C} = L \otimes_\mathbb{Q} \mathbb{C} = \prod_{\sigma: L \to \mathbb{C}} L \otimes_\sigma \mathbb{C}$$

the Haar measure $\mu_{L,\mathbb{C}}$ coming from Euclidean Haar measure on each $L \otimes_\sigma \mathbb{C} = \mathbb{C}$, where σ runs over the distinct embeddings of L into $\mathbb{C}$. Here

$$H^0(D, \mathcal{O}_D) \otimes_\mathbb{Z} \mathbb{C} = \prod_{\tau: K \to \mathbb{C}} H^0(D, \mathcal{O}_D) \otimes_{\mathcal{O}_K, \tau} \mathbb{C},$$

where the product is over the distinct embeddings of K into $\mathbb{C}$. Let $\infty \in S_\mathrm{inf}$ be the place associated to $\tau: K \to \mathbb{C}$. Evaluation of functions in $H^0(D, \mathcal{O}_D)$ at the points P_{j+1}^∞ induces a natural isomorphism

$$H^0(D, \mathcal{O}_D) \otimes_{\mathcal{O}_K, \tau} \mathbb{C} = \prod_{0 \le j < n} \mathcal{O}_{\mathscr{X}_\infty}(D_0^\infty)[P_{j+1}^\infty],$$

where D_0^∞ is the trivial divisor on $\mathscr{X}_\infty$. The metric on $\mathcal{O}_{\mathscr{X}_\infty}(D_0^\infty)[P_{j+1}^\infty]$ is just the Euclidean absolute value on the fibre of $\mathcal{O}_{\mathscr{X}_\infty}$ at P_{j+1}^∞. Thus the metrics on the $\mathcal{O}_{\mathscr{X}_\infty}(D_0^\infty)[P_{j+1}^\infty]$ give rise to the measure $\mu_{L,\mathbb{C}}$ on $H^0(D, \mathcal{O}_D) \otimes_\mathbb{Z} \mathbb{C}$, and hence also to the measure μ_L on $H^0(D, \mathcal{O}_D) \otimes_\mathbb{Z} \mathbb{R}$.

Fix $\infty \in S_\mathrm{inf}$. If we multiply the tensor product metric on

$$\bigotimes_{0 \le j < n} \mathcal{O}_{\mathscr{X}_\infty}(D_0^\infty)[P_{j+1}^\infty]$$

by the positive scalar $G(\infty)$, this multiplies the Haar measure on the real vector space $H^0(D, \mathcal{O}_D) \otimes_{\mathcal{O}_K} K_\infty$ in

$$H^0(D, \mathcal{O}_D) \otimes_{\mathcal{O}_K} \bar{K}_\infty = \prod_{0 \le j < n} \mathcal{O}_{\mathcal{X}_\infty}(D_0^\infty)[P_{j+1}^\infty]$$

by the factor $G(\infty)^{\varepsilon_\infty}$. The isometry in (4.6) now shows that the measure μ' for which (4.3) is volume exact is

$$\mu' = \left(\prod_{\infty \in S_{\inf}} G(\infty)^{\varepsilon_\infty} \right) \mu_L.$$

This is exactly the equality (4.4), so the proof is complete. □

§5. The Hodge Index Theorem

Throughout this section, $\mathcal{X}$ will be a regular curve over S such that $K = K(S)$ is algebraically closed in $K(\mathcal{X})$ and the general fibre $\mathcal{X}_K$ of $\mathcal{X}$ has positive genus.

Let D_K be a divisor of degree zero on $\mathcal{X}_K$. Then D_K determines a point $[\mathcal{O}(D_K)]$ on the Jacobian $\text{Jac}(\mathcal{X}_K)$ of $\mathcal{X}_K$. Let $\text{Height}([\mathcal{O}(D_K)])$ be the Néron–Tate height of this point. The central part of the Hodge index theorem connects $\text{Height}([\mathcal{O}(D_K)])$ to intersection theory. This was done by Hriljac in [11]–[12] by comparing Néron local heights to Arakelov theory. The same result was proved by Faltings in [8] using the Riemann–Roch theorem; we will sketch this approach below.

To apply results of Raynaud [17] on Picard functors, we make the following technical assumption:

(5.1) Let $\mathcal{X}_x$ be the fibre of $\mathcal{X}$ at a closed point x of S. Let δ_x be the greatest common divisor of the multiplicities in $\mathcal{X}_x$ of the prime Weil divisors of $\mathcal{X}$ which lie over x. Then $\delta_x = 1$ for all closed points x of S.

If, for example, there is a section $s: S \to \mathcal{X}$ of the projection $\mathcal{X} \to S$, then $\mathcal{X}$ will satisfy (5.1).

Theorem 5.1 (Hirljac [11]–[12], Faltings [8]). *Let $\mathcal{X}$ be a regular curve over S with general fibre of positive genus which satisfies (5.1), and for which K is algebraically closed in $K(\mathcal{X})$. If v is a finite or infinite place of K, let $\mathcal{X}_v$ be the fibre of $\mathcal{X}$ over v. Let V_v be the real vector space having basis the irreducible components of $\mathcal{X}_v$, where fibres over infinite v are defined to be irreducible.*

(i) *For each place v, [,] induces a negative semi-definite bilinear form on V_v. The scalar multiples of $\mathcal{X}_v$ are the only elements of V_v having self-intersection zero.*

(ii) *Let D be an Arakelov divisor of $\mathcal{X}$ which is perpendicular to every element of V_v for all v. Then D determines a degree-zero divisor D_K on the generic* fibre $\mathscr{X}_K$ of $\mathscr{X}$. One has

$$[D, D] = -2[K:Q] \text{Height}([\mathcal{O}(D_K)]).$$

(iii) *Define* $\text{Div}_{\mathbb{R}}^{\vee}(\mathscr{X})$ *to be the group of finite real linear combinations of Arakelov divisors. Let* $\text{Div}_{\mathbb{R}}^{\vee n}(\mathscr{X})$ *be the subgroup of* $C \in \text{Div}_{\mathbb{R}}^{\vee}(\mathscr{X})$ *which are numerically equivalent to* 0, *i.e. for which* $[C, C'] = 0$ *for all Arakelov divisors* C'. *Let* $n(v)$ *be the number of components of the fibre* X_v. *The bilinear form induced by* [,] *on the real vector space* $\text{Num}(\mathscr{X}) = \text{Div}_{\mathbb{R}}^{\vee}(\mathscr{X})/\text{Div}_{\mathbb{R}}^{\vee n}(\mathscr{X})$ *is nondegenerate. This form is equivalent over* $\mathbb{R}$ *to a diagonal form having one positive eigenvalue and* t *negative ones, where*

$$t = \sum_v (n(v) - 1) + \text{rank}_{\mathbb{Z}}(\text{Jac}(\mathscr{X}_K(K))),$$

the sum being over all places v *of* K.

SKETCH OF THE PROOF. Part (i) is proved by the argument of Mumford in [16], which is recalled in [8] and [7, Lemma 7.1(b)]. Before proving (ii), let us see how (iii) follows from (i) and (ii).

Let C be an effective finite horizontal Arakelov divisor, and let $\mathscr{X}_x$ be the fibre of $\mathscr{X}$ over a closed point x of S. For large enough n, $C_n = C + n\mathscr{X}_x$ has positive self-intersection, so [,] has at least one positive eigenvalue on $\text{Num}(\mathscr{X})$. By (i), every Arakelov divisor C' is equal modulo $\text{Div}_{\mathbb{R}}^{\vee n}(\mathscr{X})$ to the sum of a real multiple of C and a real linear combination of components of reducible fibres of $\mathscr{X}$ and divisors D as in part (ii). As discussed in [19], Height is a positive definite quadratic form on $\text{Jac}(\mathscr{X}_K)(K)$ modulo torsion, which is a finitely generated abelian group. This fact together with (i) and (ii) imply (iii).

To prove (ii), Faltings uses the construction of the Riemann–Roch theorem, working over S rather than over $\mathbb{C}$. Raynaud shows in [17] that because of assumption (5.1), the set of line bundles of degree $g - 1$ over $\mathscr{X}$ forms a scheme $\text{Pic}_{g-1}(\mathscr{X})$ which is locally of finite type over S. Suppose x is a closed point of S, and that C and C' are two Weil divisors on $\mathscr{X}$ of degree $g - 1$. The points on the fibre of $\text{Pic}_{g-1}(\mathscr{X})$ over x which are determined by $\mathcal{O}_{\mathscr{X}}(C)$ and $\mathcal{O}_{\mathscr{X}}(C')$ are equal iff $\mathcal{O}_{\mathscr{X}}(C)$ and $\mathcal{O}_{\mathscr{X}}(C')$ induce isomorphic invertible sheaves on $\mathscr{X}_x$; these points lie on different connected components of the fibre of $\text{Pic}_{g-1}(\mathscr{X})$ over x if C and C' have different intersection numbers with some irreducible component of $\mathscr{X}_x$. (From this one sees that there are an infinite number of connected components in the fibres of $\text{Pic}_{g-1}(\mathscr{X})$ over closed points x of S for which $\mathscr{X}_x$ is reducible. Thus $\text{Pic}_{g-1}(\mathscr{X})$ need not be of finite type over S.) Let P be the open subscheme of $\text{Pic}_{g-1}(\mathscr{X})$ obtained by removing, for each $x \in S$ such that $\mathscr{X}_x$ is reducible, all connected components of fibres of $\text{Pic}_{g-1}(\mathscr{X})$ over x except for those containing $\mathcal{O}_{\mathscr{X}}(C)$. Then by [17], P will be of finite type over S. Let D be an Arakelov divisor perpendicular to every element of V_v for all places v of K. For this to be true for archimedean v, D must have degree zero on the generic fibre $\mathscr{X}_K$. From the above discussion, we see that $\mathcal{O}_{\mathcal{X}}(C + nD)$ determines a point $[\mathcal{O}_{\mathcal{X}}(C + nD)]$ on P for each natural number n.

Define Θ to be the divisor on P defined by those line bundles having global sections, and let $\mathscr{L}$ be the universal bundle on $P \times X$. For a given point p of P, the cohomology groups of the fibre $\mathscr{L}(p)$ are finitely generated $\mathfrak{O}_K$ modules. Taking the tensor Euler characteristics of the highest exterior powers of these $\mathfrak{O}_K$ modules, we arrive by the construction of the proof of the Riemann–Roch theorem at an isomorphism

$$(5.2) \qquad \lambda(R\Gamma(\mathcal{X}, \mathscr{L})) \cong \mathcal{O}(-\Theta)$$

of line bundles on P. The divisor Θ determines the usual Θ-divisor on the general fibre of $\mathrm{Pic}_{g-1}(\mathcal{X})$, and $\mathcal{O}(-\Theta)$ is a metrized line bundle on P in this way. In the proof of the Riemann–Roch theorem, it was shown that (5.2) becomes an isomorphism of metrized line bundles if for each archimedean place ∞ of K, one multiplies the metric on $\lambda(R\Gamma(\mathcal{X}_\infty, \mathscr{L}_\infty))$ which comes from Theorem 3.1 by a positive scalar α_∞^{-1} depending only on $\mathcal{X}_\infty$.

For each integer n, let $s_n: S \to P$ be the section of $P \to S$ determined by the point $[\mathcal{O}_{\mathcal{X}}(C + nD)]$ of P. We pull back the isometry (5.2) via s_n and take the degrees of the resulting metrized line bundles on S. The degree of $s_n^*(\lambda(R\Gamma(\mathcal{X}, \mathscr{L})))$ is the sum of the Euler characteristic $\chi(\mathcal{O}_{\mathcal{X}}(C + nD))$ of $\mathcal{O}_{\mathcal{X}}(C + nD)$ and a constant c independent of n which depends on the scalars α_∞. By the Riemann–Roch theorem, this degree is

$$n^2[D, D]/2 + an + b$$

for some constants a and b. Let h_Θ be the absolute logarithmic height function on $P(K)$ which is determined by Θ. (See [19], noting that a multiple of Θ is very ample.) By the argument of [19, Prop. 7.2], the degree of $s_n^*(\mathcal{O}(\Theta))$ differs from

$$[K : \mathbb{Q}] h_\Theta([\mathcal{O}_{\mathcal{X}}(C + nD)])$$

by an amount which is bounded independently of n. By the quadratic property of the Néron–Tate height, we have

$$\mathrm{Height}([\mathcal{O}(D_K)]) = \lim_{n \to \infty} h([\mathcal{O}_{\mathcal{X}}(C + nD)])/n^2,$$

where $[\mathcal{O}(D_K)]$ is the point on $\mathrm{Jac}(\mathcal{X}_K)(K)$ determined by D. The isometry (5.2) now shows

$$[K : \mathbb{Q}] \, \mathrm{Height}([\mathcal{O}(D_K)]) = -[D, D]/2,$$

which completes the proof. □

EXERCISE

Suppose that the general fibre of $\mathcal{X}$ has genus 1. Show that there is a (possibly negative) constant c depending on $\mathcal{X}$ and K such that

$$[D_1, D_2] > c$$

for all sections D_1 and D_2 of $\mathscr{X} \to S$. (Hint: Show that there is an integer $m > 0$ such that for all D_1 and D_2, one can find a finite Arakelov divisor T supported on the reducible fibres of $\mathscr{X}$ for which $D = m(D_1 - D_2) + T$ satisfies the conditions of Theorem 5.1 (ii).)

REFERENCES

[1] Abhyankar, S. Resolutions of singularities of arithmetical surfaces, in *Arithmetical Algebraic Geometry*. Harper and Row: New York, 1965.
[2] Arakelov, S. Intersection theory of divisors on an arithmetic surface. *Izv. Akad. Nauk.*, **38** (1974), 1179–1192.
[3] Arakelov, S. Theory of intersections on the arithmetic surface. *Proceedings of the International Congress on Mathematics*, Vancouver, 1974, pp. 405–408.
[4] Artin, M. Lipman's proof of resolution of singularities for surfaces, this volume, pp. 267–287.
[5] Cantor, D. On an extension of the definition of transfinite diameter and some applications. *J. Reine. Angew. Math.*, **316** (1980), 160–207.
[6] Chinburg, T. Intersection theory and capacity theory on arithmetic surfaces. *Proc. of the Canadian Math. Soc. Summer Seminar in Number Theory*, 7. American Mathematical Society: Providence, RI, 1986.
[7] Chinburg, T. Minimal models for curves over Dedekind rings, this volume, pp. 309–326.
[8] Faltings, G. Calculus on arithmetic surfaces. *Ann. Math.*, **119**, no. 2 (1984), 387–424.
[9] Hartshorne, R. *Algebraic Geometry*. Springer-Verlag: New York, 1977.
[10] Hartshorne, R. *Residues and Duality*. Springer Lecture Notes, 20. Springer-Verlag: Heidelberg, 1966.
[11] Hriljac, P. Thesis (1982), Massachusetts Institute of Technology.
[12] Hriljac, P. Heights and Arakelov's intersection theory. *Amer. J. Math.*, **107**, no. 1 (1985), 23–38.
[13] Lang, S. *Introduction to Arakelov intersection theory*, Springer-Verlag: New York, Berlin, Heidelberg, 1988.
[14] Lichtenbaum, S. Curves over discrete valuation rings. *Amer. J. Math.*, **15**, no. 2 (1968), 380–405.
[15] Lipman, J. Rational singularities with applications to algebraic surfaces and unique factorization. *Publ. Math. I.H.E.S.*, **36** (1969), 195–279.
[16] Mumford, D. The topology of normal singularities of an algebraic surface and a criterion for simplicity. *Publ. Math. I.H.E.S.*, **9** (1961), 5–22.
[17] Raynaud, M. Specialization du functeur Picard. *Publ. Math. I.H.E.S.*, **38** (1970), 27–76.
[18] Rumely, R. *Capacity Theory on Algebraic Curves*. (To appear in Springer Lecture Notes in Mathematics.)
[19] Silverman, J. The theory of height functions, this volume, pp. 151–166.
[20] Szpiro, L. Degrée, intersections, hauteur, in *Séminaire sur les Pinceaux Arithmetiques: La Conjecture de Mordell*, L. Szpiro ed., *Asterisque*, **127**, (1985), 11–28.

CHAPTER XIII

Minimal Models for Curves over Dedekind Rings

T. CHINBURG

In this chapter we review the construction by Lichtenbaum [8] and Shafarevitch [11] of relatively minimal and minimal models of curves over Dedekind rings. We have closely followed Lichtenbaum [8]; some proofs have been skipped or summarized so as to go into more detail concerning other parts of the construction. Since the main arguments of [8] apply over Dedekind rings, we work always over Dedekind rings rather than discrete valuation rings.

§1. Statement of the Minimal Models Theorem

All rings and schemes will be assumed to be excellent and Noetherian; k will denote a field, $\mathfrak{O}$ will denote a Dedekind ring and S will denote $\text{Spec}(\mathfrak{O})$. A curve over k is a scheme F of finite type over k such that $\dim \mathcal{O}_{F,x} = 1$ for all closed points x of F. A curve $\mathscr{X}$ over $\mathfrak{O}$ is a connected normal scheme together with a morphism $\pi: \mathscr{X} \to S$ which is proper, flat and of finite type, and whose fibres are curves. By results of Abhyankhar [2] and Lipman [9], [3], one may perform a finite sequence of blow-ups and normalizations to arrive at a regular curve $\mathscr{X}'$ over $\mathfrak{O}$ together with a proper birational morphism $f: \mathscr{X}' \to \mathscr{X}$ over S. In what follows we will be concerned with regular curves over $\mathcal{O}$.

Theorem 1.1 (Lichtenbaum [8]). *A regular curve $\mathscr{X}$ over $\mathfrak{O}$ is projective over S.*

Theorem 1.1 is proved essentially by constructing an effective Weil divisor on $\mathscr{X}$ which has no components lying in the fibres of $\mathscr{X}$ over S, and which meets every irreducible component of every fibre of $\mathscr{X}$ positively. Such a divisor may be shown to be ample over S; the details may be found in [8, Section 2].

Definition 1.1. A regular curve $\mathscr{Y}$ over $\mathfrak{O}$ is a relatively minimal model of its function field $K(\mathscr{Y})$ if every proper S-birational morphism $f: \mathscr{Y} \to \mathscr{Y}'$ to a regular curve $\mathscr{Y}'$ over $\mathfrak{O}$ is necessarily an isomorphism. If all relatively minimal models in the birational equivalence class of $\mathscr{Y}$ are isomorphic over S, we say $\mathscr{Y}$ is a minimal model of $K(\mathscr{Y})$.

Definition 1.2. A surface is an integral scheme of dimension 2.

If $\mathscr{X}$ is an integral curve over $\mathfrak{O}$, and $\mathfrak{O}$ has dimension 1, then $\mathscr{X}$ is a surface in the sense of Definition 1.2.

Definition 1.3. A prime divisor E on a regular surface $\mathscr{X}$ is exceptional if there exits a proper birational morphism $\pi: \mathscr{X} \to \mathscr{Y}$ to a regular surface $\mathscr{Y}$ such that π is an isomorphism outside of E, and $\pi(E)$ is a point P of $\mathscr{Y}$ such that $\dim \mathcal{O}_{\mathscr{Y},P} = 2$. In this case we say that π is a blow-down of E on $\mathscr{X}$.

It is shown in Corollary 2.1 below that that $\mathscr{Y}$ and P determine $\mathscr{X}$ and E up to isomorphism over $\mathscr{Y}$, and that $\mathscr{X}$ and E determine $\mathscr{Y}$ up to isomorphism.

Our main object is the following result of Lichtenbaum [8, Theorem 4.4] and Shafarevitch [11, Chap. II].

Theorem 1.2 (Minimal Models Theorem). *Let $\mathfrak{O}$ be a Dedekind domain, and let $\mathscr{X}$ be a regular curve over $S = \mathrm{Spec}(\mathfrak{O})$. Assume that the fraction field $K(S)$ of $\mathfrak{O}$ is algebraically closed in $K(\mathscr{X})$. Construct a sequence of regular curves $\{\mathscr{X}(n)\}_n$ over S in the following way. Let $\mathscr{X}(0) = \mathscr{X}$. If $\mathscr{X}(n)$ has been defined, and there is an exceptional curve on $\mathscr{X}(n)$, let $\pi(n): \mathscr{X}(n) \to \mathscr{X}(n+1)$ be a blow-down of this curve. Then the sequence $\{\mathscr{X}(n)\}_n$ must be finite. The final curve $\mathscr{Y}$ in this sequence is a relatively minimal model of the common function field $K(\mathscr{X})$ of the $\mathscr{X}(n)$. Let W be the fibre of $\mathscr{X}$ over the generic point of S. If $H^1(W, \mathcal{O}_W) \neq 0$, then $\mathscr{Y}$ is a minimal model of $K(\mathscr{X})$.*

If $H^1(W, \mathcal{O}_W) = 0$ then the construction of [7, p. 416] produces examples in which $\mathscr{Y}$ is not minimal and W is isomorphic to $P^1_{K(S)}$.

As in the case of complex surfaces, the proof of the Minimal Models Theorem involves giving a factorization theorem for birational morphisms between surfaces and a numerical characterization of exceptional curves (Castelnuovo's criterion). The key step in showing that $\mathscr{Y}$ is minimal if W has genus greater than zero is to show (cf. Lemma 7.2 below) that in this case, the exceptional curves on $\mathscr{X}(n)$ are disjoint for each n, and if $\pi: \mathscr{X}(n) \to \mathscr{X}(n+1)$

is a blow-down of one such curve, the other exceptional curves on $\mathscr{X}(n)$ are sent by π to isomorphic exceptional curves on $\mathscr{X}(n+1)$.

§2. Factorization Theorem

Let $\mathscr{X}$ and $\mathscr{Y}$ be regular surfaces over a base scheme S_0. We recall that an S_0-rational map $f: \mathscr{X} \to \mathscr{Y}$ is an equivalence class of S_0-morphisms from open dense subsets of $\mathscr{X}$ to $\mathscr{Y}$, where two such morphisms are equivalent if they agree on the intersection of their domains. By [6, EGA I, Prop. 7.2.2], there is a largest dense open set U_0 of $\mathscr{X}$ where an element in the equivalence class of f is defined; U_0 will be called the domain of definition of f. In particular, f is a morphism if and only if f is defined at each $x \in \mathscr{X}$, by which we mean that an element in the equivalence class of f has x in its domain. By [6, EGA II, Prop. 7.3.5], the codimension of $\mathscr{X} - U_0$ in $\mathscr{X}$ is at least two. If f induces an isomorphism between the function fields of $\mathscr{X}$ and $\mathscr{Y}$, then f is called an S_0-birational map.

Let P be a closed point of $\mathscr{X}$ for which $\dim \mathcal{O}_{\mathscr{X},P} = 2$. We refer the reader to [7, §§II.7, V.3] and [6, EGA II, §II.8] for an account of the basic properties of the blow up $\pi: \mathscr{X}' \to \mathscr{X}$ of $\mathscr{X}$ at P. By definition, $\mathscr{X}' = \text{Proj}(\sum_{n \geq 0} I^n)$, where I is the $\mathcal{O}_\mathscr{X}$ ideal defining the closed subscheme $(P, \text{Spec}(k(P))$ of $\mathscr{X}$. The natural morphism $\pi: \mathscr{X}' \to \mathscr{X}$ is called the locally quadratic transformation of $\mathscr{X}$ with center P, or simply a locally quadratic transformation. By [6, EGA II, Defin. 8.1.3 and Prop. 8.1.4], π is projective, surjective and an isomorphism outside of $|\pi^{-1}(P)|$. In [8, Cor. 1.5] $\mathscr{X}'$ is shown to be regular. One can describe $\mathscr{X}'$ in an open neighborhood of $\pi^{-1}(P)$ in the following way. Let x and y be generators of the maximal ideal of $\mathcal{O}_{\mathscr{X},P}$. Let U be an affine open subset of $\mathscr{X}$ containing P such that x and y generate the maximal ideal of P in $\Gamma(U, \mathcal{O}_U) = A$. Then $\mathscr{X}'' = \pi^{-1}(U)$ is isomorphic to the closed subscheme of $\mathbb{P}_U^1$ defined by $ty - ux$, taking t and u as homogenous coordinates of $\mathbb{P}_U^1$. In particular, $\mathscr{X}''$ is covered by the affine patches $V_1 = \text{Spec } A[y/x]$ and $V_2 = \text{Spec } A[x/y]$.

Theorem 2.1 (Factorization Theorem [8, Theorem 1.15] and [11, Chap. I, §2]). *Let $\mathscr{X}$ and $\mathscr{X}'$ be regular surfaces and let $f_0: \mathscr{X}' \to \mathscr{X}$ be a proper birational morphism. Then $\mathscr{X}'$ is isomorphic to the scheme obtained from $\mathscr{X}$ by a finite number of successive locally quadratic transformations.*

PROOF. Since f_0 induces an isomorphism between the function fields of $\mathscr{X}$ and $\mathscr{X}'$, there is a birational map $g_0: \mathscr{X} \to \mathscr{X}'$ induced by f_0. Define $\mathscr{X}_0 = \mathscr{X}$ and assume by induction on $n \geq 1$ that $\mathscr{X}_{n-1}$ is a regular surface for which we have a birational map $g_{n-1}: \mathscr{X}_{n-1} \to \mathscr{X}'$. Since $\mathscr{X}_{n-1}$ is regular, the set of points where g_{n-1} is not defined is of codimension at least 2 in $\mathscr{X}_{n-1}$, and is thus a finite set of closed points of codimension 2 in $\mathscr{X}_{n-1}$. Suppose that P_{n-1} is such a point. Let $\pi_n: \mathscr{X}_n \to \mathscr{X}_{n-1}$ be the blow-up of $\mathscr{X}_{n-1}$ at P_{n-1}, and let $g_n: \mathscr{X}_n \to \mathscr{X}'$ be the birational map induced by g_{n-1}.

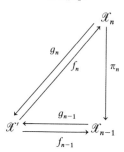

We will prove the following facts:

(2.1) Suppose $f_{n-1}: \mathscr{X}' \to \mathscr{X}_{n-1}$ is a proper birational morphism which induces the isomorphism between the function fields of $\mathscr{X}'$ and $\mathscr{X}_{n-1}$ which is the inverse of the isomorphism induced by g_{n-1}. Then f_{n-1} factors through π_n, i.e. there is a proper birational morphism $f_n: \mathscr{X}' \to \mathscr{X}_n$ such that $f_{n-1} = \pi_n \circ f_n$.

(2.2) The sequence of $\{\mathscr{X}_n, g_n\}$ constructed inductively as above is necessarily finite, i.e. there is an N such that the birational map $g_N: \mathscr{X}_N \to \mathscr{X}'$ is defined everywhere.

Since the birational map g_N in (2.2) is defined everywhere, it is a birational morphism. Each f_n constructed inductively in (2.1) is a birational morphism. Hence (2.1) and (2.2) will imply $\mathscr{X}'$ is isomorphic to $\mathscr{X}_N$, thus proving the Factorization Theorem. □

To prove (2.1) we will need the following criterion of [8, Cor. 1.13] for a birational map between integral schemes to be defined at a given point. This criterion follows readily from [6, EGA I, Prop. 6.5.1].

Proposition 2.1. *Let $\mathscr{F}$ and $\mathscr{G}$ be integral schemes over a base scheme S_0, with $\mathscr{G}$ separated and of finite type over S_0. Let π be an S_0-birational map from $\mathscr{F}$ to $\mathscr{G}$. Identify $K(\mathscr{F})$ with $K(\mathscr{G})$ via π. Then π is defined at a point x of $\mathscr{F}$ iff there is a point y of $\mathscr{G}$ such that $\mathcal{O}_{\mathscr{F},x}$ dominates $\mathcal{O}_{\mathscr{G},y}$.*

PROOF OF (2.1). Since $\mathscr{X}'$, $\mathscr{X}_{n-1}$ and $\mathscr{X}_n$ are regular and all have the same function field, $f_n: \mathscr{X}' \to \mathscr{X}_n$ exists as a birational map, and the codimension of the set of points where f_n is not defined is at least two. Hence to show that f_n is a birational morphism, it will suffice to show f_n is defined at all points z of $\mathscr{X}'$ such that $\dim \mathcal{O}_{\mathscr{X}',z} = 2$.

Suppose first that $f_{n-1}(z)$ is not P_{n-1}. Then π_n^{-1} is defined in an open neighborhood U of z. Let $V = f_{n-1}^{-1}(U)$. Then V is dense and open in $\mathscr{X}'$, and $\pi_n^{-1} \circ f_{n-1}: V \to \mathscr{X}_n$ is in the class of f_n. Hence f_n is defined at z.

Now suppose that $f_{n-1}(z) = P_{n-1}$. To simplify notation, let $\mathcal{Y} = \mathcal{X}_{n-1}$ and $P = P_{n-1}$. By Proposition 2.1, $\mathcal{O}_{\mathcal{X}',z}$ dominates $\mathcal{O}_{\mathcal{Y},P}$. Since g_{n-1} is by assumption not defined at P, $\mathcal{O}_{\mathcal{X}',z}$ must properly contain $\mathcal{O}_{\mathcal{Y},P}$. Let x and y be generators for the maximal ideal of $\mathcal{O}_{\mathcal{Y},P}$. By a Theorem of Zariski and Abhyankar [1, Theorem 3], if $\mathcal{O}_{\mathcal{X}',z}$ and $\mathcal{O}_{\mathcal{Y},P}$ are two-dimensional regular local rings with the same fraction field, and $\mathcal{O}_{\mathcal{X}',z}$ dominates and properly contains $\mathcal{O}_{\mathcal{Y},P}$, then $\mathcal{O}_{\mathcal{X}',x}$ must contain a local ring of either $\mathcal{O}_{\mathcal{Y},P}[y/x]$ or $\mathcal{O}_{\mathcal{Y},P}[x/y]$. Now the local rings of $\mathcal{O}_{\mathcal{Y},P}[y/x]$ and $\mathcal{O}_{\mathcal{Y},P}[x/y]$ are exactly the local rings of the inverse image of P under the blow-up $\pi_n \colon \mathcal{X}_n \to \mathcal{Y}$ of $\mathcal{Y}$ at P. Therefore $\mathcal{O}_{\mathcal{X}',z}$ dominates the local ring of a point of $\mathcal{X}_n$. We conclude from Proposition 2.1 that f_n is defined at z. Since $\pi_n \circ f_n = f_{n-1}$, f_{n-1} is proper and π_n is separated, one concludes that f_n is proper (cf. [7, p. 107]). □

PROOF OF (2.2). Suppose that (2.2) is false, so that the sequence $\{\mathcal{X}_n, g_n\}_n$ is infinite. Since each $\mathcal{X}_n$ is regular, the set of points S_n where g_n is not defined is finite and of codimension two in $\mathcal{X}_n$. Hence for each n, there is a point of S_n which has infinitely many of the points P_m for $m \geq n$ lying over it. Therefore we can find an infinite subsequence $\{P_{m(n)}\}$ of the $\{P_n\}$ such that the local ring $\mathcal{O}_{m(n+1)}$ of $P_{m(n+1)}$ dominates $\mathcal{O}_{m(n)}$. By another result of Zariski and Abhyankar [1, Lemma 12 and Theorem 3], the union $C = \bigcup \mathcal{O}_{m(n)}$ of an ascending sequence of two-dimensional local rings with the same fraction field, and for which each $\mathcal{O}_{m(n+1)}$ dominates the previous $\mathcal{O}_{m(n)}$, is the valuation ring of a valuation v of the common field of functions K of $\mathcal{X}'$ and the $\mathcal{X}_{m(n)}$. The birational morphism $f_{m(0)} \colon \mathcal{X}' \to \mathcal{X}_{m(0)}$ constructed in the proof of (2.1) is proper. Hence by the valuative criterion of properness, C dominates the local ring $\mathcal{O}_{\mathcal{X}',x}$ of some point x of $\mathcal{X}'$. Because $\mathcal{O}_{\mathcal{X}',x}$ is the localization of a ring of finite type over $\mathcal{O}_{m(0)}$, there is an n such that $\mathcal{O}_{m(n)}$ dominates $\mathcal{O}_{\mathcal{X}',x}$. But now by Proposition 2.1, $g_{m(n)}$ is well defined at $P_{m(n)}$, which is a contradiction. □

We now observe that to construct the $\mathcal{X}_n$ and the birational maps g_n: $\mathcal{X}_n \to \mathcal{X}'$, we made use only of the original birational map $g_0 \colon \mathcal{X} = \mathcal{X}_0 \to \mathcal{X}'$, and not of the morphisms f_n. The only point at which of f_n entered into the proof was in showing that a valuation ring C of $K(\mathcal{X}) = K(\mathcal{X}')$ which dominates the local ring of a point of $\mathcal{X}$ must also dominate the local ring of a point of $\mathcal{X}'$. Suppose now that both $\mathcal{X}$ and $\mathcal{X}'$ are regular surfaces which are proper over a base scheme S_0, but that we are given only the existence of an S_0-birational map $g_0 \colon \mathcal{X} \to \mathcal{X}'$ and not of a proper birational morphism $f_0 \colon \mathcal{X}' \to \mathcal{X}$. Then a valuation ring C of $K(\mathcal{X})$ which dominates the local ring of a point of $\mathcal{X}$ also dominates the local ring of a point of S_0. Since we have now assumed that $\mathcal{X}'$ is proper over S_0, C must dominate a local ring of a point of $\mathcal{X}'$ by the valuative criterion of properness. Thus the argument used in showing (2.2) also proves the following result.

Proposition 2.2. *Let* $g_0 \colon \mathcal{X} \to \mathcal{X}'$ *be a proper S_0-birational map between regular surfaces which are proper over a base scheme S_0. Then there is a regular* surface $\mathscr{X}_N$ and proper birational S_0-morphisms $\pi_N: \mathscr{X}_N \to \mathscr{X}$ and $g_N: \mathscr{X}_N \to \mathscr{X}'$ such that $g_N = g_0 \circ \pi_N$.

Using the Factorization Theorem, we can now clarify the uniqueness of blow-downs of exceptional curves on regular surfaces.

Lemma 2.1. *Let $\mathscr{X}$ (resp. $\mathscr{Y}$) be a regular (resp. normal) surface. Suppose $\pi: \mathscr{X} \to \mathscr{Y}$ is a proper birational morphism which blows down a prime divisor E on $\mathscr{X}$ to a point P of $\mathscr{Y}$, and which is an isomorphism off of $|E|$. Then $\pi_*(\mathcal{O}_\mathscr{X}) = \mathcal{O}_\mathscr{Y}$.*

PROOF. By [5, Chap. VI, §1, no. 3, Theorem 3, p. 92], the integral closure of $\mathcal{O}_{\mathscr{Y},P}$ in its quotient field $K = K(\mathscr{Y}) = K(\mathscr{X})$ is the intersection of all the valuation rings of K which contain $\mathcal{O}_{\mathscr{Y},P}$. Now $\mathcal{O}_{\mathscr{Y},P}$ is integrally closed in K since $\mathscr{Y}$ is normal, and since $|\pi^{-1}(P)| = |E|$, $\mathcal{O}_{\mathscr{Y},P}$ is a subring of $\mathcal{O}_{\mathscr{X},x}$ for each $x \in E$. By the valuative criterion of properness applied to π, each valuation ring of K containing $\mathcal{O}_{\mathscr{Y},P}$ has a unique center x on E. Hence $\mathcal{O}_{\mathscr{Y},P} = \bigcap \{\mathcal{O}_{\mathscr{X},x}: x \in E\}$. Therefore

$$\pi_*(\mathcal{O}_\mathscr{X}) = \mathcal{O}_\mathscr{Y},$$

since these sheaves on $\mathscr{Y}$ are equal on all stalks. □

Corollary 2.1. *With assumptions of Lemma 2.1, suppose further that $\mathscr{Y}$ is regular. Then $\mathscr{Y}$ and P determine $\mathscr{X}$ and E up to isomorphism over $\mathscr{Y}$, and $\mathscr{X}$ and E determine $\mathscr{Y}$ up to isomorphism.*

PROOF. Since E is a prime, the Factorization Theorem shows π is the blow up of $\mathscr{Y}$ at P, so $\mathscr{X}$ and E are determined up to an isomorphism over $\mathscr{Y}$. The second assertion is a consequence of $\pi_*(\mathcal{O}_\mathscr{X}) = \mathcal{O}_\mathscr{Y}$.

§3. Statement of the Castelnuovo Criterion

Definition 3.1. Let F be a complete connected curve over a field k. Let $\mathscr{L}$ be an invertible sheaf on F. Let $\pi: \bar{F} \to F$ be the canonical morphism from the normalization $\bar{F}$ of F to F. The invertible sheaf $\pi^*\mathscr{L}$ corresponds to a divisor $D = \sum n_P P$ of $\bar{F}$. Define the degree $\deg_k(\mathscr{L})$ of $\mathscr{L}$ with respect to k to be $\sum n_P [k(P) : k]$.

Definition 3.2. Let $i: E \to \mathscr{X}$ be a closed immersion of a complete integral curve E over a field k into a scheme $\mathscr{X}$. A positive Cartier divisor on $\mathscr{X}$ is a closed subscheme F of $\mathscr{X}$ such that the sheaf of ideals I which defines F is invertible. Define $i_k(E, F)$, the intersection of E and F with respect to k, to be $\deg_k(i^*I^{-1})$, where i^*I^{-1} is the induced invertible sheaf on E. If E is a positive

Cartier divisor on $\mathscr{X}$ and $k = H^0(E, \mathcal{O}_E)$, the self-intersection $E^{(2)}$ of E is defined to be $i_k(E, E)$. A Cartier divisor on $\mathscr{X}$ is a formal integral combination of positive Cartier divisors.

This terminology differs slightly from [7, §II.6]. If $\mathscr{X}$ is a regular scheme, then [7, Prop. II.6.11] shows that the notions of Cartier divisors and Weil divisors on $\mathscr{X}$ coincide.

Theorem 3.1 (Castelnuovo Criterion [8, p. 399], [11]). *Let $\mathfrak{O}$ be a Dedekind ring and let $\mathscr{X}$ be a regular curve over $S = \mathrm{Spec}(\mathfrak{O})$. A prime divisor E of $\mathscr{X}$ is exceptional in the sense of Definition 1.3 if and only if* (a) E is contained in a fibre of $\mathscr{X}$ over a closed point of S, (b) $H^1(E, \mathcal{O}_E) = 0$, and (c) $E^{(2)} = -1$. In this case E is isomorphic to $\mathbf{P}_k^1$ over the field $k = H^0(E, \mathcal{O}_E)$.

In the next section we develop the results from intersection theory needed to prove that if E is exceptional on $\mathscr{X}$, then E has the properties stated in Theorem 3.1. We also prove certain results about the infinitesimal neighborhoods of prime divisors E' satisfying conditions (a), (b) and (c) of Theorem 3.1. These results are needed to show that such divisors may be blown down to points on *regular* curves over $\mathrm{Spec}(\mathfrak{O})$. This blown-down surface is constructed (as in the case of complex surfaces) as the normalization of the image of a morphism from $\mathscr{X}$ to projective space over $\mathfrak{O}$, this morphism being associated to the tensor product of a very ample line bundle on $\mathscr{X}$ with a power of the line bundle associated to E'.

§4. Intersection Theory and Proper and Total Transforms

We will omit the proof of the following standard result, which is given in [8, Prop. 1.6].

Proposition 4.1. *Let E be a complete integral curve over a field k. Suppose that E is a positive Cartier divisor on a surface $\mathscr{X}$. Let F be a positive integral Cartier divisor on $\mathscr{X}$, and define $E \cap F = E \times_{\mathscr{X}} F$. Suppose that E and F meet properly, i.e. that $|E \cap F|$ is a finite set $\{P_1, \ldots, P_n\}$ of closed points of $\mathscr{X}$. Let $\mathfrak{O}_i$ be the local ring of P_i on $\mathscr{X}$, and let e_i and f_i be local equations for E and F at P_i. Then*

$$i_k(E, F) = \sum \dim_k(\mathfrak{O}_i/(e_i, f_i)).$$

Definition 4.1. Suppose D is an arbitrary Cartier divisor on $\mathscr{X}$, and that C is a Cartier divisor such that each component E' of C satisfies the following condition: $H^0(E', \mathcal{O}_{E'})$ is a field of finite degree over k, and E' is a complete curve over $H^0(E', \mathcal{O}_{E'})$. Extend the definition of $i_k(E, F)$ to one of $i_k(C, D)$ for all such C and D via bilinearity in C and D.

The symmetry of the formula in Proposition 4.1 has the following corollary.

Corollary 4.1. *Let $\mathcal{X}$ be a regular curve over a Dedekind ring $\mathfrak{O}$. Let k be the residue field of $\mathfrak{O}$ at a closed point x of $\mathrm{Spec}(\mathfrak{O})$. Then $i_k(E, F)$ is defined for all divisors E and F of $\mathcal{X}$ such that E has support in the fibre $\mathcal{X}_x$ of $\mathcal{X}$ over x. One has*

(a) $i_k(E, F) = \deg_k \mathcal{L}(F) \otimes \mathcal{O}_E$ *if E is a prime divisor in $\mathcal{X}_x$, where $\mathcal{L}(F)$ is the invertible sheaf on $\mathcal{X}$ defined by F.*
(b) $i_k(E, F) = 0$ *if F is a principal divisor.*
(c) $i_k(E, F) = i_k(F, E)$ *if F has support in a fibre of $\mathcal{X}$.*

Definition 4.2. Let $f: \mathcal{X} \to \mathcal{Y}$ be a proper birational morphism between regular surfaces. Suppose C is a prime divisor on $\mathcal{Y}$ corresponding to an invertible sheaf $\mathcal{L}$. The total transform $f^{-1}(C)$ of C is the divisor on $\mathcal{X}$ associated to $f^*\mathcal{L}$. Let y be the generic point of C. Define the proper transform $f^{-1}[C]$ to be the Zariski closure of $f^{-1}(y)$ with the induced reduced scheme structure. (Since f^{-1} is a morphism outside a set of codimension two on $\mathcal{Y}$, $f^{-1}[C]$ is a divisor of $\mathcal{X}$.) Extend the definition of total and proper transforms to all divisors on $\mathcal{Y}$ by linearity.

Proposition 4.2. *Let g be a local equation for the prime divisor C of Definition 4.2 at the point y of $\mathcal{Y}$. Then g is also a local equation for $f^{-1}(C)$ at any point x of $\mathcal{X}$ such that $f(x) = y$.*

PROOF. By the definition of a local equation for C, $\mathcal{O}_{C,y} = \mathcal{O}_{\mathcal{Y},y}/g$. Now $\mathcal{O}_{f^{-1}(C),x} = \mathcal{O}_{\mathcal{X},x} \otimes \mathcal{O}_{C,y}$ is equal to $\mathcal{O}_{\mathcal{X},x}/g$ since $\mathcal{O}_{\mathcal{X},x}$ is an integral domain containing $\mathcal{O}_{\mathcal{Y},y}$. □

Corollary 4.2. *The total transform from divisors on $\mathcal{Y}$ to divisors on $\mathcal{X}$ preserves linear equivalence. If k is a field and C and D are divisors on $\mathcal{Y}$ for which $i_k(C, D)$ is well defined, then $i_k(f^{-1}(C), f^{-1}(D))$ is well defined and equal to $i_k(C, D)$.*

PROOF. Suppose C and D are divisors on $\mathcal{Y}$ satisfying the conditions of Definition 4.1. Then $f^{-1}(C)$ and $f^{-1}(D)$ satisfy the same conditions for $\mathcal{X}$ because f is proper. Since the birational map $f^{-1}: \mathcal{Y} \to \mathcal{X}$ is defined outside of a set of codimension two on $\mathcal{Y}$, we can find a divisor D' linearly equivalent to D such that f^{-1} is defined at all points of D'. Now $i_k(C, D) = i_k(C, D')$ may be computed by local equations; by Proposition 4.2, $i_k(C, D) = i_k(f^{-1}(C), f^{-1}(D))$. □

§5. Exceptional Curves

5A. Intersection Properties

Throughout Section 5A we will assume that $\mathscr{X}$ and $\mathscr{Y}$ are regular surfaces, that $f: \mathscr{X} \to \mathscr{Y}$ is a birational morphism which is an isomorphism outside of the prime divisor E of $\mathscr{X}$, and that $f(E) = P$ is a point of $\mathscr{Y}$ such that $\dim \mathcal{O}_{\mathscr{Y}, P} = 2$. By the Factorization Theorem, f must then be the blow-up of $\mathscr{Y}$ at P.

Lemma 5.1. *Let D be a divisor of $\mathscr{Y}$ and let $k = k(P)$. Then $i_k(E, f^{-1}(D)) = 0$.*

PROOF. We can find a divisor D' on $\mathscr{Y}$ which is linearly equivalent to D such that D' does not contain P. Then $f^{-1}(D')$ and E are disjoint, so $i_k(E, f^{-1}(D)) = i_k(E, f^{-1}(D')) = 0$.

Lemma 5.2. *Let D be a prime divisor of $\mathscr{Y}$ passing through P. Then D has a regular point at P iff $f^{-1}(D) = f^{-1}[D] + E$ and $i_{k(P)}(E, f^{-1}[D]) = 1$. If D is regular at P then f induces a isomorphism between $f^{-1}[D]$ and D.*

PROOF. Let g be a local equation for D at P. We first suppose $f^{-1}(D) = f^{-1}[D] + E$ and $i_{k(P)}(E, f^{-1}[D]) = 1$. By Proposition 4.1, there is a unique point Q on $\mathscr{X}$ where $f^{-1}[D]$ and E intersect. Let y be a local equation for E at Q. Because g is a local equation for $f^{-1}(D)$ in $\mathcal{O}_{\mathscr{X}, Q}$, $i_{k(P)}(E, f^{-1}[D]) = 1$ iff $\mathcal{O}_{\mathscr{X}, Q}/(g, y) = k(P)$. This implies g cannot be in the square of the maximal ideal $m_P \subseteq \mathcal{O}_{\mathscr{Y}, P}$ of P on $\mathscr{Y}$. But g must be in m_P, since E is a component of $f^{-1}(D)$. Now because $\mathcal{O}_{\mathscr{Y}, P}$ is regular, the local ring $\mathcal{O}_{D, P} = \mathcal{O}_{\mathscr{Y}, P}/(g)$ of P on D must be regular. Hence $\mathcal{O}_{D, P}$ is one dimensional and normal, so f must induce an isomorphism between the local ring of P on D and the local ring of Q on $f^{-1}[D]$. Since f is an isomorphism off of E, and Q is the unique point of $f^{-1}[D]$ over P, f induces an isomorphism between $f^{-1}[D]$ and D.

We now suppose D has a regular point at P and that, as above, g is a local equation for D at P. We may choose a y in $\mathcal{O}_{\mathscr{Y}, P}$ so that $\{g, y\}$ is a system of local parameters for P. Let $U = \operatorname{Spec}(A)$ be an open affine neighborhood of P such that $\{g, y\}$ is a set of generators for the maximal ideal of P in U. Let t and u be a set of homogeneous coordinates for $\mathbb{P}_U^1$. Then $f^{-1}(U)$ is the closed subset of $\mathbb{P}_U^1$ defined by $ty - ug = 0$. On the affine patch where $u = 1$, $f^{-1}(U)$ is thus defined by $g = ty$. On this patch, y is a local equation for $E = f^{-1}(P)$, g is a local equation $f^{-1}(D)$ by Proposition 4.2, and t does not vanish on E. Hence on the patch $u = 1$, $f^{-1}(D) = (t) + E$, where the divisor (t) of t intersects E in the single point Q of $\mathscr{X}$ where $t = y = 0$ in the residue field of Q. Now $f^{-1}(D)$ and $f^{-1}[D]$ differ by at most a multiple of E, since f is an isomorphism off of E and E is prime. Since (t) does not contain E, we conclude that $(t) = f^{-1}[D]$ on the patch $u = 1$. Taking the Zariski closure of (t), we conclude that $f^{-1}(D) = f^{-1}[D] + E$. We have also shown that $f^{-1}[D]$ and E intersect in the single point Q, which has residue field $\mathcal{O}_{\mathcal{X},P}[y, g/y]/(y, g/y) = k(P)$. Hence $i_{k(P)}(E, f^{-1}[D]) = 1$ □

Proposition 5.1. *The divisor E is isomorphic to the projective line over $H^0(E, \mathcal{O}_E) = k(P)$. One has $H^1(E, \mathcal{O}_E) = 0$ and $E^{(2)} = i_{k(P)}(E, E) = -1$.*

PROOF. By the construction of the blow-up of $\mathcal{Y}$ at P, $E = f^{-1}(P)$ is a complete curve over $k(P)$ covered by two copies of the affine line over $k(P)$. Hence E is isomorphic to the projective line over $k(P) = H^0(E, \mathcal{O}_E)$, and $H^1(E, \mathcal{O}_E) = 0$.

To show $E^2 = -1$, choose a divisor D of $\mathcal{Y}$ having a regular point at P. By Lemmas 5.1 and 5.2,

$$0 = i_{k(P)}(E, f^{-1}(D))$$
$$= i_{k(P)}(E, f^{-1}[D] + E)$$
$$= 1 + i_{k(P)}(E, E).$$

Hence $i_{k(P)}(E, E) = -1$, so $H^0(E, \mathcal{O}_E) = k(P)$ implies $E^{(2)} = -1$. □

The following proposition is used in the proof of Lemma 7.2, which underlies the proof that relatively minimal models of curves over Dedekind rings are minimal if the general fibre of the curve has genus greater than zero.

Proposition 5.2. *Let C' be a complete integral curve on $\mathcal{X}$ over $k' = H^0(C', \mathcal{O}_{C'})$. Assume $C' \neq E$. Then $C = f(C')$ is a prime divisor on $\mathcal{Y}$. The degree of k' over the field $k = H^0(C, \mathcal{O}_C)$ is finite, and C is a complete integral curve over k.*

(a) *If C does not contain P, then C and C' are isomorphic and $C^{(2)} = C'^{(2)}$.*
(b) *If C contains P, then*

$$C^{(2)} \geq [k':k](C'^{(2)} + 1),$$

with equality iff C' is isomorphic to C via f and C has a regular point at P.

PROOF. Since f is proper and an isomorphism off of E, C is an irreducible curve on $\mathcal{Y}$ over k. Because f is of finite type, $[k':k]$ is finite. Clearly C is isomorphic to C' if P is not on C, and $C^{(2)} = C'^{(2)}$ in this case by Corollary 4.2. Suppose now that C contains P. Then $f^{-1}(C) = C' + mE$ for some integer $m \geq 1$. Therefore $C^{(2)} = i_k(C, C) = i_k(f^{-1}(C), C') + i_k(f^{-1}(C), mE) = i_k(f^{-1}(C), C')$ by Lemma 5.1. Now

$$i_k(f^{-1}(C), C') = i_k(C', C') + i_k(mE, C')$$
$$= [k':k](i_{k'}(C', C') + mi_{k'}(C', E)),$$

since C' is a curve over k' and $i_k(E, C') = i_k(C', E)$. Because C' and E intersect properly and nontrivially, $i_{k'}(C', E) \geq 1$, and so the inequality in part (b) holds. This inequality is an equality iff $m = 1$ and $i_k(E, C') = 1$. Lemma 5.2 now completes the proof. □

5B. Prime Divisors Satisfying the Castelnuovo Criterion

In this section we no longer make the assumptions of Section 5A. In particular, we no longer assume that E is an exceptional curve on $\mathcal{X}$.

Proposition 5.3. *Let $\mathcal{X}$ is a regular surface. Let E be a prime divisor of $\mathcal{X}$ which is a proper curve over the field $k' = H^0(E, \mathcal{O}_E)$. Suppose that $H^1(E, \mathcal{O}_E) = 0$ and $E^{(2)} = -1$. Then E is isomorphic to $P^1_{k'}$ over k'.*

PROOF. Let $\bar{E}$ be the normalization of E, and let $f: \bar{E} \to E$ be the natural morphism. Let i be the closed immersion of E into $\mathcal{X}$, and let I be the sheaf of ideals of E on $\mathcal{X}$. From $E^{(2)} = \deg_{k'} i^*(I^{-1})$, we conclude that $f^* i^* I$ is an invertible sheaf on $\bar{E}$ which corresponds to an effective divisor on $\bar{E}$ of degree 1 over k'. Hence there is a point $\bar{Q}$ on $\bar{E}$ which has residue field k' on $\bar{E}$. Since $\bar{E}$ is normal, k' must be algebraically closed in the common field of functions $K(E)$ of $\bar{E}$ and E. We also conclude that $Q = f(\bar{Q})$ must have residue field k' on E, since $k' = H^0(E, \mathcal{O}_E)$. Therefore Q is a nonsingular point of E defined over k'.

We now show that E is nonsingular. Consider the sequence

$$0 \to \mathcal{O}_E \to f_* \mathcal{O}_{\bar{E}} \to F \to 0$$

of sheaves on E. In cohomology this gives

$$0 \to k' = H^0(\mathcal{O}_E) \xrightarrow{\phi} H^0(f_* \mathcal{O}_{\bar{E}}) \to H^0(F) \to H^1(E, \mathcal{O}_E) = 0.$$

Now k' is algebraically closed in $K(E)$, and $H^0(f_* \mathcal{O}_E) = H^0(\bar{E}, \mathcal{O}_{\bar{E}})$, so ϕ is an isomorphism. Since F is supported on a finite number of closed points of E, F must be trivial, so E is nonsingular. By descent from $\bar{k}'$ to k', a nonsingular curve of genus 0 which is proper over k' and which has a rational point over k' is isomorphic to $P^1_{k'}$ over k'. □

Proposition 5.4. *With the hypotheses and notations of Proposition 5.3, let I be the $\mathcal{O}_{\mathcal{X}}$ ideal defining E, and let $E(n) = \mathrm{Spec}(\mathcal{O}_{\mathcal{X}}/I^n)$ be the closed subscheme of $\mathcal{X}$ defined by I^n. Then*

(a) $H^0(\mathcal{X}, I/I^2) = H^0(E, I/I^2)$ *is a vector space of dimension two over k'. Let T_0 and T_1 be generators of this vector space. Then $\sum H^0(\mathcal{X}, I^n/I^{n+1})$ is isomorphic to the polynomial algebra on T_0 and T_1 over k'.*
(b) $H^1(E, I^n/I^{n+1}) = 0$ *for $n \geq 0$.*
(c) *The natural maps* $H^0(\mathcal{X}, \mathcal{O}_{\mathcal{X}}/I^{n+1}) \to H^0(\mathcal{X}, \mathcal{O}_{\mathcal{X}}/I^n)$ *and* $H^0(\mathcal{X}, I^m/I^{n+1}) \to H^0(\mathcal{X}, I^m/I^n)$ *are surjective for all $n \geq m \geq 1$.*
(d) $H^0(\mathcal{X}, I/I^n)^2 = H^0(\mathcal{X}, I^2/I^n)$ *for all $n \geq 2$.*

(e) $H^1(\mathscr{X}, \mathcal{O}_\mathscr{X}/I^n) = H^1(E(n), \mathcal{O}_{E(n)}) = 0$ for all $n \geq 1$.
(f) $H^1(E(n), \mathcal{O}^*_{E(n)}) = \mathrm{Pic}(E(n)) = \mathbb{Z}$ for all $n \geq 1$.

PROOF. Let $i: E \to \mathcal{O}_\mathscr{X}$ be the closed immersion of E into $\mathscr{X}$. Since $i^*I = I/I^2$ as sheaves on E, the sheaf $I^n/I^{n+1} = (i^*I)^{\otimes n}$ on E has degree $n \deg_{k'} i^*I = n(-E^{(2)}) = n \geq 0$. By Proposition 5.1, E is isomorphic to $P^1_{k'}$ over k'. Now (a) and (b) follow from Riemann–Roch on E.

Consider the sequences

(5.1) $\quad 0 \to I^n/I^{n+1} \to \mathcal{O}_\mathscr{X}/I^{n+1} \to \mathcal{O}_\mathscr{X}/I^n \to 0,$

(5.2) $\quad 0 \to I^n/I^{n+1} \to I^m/I^{n+1} \to I^m/I^n \to 0,$

for $n \geq m \geq 1$. On taking cohomology, we see that (b) implies (c). In particular, we have an exact sequence

(5.3) $\quad 0 \to H^0(\mathscr{X}, I^n/I^{n+1}) \to H^0(\mathscr{X}, I^m/I^{n+1}) \to H^0(\mathscr{X}, I^m/I^n) \to 0.$

In part (d) it is clear that

(5.4) $\quad\quad\quad H^0(\mathscr{X}, I/I^n)^2 \subseteq H^0(\mathscr{X}, I^2/I^n)$

if $n \geq 2$, with equality if $n = 2$. Suppose now by induction that

$$H^0(\mathscr{X}, I/I^n)^2 = H^0(\mathscr{X}, I^2/I^n)$$

for some $n \geq 2$. Since $H^0(\mathscr{X}, I/I^{n+1}) \to H^0(\mathscr{X}, I/I^2)$ is surjective, we have from part (a) that

$$H^0(\mathscr{X}, I/I^{n+1})^2 \supseteq H^0(\mathscr{X}, I/I^{n+1})^n = H^0(\mathscr{X}, I^n/I^{n+1}).$$

The induction hypohtesis and (5.3) with $m = 2$ now imply

$$H^0(\mathscr{X}, I/I^{n+1})^2 \supseteq H^0(\mathscr{X}, I^2/I^n).$$

Since (5.4) holds with n replaced by $n + 1$, the induction holds for $n + 1$, and part (d) is proved.

To show (e), we have from (b) and the cohomology of (5.1) that the natural map $H^1(\mathscr{X}, \mathcal{O}_\mathscr{X}/I^{n+1}) \to H^1(\mathscr{X}, \mathcal{O}_\mathscr{X}/I^n)$ is an isomorphism for all $n \geq 1$. Since $E = P^1_{k'}$, one has $H^1(E, \mathcal{O}_E) = H^1(\mathscr{X}, \mathcal{O}_\mathscr{X}/I) = 0$, and (e) follows.

It remains to prove $\mathrm{Pic}(E(n)) = \mathbb{Z}$ if $n \geq 1$. From the surjection of sheaves $\mathcal{O}_{E(n)} \to \mathcal{O}_E$ on $E(n)$, one has exact sequences

$$0 \to N \to \mathcal{O}_{E(n)} \to \mathcal{O}_E \to 0,$$

$$0 \to M \to \mathcal{O}^*_{E(n)} \to \mathcal{O}^*_E \to 0,$$

for some sheaves N and M on $E(n)$. Since $H^0(E(n), \mathcal{O}_{E(n)}) \to H^0(E, \mathcal{O}_E)$ is surjective, $H^0(E(n), \mathcal{O}^*_{E(n)}) \to H^0(E, \mathcal{O}^*_E)$ is surjective. Thus on taking cohomology, we have exact sequences

$$0 \to H^1(E(n), N) \to H^1(E(n), \mathcal{O}_{E(n)}) \to H^1(E, \mathcal{O}_E) \to 0,$$

$$0 \to H^1(E(n), M) \to H^1(E(n), \mathcal{O}^*_{E(n)}) \to H^1(E, \mathcal{O}^*_E) \to 0.$$

In this situation, Artin proved in [4] that $H^1(E(n), N)$ and $H^1(E(n), M)$ have isomorphic Jordan–Holder series as finitely generated abelian groups. Because we have already shown $H^1(E(n), \mathcal{O}_{E(n)}) = 0$ for all n, it follows that $H^1(E(n), N) = H^1(E(n), M) = 0$. Therefore $H^1(E(n), \mathcal{O}^*_{E(n)}) = H^1(E, \mathcal{O}^*_E) = \text{Pic}(P^1_{k'}) = \mathbb{Z}$.

§6. Proof of the Castelnuovo Criterion

Let $\mathfrak{O}$ be a Dedekind ring, and let $\mathscr{X}$ be a complete regular curve over $S = \text{Spec}(\mathfrak{O})$. Suppose E is an exceptional curve on $\mathscr{X}$. A proper birational morphism of $\mathscr{X}$ to another regular curve $\mathscr{Y}$ over S must induce an isomorphism on the fibres of $\mathscr{X}$ and $\mathscr{Y}$ over the generic point of S. Hence $f(E)$ does not contain the generic point of S. Since E is irreducible, E lies in the fibre of $\mathscr{X}$ over a closed point of S. Proposition 5.1 now shows that E satisfies all the conditions of the Castelnuovo criterion.

We now suppose that E is an irreducible divisor contained in a fibre of $\mathscr{X}$ over a closed point of S, that $H^1(E, \mathcal{O}_E) = 0$ and that $E^{(2)} = -1$. We are to show that there is a proper S-morphism $\pi: \mathscr{X} \to \mathscr{Y}$ to a regular curve $\mathscr{Y}$ over S which is an isomorphism off E, and which blows down E to a point.

By Theorem 1.1, there is a divisor D on $\mathscr{X}$ which is very ample over S. By Serre's theorem, we may replace D by a suitably high multiple of D to be able to assume that $H^1(\mathscr{X}, \mathcal{O}_{\mathscr{X}}(D)) = 0$.

Let i be the immersion of E into $\mathscr{X}$, and let k' be the field $H^0(E, \mathcal{O}_E)$. Define $r = \deg_{k'} i^*\mathcal{O}_{\mathscr{X}}(D)$. Let Z be the divisor rE. Our object is to show that if $\mathscr{Y}_0$ is the image of the rational map into projective space defined by the divisor $D + Z$, then the normalization $\mathscr{Y}$ of $\mathscr{Y}_0$ is the regular curve desired.

Step 1. We first show that the rational map $\pi_0: \mathscr{X} \to \mathscr{Y}_0$ defined by $D + Z$ is defined everywhere. Clearly π_0 is defined and regular off of $|Z|$ since $\mathscr{X}$ is regular and D is very ample. By [6, EGA II, Cor. 4.4.9], a sufficient condition for π_0 to be defined on $|Z|$ is that there exist an element s of $H^0(\mathscr{X}, \mathcal{O}_{\mathscr{X}}(D + Z))$ which is nonzero at every local ring of Z.

To find such an s, let j be the natural injection of Z into $\mathscr{X}$. Since $E^{(2)} = -1$, we have $i_{k'}(E, D + Z) = 0$. Therefore $i_{k'}(Z, D + Z) = 0$. Because the natural map $\text{Pic}(Z) \to \text{Pic}(E) = \mathbb{Z}$ is an isomorphism by Proposition 5.4(f), we conclude that $j^*(\mathcal{O}_{\mathscr{X}}(D + Z))$ is isomorphic to $\mathcal{O}_Z$. Now tensor the exact sequence of sheaves

$$0 \to \mathcal{O}_{\mathscr{X}}(-Z) \to \mathcal{O}_{\mathscr{X}} \to \mathcal{O}_Z \to 0,$$

with $\mathcal{O}_{\mathscr{X}}(D + Z)$ to obtain

$$0 \to \mathcal{O}_{\mathscr{X}}(D) \to \mathcal{O}_{\mathscr{X}}(D + Z) \to j^*(\mathcal{O}_{\mathscr{X}}(D + Z)) = \mathcal{O}_Z \to 0.$$

Since $H^1(\mathscr{X}, \mathcal{O}_{\mathscr{X}}(D)) = 0$, restriction to Z induces a surjection $H^0(\mathscr{X}, \mathcal{O}_{\mathscr{X}}(D+Z)) \to H^0(Z, \mathcal{O}_Z)$. We may thus chose s to be any element of $H^0(\mathscr{X}, \mathcal{O}_{\mathscr{X}}(D + Z))$

which restricts to the identity element of $H^0(Z, \mathcal{O}_Z) = H^0(Z, j^*(\mathcal{O}_{\mathcal{X}}(D + Z)))$. (Notice that this s will not in general be the identity.) We have now shown that $\pi_0: \mathcal{X} \to \mathcal{Y}_0$ is a projective birational morphism and an isomorphism off of $|Z|$.

Step 2. Suppose $Z_0 = \pi_0(|Z|)$ is a curve on $\mathcal{Y}_0$. The restriction of $\mathcal{O}_{\mathcal{X}}(D + Z)$ to Z would then be ample. This contradicts the fact proved above that this restriction is isomorphic to $\mathcal{O}_Z$, which has degree zero and is thus not ample. Since $|Z|$ is connected, and π_0 is an isomorphism off of $|Z|$, π_0 must map $|Z|$ to a point P_0 of $\mathcal{Y}_0$. Since $\mathcal{X}$ is normal, π_0 factors through the normalization $\mathcal{Y}$ of $\mathcal{Y}_0$ of $\mathcal{Y}$ via a morphism $\pi: \mathcal{X} \to \mathcal{Y}$. Because $|\pi_0^{-1}(P_0)| = |Z|$, it follows that there is a unique point P of $\mathcal{Y}$ over P_0, and that $\pi(|Z|) = P$.

Step 3. We must show $\mathcal{Y}$ is regular to complete the proof. For this it will suffice to prove that $\mathcal{O}_{\mathcal{Y}, P}$ is a regular local ring.

Let m_P be the maximal ideal of $\mathcal{O}_{\mathcal{Y}, P}$. By the Theorem on Formal Functions [6, EGA II, Cor. 4.2.4], one has an isomorphism

$$((\pi_* \mathcal{O}_{\mathcal{X}})_P)^\wedge = \varprojlim_n \Gamma(\pi^{-1}(P), \mathcal{O}_{\mathcal{X}}/m_P^n \mathcal{O}_{\mathcal{X}}),$$

where the left-hand side is the m_P-adic completion of the stalk. Recall that I denotes the ideal sheaf of E. The sets of ideals $\{m_P^n \mathcal{O}_{\mathcal{X}}\}_n$ and $\{I^n\}_n$ are cofinal in the collection of all $\mathcal{O}_{\mathcal{X}}$ ideals J for which $\mathcal{O}_{\mathcal{X}}/J$ has support on $|E| = |\pi^{-1}(P)|$. By Lemma 2.1, $\mathcal{O}_{\mathcal{Y}, P} = (\pi_* \mathcal{O}_{\mathcal{X}})_P$. Therefore

$$\mathcal{O}_{\mathcal{Y}, P}^\wedge = ((\pi_* \mathcal{O}_{\mathcal{X}})_P)^\wedge = \varprojlim_n \Gamma(E, \mathcal{O}_{\mathcal{X}}/I^n).$$

It will be enough to show $\mathcal{O}_{\mathcal{Y}, P}^\wedge$ is a regular local ring. Let $E(n) = \mathrm{Spec}(\mathcal{O}_{\mathcal{X}}/I^n)$. The following was shown in Proposition 5.4(c):

(6.1) The natural map from $A_n = H^0(E(n), \mathcal{O}_{E(n)}) = \Gamma(\mathcal{X}, \mathcal{O}_{\mathcal{X}}/I^n)$ to A_{n-1} is surjective for all $n \geq 1$.

Since $\mathcal{O}_{\mathcal{Y}, P}^\wedge = \varprojlim A_n$, this gives for each n an exact sequence

(6.2) $$0 \to B_n \to \mathcal{O}_{\mathcal{Y}, P}^\wedge \to A_n \to 0$$

for some $\mathcal{O}_{\mathcal{Y}, P}^\wedge$ ideal B_n. Here if $m \leq n$ then

(6.3) $$\begin{aligned} B_m/B_n &= \ker\{A_n \to A_m\} \\ &= \ker\{H^0(\mathcal{X}, \mathcal{O}_{\mathcal{X}}/I^n) \to H^0(\mathcal{X}, \mathcal{O}_{\mathcal{X}}/I^m)\} \\ &= H^0(\mathcal{X}, I^m/I^n). \end{aligned}$$

By (6.3) and Proposition 5.4(d), we have

(6.4) $$B_1^2 = B_2 \mod B_n$$

if $n \geq 2$. Since $\mathcal{O}_{\mathcal{Y}, P}^\wedge = \varprojlim \mathcal{O}_{\mathcal{Y}, P}/B_n$, we conclude that

(6.5) $$B_2 = B_1^2.$$

Because $A_1 = H^0(E, \mathcal{O}_E) = k'$ is the residue field of P on $\mathcal{Y}$, $B_1 = m_P$. Thus (6.5) and (6.3) give

$$m_P/m_P^2 = B_1/B_1^2 = B_1/B_2 = H^0(\mathcal{X}, I/I^2).$$

By Proposition 5.4(a) the last cohomology group in this equality is a vector space of dimension two over the residue field k' of P in $\mathcal{O}_{\hat{\mathcal{Y}},P}$. Hence $\mathcal{O}_{\hat{\mathcal{Y}},P}$ is regular, and the proof is complete.

§7. Proof of the Minimal Models Theorem

Lemma 7.1. *Let $\mathfrak{D}$ be a Dedekind ring and let $\mathcal{X}$ be a regular curve over $S = \text{Spec}(\mathfrak{D})$. Assume that the fraction field $K(S)$ of $\mathfrak{D}$ is algebraically closed in $K(\mathcal{X})$. Then $\mathcal{X}$ has the following properties.*

(a) *The fibres of $\mathcal{X}$ are connected.*
(b) *Let x be a closed point of S, with residue field $k = k(x)$. Let V be the real vector space with basis the set $\{F_i\}$ of irreducible components of the fibre $\mathcal{X}_x$ of $\mathcal{X}$ over x. One has $i_k(\mathcal{X}_x, D) = 0$ for all $D \in V$. The pairing on $V/\mathbb{R}\mathcal{X}_x$ which is induced by $i_k(\ ,\ )$ is negative definite.*
(c) *The fibre of $\mathcal{X}$ over the generic point of S is geometrically irreducible.*
(d) *The exceptional divisors on $\mathcal{X}$ lie in reducible fibres of $\mathcal{X}$. The number $\delta(\mathcal{X})$ of irreducible divisors of $\mathcal{X}$ which lie in reducible fibres of $\mathcal{X}$ is finite.*

PROOF. By the definition of a regular curve over S, $\mathcal{X}$ is connected, and there is a flat proper morphism $\pi: \mathcal{X} \to S$ which is of finite type. By Grothendieck's Connectedness Theorem [6, EGA III, Cor. 4.3.2], to show that $\mathcal{X}$ has connected fibres, it will suffice to show that $\pi_*(\mathcal{O}_\mathcal{X}) = \mathcal{O}_S$. Since π is flat, $\pi_*\mathcal{O}_\mathcal{X}$ is the sheaf on S which is associated to the torsion-free $\mathfrak{D}$-module $H^0(\mathcal{X}, \mathcal{O}_\mathcal{X})$. Thus it will suffice to prove that $H^0(\mathcal{X}, \mathcal{O}_\mathcal{X})$ is a rank one $\mathfrak{D}$-module. Since $\mathcal{X}$ is integral, $H^0(\mathcal{X}, \mathcal{O}_\mathcal{X})$ is a submodule of $H^0(\mathcal{X}, K(\mathcal{X}))$, so it will be enough to show $H^0(\mathcal{X}, K(\mathcal{X}))$ is a rank one $K(S)$-module. Now π is proper and $\mathcal{X}$ is normal, so $H^0(\mathcal{X}, K(\mathcal{X}))$ is the algebraic closure of $K(S)$ in $K(\mathcal{X})$. By assumption, this closure is $K(S)$, so (a) is proved.

With the notations of part (b), let $\mathcal{X}_x = \sum m_i F_i$ in the vector space V for some positive integers m_i. Suppose $D = \sum r_j m_j F_j$ is an element of V for some real numbers r_j. To prove (b), we must show $i_k(D, D) \leq 0$, with equality iff all of the r_j are equal.

Since some multiple of $\mathcal{X}_x$ is principal, we have $0 = i_k(m_j F_j, \mathcal{X}_x)$ for all j. A short calculation using this and the bilinearity of $i_k(\ ,\ )$ shows

(7.1) $$i_k(D, D) = -(1/2)\sum_{i,j}(r_j - r_i)^2 i_k(m_j F_j, m_i F_i).$$

If $i \neq j$, then $i_k(m_j F_j, m_i F_i) \geq 0$, with equality iff F_j and F_i do not intersect. Hence (7.1) shows $i_k(D, D) \leq 0$, with equality iff $r_i = r_j$ whenever F_i and F_j

intersect. Since $\mathscr{X}_x$ is connected, $i_k(D, D) = 0$ implies all the r_j are equal. (Compare Mumford [10].)

Let $\mathscr{X}_K = \mathscr{X} \times_S \text{Spec}(K(S))$ be the fibre of $\mathscr{X}$ over the generic point of S. Since $K(S)$ is algebraically closed in $K(\mathscr{X})$, it follows from [12, Theorem 40, p. 197 and Cor. 2, Theorem 38, p. 195] that every zero-divisor of $K(X) \otimes_{K(S)} \overline{K(S)}$ is nilpotent. Therefore $\mathscr{X}_K$ is geometrically irreducible.

Since an irreducible fibre of $\mathscr{X}$ has self-intersection 0, the exceptional curves on $\mathscr{X}$ must lie on reducible fibres. By [6, EGA IV, p. 9.7.8], the set of points x of S where the fibre of $\mathscr{X}$ over x is geometrically irreducible is a closed set. Hence there are only finitely many such x, since $\mathfrak{O}$ is a Dedekind ring. Therefore $\delta(\mathscr{X})$ is finite. □

Corollary 7.1. *Let $\mathscr{X}$ and S be as in Lemma 7.1. Construct a sequence $\{\mathscr{X}(n)\}_n$ of regular curves over S in the following way. Let $\mathscr{X}(0) = \mathscr{X}$. If $\mathscr{X}(n)$ has been defined, and there is an exceptional curve on $\mathscr{X}(n)$, let $\pi_n: \mathscr{X}(n) \to \mathscr{X}(n+1)$ be a blow-down over S of one such curve. Then the sequence $\{\mathscr{X}(n)\}_n$ is necessarily finite. If $\mathscr{Y}$ is the last term in the sequence, then $\mathscr{Y}$ is a relatively minimal model for $K(\mathscr{X})$.*

PROOF. The number $\delta(\mathscr{X}(n))$ of irreducible divisors of $\mathscr{X}(n)$ which lie in reducible fibres must decrease as n increases. Since $\delta(\mathscr{X}(0))$ is finite, the sequence $\{\mathscr{X}(n)\}_n$ must be finite. The final term $\mathscr{Y}$ in this sequence can have no exceptional curves, so $\mathscr{Y}$ is a relatively minimal model of $K(\mathscr{X})$ by the Factorization Theorem. □

The key to showing that the $\mathscr{Y}$ of Corollary 7.1 is a minimal model if the generic fibre of $\mathscr{X}$ has genus greater than 0 is the following lemma, which is Proposition 4.3 of [8].

Lemma 7.2. *Let $\mathscr{X}$ and S be as in Lemma 7.1. Let $f: \mathscr{X}' \to \mathscr{X}$ be a birational S-morphism from a regular curve $\mathscr{X}'$ over S to $\mathscr{X}$ which is the blow-down of an exceptional curve E on $\mathscr{X}'$. Let W be the generic fibre of $\mathscr{X}$, and assume that $H^1(W, \mathcal{O}_W) \neq 0$. Suppose that C' is an exceptional curve on $\mathscr{X}'$. Then either $C' = E$, or $C = f(C')$ is an exceptional curve in $\mathscr{X}$ which does not contain P.*

PROOF. Suppose that $C' \neq E$. Then C is a curve on $\mathscr{X}$ which is birational to C'. If C does not contain P, then C and C' are isomorphic, and $C^{(2)} = C'^{(2)}$ by Proposition 5.2. Therefore C is exceptional in this case by the Castelnuovo criterion.

Suppose now that C contains P. Because $C'^{(2)} = -1$, Proposition 5.2 shows $C^{(2)} \geq 0$. Therefore C must be a rational multiple of a fibre of $\mathscr{X}$ over a closed point of S by Lemma 7.1(b). Hence $C^2 = 0$. From Proposition 5.2(b), we conclude that f induces an isomorphism between C' and C. Because C' is an exceptional curve, C is thus isomorphic to P_k^1, where $k = H^0(C', \mathcal{O}_{C'}) = H^0(C, \mathcal{O}_C)$.

Let I_C be the sheaf of ideals defining C on $\mathscr{X}$. From the sequence

$$0 \to I_C^m/I_C^{m+1} \to \mathcal{O}_\mathscr{X}/I_C^{m+1} \to \mathcal{O}_\mathscr{X}/I^m \to 0$$

and the fact that C is isomorphic to P_k^1, one has by induction on m that $H^1(\mathscr{X}, \mathcal{O}_\mathscr{X}/I^m) = 0$ for all $m > 0$. Since C is a rational multiple of a fibre of $\mathscr{X}$ over a closed point of S, there is an m and a nonzero $t \in \mathfrak{O}$ such that $I^m = t\mathcal{O}_\mathscr{X}$. From the exact sequence

$$H^1(\mathscr{X}, \mathcal{O}_\mathscr{X}) \xrightarrow{t} H^1(\mathscr{X}, \mathcal{O}_\mathscr{X}) \to H^1(\mathscr{X}, \mathcal{O}_\mathscr{X}/I^m) = 0,$$

we conclude that $H^1(\mathscr{X}, \mathcal{O}_\mathscr{X})$ is a torsion $\mathfrak{O}$-module. Let K be the quotient field of $\mathfrak{O}$. Then $H^1(W, \mathcal{O}_W) = H^1(\mathscr{X}, \mathcal{O}_\mathscr{X}) \otimes_\mathfrak{O} K = 0$, contradicting the assumption that W has nonzero genus. □

COMPLETION OF THE PROOF OF THE MINIMAL MODELS THEOREM. Suppose $\mathscr{Y}$ and $\mathscr{Y}'$ are relatively minimal models of $K(\mathscr{X})$, where we now assume that the general fibre of $\mathscr{X}$ has positive genus. In view of Proposition 2.2, there is a regular curve $\mathscr{X}'$ over S for which there are proper S-birational morphisms $\pi: \mathscr{X}' \to \mathscr{Y}$ and $\pi': \mathscr{X}' \to \mathscr{Y}'$. We may further suppose that $\mathscr{X}'$ has been chosen so that the number $\delta(\mathscr{X}')$ or irreducible divisors contained in reducible fibres of $\mathscr{X}'$ is minimal. We will now show that $\mathscr{X}'$ can have no exceptional curves. It will then follow from the Factorization Theorem that π and π' are isomorphisms, which will complete the proof.

Suppose that $\mathscr{X}'$ contains an exceptional curve E. By the Factorization Theorem, we can factor each of π and π' into products $\pi_1 \ldots \pi_n$ and $\pi'_1 \ldots \pi'_m$ of locally quadratic transformations. By Lemma 7.2, each π_i and each π'_j either blow down E to a point or send E to an isomorphic exceptional curve, which we will identify with E. Since $\mathscr{Y}$ and $\mathscr{Y}'$ contain no exceptional curves, E must be the exceptional curve associated to some π_i and to some π_j. The exceptional curves associated to $\pi_1, \ldots, \pi_{i-1}$ are disjoint from E by Lemma 7.2. Hence we may rearrange the $\pi_1, \ldots, \pi_i$ to be able to assume that E is the exceptional curve of π_1. Similarly, we can assume E is the exceptional curve of π'_1. But now π and π' factor through the range $\mathscr{X}(1)$ of π_1. Since $\delta(\mathscr{X}(1)) < \delta(\mathscr{X}')$, this contradicts the assumption that $\mathscr{X}'$ was chosen to minimize $\delta(\mathscr{X}')$, so the proof is complete.

REFERENCES

[1] Abhyankar, S. On the valuations centered in a local domain, *Amer. J. Math.*, **78** (1956), 321–348.
[2] Abhyankar, S. Resolution of singularities of arithmetical surfaces, in *Arithmetical Algebraic Geometry*. Harper and Row: New York, 1965.
[3] Artin, M. Lipman's proof of resolution of singularities for surfaces, this volume, pp. 267–287.
[4] Artin, M. Some numerical criteria for contractibility of curves on algebraic surfaces. *Amer. J. Math.*, **84** (1962), 485–496.
[5] Bourbaki, N. *Algèbre Commutative*. Eléments de Mathematiques, 27, 28, 30, 31. Hermann: Paris, 1961–65.

[6] Grothendieck, A. Eléments de géométrie algébrique (EGA), I–IV. *Publ. Math. I.H.E.S.*, **4, 8, 11, 17, 20, 24, 28, 32** (1960–67).
[7] Hartshorne, R. *Algebraic Geometry*. Springer-Verlag: New York, 1977.
[8] Lichtenbaum, S. Curves over discrete valuation rings, *Amer. J. Math.*, **25**, no. 2 (1968), 380–405.
[9] Lipman, J. Rational singularities with applications to algebraic surfaces and unique factorization. *Publ. Math. I.H.E.S.*, **36** (1969), 195–279.
[10] Mumford, D. The topology of normal singularities of an algebraic surface and a criterion for simplicity. *Publ. Math. I.H.E.S.*, **9** (1961), 5–22.
[11] Shafarevitch, I. *Lectures on Minimal Models and Birational Transformations of Two-dimensional Schemes*. Tata Institute: Bombay, 1966.
[12] Zariski, O. and Samuel, P. *Commutative Algebra*, Vol. I. Van Nostrand: Princeton, NJ, 1958.

CHAPTER XIV

Local Heights on Curves

BENEDICT H. GROSS

In this paper we will review the theory of local heights on curves and describe its relationship to the global height pairing on the Jacobian. The local results are all special cases of Néron's theory [9], [10]; the global pairing was discovered independently by Néron and Tate [5]. We will also discuss extensions of the local pairing to divisors of arbitrary degree and to divisors which are not relatively prime. The first extension is due to Arakelov [1]; the second is implicit in Tate's work on elliptic curves [12]. I have also included several sections of examples which illustrate the general theory.

I would like to thank R. Rumley, J. Tate, and D. Zagier for their help.

§1. Definitions and Notations

Let k be a field, and let X be a complete, non-singular, geometrically irreducible curve of genus g which is defined over k and has a k-rational point. Let $\mathrm{Div}(X/k)$ denote the subgroup of divisors on X which are rational over k. If $a = \sum m_x(x)$ is a divisor, we let $|a| = \{x : m_x \neq 0\}$ denote its support and $\deg a = \sum m_x$ denote its degree. Let $\mathrm{Div}^0(X/k)$ denote the subgroup of divisors of degree zero and $P(X/k)$ the subgroup of principal divisors—those of the form $a = \mathrm{div}(f)$ with $f \in k(X)^*$.

Let J be the Jacobian of X; this is an abelian variety of dimension g which is defined over k. For any extension field H of k we have $J(H) = \mathrm{Div}^0(X/H)/P(X/H)$.

§2. Néron's Local Height Pairing

In this section, we assume that $k = k_v$ is a locally compact field. Let dx be a Haar measure on k_v^+ and define the valuation homomorphism $|\ |_v \colon k_v^* \to \mathbf{R}_+^*$ by $\alpha^*(dx) = |\alpha|_v \cdot dx$. If k_v is archimedean, we have

(2.1) $$|\alpha|_v = \begin{cases} \alpha \cdot \operatorname{sign} \alpha = |\alpha| & \text{if } k_v = \mathbf{R}, \\ \alpha \cdot \bar{\alpha} = |\alpha|^2 & \text{if } k_v \cong \mathbf{C}. \end{cases}$$

If k_v is non-archimedean, let $\mathcal{O}_v$ denote the ring of integral elements and π_v a uniformizing parameter in $\mathcal{O}_v$. Normalize the valuation v on k_v^* so that $v(\pi_v) = 1$. The residue field $F_v = \mathcal{O}_v/(\pi_v)$ is finite, with q_v elements, and

(2.2) $$|\alpha|_v = q_v^{-v(\alpha)}.$$

If X is a curve over k_v, the set of k_v-rational points $X(k_v)$ has the natural structure of a topological space, which we will assume is non-empty. We let $Z^0(X/k_v)$ be the elements of degree zero in the free abelian group on $X(k_v)$, viewed as a subgroup of $\operatorname{Div}^0(X/k_v)$. We say two divisors a and b are relatively prime if $|a| \cap |b|$ is empty.

Suppose f is a function on X over k_v whose divisor is relatively prime to $a = \sum m_x(x)$ in $Z^0(X/k_v)$. We define $f(a)$ in k_v^* by the formula $f(a) = \prod f(x)^{m_x}$; this depends only on $b = \operatorname{div}(f)$, as a has degree zero. The following basic result is due to Néron [10, Chap. II, Theorems 3 and 4]; we will give a different proof in Section 3.

Proposition 2.3. *There is a unique function $\langle a, b \rangle_v$ on relatively prime divisors $a \in Z^0(X/k_v)$, $b \in \operatorname{Div}^0(X/k_v)$ with values in $\mathbf{R}$ which satisfies the following four properties*:

(1) $\langle a, b \rangle_v + \langle a, c \rangle_v = \langle a, b + c \rangle_v$.
(2) $\langle a, b \rangle_v = \langle b, a \rangle_v$ whenever $b \in Z^0(X/k_v)$.
(3) $\langle a, \operatorname{div}(f) \rangle_v = \log |f(a)|_v$.
(4) *Fix b and a point $x_0 \in X(k_v) - |b|$. Then the function $X(k_v) - |b| \to \mathbf{R}$, defined by $x \to \langle (x) - (x_0), b \rangle_v$ is continuous.*

If H is a finite extension of $k = k_v$ and a and b are relatively prime divisors of degree zero over k with $a \in Z^0(X/k_v)$, we have

(2.4) $$\langle a, b \rangle_H = [H:k] \langle a, b \rangle_k.$$

If a and b are pointwise rational over H and σ is any automorphism of H over k, we have

(2.5) $$\langle a^\sigma, b^\sigma \rangle_H = \langle a, b \rangle_H.$$

This suggests an extension of the pairing $\langle a, b \rangle_v$ to relatively prime divisors in $\operatorname{Div}^0(X/k_v)$; we choose a finite extension H where a becomes pointwise rational and define

$$\langle a, b \rangle_v = \frac{1}{[H:k]} \langle a, b \rangle_H.$$

The local pairing is functoral for morphisms in the following sense. Let Y be another curve and $T \subset X \times Y$ a correspondence which is rational over k_v. Then for $a \in \text{Div}^0(X/k_v)$ and $b \in \text{Div}^0(Y/k_v)$ we have

(2.6) $$\langle a, T^*b \rangle_X = \langle T_*a, b \rangle_Y,$$

whenever both sides are defined.

§3. Construction of the Local Height Pairing

In this section, we will present a proof of Proposition 2.3 which does not use the theory of Néron models for abelian varieties, but works directly on the curve. To show a local pairing exists, we will construct it explicitly by using potential theory at the archimedean places and intersection theory at the non-archimedean places. The uniqueness is much easier to demonstrate. Indeed, the difference of any two local pairings satisfying the four properties of (2.3) would give a symmetric, bilinear pairing $J(k_v) \times J(k_v) \to \mathbf{R}$ which is continuous in the first variable when the second is fixed. This must be the trivial pairing, as $J(k_v)$ is compact and $\mathbf{R}$ contains no non-trivial compact subgroups.

We begin with the archimedean case; by (2.4) there is no loss in generality in assuming that $k_v = \mathbf{C}$. Then $M = X(K_v)$ is a Riemann surface. A meromorphic differential ω on M is said to be of the third kind if $\text{ord}_x(\omega) \geq -1$ for all $x \in M$. The divisor $\text{Res}(\omega) = \sum_x \text{res}_x(\omega) \cdot (x)$ with complex coefficients has degree zero by the global residue theorem. Conversely, every a in $\text{Div}^0(X/k_v)$ has the form $a = \text{Res}(\omega)$ for a differential ω of the third kind on M; this is a simple consequence of the Riemann–Roch theorem. The differential ω is unique up to the addition of differential which is holomorphic on M; since the real parts of the periods of a holomorphic differential determine it precisely and may be chosen arbitrarily, we may normalize $\omega = \omega_a$ uniquely by insisting that its periods be purely imaginary. If $a = \text{div}(f)$ is principal, then $\omega_a = df/f$ is the associated normalized differential of the third kind.

The differential ω_a arises naturally in the theory of mixed Hodge structures on open curves. Let $S = |a|$ and put $U = M - S$. We have an exact sequence

(3.1) $$0 \to H^1(M) \to H^1(U) \to H^0(S)(-1) \to H^2(M) \to 0$$

in the rational (Betti) cohomology, as well as in the complex (de Rham) cohomology theory $H_{\text{DR}} = H \otimes \mathbf{C}$. This gives the weight filtration on $H^1(U)$, where $W_1 = H^1(M)$. The Hodge filtration of $H^1_{\text{DR}}(U)$ is given by taking Fil^1 to be the image of the meromorphic differentials $\Omega^1(U)$ of the third kind on

M with poles contained in S. Clearly $\Omega^1(U) \cap H^1_{DR}(M) = \Omega^1(M)$ consists of the differentials of the first kind on M; the desired splitting of the sequence

(3.2) $$0 \to \Omega^1(M) \to \Omega^1(U) \xrightarrow{\text{Res}} \text{Div}^0(S) \to 0$$

is obtained by taking a to the unique element ω_a in $\Omega^1(U) \cap \overline{\Omega^1(U)}$ which maps to a in $\text{Div}^0(S)$. We emphasize that complex conjugation of $\Omega^1(U)$ takes place in $H^1_{DR}(U) = H^1(U) \otimes \mathbf{C}$. This is the canonical real splitting of a mixed Hodge structure with two weights [2, pp. 36–37].

Since the periods of ω_a are purely imaginary, the differential

(3.3) $$\omega_a + \bar{\omega}_a = dg_a$$

is exact. Here g_a is a harmonic function on $M - |a|$ which is well determined up to the addition of a constant function. This is the Green's function associated to the divisor a; if $a = \text{div}(f)$ is principal then $g_a = \log|f|_v$. In general, if m_x is the order of x in a, then $g_a - m_x \log|z|_v$ is harmonic near x, where z is any uniformizing parameter there. In the sense of distributions and $(1,1)$ currents, g_a is the solution of the differential equation:

(3.4) $$\partial \bar{\partial} g_a = -2\pi i \delta_a,$$

where δ_a is the $(1,1)$-current which represents the evaluation of $(0,0)$-forms at a.

Now assume $b = \sum m_y(y)$ is a divisor of degree zero on $X(k_v)$ which is relatively prime to a. We define

(3.5) $$\langle a, b \rangle_v = \sum m_y g_a(y).$$

This is independent of the choice of g_a, as $\sum m_y = 0$. It satisfies the four properties of Néron's local pairing; for example, the symmetry $\langle a, b \rangle_v = \langle b, a \rangle_v$ follows from Green's theorem.

We now consider the non-archimedean case. Let $\mathcal{O}_v$ denote the ring of integers of k_v and let $\mathscr{X}$ be a regular model for X over $\mathcal{O}_v$; this exists by the desingularization theory for arithmetic surfaces. For an account of that theory, and proofs of the results we will need below, see Lichtenbaum [7] or Shafarevitch [11]. Let $\mathscr{F} = \sum a_i \mathscr{F}_i$ denote the special fibre $\mathscr{X} \otimes F_v$, where the $\mathscr{F}_i$ are the irreducible components over F_v and $a_i \geq 1$ are their multiplicities.

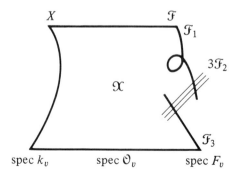

It is well known that we may define an intersection product $(A \cdot B)$ for two divisors A and B which intersect property on $\mathscr{X}$. For divisors $\sum m_i \mathscr{F}_i$ concentrated in the special fibre, we can also define a self-intersection using the rule: $(\mathscr{F} \cdot \mathscr{F}_i) = 0$ for all i. On the quotient group $\sum_i \mathbf{Z}\mathscr{F}_i/\mathbf{Z}\mathscr{F}$ this pairing is negative definite.

Let a be a divisor of degree zero in $Z^0(X/k_v)$, and let A be an extension to a *rational* divisor on $\mathscr{X}$ which satisfies

(3.6) $$(A \cdot \mathscr{F}_i) = 0 \quad \text{for all } i.$$

This exists by the non-degeneracy of the pairing (over $\mathbf{Q}$) on the special fibre, which also shows that A is unique up to the addition of a rational multiple of $\mathscr{F}$. Let b be a divisor of degree zero which is relatively prime to a, and B any extension of b to $\mathscr{X}$. We define

(3.7) $$\langle a, b \rangle_v = -(A \cdot B)\log q_v,$$

where $q_v = \text{Card}(F_v)$. This pairing satisfies the four properties of Néron's local pairing.

§4. The Canonical Height

In this section, we assume k is a global field, and for each place v we normalize the valuation $|\ |_v$ on the completion k_v as in Section 2. For any $\alpha \in k^*$ we have the product formula

(4.1) $$\prod_v |\alpha|_v = 1.$$

Assume that the genus g of X is positive, and let W be the image of $\text{Sym}^{g-1} X$ in $\text{Pic}^{g-1}(X)$. Let $\kappa \in \text{Pic}^{g-1}(X)$ be a theta-characteristic; that is, a divisor class of degree $g - 1$ over $\bar{k}$ with 2κ linearly equivalent to the canonical class of X. The theta-divisor $\Theta = W - \kappa$ is symmetric and ample in J; furthermore, the class of the divisor $2(\Theta)$ in $\text{Pic}(J)$ is independent of the choice of κ and is defined over k. The sections of the associated line bundle L on J give rise to the Kummer embedding

(4.2) $$\phi: J/\pm 1 \to \mathbf{P}(H^0(J, L)).$$

The image has degree $2^{g-1}g!$ in a projective space of dimension $2^g - 1$.

To define the height function $\hat{h} = \hat{h}_L: J(k) \to \mathbf{R}$, we fix an isomorphism: $\mathbf{P}(H^0(J, L)) = \mathbf{P}^{2^g-1}$ and for any point $z = (z_1, \ldots, z_{2^g})$ in $\mathbf{P}^{2^g-1}(k)$ define the naive height of z by

(4.3) $$h(z) = \sum_v \max_i \{\log |z_i|_v\}.$$

Tate [4, Chap. 5] proved that for $\alpha \in J(k)$ the limit

(4.4) $$\hat{h}(\alpha) = \lim_{n \to \infty} \frac{h \circ \phi(n\alpha)}{n^2}$$

exists, is independent of the basis of sections chosen for L, and defines a positive-definite quadratic form $\hat{h}\colon J(k)\otimes \mathbf{R}\to \mathbf{R}$.

In particular, the function $\langle\ ,\ \rangle\colon J(k)\times J(k)\to \mathbf{R}$ defined by

(4.5) $$\langle \alpha, \beta\rangle = \tfrac{1}{2}(\hat{h}(\alpha+\beta) - \hat{h}(\alpha) - \hat{h}(\beta))$$

is bi-additive. To prove this, one identifies this pairing with the canonical height associated to the Poincaré divisor on $J\times J^\vee$ using the isomorphism $J\xrightarrow{\sim} J^\vee$ defined by $\alpha\mapsto$ class of $(\Theta-\alpha)-(\Theta)$ [4]. Néron [10, Chap. II, Theorems 8 and 10] gave a formula for this pairing in terms of his local symbols, which we now recall.

Suppose that $\alpha = \mathrm{class}(a)$ and $\beta = \mathrm{class}(b)$, where a and b are relatively prime divisors of degree zero over k. For each place v we let $\langle a, b\rangle_v$ denote the value of the local pairing described in Section 3. These values are zero for all but a finite number of places of k, and their sum depends only on α and β, by property 3 of (2.3) and the product formula (4.1). Néron's formula is

(4.6) $$\sum_v \langle a, b\rangle_v = \langle \alpha, \beta\rangle.$$

Finally, we remark that the functorial properties of the local pairing all hold for the global height pairing. If H is a finite extension of k and σ is any automorphism of H over k, we have

(4.7) $$\langle \alpha, \beta\rangle_H = [H:k]\langle \alpha, \beta\rangle_k,$$

(4.8) $$\langle \alpha^\sigma, \beta^\sigma\rangle_H = \langle \alpha, \beta\rangle_H.$$

If $T\subset X\times Y$ is a correspondence over k and $t_*\colon J_X\to J_Y$, $t^*\colon J_Y\to J_X$ are the corresponding homomorphisms of the Jacobians we have

(4.9) $$\langle \alpha, t^*\beta\rangle = \langle t_*\alpha, \beta\rangle.$$

§5. Local Heights for Divisors with Common Support

In view of (4.6) it is desirable to have an extension of the local pairing to divisors a and b of degree zero on X which are not relatively prime. At the loss of some functorality, we may accomplish this as follows.

At each point x in the common support, chose a basis $\partial/\partial t$ for the tangent space and let z be a uniformizing parameter with $\partial z/\partial t = 1$. Any function $f\in k_v(X)^*$ then has a well-defined "value" at x:

(5.1) $$f[x] = \frac{f}{z^m}(x) \quad \text{in } k_v^*,$$

where $m = \mathrm{ord}_x f$. This depends only on the choice of $\partial/\partial t$, not on z. Clearly we have

(5.2) $$fg[x] = f[x]\cdot g[x].$$

To pair a with b, we may find a function f on X such that $b = \text{div}(f) + a'$, where a' is relatively prime to a. We then define

(5.3) $$\langle a, b \rangle_v = \log|f[a]|_v + \langle a, a' \rangle_v,$$

where $f[a] = \prod f[x]^{m_x}$ is defined via (5.1) and $\langle a, a' \rangle_v$ as in Section 3. By property 3 of (2.3), this definition is independent of the choice of f used to move b away from a. The pairing (5.3) depends on the choice of tangent vectors, but is unchanged if $\partial/\partial t$ is multiplied by any $\alpha \in k_v^*$ with $|\alpha|_v = 1$.

The formulas given in Section 3 for Néron's local pairing can be modified to give the extension defined by (5.3). Formula (5.3) continues to hold, provided we define $g_a(x)$ to be the value of $g_a - m_x \log|z|_v$ at x, where z is a local parameter with $|\partial z/\partial t|_v = 1$. In the non-archimedean case, it is convenient to fix $\partial/\partial t$ up to a unit by insisting that it be a generator of the $\mathcal{O}_v$-module $T_\Delta \mathscr{X}$, where Δ is the section corresponding to the point x. Formula (3.7) continues to hold, provided we adopt the convention that $(\Delta \cdot \Delta) = 0$.

Finally, the global formula (4.6) holds for *any* choice of global tangent vector $\partial/\partial t$, by the product formula.

§6. Local Heights for Divisors of Arbitrary Degree

Arakelov has described an extension of the archimedean local pairing to relatively prime divisors of arbitrary degree. Again, this is only accomplished at the loss of some functorality. Assume $k_v \simeq \mathbf{C}$ and let $d\mu$ be a positive real analytic 2-form on $M = X(\mathbf{C})$ with $\int_M d\mu = 1$. We remark that the choice of such a volume form is equivalent to the choice of a Riemannian structure in the conformal equivalence class determined by the complex structure on M.

Proposition 6.1. *For each divisor a on X there is a unique real analytic function $g_a: M - |a| \to \mathbf{R}$ such that:*

(1) *$g_a - m_x \log|z|_v$ is real analytic near x, where z is any uniformizing parameter at x and m_x is the multiplicity of x in a,*
(2) *g_a satisfies the differential equation:*

$$\partial\bar{\partial} g_a = 2\pi i (\deg(a) \cdot d\mu - \delta_a),$$

where δ_a is the (1, 1)-current which represents the evaluation of (0, 0) forms at a.
(3) $\int_M g_a \, d\mu = 0$.

The system of functions $\{g_a\}_{a \in \text{Div}(x)}$ satisfy the further identities:

(4) $g_{a+b} = g_a + g_b$.
(5) *$g_a(b) = g_b(a)$ whenever a and b are disjoint.*

When a has degree zero, g_a is the unique solution of (3.3) which satisfies condition (3). In general, the existence of a function g_a satisfying (1) and (2) follows from the theory of elliptic partial differential equations on M. Such a function is unique up to a global harmonic function on M, which must be a constant, and this constant is normalized by condition (3). Condition (4) is immediate, and condition (5) follows from Green's theorem.

We now define the function of two variables by

(6.2) $$G(x, y) = g_{(x)}(y) = g_{(y)}(x).$$

This is symmetric with a logarithmic pole on the diagonal and satisfies the differential equation

(6.3) $$\frac{\partial^2}{\partial x\, \partial \bar{x}} G(x, y) = 2\pi i (d\mu - \delta_y)$$

as a function of x. We have the formula

(6.4) $$\langle \sum m_x(x), \sum m_y(y) \rangle_v = \sum m_x m_y G(x, y)$$

for the local height pairing on relatively prime divisors of degree zero.

In many cases, it is convenient to permit the measure μ to have singularities on M; then (6.4) must be interpreted in the limit if G has a singularity at x or y. Condition (3) must also be dropped if g_a is not integrable on M.

§7. Local Heights on Curves of Genus Zero

We will now illustrate the ideas of Sections 5 and 6 when X has genus zero. Let x_∞ be a fixed point on $X(k_v)$ and let $\partial/\partial t$ be a non-zero tangent vector at x_∞. Choose an isomorphism

$$f: X \xrightarrow{\sim} \mathbf{P}^1,$$

$$x_\infty \mapsto \infty,$$

such that the uniformizing parameter $z = 1/f$ at x_∞ satisfies $\partial z/\partial t = 1$. Then f is determined up to the addition of a constant and has the "value": $f[x_\infty] = 1$. Let $x \neq y$ be two points of X which are not equal to x_∞ and put $a = (x) - (x_\infty)$ and $b = (y) - (x_\infty)$. We then find

(7.1) $$\langle a, b \rangle_v = \log |f(x) - f(y)|_v.$$

Indeed, $g = f - f(y)$ is a function on X with $\operatorname{div}(g) = b$, and $g(a) = g(x)/g[x_0] = f(x) - f(y)$.

Now suppose k_v is archimedean, and fix a uniformization

$$k_v \xrightarrow{\pi} X(k_v) - \{x_\infty\}$$

such that $\pi(z) = x$ and $\pi(w) = y$. Then

(7.2) $$G(x, y) = \log |z - w|_v$$

is a Green's function associated to the (1, 1)-current $d\mu = \delta_{x_\infty}$. The height pairing of $a = (x) - (x')$ with $b = (y) - (y')$ is then

$$\langle a, b \rangle_v = \log \left| \frac{(z-w)(z'-w')}{(z-w')(z'-w)} \right|_v$$

independent of the uniformization π.

§8. Local Heights on Elliptic Curves

We will now illustrate the ideas of Sections 5 and 6 when X has genus 1 and a rational point x_0, so X is an elliptic curve with x_0 taken as the origin. We begin with the pairing of divisors of degree zero with common support; note that once we choose a non-zero tangent vector $\partial/\partial t$ at x_0, it determines (by parallel translation) a non-zero tangent vector at every point of X.

First suppose k_v is non-archimedean, and let $\mathscr{X}$ be the *minimal* regular model for X over $\mathcal{O}_v$ (so $\mathscr{X}$ is the compactification of the Néron minimal model over $\mathcal{O}_v$). Fix a basis $\partial/\partial t$ for the $\mathcal{O}_v$-module $T_{\Delta_0}\mathscr{X}$, where Δ_0 is the identity section; this is dual to a Néron differential ω. Let $x \neq y$ be two points on X whose associated sections Δ_x and Δ_y reduce to the identity component $\mathscr{F}_0$ in the special fibre. Then for the divisors $a = (x) - (x_0)$ and $b = (y) - (x_0)$ we have the formulas

(8.1)
$$\langle a, b \rangle_v = \{(\Delta_x \cdot \Delta_0) + (\Delta_y \cdot \Delta_0) - (\Delta_x \cdot \Delta_y)\} \log q_v,$$
$$\langle a, a \rangle_v = 2(\Delta_x \cdot \Delta_0) \log q_v,$$

in agreement with Tate [12].

Now suppose k_v is archimedean, and fix a uniformization:

(8.2)
$$\mathbf{C}/\Lambda \overset{\pi}{\rightleftarrows} X(\bar{k}_v) = M.$$

Then $\pi_*(\partial/\partial z) = \partial/\partial t$ is a non-zero tangent vector at x_0. Let $\sigma(z)$ be the Weierstrass σ-function of the lattice Λ, and let $u(z)$ be the **R**-linear map taking the periods (ω_1, ω_2) of $\wp(z)$ to the quasi-periods (η_1, η_2) of $\zeta(z)$. The Klein function

(8.3)
$$k(z) = \Delta(\Lambda)^{1/12} e^{-zu(z)/2} \sigma(z)$$

is real analytic with a simple zero at each lattice point. It is not periodic, but its multipliers $k(z + \omega)/k(z)$ all have absolute value 1. Let z and w be points in **C** with $\pi(z) = x$ and $\pi(w) = y$; then

(8.4)
$$G(x, y) = \log |k(z - w)|_v$$

is the Green's function associated to the unique multiple $c\, dx\, dy = d\mu$ of Lebesque measure on **C** such that the volume of any fundamental domain for Λ with respect to $d\mu$ is equal to 1. (If $\Lambda = \mathbf{Z}\omega_1 + \mathbf{Z}\omega_2$ we have $|\omega_1 \bar{\omega}_2 - \omega_2 \bar{\omega}_1| = 2/c$.)

§9. Green's Functions on the Upper Half-Plane

We now discuss the Green's function $G(x, y)$ on complex curves X uniformed by the upper half-plane $\mathfrak{H} = \{z = x + iy : y > 0\}$. Fix a uniformization

(9.1) $$\Gamma | \mathfrak{H} \xrightarrow{\pi} X(\mathbf{C}) - S,$$

where Γ is a discrete subgroup of $\mathrm{PSL}_2(\mathbf{R})$ and S is a finite set of cusps on X corresponding to the parabolic conjugacy classes in Γ. For each $x \in X(\mathbf{C}) - S$, let z be a point in $\mathfrak{H}$ with $\pi(z) = x$, and let $e_x = \mathrm{Card}(\Gamma_z)$ be the order of the cyclic stabilizer subgroup. For $x \in S$ we let $e_x = \infty$. The rational number $\chi = 2 - 2g - \sum(1 - (1/e_x))$ is negative, and $-2\pi\chi$ is the volume of $\Gamma | \mathfrak{H}$ with respect to the invariant $(1, 1)$-form $d\mu = dx\, dy/y^2$.

The function

(9.2) $$g_1(z, w) = \log \left| \frac{z - w}{\bar{z} - w} \right|_v$$

is harmonic on $\mathfrak{H} \times \mathfrak{H}$ minus the diagonal, with a logarithmic pole as $z \mapsto w$ and satisfies $g_1(\gamma z, \gamma w) = g_1(z, w)$. Unfortunately, the average of this function over the group Γ gives a divergent series. Instead, it is best to look for a real-analytic function $g_s(z, w)$ on $\mathfrak{H} \times \mathfrak{H}$ minus the diagonal which satisfies

(9.3) $$\begin{cases} g_s(\gamma z, \gamma w) = g_s(z, w) & \text{for all } \gamma \in \mathrm{PSL}_2(\mathbf{R}), \\ g_s(z, w) \sim \log|z - w|_v & \text{as } z \mapsto w, \\ \Delta g_s(z, w) = s(s-1)g_s(z, w) & \text{where } \Delta = y^2(\partial/\partial x^2 + \partial/\partial y^2) \text{ is the hyperbolic Laplacian in the } z \\ & \text{variable and } w \text{ is fixed.} \end{cases}$$

The first property implies that $g_s(z, w)$ is a function of the hyperbolic distance

$$r = d(z, w) = \cosh^{-1}\left(\frac{|z - w|^2}{2 \operatorname{Im} z \operatorname{Im} w} + 1 \right).$$

Writing $g_s(z, w) = f(\cosh r)$ we find that f satisfies the Legendre differential equation

(9.4) $$(1 - u^2) \frac{d^2 f}{du^2} - 2u \frac{df}{du} + s(s-1)f = 0.$$

The unique solution with the correct pole on the diagonal and slow growth at infinity is the function

$$f(u) = -2Q_{s-1}(u) = -2 \int_0^\infty (u + \sqrt{u^2 - 1} \cosh t)^{-s} dt \quad \text{for } u > 1$$

[6, Chap. 7].

Hence we may take

$$g_s(z, w) = -2Q_{s-1}(\cosh d(z, w))$$
(9.5)
$$= -2Q_{s-1}\left(\frac{|z - w|^2}{2 \operatorname{Im} z \operatorname{Im} w} + 1\right);$$

then $g_1(z, w)$ is the function in (9.2). The advantage in using this generality is that for $\operatorname{Re}(s) > 1$, the series

(9.6) $$G_s(z, w) = \sum_\Gamma g_s(z, \gamma w)$$

is convergent and defines a symmetric bi-Γ-invariant function with a logarithmic pole on the diagonal which satisfies $\Delta G_s(z, w) = s(s - 1)G_s(z, w)$ for fixed w. It can be meromorphically continued to the half-plane $\operatorname{Re}(s) \geq \frac{1}{2}$, and has a simple pole at $s = 1$ with residue $2/\chi$ independent of z and w [3].

The function

(9.7) $$G(x, y) = \lim_{s \to 1}\left\{G_s(z, w) - \frac{\text{Residue } G_s}{s(s - 1)}\right\}$$

is the Green's function associated to the measure $d\mu/-2\pi\chi$ on X, where $\pi(z) = x$ and $\pi(w) = y$. When X has cusps, the height pairing for divisors with cuspidal support must be computed via a limiting procedure, as $d\mu$ has singularities on S.

§10. Local Heights on Mumford Curves

In this section, we will describe the local height pairing on Mumford curves over the non-archimedean local field k_v. Our basic reference for the function theory used below is Manin and Drinfel'd [8].

Let Γ be a Schottky group in $\operatorname{PGL}_2(k_v)$, and let Σ denote its limit set—the closure of its fixed points on $\mathbf{P}^1(k_v)$. The space $\Omega = \mathbf{P}^1(k_v) - \Sigma$ admits a properly discontinuous, rigid analytic action of Γ and we have a uniformation

(10.1) $$\Gamma\backslash\Omega \xrightarrow{\pi} X(k_v)$$

for some Mumford curve X over k_v. The number of independent generators for the free group Γ is equal to the genus of X.

Let a and b be relatively prime divisors of degree zero on X over k_v, and let A and B denote any liftings of these divisors to divisors on Ω. Let w_A be a function on $\mathbf{P}^1$ with $\operatorname{div}(w_A) = A$; for any z not in the Γ-orbit of A in Ω and any $g \in \Gamma$ we define

(10.2) $$\mu_A(g) = \prod_\Gamma \frac{w_A(\gamma g z)}{w_A(\gamma z)}.$$

This product converges to a limit in k_v^* which is independent of z, and the resulting map $\mu_A: \Gamma \to k_v^*$ is a group homomorphism.

The fundamental theorem on the positive-definiteness of the polarization pairing $\langle\ ,\ \rangle: \Gamma^{ab} \times \Gamma^{ab} \to \mathbf{Z}$ allows one to choose a *rational* divisor A representing a such that μ_A takes values in the *units* of k_v^*. Such a representative A will be called unitary; for unitary A we have the formula

(10.3) $$\langle a, b\rangle_v = \sum_\Gamma \log|w_A(\gamma B)|_v$$

as the sum on the right-hand side defines a pairing satisfying all the properties of Proposition 2.3.

For X of genus zero, $\Gamma = (1)$ and (10.3) gives the logarithm of the cross-ratio, as in Section 7. For X of genus one, Γ is a free group on one generator; conjugating by an automorphism of $\mathbf{P}^1$, we may assume Γ is generated by $\gamma = \begin{pmatrix} q & 0 \\ 0 & 1 \end{pmatrix}$ with $|q|_v < 1$. Then $\Sigma = \{0, \infty\}$, $\Omega = k_v^*$, and (10.1) is Tate's parametrization: $k_v^*/q^{\mathbf{Z}} \simeq X(k_v)$ taking 1 (mod $q^{\mathbf{Z}}$) to the origin for the group law on X. Let

(10.4) $$\theta(t) = (1-t)\prod_{n\geq 1}(1-q^n t)(1-q^n t^{-1})$$

and suppose A and B are two divisors of degree zero on Ω with images a and b on X. If $A = \sum m_i(\alpha_i)$, then A is a unitary representation for a if and only if $\alpha = \prod \alpha_i^{m_i}$ is a unit in k_v^*. Since the function

(10.5) $$\theta_A(t) = \prod \theta(t/\alpha_i)^{m_i} \quad \text{satisfies}$$
$$\theta_A(qt) = \alpha\theta_A(t),$$

we see that $\log|\theta_A(t)|_v$ is a function on $X(k_v)$ if A is unitary. In this case, some elementary manipulation shows that (10.3) is equivalent to Tate's formula [12]:

(10.6) $$\langle a, b\rangle_v = \log|\theta_A(B)|_v.$$

We remark that a will have an integral unitary representatiave A if and only if the class of a lies in the connected component of the origin in $J(k_v)$.

For general Mumford curves of genus $g \geq 2$, the unitary representatives A may be hard to find. But the following case seems interesting. Assume Γ has limit set $\Sigma = \mathbf{P}^1(k_v)$, but view it as a subgroup of $\mathrm{PGL}_2(K_v)$ where K_v is a separable quadratic extension of k_v. Then $\Omega = \mathbf{P}^1(K_v) - \mathbf{P}^1(K_v)$ and $\Gamma/\Omega \simeq X(K_v)$; such groups might be considered of Fuchsian type. Let $z \mapsto \bar{z}$ denote the non-trivial automorphism of K_v over k_v; for any divisor A on Ω of the form $A = A_1 - \bar{A}_1$ we find $\mu_A(g) = \mu_A(g)^{-1}$, so A is unitary. For example, if $a = (x) - (\bar{x})$ and $b = (y) - (\bar{y})$ we find

(10.7) $$\langle a, b\rangle_v = 2\sum_\Gamma \log\left|\frac{z - \gamma w}{\bar{z} - \gamma w}\right|_v,$$

where z and w are points of Ω with $\pi(z) = x$, $\pi(w) = y$.

References

[1] Arakelov, S. J. Theory of intersections on an arithmetic surface. *Proceedings of the International Congress on Mathematics*, Vancouver, 1974, pp. 405–408.

[2] Griffiths, P. and Schmid, W. Recent developments in Hodge theory. *Proceedings of the International Colloquium on Discrete Subgroups*, Bombay, 1973, pp. 31–127.

[3] Hejhal, D. *The Selberg Trace Formula for* PSL $(2, \mathbf{R})$, Vols. I, II. Springer Lecture Notes, 548, 1001. Springer-Verlag: New York, 1978, 1983.

[4] Lang, S. *Fundamentals of Diophantine Geometry*. Springer-Verlag: New York, 1983.

[5] Lang, S. Les formes bilinéares de Néron et Tate. *Séminaire Bourbaki*, Éxposé 274, 1964.

[6] Lebedev, N. N. *Special Functions and Their Applications*. Dover: New York, 1972.

[7] Lichtenbaum, S. Curves over discrete valuation rings. *Amer. J. Math.*, **90** (1968), 380–405.

[8] Manin, Y. and Drinfel'd, V. Periods of p-adic Schottky groups. *J. Crelle*, **262/263** (1973), 239–247.

[9] Néron, A. Modèles minimaux des variétés abéliennes sur les corps locaux et globaux. *Publ. Math. I.H.E.S.*, **21** (1964), 361–482.

[10] Néron, A. Quasi-fonctions at hauteurs sur les variétés abéliennes. *Ann. Math.*, **82** (1965), 249–331.

[11] Shafarevitch, I. *Lectures on Minimal Models and Birational Transformations of Two-dimensional Schemes*. Tate Institute: Bombay, 1966.

[12] Tate, J. Letter to J.-P. Serre, 21 June 1968. (Some of the results in this letter have been reproduced by S. Lang, *Elliptic Curves Diophantine Analysis*. Grundlehren der Mathematischen Wissenschaften, 231. Springer-Verlag: New York, 1978, Chapter I, §§7–8, Chapter III, §§4–5, Chapter IV, §6.)

CHAPTER XV

A Higher Dimensional Mordell Conjecture

PAUL VOJTA

Faltings' long awaited proof of the Mordell conjecture completes, roughly speaking, the question of whether a given curve has only finitely many integral or rational points. Indeed, if a complete curve has genus $g \geq 2$, then it has finitely many rational points; any affine curve whose projective closure is a curve of genus at least two will, *a fortiori*, have only finitely many integral points. A curve of genus 1 is an elliptic curve; it will have infinitely many rational points over a sufficiently large ground field, but no affine subvariety has an infinite number of integral points. Finally, a curve of genus zero is, after a base change, the projective line, which has an infinite number of rational points; affine subvarieties omitting at most two points will have infinitely many integral points over a sufficiently large ring; but affine subvarieties omitting at least three points will have only finitely many integral points. Thus the answer to the finiteness question is given entirely by the structure of the curve over the complex numbers.

A natural next step, then, is to ask whether such a classification holds for higher dimensional varieties. The purpose of this paper is to present a conjectural answer to this question. The motivation for this conjecture comes from the field of complex analysis: all of the above theorems for curves have analogues for holomorphic maps to Riemann surfaces. For example, there are nonconstant meromorphic maps to the Riemann sphere which omit n points if and only if $n < 3$ (the Borel lemma), and no nonconstant holomorphic maps to a Riemann surface of genus at least two (Picard's theorem). Thus a curve admits infinite sets of integral or rational points if and only if the corresponding complex manifold admits at least one nonconstant holomorphic map.

On the analytic side, the proofs of the theorems all use Nevanlinna theory, which we will introduce in Section 1. On the number-theoretic side, the statements on integral points can be proved by diophantine approximations, using Roth's theorem. Although they come from entirely different fields of mathematics, Osgood observed [9], that these two fields have a startling similarity. This similarity will be described in Section 2.

Section 3 presents the main conjecture. It presents a description of Nevanlinna theory in the higher dimensional case, due to Stoll and Griffiths. The main theorem, when translated into number theory, will be the conjecture of this paper.

This conjecture implies many outstanding conjectures regarding integral and rational points. This includes two conjectures of Lang, Shafarevich and Hall, and a question posed by Bombieri. Except for Hall's conjecture, we show how these follow in Section 4. The deduction of Hall's conjecture is fairly lengthy; details will be published elsewhere.

In Section 5 we compare methods in Griffiths' proof of the theorem in Nevanlinna theory, with the methods of Faltings' proof.

I owe many thanks to Serge Lang and Joe Silverman for their helpful comments during the preparation of this paper.

§1. A Brief Introduction to Nevanlinna Theory

Nevanlinna theory is a generalization of the observation of Hadamard that, for a holomorphic function f,

(1.1)   (number of zeros in $|z| < r$) $\leq \log \max(|f(z)|: |z| \leq r) + O(1)$,

where the constant in $O(1)$ depends on f but not on r. This inequality is less than ideal in two ways. Although it is still true for meromorphic functions, it does not give an upper bound for the number of zeros, since the right-hand side is infinite. Moreover, it is necessarily a strict inequality even for entire functions; e.g. $f(z) = e^z$ has no zeros but grows rapidly as $|z|$ increases.

Nevanlinna improved (1.1) by eliminating both of the deficiencies mentioned above. He did this by defining the characteristic functions to replace each side of (1.1). The counting function is a weighted sum of the number of zeros: if $n(a, r)$ is the number of zeros of $f(z) - a$ inside the circle of radius r (counted with multiplicity), then the counting function is

$$N(a, r) = \int_0^r (n(a, r) - n(a, 0)) \frac{dr}{r}$$
$$= \sum_{|z| < r} \operatorname{ord}_z^+ (f - a) \log \frac{r}{|z|}.$$

In the second equation, ord is the order of vanishing of $f - a$ at z; the symbol ord$^+$ means max(0, ord). If $n(\infty, r)$ counts the number of poles inside the circle of radius r, then one can define $N(\infty, r)$ similarly.

To replace the right-hand side of (1.1), Nevanlinna defined the characteristic function of f, $T(r)$, as

$$T(r) = \frac{1}{2\pi} \int_0^{2\pi} \log^+ |f(re^{i\theta})| \, d\theta + N(\infty, r).$$

A simple calculation shows that $T(r)$ is defined even when f has a pole on the circle $|z| = r$. With these definitions (1.1) now becomes the Nevanlinna inequality $N(0, r) \leq T(r) + O(1)$, which is valid (and nontrivial) for meromorphic functions in general.

Still, this inequality is strict for functions such as e^z. To eliminate the strictness, we note that while e^z has no zeros, it does have many values very close to zero. With this in mind, we define the proximity function

$$m(a, r) = \frac{1}{2\pi} \int_0^{2\pi} \log^+ \left| \frac{1}{f(re^{i\theta}) - a} \right| d\theta.$$

Adding this term to the left-hand side of the Nevanlinna inequality turns it into an equality, called the First Main Theorem of Nevanlinna theory:

(1.2) $$m(a, r) + N(a, r) = T(r) + O(1).$$

This is actually a fairly straightforward consequence of Jensen's formula:

$$\log |c_\lambda| = \frac{1}{2\pi} \int_0^{2\pi} \log |f(re^{i\theta})| \, d\theta + N(\infty, r) - N(0, r).$$

where c_λ is the leading coefficient in the Laurent expansion of zero.

Thus the First Main Theorem gives an upper bound for the number of zeros of a function. The much deeper Second Main Theorem gives an upper bound for $N(a, r)$. Because of functions such as e^z, such a theorem must consider several values of a:

$$\sum_{i=1}^n m(a_i, r) < 2T(r) - N_1(r) + O(\log(rT(r))). \qquad //$$

Following Weyl, the symbol // means that the inequality holds except for a set of r of finite measure. The term $N_1(r)$ counts ramification of f in the same manner as $N(a, r)$ counts zeros: in particular, it is nonnegative and can be dropped. Dividing this equation by $T(r)$ and taking $\liminf_{r \to \infty}$ gives

(1.3) $$\sum_{a \in C} \delta(a) \leq 2,$$

where

$$\delta(a) = \lim_{r \to \infty} \frac{m(a, r)}{T(r)}.$$

The number $\delta(a)$ is called the defect of a; equation (1.3) is called the Nevanlinna defect relation.

By the First Main Theorem, $0 \le \delta(a) \le 1$; by the Second Main Theorem, $\delta(a)$ is almost always zero. What this means is that, for almost all a, $f^{-1}(a)$ is as large as (1.2) allows it to be; other values of a are called deficient; and $\delta(a)$ measures how deficient they are. If f never attains the value a, then $\delta(a) = 1$.

§2. Correspondence with Number Theory

In this section we set up a dictionary with which to translate the theorems and definitions of the previous section into number theory. Under this correspondence, all of the theorems of the last section translate into well-known theorems or conjectures of number theory.

To fix notation, let k be a number field and let S be a finite set of places v containing the set S_∞ of infinite places. We use the normalized absolute value $\| \ \|_v$, defined as in [10]:

$$\|z\|_v = |z| \quad \text{if } k_v = \mathbb{R};$$
$$\|z\|_v = |z|^2 \quad \text{if } k_v = \mathbb{C}; \text{ and}$$
$$\|p\|_v = p^{-ef} \quad \text{if } v \text{ is } p\text{-adic and } [k_v : \mathbb{Q}_p] = ef.$$

The product formula can then be written without multiplicities:

(2.1) $$\prod_v \|b\|_v = 1.$$

Now, as suggested in the introduction, let a meromorphic function correspond to an infinite set $\{b\} \subseteq k$. This is done by letting the function f parametrize an infinite family of functions on the unit disk, parametrized by r, by contraction. On the boundary of the disk, we can take the absolute value, giving an archimedean absolute value $|f(re^{i\theta})|$ which can correspond to the absolute values $\|b\|_v$ for $v \in S$. On the interior of the disk, one can take the order of vanishing of a function $\mathrm{ord}_z(f)$, which can correspond to the order, relative to a prime ideal, of $b \in k$: $\mathrm{ord}_p(b)$. Finally, it will soon be clear that the term $\log(r/|z|)$ should correspond to $[k_v : \mathbb{Q}_p]\log p$ if v is a finite place lying over a rational prime p. Table 1 summarizes these identifications.

Having made these identifications, it is now possible to translate the definitions and theorems of the preceding section. We start with the counting function: let $a \in k$; then

$$N(a, b) = \sum_{v \notin S} [k_v : \mathbb{Q}_p]\log p \cdot \mathrm{ord}_v^+(b - a)$$

$$= \sum_{v \notin S} \log^+ \left\| \frac{1}{b - a} \right\|_v.$$

A Higher Dimensional Mordell Conjecture

Table 1. The Dictionary in the One-Dimensional Case.

Nevanlinna Theory	Roth's Theorem
$f: \mathbb{C} \to \mathbb{C}$ nonconstant	$\{b\} \subseteq k$ infinite
r	b
θ	$v \in S$
$z \in \mathbb{C}, |z| < r$	$v \notin S$
$|f(re^{i\theta})|$	$\|b\|_v, v \in S$
$\operatorname{ord}_z f$	$\operatorname{ord}_v f, v \notin S$
$\log \dfrac{r}{|z|}$	$\log N_v$
Characteristic function	Logarithmic height
$T(r) = \dfrac{1}{2\pi} \int_0^{2\pi} \log^+ |f(re^{i\theta})| \, d\theta + N(\infty, r)$	$h(b) = \dfrac{1}{[k:\mathbb{Q}]} \sum_v \log^+ \|b\|_v$
Proximity function	
$m(a, r) = \dfrac{1}{2\pi} \int_0^{2\pi} \log^+ \left| \dfrac{1}{f(re^{i\theta}) - a} \right| d\theta$	$m(a, b) = \dfrac{1}{[k:\mathbb{Q}]} \sum_{v \in S} \log^+ \left\| \dfrac{1}{b - a} \right\|_v$
Counting function	
$N(a, r) = \sum_{z_k} \log \dfrac{r}{|z_k|}$	$N(a, b) = \dfrac{1}{[k:\mathbb{Q}]} \sum_{v \notin S} \log^+ \left\| \dfrac{1}{b - a} \right\|_v$
First Main Theorem	
$N(a, r) + m(a, r) = T(r) + O(1)$	$N(a, b) + m(a, b) = h(b) + O(1)$
Defect	
$\delta(a) = \liminf\limits_{r \to \infty} \dfrac{m(a, r)}{T(r)}$	$\delta(a) = \liminf\limits_b \dfrac{m(a, b)}{h(b)}$
Defect relation $\sum\limits_{a \in \mathbb{C}} \delta(a) \leq 2$	$\sum\limits_{a \in \mathbb{Q}} \delta(a) \leq 2$ Roth's theorem
Jensen's theorem	Artin–Whaples product formula
$\log|c_\lambda| = \dfrac{1}{2\pi} \int_0^{2\pi} \log|f(re^{i\theta})| \, d\theta + N(\infty, r) - N(0, r)$	$\sum\limits_v \log \|b\|_v = 0$

For the proximity and characteristic functions, the integrals correspond to the sums:

$$m(a, b) = \sum_{v \in S} \log^+ \left\| \dfrac{1}{b - a} \right\|_v,$$

$$h(b) = \sum_v \log^+ \|b\|_v.$$

Notice that the characteristic function has been written $h(b)$ instead of $T(b)$; this is because it is the height function defined in [10].

Now consider Jensen's formula. The term $\log|c_\lambda|$ cannot be translated, but the right-hand side becomes

$$\sum_{v \in S} \log \|b\|_v + \sum_{v \notin S} \log^+ \|b\|_v - \sum_{v \notin S} \log^+ \left\|\frac{1}{b}\right\|_v,$$

which reduces to the left-hand side of the product formula

$$\sum_v \log \|b\|_v = 0.$$

Similarly, the First Main Theorem reduces to

$$h\left(\frac{1}{b-a}\right) = h(b) + O(1),$$

which is an exercise using the definition of height, the product formula (giving $h(1/x) = h(x)$ for $x \neq 0$), and the triangle inequality (giving $h(x+y) \leq h(x) + h(y) + \log 2$).

This leaves the Second Main Theorem and the defect relation. We claim that the defect relation is equivalent to Roth's theorem, which is the following.

Theorem 2.2. *Let k, S, and $\| \ \|$ be as above. For each $v \in S$ let α_v be an algebraic number in k_v; let $\varepsilon > 0$ and $c > 0$ be constants. Then only finitely many $b \in k$ satisfy*

(2.3) $$\prod_{v \in S} \min(1, \|b - \alpha_v\|_v) < \frac{c}{H(b)^{2+\varepsilon}}.$$

Expressing (2.3) logarithmically and taking the contrapositive, this theorem is equivalent to saying that for all infinite sets of elements $b \in k$

$$\sum_{v \in S} \log^+ \left\|\frac{1}{b - \alpha_v}\right\|_v < (2 + \varepsilon)h(b) + O(1).$$

By enlarging k, we may assume the α_v to lie in k; then a few more manipulations place this in the form

(2.4) $$\sum_{i=1}^n m(a_i, b) < (2 + \varepsilon)h(b) + O(1).$$

Defining as before, $\delta(a) = \varliminf m(a,b)/h(b)$, this is equivalent to the defect relation

$$\sum_{a \in k} \delta(a) \leq 2.$$

From (2.4) it follows that the Second Main Theorem corresponds to a conjectural refinement of Roth's theorem which uses the condition

$$\prod_{v \in S} \min(1, \|b - \alpha_v\|_v) < \frac{C}{H(b)^2 h(b)^{c'}},$$

instead of (2.3). This conjecture is due to Lang [7], with $c' = 1 + \varepsilon$.

§3. Higher Dimensional Nevanlinna Theory

In this section we introduce higher dimensional Nevanlinna theory and consider how it relates to number theory. The higher dimensional analogue of the defect relation gives a new conjecture which generalizes Roth's theorem to higher dimensions. The relation is due to Stoll [11]; this section follows Griffiths [4].

One of the major advances in Nevanlinna theory since its appearance has been the introduction of differential-geometric methods. In particular, both the proximity function $m(a, r)$ and the characteristic function $T(r)$ have new expressions in terms of metrics defined on line bundles. This corresponds well with the use of metrics in Arakelov theory.

To fix notation, let V be a projective algebraic variety of dimension n and let D be an effective divisor on V, with associated line bundle $[D]$. Let s be a section of $[D]$ whose divisor is D. On the analytic side, assume $f: \mathbb{C}^n \to V$ is an equidimensional holomorphic map which is nondegenerate (i.e. its Jacobian determinant does not vanish identically). The domain can again be written in polar coordinates, using $r = |z_1|^2 + \cdots + |z_n|^2$ and $\theta \in \partial B[1]$ the boundary of the ball of radius 1. On the number-theoretic side, let S and k be as before; assume V, D, etc. to be defined over k. Then let $P = \{P\} \subseteq V(k)$ be a set of points which is nondegenerate; i.e. dense in the Zariski topology.

The easiest definition is the proximity function. Instead of being defined relative to an element $a \in \mathbb{C}$, it is defined relative to a divisor D of V. Letting $|\ |$ be a metric on $[D]$, we have

$$m(D, r) = \int_{\partial B[r]} -\log|s(f(z))| \, d\sigma,$$

where $d\sigma$ is a rotationally invariant measure on $\partial B[r]$ normalized to have $\int_{\partial B[r]} d\sigma = 1$. In the number-theoretic context, this translates into

$$m(D, P) = \sum_{v \in S} -\log\|s(P)\|_v$$
$$= \sum_{v \in S} d_{D,v}(P),$$

where $d_{D,v}$ is the logarithmic distance function defined in Section VIII of [10]. (These distance functions are also known as Weil functions.) If $S = S_\infty$, then the above is the component of the height at infinity in the definition of Arakelov theory.

As before, the counting function can be written as a sum over the components of f^*D. The exact formula appears in [4]; its number-theoretic equivalent is

(3.1) $$N(D, P) = \sum_{v \notin S} d_{D,v}(P).$$

The Nevanlinna characteristic function also has a fairly complicated definition, using Chern classes; it seems to have no obvious translation as in Section 2. Therefore we merely define the height to be

$$h_D(P) = \sum_v d_{D,v}(P),$$

so that it automatically satisfies the First Main Theorem and, by [10, Theorem 3.1], is independent of the choice of D within its linear equivalence class. This is consistent with the picture in Nevanlinna theory, in which the characteristic function (height) is defined in terms of the line bundle $[D]$, so that it is automatically independent of linear equivalence; then by the First Main Theorem, the sum $N(D, r) + m(D, r)$ is independent of the choice of D within its linear equivalence class, up to a bounded function.

As in the one variable case, there is a simple expression for the defect:

$$\delta(D) = \varliminf \frac{m(D, -)}{h_D(-)}.$$

As before, $0 \leq \delta(D) \leq 1$. In the number-theoretic case this limit is taken with respect to Zariski-open subsets of V. In other words, we let $\delta(D) = A$, where A is the smallest real number such that for all $\varepsilon > 0$ and all Zariski-open subsets U of V, there is a point $P \in \{P\} \cap U$ for which $m(D, P)/h_{[D]}(P) < A + \varepsilon$.

Theorem 3.2 (Second Main Theorem). *Let D be a normal crossings divisor on V (i.e. it is locally of the form $z_1 z_2 \ldots z_i = 0$). Let K denote the canonical bundle on V. Then:*

(a) *If D is ample, then $\delta(D) \leq \varliminf (T_{-K}(r)/T_D(r))$.*
(b) *If D is effective, A is ample, and $\varepsilon > 0$, then*

$$m(D, r) + T_K(r) \leq \varepsilon T_A(r). \qquad //$$

This leads naturally to

Conjecture 3.3. *In the number field case, with $V, D, K,$ and $\{P\} \subseteq V(k)$ as above*

(a) *If D is ample, then*

$$\delta(D) \leq \varliminf \frac{h_{-K}(P)}{h_D(P)}$$

$$\leq \inf\{p/q \in \mathbb{Q} | q > 0, pD + qK \text{ ample} + \text{effective}\}.$$

(b) *If D is effective, A is ample, and $\varepsilon < 0$, then $m(D, P) + h_K(P) \leq \varepsilon h_A(P)$ for all P outside of some Zariski-closed subset $Z(\varepsilon)$ of V.*

Remarks. (a) In case $V = \mathbb{P}^1$ and D has n distinct points, then $D \sim O(n)$ and $K = O(-2)$, so the right-hand side is $2/n$. Multiplying by n and writing $D = \Sigma [a_i]$, we have

$$\sum_{i=1}^{n} \delta(a_i) = \sum \varlimsup \frac{m(a_i, b)}{h(b)}$$

$$\leq \varlimsup \sum \frac{m(a_i, b)}{h_{O(1)}(b)}$$

$$= n \varlimsup \frac{m(D, b)}{h_D(b)}$$

$$= n\delta(D) \leq 2.$$

Thus the above conjecture contains Roth's theorem.

(b) In general, (3.3) reduces to a statement in diophantine approximations. However, because of the "Zariski limit" used in defining $\delta(D)$, the conclusion will always be that the inequality holds outside of some Zariski-closed subset. This exceptional set can sometimes be eliminated in special cases, but examples in the next section will show that in general it is necessary.

§4. Consequences of the Conjecture

Conjecture 3.3 implies many conjectures in number theory that have already been posed. In this section we examine conjectures of Shafarevich, and Lang, and Hall, and a question posed by Bombieri.

Consider first the question posed by Bombieri: On a variety V of general type, are all sets of rational points degenerate? Conjecture 3.3 implies that the answer is "yes." To see this, we need a definition and a lemma.

Definition. Let L be a line bundle on V. Then the dimension of L is the smallest integer i such that $l(L^{\otimes n}) = O(n^i)$. If $L^{\otimes n}$ never has any global sections, we set $i = -\infty$. If $i = \dim V$, then we say that L is maximal dimensional.

Remarks. (a) By [6], we always have dim $L \leq \dim V$.

(b) If K is the canonical bundle on V, then dim K is the Kodaira dimension of V.

(c) If D is a divisor, then dim$[D]$ is independent of the multiplicities of the components of D.

(d) If $f: V \to W$ is a morphism and L is a line bundle on W, then dim $f^*L = \dim L$.

Lemma 4.1 (Kodaira). *Let L be a maximal dimensional line bundle on V, let A be an ample line bundle on V. Then for some $n > 0$, $h^0(L^{\otimes n} \otimes A^{-1}) > 0$.*

In particular

$$nh_L(P) > h_A(P) + O(1)$$

outside of some Zariski-closed subset of V.

Now let V be of general type. Then K is of maximum dimension and Lemma 4.1 applies with $L = K$. But this contradicts Conjecture 3.3b when $D = 0$ and $\varepsilon < 1/n$, unless $V(k)$ is degenerate. This implies that the answer to Bombieri's question is yes. In general, a variety of general type may contain rational curves, so the condition on the exceptional subset is necessary.

For curves, the above discussion implies that if a curve has genus $g \geq 2$, then $\deg K = 2g - 2 > 0$, so K is ample, hence of maximal dimension. Then $V(k)$ is degenerate, hence finite since V is a curve. This is Mordell's conjecture.

Before discussing the conjectures of Shafarevich and Lang, it is necessary to digress momentarily to define integral points. Naively, assume that an affine variety is given by equations $f_i(X_1, \ldots, X_n) = 0$, $1 \leq i \leq m$. In the language of schemes, an integral point is a map,

of schemes over Spec $\mathbb{Z}$. In the present case, however, we are considering complete varieties defined over a field k. Therefore let U be the variety Spec $k[X]/(f)$ and let V be its completion, with D the divisor $V - U$. The functions $\{1, X_1, \ldots, X_n\}$ form a basis for $\mathscr{L}(D)$, a point $P \in V(k)$ is integral if its images under these functions are integers. This is the definition used by Lang [8]. For our purposes, we note that

$$d_{D,v}(P) = -\log \max(1, \|X_1\|_v, \ldots, \|X_n\|_v)$$

is a distance function for D at any valuation v; therefore, if P is an integral point relative to the above situation, then

$$N(D, P) = \sum_{v,s} -\log \max(1, \|X_1(P)\|_v, \ldots, \|X_n(P)\|_v)$$

$$= 0.$$

since all $X_i(P)$ are S-integers. In particular, this means that:

Lemma 4.2. *If $\mathscr{P} = \{P\}$ is a set of D-integral points of V, then $\delta(D) = 1$.*

This is similar to the analytic situation in which the image of f misses D. Then again $N(D, r) = 0$ and $\delta(D) = 1$.

Conjecture 3.3, as it relates to integral points, can best be stated with a few definitions.

Let D be a normal crossing divisor on a complete nonsingular variety V.

Definition. The logarithmic canonical divisor of the variety $V \backslash D$ is the sheaf of maximal degree differential forms on $V \backslash D$ which have at most simple poles at D; i.e. in an affine open set U, a form has logarithmic singularities along D if it can be written as $f dx_1 \wedge \cdots \wedge dx_n$ with $x_i \in O(U)$, $n = \dim V$, and f has at most simple poles along D. Thus $K_{V \backslash D} = K_V \otimes [D]$.

Definition. A quasi-projective variety is of logarithmic general type if it can be written as $V \backslash D$ (as above), and $K_{V \backslash D}$ is of maximal dimension on V.

The above definition is independent of the representation of the choice of V and D (which may vary by blow-ups on D) (see [6]).

Now Conjecture 3.3 implies

Conjecture 4.3. *If $V \backslash D$ is a variety of logarithmic general type, then all sets of D-integral points of V are degenerate.*

This follows from part (b) of Conjecture 3.3 and from Lemma 4.1 in the same manner as before, i.e. since $N(D, P) = 0$, then (3.3b) becomes,

$$h_D(P) + h_K(P) \leq \varepsilon h_A(P);$$

yet $D + K > (1/n)A$ by Lemma 4.1.

Now $A_{g,n}$ classifies principally polarized abelian varieties and S-integral points correspond to abelian varieties with good reduction outside S. But since $A_{g,n}$ is of logarithmic general type [1], we have the fact that S-integral points lie in some Zariski-closed subset of $A_{g,n}$. The Shafarevich conjecture is that this subset has dimension zero; thus it is partially implied by Conjecture 4.3.

Consider now

Conjecture 4.4 (Lang). *Let D be an ample divisor on an abelian variety A. Then all sets of D-integral points on A are finite.*

The conclusion of this conjecture is stronger than that of Conjecture 4.3; however, disregarding this discrepancy, the above conjecture follows even when D is not a normal crossings divisor. Indeed, let $\pi: V \to A$ denote a blowing-up of A such that $\pi^* D$ has normal crossings. Then since D is ample, it is of maximum dimension; thus $V \backslash \pi^* D$ is of logarithmic general type. Thus Lang's conjecture follows from Conjecture 4.3 up to Zariski-closed subsets (as above).

Finally, we consider Hall's conjecture [5]. In its original form (over $\mathbb{Z}$), it was posed as

$$|y^2 - x^3| \gg x^{1/2 - \varepsilon}, \quad \text{if } y^2 \neq x^3.$$

Work over function fields has suggested that the correct formulation over number fields should be

(4.5) $\quad h(y^2 - x^3) > (\frac{1}{6} - \varepsilon)h([x^3, y^2, 1]) - O(1),$

where x and y are integers. This latter formula is a consequence of Conjecture 3.3. Let D be the divisor $(Y^2 Z - X^3)$ on $\mathbb{P}^2$. The support of this divisor has two singular points. When sufficiently blown-up, the pull-back of D has normal crossings. Applying the conjecture to the support of D on the blow-up then gives an inequality which implies (4.5) when x and y are integers. The exceptional Zariski-closed subset Z can be eliminated in this case that $\mathbb{G}_m$ acts on $\mathbb{P}^2$ by

$$u \cdot (x, y) = (u^2 x, u^3 y),$$

preserving (4.5) up to a constant (depending on u). Thus any curve in Z must be fixed by $\mathbb{G}_m$, and $\mathbb{G}_m$-orbits obey (4.5). Any point of Z can be absorbed into the $O(1)$ term. The explicit details of the blowings-up are rather lengthy; they will be published elsewhere.

§5. Comparison with Faltings' Proof

Since the statements of the theorems become so similar when viewed in this way, it is natural to ask whether the proof of Mordell's conjecture is at all similar to the proof of Picard's theorem. The answer is no: the proof of Picard's theorem makes no reference to abelian varieties. However, the Shafarevich conjecture is also a consequence of the conjecture, and some aspects of Griffiths' proof of Theorem 9 appear also in Faltings' proof.

Indeed, let V be the toroidal compactification $\bar{A}_{g,n}$ and let $A_{g,n} = V \backslash D$. Then D is a normal crossings divisor and $K_V + D$ is of maximal dimension.

The first step in Griffiths' proof is to construct a metric on this line bundle, which has at most logarithmic singularities at D. Later he shows that the characteristic function relative to this "pseudo-metric" does not differ substantially from the characteristic function relative to any metric defined on all of V ([2, Lemma 5.15]).

Faltings does the same procedure for a line bundle which is not the same as $K_V \otimes [D]$, but is proportional. Indeed, using the notation of [3], the height of an abelian variety is defined relative to a hermitian metric on

$$\omega = \Lambda^g W.$$

As is shown in [3], the second symmetric power of W is isomorphic to the sheaf of differential forms with logarithmic poles at D:

$$S^2(W) \cong \Omega^1_V[D].$$

But $K_V \otimes [D]$ is the bundle of top degree differential forms with logarithmic singularities at D, so

$$\Lambda^{g(g+1)/2} \Omega^1_V[D] = K_V \otimes [D].$$

Thus, a little linear algebra shows that

$$\omega^{g+1} \cong K_V \otimes [D].$$

The metric on ω that Faltings uses also has logarithmic singularities at D; he also uses an equivalent of Carlson–Griffiths' Lemma 5.15, which appears implicitly in Proposition 8.2 of [10].

REFERENCES

[1] Baily, W. L. and Borel, A. Compactification of arithmetic quotients of bounded symmetric domains. *Ann. Math.*, **84** (1966), 442–528.
[2] Carlson, J. and Griffiths, P. A. A defect relation for equidimensional holomorphic mappings between algebraic varieties. *Ann. Math.*, **95** (1972), 557–584.
[3] Faltings, G. Arakelov's theorem for abelian varieties. *Invent. Math.*, **73** (1983), 337–347.
[4] Griffiths, P. A. *Entire Holomorphic Mappings in One and Several Variables: Herman Weyl Lectures.* Princeton University Press, Princeton, NJ, 1976.
[5] Hall, M., Jr., The diophantine equation $x^3 - y^2 = k$, in *Computers in Number Theory*, (A.O.C. Atkin and B. J. Birch, eds.). Academic Press, London, 1971, pp. 173–198.
[6] Iitaka, S. *Algebraic Geometry.* Graduate Texts in Mathematics, 76. Springer-Verlag: New York, Heidelberg, Berlin, 1982.
[7] Lang, S. *Introduction to Diophantine Approximations.* Addison-Wesley: Reading, MA, 1966, p. 71.
[8] Lang, S. *Fundamentals of Diophantine Geometry.* Springer-Verlag: New York, Heidelberg, Berlin, 1983.
[9] Osgood, C. F. A number-theoretic differential equations approach to generalizing Nevanlinna theory. *Indian J. Math.*, **23** (1981), 1–15.
[10] Silverman, J. H. The theory of height functions, this volume, pp. 151–166.
[11] Stoll, W. *Value Distribution on Parabolic Spaces.* Lecture Notes in Mathematics, 600. Springer-Verlag: New York, Heidelberg, Berlin, 1973.
[12] Vojta, P. A. Integral points on varieties. Thesis, Harvard University (1983).

Notation

$\omega_{X/B}$	relative dualizing sheaf, 5
π	absolute Galois group $\mathrm{Gal}(\bar{K}/K)$, 9
$T_l(A)$	Tate module, 9
$\omega_{A/S}$	the line bundle $s^*(\Omega^g_{A/S})$, 11
$\overline{\mathfrak{M}}_g$	moduli stack of stable curves, 11
$\overline{M}_g$	moduli variety of stable curves, 11
$\overline{\mathfrak{A}}_g$	moduli stack of abelian varieties, 11
$\overline{A}_g$	moduli variety of abelian varieties, 11
$h(A)$	moduli-theoretic height of an abelian variety, 14
$\mathbb{G}_a$	additive group scheme, 30
$\mathbb{GL}(n)$	general linear group scheme, 30
$\mathbb{G}_m$	multiplicative group scheme, 30
$\mathbb{SL}(n)$	special linear group scheme, 30
μ_r	roots of unity group scheme, 30
α_{p^r}	roots of zero group scheme, 30
$\mathcal{H}$	constant group functor, 31
$\mathcal{G}^0$	connected component of the identity, 40
$\mathcal{G}^{\mathrm{et}}$	étale quotient, 40
$\mathrm{Hilb}^r(G/S)$	Hilbert scheme, 41
G^D	Cartier dual of G, 46
$\mathbb{W}_t$	Witt group scheme, 46
$\mathbb{F}_S(G)$	Frobenius group scheme, 49
$\mathbf{f}_{G/S}$	Frobenius morphism, 49
$\hat{F}$	formal completion of F, 57
$\mathrm{Spf}\, A$	formal spectrum of A, 57
$\hat{\otimes}$	complete tensor product, 58
$T_p(X)$	Tate module of a group scheme or formal group, 65
$\Phi_p(G)$	Tate comodule of a p-divisible group, 65
$\mathbb{Z}_p(1)$	Tate module $T_p(\mathbb{G}_m)$, 67
$\mathbb{Z}_p(n)$	n-fold Tate twist, 67
$\rho_{\mathbb{C}}$	complex representation of complex tori, 80
$\rho_{\mathbb{Z}}$	rational representation of complex tori, 80
$\deg(\phi)$	degree of an isogeny of complex tori, 81
$\tilde{\phi}$	dual isogeny, 82
$\mathrm{End}(T)$	endomorphism ring of a complex torus, 82
$\mathrm{End}_0(T)$	endomorphism algebra of a complex torus, 82

$\text{char}(\phi, x)$	characteristic polynomial of an endomorphism, 82
$\mathcal{M}(T)$	field of meromorphic functions on torus T, 83
$\wp$	Weierstrass $\wp$-function, 84
$H(u, v)$	Hermitian form, 84
$S(u, v)$	symmetric Hermitian form, 84
$E(u, v)$	antisymmetric Hermitian form, 84
$\mathcal{D}(T)$	group of theta functions modulo trival theta functions, 88
$\text{Pf}(G)$	the Pfaffian of G, 89
$\hat{A}$	dual abelian manifold, 93
$\mathcal{A}_d$	space of principally polarized abelian manifolds, 97
$\mathcal{L}(D)$	the invertible sheaf defined by D, 104
p, q	the projections $p: V \times W \to V$ and $q: V \times W \to W$, 104
$T_v(V)$	tangent space to a variety V at the point v, 104
$L(D)$	space of global sections to $\mathcal{L}(D)$, 112
$\|D\|$	complete linear system containing D, 112
$T_l A$	Tate module of abelian variety A, 114
$\text{Pic}^0(A)$	subgroup of Picard group of A, 118
$\check{A}$	dual abelian variety, 118
$\text{NS}(A)$	Néron-Severi group of A, 124
$\pi_1^{\text{et}}(A, 0)$	étale fundamental group of A, 129
$\bar{e}_m$	Weil pairing on A_m, 131
e_l	Weil pairing on $T_l A$, 132
$\check{\mathcal{A}}$	dual abelian scheme, 149
$M_{g,d}$	moduli space of abelian varieties, 149
M_K	set of absolute values on K, 151
$\|\cdot\|_v$	normalized absolute value, 151
$H_K(P)$	height of P relative to K, 151
$H(P)$	absolute height of P, 151
$\mathcal{H}(V)$	functions on V modulo bounded functions, 154
$h_\mathcal{L}$	height attached to the sheaf $\mathcal{L}$, 154
$\hat{h}_\mathcal{L}$	canonical height attached to the sheaf $\mathcal{L}$, 158
$\text{Pic}(S)$	Picard group of a scheme, 167
$\text{Pic}^0(S)$	Picard group of degree 0 invertible sheaves on a scheme, 167
$P_C^0(T)$	group of families of invertible sheaves on C, 168
$\mathcal{L}^P$	the invertible sheaf $\mathcal{L}(\Delta - C \times P - P \times C)$, 171
f^P	a map $C \to J$ associated to $\mathcal{L}^P$, 171
J^{an}	analytic Jacobian variety, 173
S_r	symmetric group on r letters, 174
$V^{(r)}$	r^{th} symmetric power of V, 174
$\text{Div}_C^r(T)$	set of relative effective Cartier divisors of degree r, 178
$P_C^r(T)$	classes of invertible sheaves of degree r on $C \times T$, 179
$f^{(r)}$	map of symmetric power $C^{(r)}$ to Jacobian variety, 182
W^r	image of symmetric power $C^{(r)}$ in Jacobian variety, 182
Θ	theta divisor on Jacobian variety, 186
Θ^-	image of Θ under $(-1)_J$, 186
Θ_a	translate of Θ by a, 186

NOTATION

$P_C^r(T)$	classes of invertible sheaves of degree r on $C \times_S T$, 192
$\mathrm{Grass}_n^{\mathcal{E}}$	Grassmann functor, 193
$G_n^{\mathcal{E}}$	Grassmann scheme, 194
$\pi_1(V, P)$	étale fundamental group, 195
$M_{m,n}$	group scheme of $m \times n$ matrices, 231
GL_n	general linear group scheme, 231
Sp_{2g}	symplectic group scheme, 231
$\Gamma_g(n)$	principal congruence group scheme of level n, 232
$C_r, \bar{C}_r$	cones of positive definite matrices, 232
$\mathfrak{H}_g$	Siegel upper half space, 232
h_X	contravariant functor attached to a scheme X, 232
$[n]_X$	multiplication by n map, 233
$X_{[n]}$	kernel of multiplication by n map, 233
$\mathbf{A}_g$	functor classifying principally polarized abelian schemes, 233
$\mathbf{A}_{g,d,n}$	functor classifying abelian schemes with polarization degree and level structure, 233
$\mathbf{A}_{g,n}^*$	functor classifying abelian schemes with level structure obeying Weil pairing, 234
$\mathcal{A}_{g,d,n}$	coarse moduli scheme of $\mathbf{A}_{g,d,n}$, 234
$\mathcal{X}_g$	universal holomorphic family of abelian varieties, 235
$\mathcal{X}_{g,\Gamma}$	quotient of $\mathcal{X}_g$ by Γ, 235
$R_k(\Gamma)$	space of Siegel modular forms of weight k, 237
$\mathcal{F}_g(u)$	Siegel subset of $\mathfrak{H}$, 239
M_K	set of absolute values on K, 253
$\mathbb{H}$	upper half plane, 254
$\Upsilon_{E/K}$	unstable minimal discriminant of E/K, 257
$[Z \cdot C]$	proper intersection number, 271
$[D_1, D_2]$	Arakelov intersection index, 290
$[D_1, D_2]_{\mathrm{fin}}$	finite part of Arakelov intersection index, 290
$[D_1, D_2]_{\mathrm{inf}}$	infinite part of Arakelov intersection index, 290
$G(P, z)$	Green's function on a Riemann surface, 291
$g(P, Q)$	logarithmic Green's function on a Riemann surface, 291
$[\mathcal{L}, \mathcal{L}']$	Arakelov intersection index of metrized line bundles, 294
$\mathrm{Div}^\wedge(\mathcal{X})$	group of Arakelov divisors, 294
$\mathrm{P}^\wedge(\mathcal{X})$	group of principal Arakelov divisors, 294
$\lambda(R\Gamma(X, L))$	the space $\mathrm{Hom}_{\mathbb{C}}(\Lambda(H^1(X, L)), \Lambda(H^2(X, L)))$, 295
μ_K	Haar measure on $\mathfrak{O}_K \otimes \mathbb{R}$, 299
$\mathrm{Num}(\mathcal{X})$	space of numerical equivalence classes of Arakelov divisors, 305
$\deg_k(L)$	degree of L with respect to k, 314
$i_k(E, F)$	intersection index on a surface, 314
$\|a\|$	the support of a divisor a, 327
$\mathrm{Div}^0(X/k)$	divisors of degree 0 defined over k, 327
$Z^0(X/k_v)$	divisors of degree 0 supported on $X(k_v)$, 328
$f(a)$	value of a function at a divisor, 328
$\langle a, b \rangle_v$	Neron local height pairing, 328

$n(a,r)$	number of zeros of $f(z)-a$ inside $|z|<r$, 342
$N(a,r)$	Nevanlinna counting function, 342
ord^+	maximum of 0 and ord, 342
$T(r)$	Nevanlinna characteristic function, 343
$m(a,r)$	Nevanlinna proximity function, 343
$//$	inequality that holds off of a set of finite measure, 343
$\delta(a)$	the defect of f at a, 343
$K_{V/D}$	logarithmic canonical divisor $K_V \otimes [D]$, 351

Index

Abel-Jacobi theorem, 96, 173
Abelian cover,
 of a curve, 195
 of a variety, 195
Abelian group, descent lemma, 158
Abelian integral, 173
Abelian manifold, 86
 ample divisor, 88
 automorphism group is finite, 95
 Cartier divisor, 86
 complex torus need not be an, 99
 dual, 93, 94
 endomorphism algebra is semi-simple, 86
 generic is simple, 99
 Lefschetz embedding theorem, 91
 Néron-Severi group, 92
 period matrix, 97
 periodic divisor, 86
 Picard group, 92
 Poincaré reducibility theorem, 86
 polarized, 95
 principally polarized, 96
 Riemann relations, 97
 Riemann-Roch theorem, 91
 Rosati involution, 94
 Siegel upper half space, 98
 space of principally polarized, 97
 theta function, 87
Abelian scheme, 35, 145, 232
 abelian variety extends to, 148
 dual, 149
 functor classifying principally polarized, 233
 group law is commutative, 146
 Hodge-Tate decomposition, 73

 is Néron model of generic fiber, 215
 kernel of multiplication-by-n, 147
 K/k image, 147
 K/k trace, 147
 morphism is homomorphism, 146
 multiplication-by-n map, 147, 233
 Néron model, 148
 p-divisible group of, 62
 Poincaré sheaf, 233
 polarization, 149, 233
 principally polarized, 233
 semi-, 220
 Tate module, 72
 theorem of the cube, 233
Abelian variety, 103–150,
 See also Abelian manifold, Abelian scheme, Complex torus
 anti-symmetric sheaf on, 111, 157
 automorphism group is finite, 139
 canonical height, 158
 characteristic polynomial of an isogeny, 125
 cohomology non-zero in one dimension, 127
 cohomology over $\mathbb{C}$, 128
 complete group variety is, 104
 complex, 83
 complex torus is an, 85
 degeneration of a family of, 246
 degree is a polynomial function, 123
 degree of an isogeny, 115
 degree of multiplication-by-n, 115
 direct factor, 142
 divisor ample iff dim $K_{\mathcal{L}} = 0$, 117
 dual,
  See Dual abelian variety

Abelian variety (*continued*)
  dual exact sequence, 120
  dual isogney, 120
  e_l pairing, 132
  $\bar{e}_m$ pairing, 131
  $\text{End}(A) \otimes \mathbb{Z}_l \hookrightarrow \text{End}(T_l A)$, 124
  endomorphism, 20
  endomorphism algebra, 121
  endomorphism algebra is skew field, 140
  endomorphism algebra of generic, 99
  étale cohomology, 128, 129
  étale cover is abelian variety, 130
  étale fundamental group, 129, 130
  euler characteristic of a sheaf, 127
  extension by $\mathbb{G}_m$, 121
  extension to an abelian scheme, 148
  family of, 145
  field of definition, 146
  field of definition of an isomorphism, 149
  finitely many direct factors, 142
  finitely many of bounded height, 248
  finitely many over a finite field, 127, 142, 143
  finitely many polarizations of degree d, 140
  finitely many with good reduction outside S, 204
  finiteness theorems for, 9–27
  Frobenius morphism, 143
  generic is simple, 99
  good reduction of, 149
  good reduction outside S, 204
  group law is commutative, 105
  has model over a number field, 234
  has symmetric ample invertible sheaf, 114
  height after isogeny, 18
  height and multiplication-by-n map, 157
  height on, 156
  $\text{Hom}(A, B) \otimes \mathbb{Z}_l$ injects into $\text{Hom}(T_l A, T_l B)$, 123
  integral points on, 351
  is projective, 113
  is quotient of a Jacobian variety, 198
  isogeny, 114
  isognenous to principally polarized, 136, 233
  Jacobian,
    See Jacobian variety
  kernel of multiplication-by-n, 116, 233
  kernel of multiplication-by-p, 116
  Lang conjecture, 351
  l-divisible group, 18
  L-series of, 22
  map from $C \times C$ to, 186
  map from curve factors through Jacobian, 185
  map to Picard group, 112
  modular height, 14, 247
  moduli space of, 7, 149, 128
  moduli space over $\mathbb{C}$, 235
  moduli space over $\mathbb{Z}$, 248
  Mordell-Weil theorem, 2, 149, 158, 160
  morphism is homomorphism and translation, 105
  multiplication-by-n map, 111
  Néron model,
    See Néron model
  Néron-Severi group, 124
  Néron-Severi group is a functor, 126
  Néron-Tate height, 158
  number of points over a finite field, 143, 145
  of bounded height, 17
  of dimension one, 34
  over $\mathbb{C}$, 79–101
  pairings on, 131
  Picard group (Pic^0), 118
  Poincaré reducibility theorem, 86, 122
  polarization, 6, 126

INDEX

Abelian variety (*continued*)
 principally polarized, 127
 rational map from $\mathbb{P}^1$ to, 107
 rational map from group variety to, 107
 rational map from unirational variety to, 107
 rational map to, 105–107
 resolution by Jacobian varieties, 200
 Riemann hypothesis over finite field, 143
 Riemann-Roch theorem, 127
 Rosati involution, 137
 semi-, 10
 semisimple action on Tate module, 21
 Shafarevich conjecture implies Mordell conjecture, 204
 simple, 122
 symmetric sheaf on, 111, 157
 Tate module, 116
 Tate module is dual to H^1_{et}, 129
 theorem of the cube, 111, 156
 theorem of the square, 112, 228
 trace of an endomorphism, 201
 trace of an isogeny, 125
 universal, 11, 12
 universal holomorphic family, 235
 very ample linear system on, 112
 weak Mordell-Weil theorem, 148, 160
 Weil pairing,
 See Weil pairing
 with complex multiplication, 85
 with good reduction outside S, 22, 351
 Zarhin's trick, 136
 zeta function, 143, 145
Abhyankhar, S., 309
Absolute height, 152
Absolute value, 151
 with logarithmic singularities, 165
Additive group scheme, 30
Adjoint representation, 80
Adjunction formula, 263, 274, 282

arithmetic, 301
Admissible hermitian line bundle, 293
Admissible line bundle,
 attached to an Arakelov divisor, 293
 determines metric on $\lambda(R\Gamma(X, L))$, 296
 divisor attached to a section, 293
 Euler characteristic, 300
 on an arithmetic surface, 293
Admissible metric on Ω^1_X, 295
Affine group scheme, 33
Affine group variety, 104
Albanese variety, 185
Albert, A.A., 95
Algebra
 formal spectrum of, 57
 profinite, 57
Algebraic curve, Jacobian variety of, 96
Algebraic group
 arithmetic subgroup, 141
 linear, 104
 reductive, 141
Algebraic space, 219
 dual abelian scheme is, 149
Algebraic stack, 233, 249
Algebraically equivalent sheaves, 156
Algebraically equivalent to 0, 92
Alternating form
 Pfaffian of, 89
 symplectic basis, 90
Alternating Riemann form, 85
Ample divisor, 88
 on a Néron model, 227
Ample linear system, 200
Ample relative dualizing sheaf, 4
Ample sheaf
 pullback is ample, 198
 symmetric on abelian variety, 114
Analytic group, 65
Analytic space, prolongation of, 241
Anti-symmetric sheaf, 111, 157
Arakelov, S., 5, 14, 289, 301
Arakelov divisor, 289

Arakelov divisor (*continued*)
  admissible line bundle attached to, 293
  group of, 294
  horizontal, 289
  intersection index, 290
  numerically equivalent to 0, 305
  of a function, 290
  vertical, 290
Arakelov intersection index, 263, 290
  behavior under base extension, 292
  canonical height and, 305
  capacity and, 291
  equals degree of pull-back, 294
  invariant under linear equivalence, 291
  of metrized line bundles, 294
  positive semi-definite on fiber, 304
Arakelov intersection theory, 289–307, 347
Arithmetic adjunction formula, 301
Arithmetic compactification, 249
Arithmetic curve,
  See Arithmetic surface
Arithmetic Hodge index theorem, 304
Arithmetic Riemann-Roch theorem, 300
Arithmetic subgroup, 141
Arithmetic surface
  adjunction formula, 301
  admissible line bundle, 293
  blow-down of divisor, 310
  Castelnuovo criterion, 310, 315
  exceptional divisor, 310
  factorization theorem for rational maps, 311, 324, 325
  finitely many exceptional divisors, 323
  hermitian metric on $\lambda(R\Gamma(X, L))$, 296
  Hodge index theorem, 304
  intersection index, 314
  intersection properties of exceptional curves, 316
  intersection theory, 315, 331
  local height function on, 330
  locally quadratic transformation, 311
  metrized line bundle, 293
  principal divisor, 316
  proper transform of a divisor, 316
  regular implies projective, 309
  regular model, 330
  resolution of singularities, 292, 267–287
  Riemann-Roch theorem, 300
  total transform of a divisor, 316
  uniqueness of blow-down, 314
Arithmetic-geometric mean, 254
Artin, M., 233
Artin-Hasse exponential, 47
Artin-Schrier map, 47
Artin-Whaples product formula, 345
Automorphism group,
  of polarized abelian manifold, 95
  of polarized abelian variety, 139

Baily-Borel compactification, 239
  desingularization of, 243
Barbarisms, 57
Base change,
  direct image commutes with, 108
  Euler characteristic and, 300
  of a group scheme, 32
Basis, symplectic, 90
Bertini's theorem, 199
Bilinear form attached to canonical height, 158
Birational cover of a surface, 268
Birational group, 190, 217
  determines a group scheme, 217
  determines a group variety, 191
  symmetric power of curve is, 190
Birational map,
  criterion to be defined at a point, 312
  factorization of, 311, 324, 325
Blow-down
  of a divisor, 310
  uniqueness of, 314
Blow-up, 267, 281

Bombieri conjecture, 349
Borel lemma, 341
Bounded height, 16
Brauer group, map from Jacobian
  variety, 169

Canonical bundle, admissible metric
  on, 295
Canonical divisor, logarithmic, 351
Canonical exact sequence,
  of a formal group, 58
  of a group scheme, 40, 43
Canonical height, 158, 331
  Arakelov intersection index and,
    305
  functorial properties, 332
  is positive definite, 158, 305
  is sum of local heights, 332
  local,
    See Local height
  lower bound for, 263
  on a Jacobian, 304
  on an elliptic curve, 263
  pairing, 331
Canonical map from a curve to its
  Jacobian, 171
Capacity, 291
Cartan, H., 241
Cartier divisor, 86, 103
  closed subscheme is, 178
  degree of, 177
  effective, 175
  fibers of, 177
  pull-back of, 177
  relative effective, 176
  split, 178
Cartier dual, 46
  of roots of unity, 232
  p-divisible group, 63
Cartier, P., 44
Castelnuovo criterion, 310, 315, 324
  prime divisors satisfying, 319
  proof, 321
Category
  of schemes, 232
  of sets, 232

Cayley form, 128
Cebotarev densisty theorem, 23
Center, of a prime divisor, 229
Character
  cyclotomic, 19, 67
  good, 54
  twist by, 67
Characteristic function, 343, 345
  higher dimensional, 348
Characteristic p lift of group scheme,
  44
Characteristic polynomial
  as a degree, 83
  of an endomorphism, 82
  of an isogeny, 125
Characteristic 0, group scheme in, 44
Chern class, 271, 348
Chinese remainder theorem, 277
Chow, W.-L., 184
Chow form, 128
Closed immersion, of varieties, 172
Closed subscheme, is relative divisor
  iff finite and flat, 178
CM, 85
Coarse moduli scheme, 234
  exists for $\mathbf{A}_{g,d,n}$, 234
Coarse moduli space, 150
Cohen-Macauly surface, 268
Coherent sheaf, direct image is co-
  herent, 108
Cohomology
  Betti, 329
  de Rham, 329
  dimension is upper semicontinu-
    ous, 108
  of schemes, 108
  varies in flat family, 108
Cokernel, 35
Commutative group law, 105, 146
Commutative group scheme, finite,
  See Finite group scheme
Compact Lie group, 80
Compactification
  Baily-Borel, 239, 243
  Satake, 238
  toroidal, 243

Complete group variety, 104
Completion, formal, 57
Complex Lie group, 80
Complex multiplication, 85
Complex representation, 80
Complex torus, 79
  alternating Riemann form, 85
  ample divisor, 88
  characteristic polynomial of an endomorphism, 82
  complex representation, 80
  dual isogeny, 81
  endomorphism algebra is semisimple, 82, 86
  endomorphism ring, 81, 82
  field of meromorphic functions, 83
  Hermitian Riemann form, 85
  holomorphic 1-forms, 80
  is abelian variety, 85
  isogeny, 81
  Jacobian variety as a, 173
  l-adic representation, 83
  Lefschetz embedding theorem, 91
  mappings between, 80
  not an abelian manifold, 99
  of dimension one, 84
  rational representation, 80
  Riemann form, 85
  Riemann-Roch theorem, 91
  semisimple, 82
  simple, 82
  Tate module, 83
  theta function, 87
Connected component
  of a group scheme, 219
  of the identity, 40, 43, 58
Connectedness theorem, 323
Conormal bundle, 284
Constant group functor, 31
Correspondence, divisorial, 168, 185
Counting function, 342, 345
  higher dimensional, 348
Cousin's theorem, 86
Cube, theorem of the, 110, 156
Cup product, 128
Curvature form, 292

Curve,
  See also Algebraic curve
  abelian cover of, 195
  arithmetic,
    See Arithmetic surface
  arithmetic adjunction formula, 301
  arithmetic Hodge index theorem, 304
  arithmetic Riemann-Roch theorem, 300
  canonical map to Jacobian, 171
  cohomology $H^1(C, \mathcal{O}_C)$ is tangent space of Jacobian, 171
  cohomology map $H^1(J, \mathbb{Z}_l) \to H^1(C, \mathbb{Z}_l)$ is isomorphism, 197
  cohomology map $H^1(J, \mathcal{O}_J) \to H^1(C, \mathcal{O}_C)$ is isomorphism, 196
  construction of Jacobian variety, 179–182
  covering ramified exactly over P, 197
  coverings obtained from Jacobian, 195
  de Rham cohomology, 329
  degree of a divisor, 167
  degree of an invertible sheaf, 167
  degree of an invertible sheaf on a singular, 192
  determined by polarized Jacobian, 201
  diagonal of $C \times C$, 171, 186
  differential form of the third kind, 329
  differentials on symmetric power, 183
  divisorial correspondence induces homomorphism of Jacobians, 185
  étale cover of, 195
  étale fundamental group, 195
  genus of is dimension of Jacobian, 171
  good reduction implies Jacobian has good reduction, 204
  good reduction of covering ramified exactly over P, 197

Curve (*continued*)
  group of degree 0 invertible sheaves, 167
  holomorphic differentials come from Jacobian, 172, 183
  hyperelliptic, 201
  image is transversal to theta divisor, 188
  integral points on, 2, 3, 264
  Jacobian scheme, 193
  Jacobian variety is birational to $C^{(g)}$, 182
  local height, 327–339, *See also* Local height
  map from $C \times C$ to abelian variety, 186
  map from symmetric power to Jacobian is a morphism, 191
  map to abelian variety factors through Jacobian, 185
  minimal model over Dedekind ring, 309
  mixed Hodge structure, 329
  modulus on a, 193
  Néron model of Jacobian, 219
  number of points over a finite field, 200
  period of, 170
  Picard variety, 196
  pointed, 168
  pull-back of theta divisor identities, 187
  regular implies projective, 309
  regular model, 330
  relatively minimal model, 310
  Riemann hypothesis over finite field, 200
  Riemann-Roch theorem, 167, 173, 181–184, 189, 190, 295, 298, 320, 329
  Shafarevich conjecture, 198
  Shafarevich conjecture implies Mordell conjecture, 198
  Siegel's theorem, 2, 264
  summary of divisorial correspondences, 189
  symmetric power, 174
  symmetric power is birational group, 190
  symmetric power is Hilbert scheme, 179
  symmetric power is non-singular, 174
  symmetric power maps to Jacobian, 182
  symmetric power represents the functor $\text{Div}_C^r(\cdot)$, 179
  Torelli's theorem, 201–208
  with good reduction outside S, 25, 198
  zeta function, 200, 201
Cusp, 237
Cyclotomic character, 19, 25, 67
  Tate twist, 67

De Franchis theorem, 198
De Rham cohomology, 329
Dedekind ring, minimal model of curve over, 309
Defect
  higher dimensional, 348
  of set of integral points, 350
Defect relation, 343, 345
  analogous to Roth's theorem, 346
Deformation,
  infinitesimal, 4
  of a group scheme, 45
Degeneration of a family of abelian varieties, 246
Degree
  finitely many points of bounded height and, 152, 155
  invertible sheaves of zero, 167
  is a polynomial function, 123
  of a divisor on a curve, 167
  of a line bundle, 314
  of a metrized line bundle, 161
  of an invertible sheaf on a curve, 167
  of an invertible sheaf on a singular curve, 192
  of an isogeny, 115

Degree (*continued*)
  of modular parametrization, 261, 262
  of multiplication-by-n, 115
Deligne, P., 52
Derived category, 268
Descent datum, 170
Descent lemma, 158
Descent theory, 170, 191
Descent, infinite, 2
Diagonal of $C \times C$, 171, 186
Differential form,
  of the third kind, 329
  on a curve, 172, 183
  on a Jacobian, 172, 183
  on symmetric power of a curve, 183
  relative, 4
Dimension
  maximal, 349
  of a formal Lie group, 59
  of a line bundle, 349
Diophantine approximation, 2
Direct factor of an abelian variety, 142
Direct image
  commutes with flat base change, 108
  of a coherent sheaf is coherent, 108
Discrete subgroup, 80
Distance function, 163
  bounded by height, 164
  logarithmic, 164, 347
Divisor,
    See also Cartier divisor
  algebraically equivalent to 0, 92
  ample, 88
  Arakelov, 289
  Arakelov intersection index, 290
  attached to a section of a metrized line bundle, 293
  blow-down, 310
  Cartier, 86, 103
  degree of, 167
  effective, 347

  exceptional, 310
  Green's function associated to, 330, 333
  horizontal, 289
  intersection index preserved by total transform, 316
  linear equivalence preserved by total transform, 316
  linear system of, 184
  modulus is an effective, 193
  numerically equivalent to 0, 305
  of degree 2 in $\mathbb{P}^2$, 284
  Ω minimal prime, 224
  on a surface, 270
  periodic, 86
  positive, 88
  principal, 92
  proper transform of, 316
  relatively prime, 328
  smooth prime, 223
  support of, 327
  total transform of, 316
  value of a function at, 328
  vertical, 289
  Weil, 103
Divisorial correspondence, 168, 185
  summary of various, 189
Double point, 281, 283
Dual abelian manifold, 93, 94
Dual abelian scheme, 149
  polarization is an isogeny to, 149, 233
Dual abelian variety, 117–120
  construction of, 119
  definition of, 118
  double dual, 119
  dual exact sequence, 120
  dual isogeny, 120
  identification with sheaf Ext, 121
  Poincaré sheaf on, 118, 186
  polarization is an isogeny to, 126
  tangent space, 118
Dual exact sequence, 120
Dual isogeny, 81, 120
Duality
  for proper maps, 268

Duality (*continued*)
  of Jacobian variety, 185
Dualizing sheaf, 10, 247
  ampleness of, 4
  blow-up of, 281
  on a surface, 268

Effective divisor, 347
  Cartier, 175
Eisenstein series, 238
Elliptic curve, 34
  canonical height, 263
  connected component of Néron model, 255
  degree of modular parametrization, 261, 262
  discriminant function, 254
  finitely many of bounded height, 259
  Hall conjecture, 352
  height of, 254
  inequalities for modular height, 258
  integral points on, 264
  integral points on $y^2 = x^3 + D$, 352
  j-invariant, 254, 257
  Klein function, 335
  local height, 335, 338
  minimal discriminant, 254
  minimal regular model, 335
  minimal Weierstrass equation, 255
  modular, 260
  modular height, 254
  Mordell-Weil theorem, 264
  Néron differential, 261
  Néron model, 218, 255, 263
  Néron-Severi group of, 127
  quasi-minimal equation, 264
  Siegel's theorem, 264
  Tate parametrization, 338
  twist, 260
  unstable minimal discriminant, 257, 258
  Weierstrass equation, 254
  Weierstrass functions, 335

Elliptic function, 84, 335
Embedding theorem of Lefschetz, 91
Endomorphism algebra
  degree is a polynomial function, 123
  of an abelian variety, 121
  of a complex torus, 82
  of a generic abelian variety, 99
  of simple abelian variety is skew field, 140
  Rosati involution, 94
  semi-simple, 86, 121
Endomorphism, 20–22
  characteristic polynomial of, 82
étale cohomology
  cup product, 129
  map $H^1(J, \mathbb{Z}_l) \to H^1(C, \mathbb{Z}_l)$ is isomorphism, 197
  of an abelian variety, 128, 129
étale cover
  of a curve, 195
  of a Jacobian variety, 195
  of abelian variety is an abelian variety, 130
étale formal group, 58
étale fundamental group, 129, 130, 195
  maximal abelian quotient, 195
  of a curve, 195
  of a Jacobian, 195
  of a variety, 196
étale group scheme, 40, 43, 51
étale morphism, multiplication-by-n is, 115, 147
étale site, large, 170
étale topology, 226
Euclidean Haar measure, 299
Euler characteristic, 295
  behavior under base change, 300
  of a sheaf, 108, 127
  of admissible line bundle, 300
  of module with Haar measure, 299
  of $\mathfrak{O}_K$-module, 300
Exact sequence, dual, 120
Excellent scheme, 269
Exceptional curve, 314 316

368 INDEX

Exceptional curve (*continued*)
  has self-intersection −1, 316
  intersection properties, 316
  is isomorphic to $\mathbb{P}^1$, 316
  uniqueness of blow-down, 314
Exceptional divisor
  finitely many, 323
  on an arithmetic surface, 310
Exponential
  Artin-Hasse, 47
  truncated, 47
Ext group, identification with Pic^0, 121

Factorization theorem, 219, 223
  for rational maps, 311, 324, 325
Faithfully flat morphism, 36
Faltings, G., 73, 165, 289, 296, 300
Faltings height,
  See Modular height
Family
  of invertible sheaves, 168
  trivial, 110
Feit-Thompson theorem, 42
Field extension, regular, 146
Field of definition
  of an abelian variety, 146
  of an isomorphism of abelian varieties, 149
Fine moduli scheme, 234
  does not exist for $\mathbf{A}_g$, 234
  exists for $\mathbf{A}_{g,d,n}$ if $n \geq 3$, 234
Fine moduli space, 150
Finite field
  finitely many abelian varieties over a, 127, 142, 143
  number of points on a curve, 200
  number of points on an abelian variety, 143, 145, 200
Finite group scheme, 37
  canonical exact sequence, 40, 43
  Cartier dual, 46
  commutative, 45–56
  connected component of identity, 40, 43
  constant, 232
  étale, 51
  étale quotient, 40, 43
  extending from K to R, 76
  Feit-Thompson theorem for, 42
  Frobenius, 49
  Frobenius height of, 50
  Frobenius morphism, 49
  in characteristic 0, 44
  is killed by m, 52
  kernel of multiplication on abelian scheme, 147
  kernel of multiplication on abelian variety, 116, 233
  lift from characteristic p, 44
  non-commutative, 45
  of prime order, 53
  of prime order is commutative, 50
  of type $(p, p, \ldots, p)$, 76
  Oort-Tate classification, 48, 53
  Raynaud classification, 53
  reduced, 44
  roots of unity, 30, 45, 46, 116, 232
  roots of zero, 30, 45, 48, 116
  Sylow theorems, 40
  Witt, 46
Finite morphism,
  intersection index for, 115
  multiplication-by-n is, 147
  pullback of hyperplane is connected, 199
Finite scheme, very, 57
First main theorem of Nevanlinna theory, 343, 345
  higher dimensional, 348
Fixed point formula, 145
Flat family, cohomology varies in, 108
Flat metric, 295
Flat morphism, multiplication-by-n is, 147
Form
  alternating Riemann, 85
  Hermitian, 84
  Hermitian Riemann, 85
  Pfaffian of, 89
  Riemann, 85

Form (*continued*)
    symplectic basis for alternating, 90
Formal completion, 57
Formal functions theorem, 322
Formal functor, 57
Formal group, 29, 56–60
    canonical exact sequence, 58
    connected component of identity, 58
    étale quotient, 58
    functor, 57
    p-divisible,
        See p-divisible group
    smooth, 59
    Tate module of, 65
Formal Lie group, 59
    dimension, 59
Formal scheme, 58
Formal spectrum, 57
Fourier series, 90
fpqc topology, 36
Frobenius group scheme, 49
Frobenius height, 50
Frobenius morphism, 47, 49, 143
    Rosati involution and, 144
Frobenius' theorem on Riemann forms, 90
Function field, prime divisor, 230
Functor
    attached to a scheme, 232
    constant group, 31
    formal, 57
    Néron-Severi group is a, 126
    representable, 31, 234
Fundamental group, étale, 129, 130, 195

Gabber, O., 27
Galois cover of a variety, 195
Gauss-Manin connection, 4
General linear group scheme, 30, 231
General type
    logarithmic, 351
    rational points on a variety of, 349, 350
    surface of, 4
    variety of, 350
Generalized Jacobian variety, 193, 197
    of projective line minus two points, 197
Generic abelian variety
    endomorphism ring of, 99
    is simple, 99
Generic point of a group scheme, 33
Genus formula, 197, 198, 285
Geometric invariant theory, 235
Good character, 54
Good reduction
    of abelian variety, 149
    of covering ramified exactly over P, 197
    of curve, 198
    outside S, 22, 25, 351
    Shafarevich conjecture, 198
Grassmann scheme, 193
Grauert, H., 4
Green's function, 291, 330, 333
    differential equation satisfied by, 330, 333
    on $\mathbb{P}^1$, 291
    on upper half plane, 336
Green's metric, 293, 296
Griffiths, P., 347
Gross, B., 42
Grothendieck, A., 58, 193
Grothendieck connectedness theorem, 323
Grothendieck topology, 36
Group
    analytic, 65
    birational, 190, 217
    birational determines a group scheme, 217
    birational determines a group variety, 191
    formal, 29, 56–60
    p-divisible,
        See p-divisible group
    reductive, 141
Group functor, formal, 57

INDEX

Group scheme, 29, 31,
    See also Finite group scheme
  additive, 30
  affine, 33
  base extension of, 32
  birational, 217
  cokernel of a homomorphism, 35
  connected component of identity, 40, 43, 219
  constant, 232
  deformation, 45
  determined by a birational group, 217
  diagonal map, 32
  étale, 33
  étale quotient, 40, 43
  finite, 37, 115
  finite and étale, 38
  formal, 58
  Frobenius morphism, 49
  general linear, 30, 231
  generic point, 33
  homomorphism, 32
  Hopf algebra associated to, 33
  identity section, 32
  illustration of a, 33
  inverse map, 32
  kernel of a homomorphism, 35
  kernel of isogeny is finite, 115
  multiplication map, 32
  multiplicative, 30, 62, 121, 232
  non-affine, 33
  non-commutative, 45
  of matrices, 231
  of prime order, 53
  of prime order is commutative, 50
  over a ring, 32
  principal congruence, 232
  quotient, 37, 38
  rational map to, 214
  reduced, 44
  roots of unity, 30, 45, 46, 116, 232
  roots of zero, 30, 45, 48, 116
  semi-abelian, 220
  special linear, 30
  subgroup, 32
  Sylow theorems, 40
  symplectic, 231
  Tate module, 65
  trivial, 32
  Witt, 46
Group variety, 104
  affine, 104
  birational, 190, 217
  complete, 104
  desecent theory, 191
  determined by a birational group, 191
  is non-singular, 104
  rational map to, 214
  rational map to abelian variety, 107
  translation map, 104

Haar measure, 299, 3228
Hadamard theorem, 342
Hall conjecture, 351
Height, 2, 14–17, 151–166
  absolute, 152
  associated to a sheaf, 153–155
  behavior under isogeny, 18
  canonical, 158
  canonical on a Jacobian, 304
  distance function bounded by, 164
  extension formula for, 152
  finitely many points of bounded, 152, 155, 165
  functoriality of, 155
  is ≥ 1, 152
  is quasi-quadratic, 156
  linear, 157
  local, 304,
    See also Local height
  logarithmic, 345
  logarithmic singularity, 249, 258
  metrized line bundles and, 161–166
  modular, 14, 247,
    See also Modular height
  multiplication-by-n map and, 157
  Néron-Tate, 158
  Nevanlinna theory analogue, 345
  of a p-divisible group, 60
  on abelian group, 158

Height (*continued*)
  on abelian variety, 156
  on projective space, 151, 331
  parallelogram law, 157
  points of bounded, 16
  positivity, 156
  quasi-equivalence, 156
  relative to a morphism, 153
  theorem of the cube and, 156
  with logarithmic singularities, 15
Height Machine, 155
Henselian ring, 213
Henselian scheme, 213
Henselization, strict, 37
Hermite-Minkowski theorem, 17
Hermitian form, 84
  associated to a theta function, 88
Hermitian line bundle, 292
  admissible, 293
Hermitian metric, 290
  on $\lambda(R\Gamma(X, L))$, 296
  on a vector space, 295
Hermitian Riemann form, 85
Higher direct image,
  See Direct image
Hilbert scheme, 41, 193, 235
  symmetric power of a curve is, 179
Hodge decomposition, p-adic, 73
Hodge index theorem, arithmetic, 304
Hodge structure, 329
Hodge theory, 247
Hodge-Tate decomposition, 71, 73
Holomorphic 1-form, on complex torus, 80
Holomorphic function, number of zeros, 342
Holomorphic functions theorem, 273
Holomorphic map,
  equidimensional, 347
  non-degenerate, 347
Holomorphic theta function, 90
Homogeneous polynomial function, 122
Homomorphism of group schemes, 32

Hopf algebra, 33, 130
Horizontal divisor, 289
Hriljac, P., 289, 301
Hurwitz genus formula, 197, 198, 285
Hyperbolic distance, 336
Hyperbolic Laplacian, 336
Hyperelliptic curve, 201

Idempotents, 142
Image, K/k of abelian scheme, 147
Immersion, 172
Induced reduced scheme structure, 316
Infinite descent, 2
Integral points,
  have defect 1, 350
  on an abelian variety, 351
  on a curve, 2
  on a variety, 350
  on a variety of logarithmic general type, 351
Integral scheme, 103
Intersection index, 314
  Arakelov, 290
  preserved by total transform, 316
  is negative semidefinite, 272, 323, 331
Intersection number, proper, 271
Intersection theory, 115, 315
Invertible sheaf
  degree of, 167, 192
  descent datum, 170
  family of, 168
  on a scheme, 167
Involution, Rosati, 94, 137
Isogeny, 17–20, 114
  between complex tori, 81
  characteristic polynomial of, 125
  degree is polynomial, 123
  degree of, 81, 115
  dual, 81, 120
  is equivalence relation, 81
  is finite, flat, and surjective, 115
  kernel is finite group scheme, 115
  trace of an, 125

Isotropy subgroup, 236

Jacobi decomposition, 239
Jacobian scheme, 193
  construction of, 194
  generalized relative to m, 193
Jacobian variety, 2, 96, 167–212
  See also Jacobian scheme
  Abel-Jacobi theorem, 173
  abelian variety is a quotient of, 198
  as Albanese variety, 185
  as a complex torus, 173
  autoduality, 185
  canonical height, 304
  canonical map from curve to, 171
  canonical polarization, 189
  Chow's construction, 184
  cohomology map $H^1(J, \mathbb{Z}_l) \to H^1(C, \mathbb{Z}_l)$ is isomorphism, 197
  cohomology map $H^1(J, \mathcal{O}_J) \to H^1(C, \mathcal{O}_C)$ is isomorphism, 196
  construction of, 179–182
  correspondence with pointed curve, 168
  curve coverings obtained from, 195
  definition of, 168
  determines curve, 201
  dimension of, 171
  divisorial correspondences, 189
  embedding of curve into, 295
  étale cover of, 195
  étale fundamental group, 195
  family of, 193
  flat metric on, 295
  generalized, 197
  generalized relative to m, 193
  has good reduction if curve has good reduction, 204
  holomorphic differentials come from curve, 172, 183
  homomorphisms associated to divisorial correspondences, 185
  is birational to $C^{(g)}$, 182, 184
  is complete, 191
  is self-dual, 186

Kummer embedding, 331
  map from $C^{(g)}$ is a morphism, 191
  map from curve to abelian variety factors through, 185
  map to Brauer group, 169
  Néron model, 219
  of $\mathbb{P}^1$ is trivial, 172
  of a family of curves, 192
  of projective line minus two points, 197
  over $\mathbb{C}$, 173
  Picard variety, 196
  Poincaré divisor, 332
  principal polarization, 189
  pull-back of theta divisor identities, 187
  represents P_C^0 functor, 168
  represents P_C^0 on large étale site, 170
  resolution of abelian variety by, 200
  Riemann form on, 173
  Riemann hypothesis over finite field, 200
  self-intersection of theta divisor, 189
  symmetric power of curve maps to, 182
  tangent space is $H^1(C, \mathcal{O}_C)$, 171
  theta divisor, 186, 331
  Torelli's theorem, 201–208
  uniqueness, 168
  universal property of, 192
  Weil's construction, 183, 189
  when $C(k)$ is empty, 169
Jensen formula, 343, 345
  analogous to product formula, 346
j-invariant, 254
  q-expansion, 257

Katz, N., 27
Kernel, group scheme, 35
Klein function, 335
Kodaira, K., 218
Kodaira dimension, 349
Kodaira-Spencer class, 6

Koecher's principle, 237, 243
Krull intersection theorem, 44
Kummer theory, 196
Künneth formula, 129

l-adic representation, 83
Lang, S., 263
Lang conjecture, 263
   Vojta conjecture implies, 351
Laplacian, 336
Large étale site, 170
Lattice, 80
   in a reductive group, 141
l-divisible group, 18
Lefschetz embedding theorem, 91
Lefschetz fixed point formula, 145
Lefschetz trace formula, 200
Legendre differential equation, 336
Leibnitz rule, 44
Leray spectral sequence, 4
Lichtenbaum, S., 309, 310
Lie group,
   adjoint representation, 80
   compact, 80
   complex, 59, 80
   formal, 59
Lift of a group scheme, 44
Line bundle,
   admissible hermitian, 293
   curvature form, 292
   degree wrt k, 314
   dimension of, 349
   Green's metric, 293, 296
   hermitian, 292
   metrized,
      See Metrized line bundle
   of maximal dimension, 349
   v-adic metric on, 162
Linear algebraic group, 104
Linear equivalence, preserved by total transform, 316
Linear system, 112, 184
   ample, 200
   very ample, 112
Lipman, J., 267, 309
Local domain, prime divisor, 229

Local height, 328, 347
   archimedean, 329, 337
   behavior under field extension, 328
   canonical height is sum of, 332
   construction of, 329
   for divisors of arbitrary degree, 333
   for divisors with common support, 332
   is bilinear, 328
   is continuous, 328
   is functorial for morphisms, 329
   is symmetric, 328
   non-archimedean, 330, 333, 337
   on curves of genus 0, 334, 338
   on elliptic curves, 335, 338
   on Mumford curves, 337
   properties of, 328
   uniqueness of, 329
   value at a principal divisor, 328
Locally free sheaf, Grassmann scheme of, 193
Locally quadratic transformation, 311
Logarithmic canonical divisor, 351
Logarithmic distance function, 347
Logarithmic general type
   $A_{g,n}$ is of, 351
   variety, 351
Logarithmic height, 345
Logarithmic singularity, 15, 163, 249, 258
   absolute value with, 165
   metrized line bundle with, 165
   of modular height, 249, 258
L-series, 22

Mahler, K., 3
Manifold, abelian, 86
Manin, Yu., 4, 261
Matrices, Siegel upper half space of, 98
Matrix
   group scheme of, 231
   Jacobi decomposition, 239
Mattuck, A., 280
Maximal dimension, 349

Mazur, B., 261
Metric
    logarithmic singularities of a, 15
    on a line bundle, 162
Metrized Euler characteristic, 295
Metrized line bundle, 14, 161, 248, 255, 292
    admissible, 293
    Arakelov intersection index, 294
    curvature form, 292
    degree of, 14, 161, 248
    degree related to height, 162
    divisor attached to a section, 293
    Euler characteristic of, 300
    on a variety, 161
    on an arithmetic surface, 293
    on Spec(R), 161
    with logarithmic singularities, 165
Minimal discriminant, 254
    unstable, 257, 258
Minimal model, 310
    of curve over Dedekind ring, 309
    relatively, 310
Minimal models theorem, 310
    proof, 323
Minimal Weierstrass equation, 255
Mixed Hodge structrue, 329
Model of a variety, 213
Modification of a surface, 268
Modular elliptic curve, 260
Modular form
    geometric interpretation, 238
    of weight 2, 261
    of weight 12, 254
    of weight k, 237, 242
    Petersson norm, 261
    q-expansion principle, 243
    Siegel, 236, 237
    space of is finitely generated, 237
    with rational Fourier coefficients, 243
Modular height, 14, 247
    behavior under base change, 248
    degree of modular parametrization and, 261
    finitely many elliptic curves of bounded, 259
    geometric, 248
    has logarithmic singularities, 249, 258
    inequalities for, 258
    measures arithmetic complexity, 254
    of an elliptic curve, 253–265
Modular parametrization, degree of, 261, 262
Module
    Haar measure on, 299
    reflexive, 268
Moduli functor
    classifying abelian schemes with level structure obeying Weil pairing, 234
    classifying abelian schemes with polarization degree and level structure, 233
    classifying principally polarized abelian schemes, 233
Moduli scheme
    coarse, 234
    constructed via algebraic theta functions, 235
    exists for $\mathbf{A}_{g,d,n}$, 234
    fine, 234
Moduli space
    as quotient $\mathfrak{H}/\Gamma_g(n)$, 236
    for abelian varieties, 7, 11, 149, 235, 236
    for curves, 11
    of logarithmic general type, 351
    over $\mathbb{C}$, 235
    over $\mathbb{Z}$, 10
    transcendental uniformization, 235
Moduli stack
    of abelian varieties, 11
    of stable curves, 11
Modulus
    generalized Jacobian relative to, 193
    on a curve, 193
Mordell conjecture, 2, 9, 195, 350

Mordell conjecture (*continued*)
  higher dimensional analogues, 341
  implied by Shafarevich conjecture, 198, 204
  over function fields, 4
  proof of, 25
Mordell-Weil theorem, 2, 158, 160, 264
  weak, 148
Morphism
  faithfully flat, 36
  Frobenius, 143
  height defined by a, 153
  of abelian schemes, 146
  of abelian varieties, 105
  quasi-compact, 36
  symmetric, 174
Multiplication-by-n
  degree of, 115
  is étale, 115, 147
Multiplicative group, 197
Multiplicative group scheme, 30, 121, 232
  Néron model, 215
  p-divisible group of, 62
Multiplicity of a rational singular point, 280, 281
Mumford, D., 243, 305
Mumford curve, 337

Nakai's criterion, 5
Nakayama's lemma, 23, 50, 52, 109, 172
Néron local height,
  See Local height
Néron, A., 158, 213
Néron measure of singularity, 225
Néron model, 148, 213–230, 247
  ample divisor on, 227
  closed fiber, 214
  closed fiber of Jacobian, 220
  cohomology $H^0(A, L)$ is finite, 227
  connected component, 220, 255
  definition of, 214
  existence over Dedekind domain, 228
  is of finite type, 214
  K-valued points extend iff, 215
  of a Jacobian, 219
  of a product, 214
  of an elliptic curve, 218, 255, 263
  of multiplicative group, 215
  of semi-abelian variety, 216
  over strictly local ring, 223
  Picard functor has same connected component, 219
  projective embedding, 227
  properties and examples, 214–220
Néron-Severi group, 124
  divsibility in, 134
  fixed by Rosati involution, 137
  is a functor, 126
  is finitely generated, 92
  is free of finite rank, 124
  of an abelian manifold, 92
  of an elliptic curve, 127
  pull-backs in, 133
Néron-Tate height,
  See Canonical height
Nevanlinna characteristic function, 343, 345, 348
Nevanlinna counting function, 342, 345, 348
Nevanlinna defect, 343, 345, 348
Nevanlinna proximity function, 343, 345, 347
Nevanlinna theory, 342
  correspondence with number theory, 344
  defect relation, 343, 345
  first main theorem, 343, 345, 348
  higher dimensional, 347
  second main theorem, 343, 348
Non-commutative group scheme, 45
Non-degenerate theta function, 88
Non-degenerate set of points, 347
Nonsingular scheme, 213
Norm, reduced, 126
Normal law of composition, 190, 217
Normalized theta function, 89

Number theory, correspondence with Nevanlinna theory, 344
Numerical equivalence of divisors, 305

Obstruction, to lifting a polarization, 127
Oort-Tate theorem, 53
Ordinary p-divisible group, 77
Ordinary reduction of a p-divisible group, 62

Parallelogram law for height, 157
Parshin, A.N., 4
Parshin's trick, 4
p-divisible group, 29, 60–78
   action of Galois, 77
   Cartier dual, 63
   connected, 62
   dimension determined by Tate module, 71
   étale, 62
   examples, 61
   extending from K to R, 76
   height, 60
   Hodge-Tate decomposition, 71
   of $\mathbb{G}_m$, 62
   of an abelian scheme, 62
   ordinary, 77
   ordinary reduction, 62
   Tate conjecture for, 74
   Tate module, 65, 71
Period
   lattice, 295
   matrix, 97
   of a curve, 170
Periodic divisor, 86
Petersson norm, 261
Pfaffian, 89
π group, 37
Picard functor, 219, 304
Picard group, 92, 118
   degree 0 part, 167
   Height Machine and the, 155
   identification with $\text{Ext}^1(\cdot, \mathbb{G}_m)$, 121
   isomorphic to Jacobian variety, 168
   of an abelian variety, 112
   of a scheme, 167
   of a variety, 196
Picard theorem, 341
Poincaré divisor, 332
Poincaré divisor theorem, 87
Poincaré duality, 173
Poincaré lemma, formal, 69
Poincaré reducibility theorem, 86, 122
Poincaré sheaf, 118, 186, 233
Poincaré-Eisenstein series, 242, 249
Point
   non-degenerate set of, 347
   on a surface, 267
   singular, 267
Pointed curve
   correspondence with Jacobian variety, 168
   divisorial correspondence between, 185
Pointed scheme, 168
   divisorial correspondence between, 168
Polarization, 6, 95, 126, 149, 233
   canonical on Jacobian, 189
   degree of a, 126, 149, 233
   finitely many of degree d, 140
   isogenous to abelian variety with principal, 136, 233
   obstruction to lifting, 127
   principal, 127, 189, 233
   pull-back of, 135
   Zarhin's trick, 136
Polarized abelian manifold, 95
   automorphism group is finite, 95
Polarized abelian scheme, 149, 233
Polarized abelian variety, 127
   automorphism group is finite, 139
   finitely many over a finite field, 127, 142, 143
Polynomial function,
   degree is a, 123
   on a vector space, 122

Positive definite quadratic form, canonical height is, 158
Positive divisor, 88
Positivity
  of height, 156
  of proper intersection number, 272
Power series ring, separable quotient, 277
Prime divisor
  becomes first kind after blow-up, 230
  center of, 229
  of the first kind, 229
  of a function field, 230
  of a local domain, 229
  of a scheme, 229
  Ω minimal, 224
  smooth, 223
Principal Arakelov divisor, 294
Principal congruence group scheme, 232
Principal divisor, 92
  has zero intersection index, 316
  local height at a, 328
Principal polarization, 96, 127, 233
  on Jacobian variety, 189
Principally polarized abelian manifold, 96
  space of, 97
Product formula, 331, 344, 345
  analogous to Jensen formula, 346
Profinite algebra, 57
Projection formula, 199
Projective embedding, of a Néron model, 227
Projective line
  Green's function, 291
  Jacobian is multiplicative group, 197
  Jacobian is trivial, 172
  local height on, 334, 338
  rational map to abelian variety, 107
Projective plane, divisor of degree 2 on, 284
Projective space
  absolute height on, 152

  finitely many points of bounded height, 152
  height on, 151, 331
Proper intersection number, 271
  formal properties, 272
Proper map, duality for, 268
Proper transform, 316
Properness, valuative criterion of, 191, 230
Proximity function, 343, 345, 347
Purity theorem, 196

q-curve, 338
q-expansion principle, 243
Quadratic character, associated to a theta function, 89
Quadratic form, equivalence classes of, 141
Quadratic function, canonical height is, 158
Quadratic transformation, 311
Quasi-compact morphism, 36
Quotient group scheme, construction of, 38
Quotient sheaf, is representable, 37

Ramified cover, of curve over one point, 197
Rational double point, 281, 283
Rational function, value at a divisor, 328
Rational map, 105
  defined at, 106
  to a group scheme, 214
  to a group variety, 214
Rational representation, 80
Rational singular point, multiplicity, 280, 281
Rational singularity, 268, 274, 278, 281
Raynaud, M., 54, 76, 213, 216, 261
Raynaud's Theorem, 55
Reduced group scheme, 44
Reduced norm, 126
Reduced representation, 126

Reduced trace, 126
Reducibility theorem, Poincaré, 86, 122
Reductive group, 141
Reflexive module, 268
Regular field extension, 146
Relative differential, 4
Relative dualizing sheaf, 10, 247
   ampleness of, 4
Relative effective Cartier divisor, 176
   closed subscheme is, 178
   degree of, 177
   fibers of, 177
   pull-back of, 177
   split, 178
Relative Picard functor, 219
Relatively minimal model, 310
Relatively prime divisors, 328
Remmert-Stein theorem, 242
Representable functor, 31
Representation
   adjoint, 80
   complex, 80
   l-adic, 83
   of a reductive group, 141
   rational, 80
   Siegel modular form attached to, 237
Resolution of singularities,
   of a surface, 267–287
   of an arithmetic surface, 267–287, 292
Riemann form, 85, 173
   associated to a positive divisor, 88
   associated to a theta function, 88
   Frobenius' theorem, 90
   non-degenerate, 85
   positive, 85
Riemann hypothesis, 95
   for abelian variety over finite field, 143
   for curve over finite field, 200
Riemann relations, 97
Riemann-Roch theorem, 90, 127, 173, 181–184, 189, 190, 280, 282, 295, 298, 300, 320, 329
   for complex torus, 91
   on a curve, 167
Riemann surface
   admissible metric on Ω_X^1, 295
   archimedean local height, 329
   de Rham cohomology, 329
   differential form of the third kind, 329
   elliptic partial differential equation on, 334
   embedding into Jacobian, 295
   Green's function, 291, 330, 333
   Green's metric on a line bundle, 293, 296
   hermitian metric, 290
   Jacobian variety of, 96
   line bundle has admissible metric, 293
   metrized Euler characteristic, 295
   metrized line bundle, 292
   period lattice, 295
   Riemann-Roch theorem, 295, 329
   volume form, 294
Rigid analysis, 337
Rigidity theorem, 104, 145
Ring
   henselian, 213
   smooth extension of, 213
   strictly local, 213
Roots of unity group scheme, 30, 45, 116, 232
   Cartier dual of, 46, 232
Roots of zero group scheme, 30, 45, 116
   Cartier dual of, 48
Rosati involution, 94, 137
   Frobenius morphism and, 144
Roth, K., 2
Roth's theorem, 345
   analogous to defect relation, 346
   is Vojta's conjecture for $\mathbb{P}^1$, 349

Satake compactification, 236, 238
   model over $\mathbb{Z}$, 249
   moduli meaning of boundary points, 246

Index

Satake topology, 240
Scheme
  abelian, 35, 145, 232
  category of, 232
  cohomology of, 108
  effective Cartier divisor, 175
  excellent, 269
  formal, 58
  functor associated to, 232
  Grassmann, 193
  henselian, 213
  integral, 103
  invertible sheaf, 167
  Jacobian, 193
  nonsingular, 213
  normal law of composition on, 217
  Picard group, 167
  pointed, 168
  prime divisor, 229
  prime divisor becomes first kind after blow-up, 230
  relative effective Cartier divisor, 176
  rigidity theorem for, 145
  seesaw principle, 109
  Stein factorization theorem, 219, 223
  strictly local, 213
  trivial family of, 110
  very finite over S, 57
Schottky group, 337
Second main theorem of Nevanlinna theory, 343
  higher dimensional, 348
  number field analogue, 348
Seesaw principle, 109
Segre embedding, 155
Semiabelian group scheme, 220
Semiabelian variety, 10
  degeneration of a family of, 246
  family of over toroidal compactification, 247
  moduli point of a, 246
  Néron model, 216
Semicontinuity, of cohomology, 108
Semisimple action, 21

Semisimple algebra, action on idempotents, 142
Semisimple complex torus, 82
Separable quotient, 277
Serre's theorem, 321
Set, category of, 232
Shafarevich, I., 310
Shafarevich conjecture, 4, 9, 198
  implies Mordell conjecture, 198, 204
  proof of, 23
  Vojta conjecture implies, 351
Sheaf
  algebraically equivalent give quasi-equivalent heights, 156
  anti-symmetric, 111, 157
  coherent, 108
  cokernel, 36
  descent datum, 170
  Euler characteristic of, 108, 127
  Grassmann scheme of locally free, 193
  Hom, 46
  projection formula, 199
  pullback of ample is ample, 198
  quotient is representable, 37
  skyscraper, 296
  symmetric, 111, 157
  theorem of the cube, 110, 156
  theorem of the square, 112
Siegel modular form, 236, 237
  geometric interpretation, 238
  of weight k, 237
  space of is finitely generated, 237
Siegel subset, 239
  topology on, 240
Siegel upper half space, 98, 232
  arithmetic quotient is compact and Hausdorff, 241
  boundary component of, 240
  canonical bundle, 238
  complex compactification, 239
  cotangent bundle, 238
  moduli meaning of boundary points, 246
  quotient by Γ, 236

380 INDEX

Siegel upper half space (*continued*)
  Satake compactification, 238
  Satake topology, 240
  stratification of, 241
  topological extension of, 240
  toroidal compactification, 243
Siegel's theorem, 2, 264
Simple abelian variety, 122
Simple algebra
  reduced norm, 126
  reduced trace, 126
Simple complex torus, 82
Singular curve, degree of a sheaf on, 192
Singular point
  multiplicity, 280
  on a surface, 267
Singularity
  Néron's measure of, 225
  rational, 268, 274, 278, 281
  resolution of on a surface, 267–287
Site, 36
Skew field, endomorphism algebra of simple abelian variety is, 140
Skyscraper sheaf, 296
Smooth extension of rings, 213
Smooth formal group, 59
Smooth prime divisor, 223
Special linear group scheme, 30
Spectral sequence, 4, 274
Spectrum, formal, 57
Split relative effective Cartier divisor, 178
Square, theorem of the, 112
Stack, 233, 249
Stein factorization theorem, 219, 223
Stoll, W., 347
Strict Henselization, 37
Strictly local ring, 213
Strictly local scheme, 213
Subgroup scheme, 32
Sublinearity, of proper intersection number, 272
Support of a divisor, 327
Surface
  adjunction formula, 274, 282
  arithmetic,
    See Arithmetic surface
  arithmetic adjunction formula, 301
  arithmetic Hodge index theorem, 304
  arithmetic Riemann-Roch theorem, 300
  birational cover, 268
  blow-up of, 267
  conormal bundle to a curve, 284
  definition of a, 267
  divisor, 270
  dualizing sheaf, 268
  modification of, 268
  multiplicity of a rational singular point, 280, 281
  normal implies Cohen-Macaulay, 268
  of general type, 4
  proper intersection number, 271
  rational double point, 281, 283
  rational singularity, 268, 274, 278, 281
  resolution of singularities, 267–287
  vanishing theorem, 273, 275
Swinnerton-Dyer, H.P.F., 261
Sylow theorems, 40
Symmetric domain,
  arithmetic quotient of a, 239
  Baily-Borel compactification, 239, 243
Symmetric function, fundamental theorem on, 175
Symmetric group, 174
Symmetric morphism, 174
Symmetric power
  differentials on, 183
  map to Jacobian is a morphism, 191
  of a curve, 174
  of a curve is birational group, 190
  of a curve is Hilbert scheme, 179
  of a curve is non-singular, 174
  of a curve maps to Jacobian, 182
  of a curve represents the functor $\text{Div}_C^r(\cdot)$, 179

Symmetric power (*continued*)
   of a variety, 174
Symmetric sheaf, 111, 157
   ample on abelian variety, 114
Symplectic basis, 90
Symplectic group, 231
   principal congruence subgroup, 232
   isotropy subgroup, 236
Szpiro, L., 6

Tai, Y.S., 246
Tangent space, 104
   of a Jacobian variety, 171
   to dual abelian variety, 118
Tate, J., 6, 74, 158
Tate algorithm, 254
Tate comodule, 65
Tate conjecture, 6, 9, 74
Tate curve, 338
Tate module, 9, 83, 116
   action of Galois, 116
   determines dimension of G, 71
   semisimple action of Galois on, 21
   Weil pairing on, 132
Tate twist, 19, 67
Taylor expansion, 227
Teichmüller character, 51
Tensor product, complete, 58
Theorem of the cube, 110, 156, 233
Theorem of the square, 112, 228
Theta constants, 235
Theta divisor, 186, 331
   degree of self-intersection ($\Theta^g = g!$), 189
   image under $(-1)_J$, 186
   is transversal to image of curve, 188
   pull-back identities, 187
   translation by a, 186
Theta function, 87
   associated Hermitian form, 88
   associated quadratic character, 89
   associated Riemann form, 88
   dimension of space of holomorphic, 90
   Fourier expansion, 90
   Frobenius' theorem, 90
   holomorphic, 90
   non-degenerate, 88
   normalized, 88
   of type (L, J), 87
   trivial, 88
   Weierstrass σ function, 92
Thue, A., 2
Topology
   Grothendieck, 36
   Satake, 240
Torelli's theorem, 5, 25, 96, 201–208
   proof of, 204–208
Toroidal compactification, 243
   arithmetic version, 246
   family of semi-abelian varieties over, 247
   local coordinates, 243
   model over $\mathbb{Z}$, 247
   moduli meaning of boundary points, 247
   universal property, 245
Torus,
   See Complex torus
Total transform, 316
   preserves intersection index, 316
   preserves linear equivalence, 316
Trace,
   formula of Lefschetz, 200
   K/k of abelian scheme, 147
   of an endomorphism, 201
   of an isogeny, 125
   reduced, 126
Transport of structure, 94
Trivial family, 110
Trivial group scheme, 32
Trivial theta function, 88
Truncated exponential, 47
Twist
   of an elliptic curve, 260
   Tate, 67

Unirational variety, 107
   rational map to abelian variety, 107

Unstable minimal discriminant, 257, 258
Upper half plane, Green's function on, 336
Upper semicontinuity of cohomology, 108

Valuative criterion of properness, 191, 230
Vanishing theorem, 273, 275
Variety, 103
  abelian cover of, 195
  Cayley form, 128
  Chow form, 128
  closed immersion of, 172
  defect of set of integral points, 350
  descent datum, 170
  descent theory, 170, 191
  distance function on, 164
  étale fundamental group, 196
  finitely many points of bounded height, 155, 165
  Frobenius morphism, 143
  Galois cover of, 195
  group, 104
  height attached to a sheaf, 153–155
  Height Machine, 155
  height relative to a morphism, 153
  integral points, 350
  integral points on logarithmic general type, 351
  intersection theory on, 115
  Kodaira dimension, 349
  logarithmic canonical divisor, 351
  logarithmic general type, 351
  metrized line bundle on, 161
  model of over R, 213
  Mordell conjecture analogue on, 341
  Nevanlinna theory, 347
  non-degenerate set of points, 347
  normal law of composition on, 190, 217
  of general type, 349, 350
  Picard group, 196
  pullback of ample sheaf is ample, 198
  pullback of hyperplane by finite map is connected, 199
  rational map, 105
  rational points on general type, 349, 350
  rigidity theorem for, 104
  seesaw principle, 109
  semiabelian, 10
  symmetric power is singular, 175
  symmetric power of, 174
  tangent space, 104
  theorem of the cube, 110
  trivial family of, 110
  unirational, 107
  zeta function, 145
Vector space
  hermitian metric on, 295
  polynomial function on, 122
  volume form on, 295
Vertical divisor, 289
Very finite scheme, 57
Vojta conjecture, 348
  consequences of, 349–352
  for $\mathbb{P}^1$, 349
  Hall conjecture implied by, 352
  Lang conjecture implied by, 351
  Shafarevich conjecture implied by, 351
Volume form, 294
  on a vector space, 295

Wahl, J., 281
Weierstrass σ function, 92
Weierstrass $\wp$ function, 84, 255
Weierstrass elliptic functions, 335
Weierstrass equation, 254
  minimal, 255
Weil, A., 183, 189, 217
Weierstrass points, 5
Weil conjectures, 20, 23
Weil curve,
  See Modular elliptic curve
Weil divisor, 103, 270
Weil function, 347

Weil pairing, 131, 234
   compatibility for, 131
   functorial properties, 132
   on Tate module, 132
   over $\mathbb{C}$, 132
Witt group scheme, 46
   Artin-Hasse exponential, 47
   Frobenius map, 48
Witt vector, 197

Yoneda's lemma, 31

Zarhin, J.G., 6, 74
Zarhin's trick, 6, 136
Zariski, O., 270
Zariski limit, 349
Zariski topology, non-degenerate set of points for, 347
Zariski's Main Theorem, 120, 218, 220, 222, 228, 270
Zeta function
   of a curve, 200, 201
   of a variety, 145
   of an abelian variety, 143, 145